RECOGNITION OF ANCIENT SEDIMENTARY ENVIRONMENTS

RECOGNITION OF ANCIENT SEDIMENTARY ENVIRONMENTS

Edited by
J. Keith Rigby
Wm. Kenneth Hamblin

SOCIETY OF ECONOMIC PALEONTOLOGISTS
AND MINERALOGISTS

Special Publication No. 16
February, 1972

A Publication of

The Society of Economic Paleontologists and Mineralogists

a division of

The American Association of Petroleum Geologists

PREFACE

This volume contains a series of papers presented as part of a symposium organized by the Research Committee of the S.E.P.M. and held in Dallas, Texas, April 15, 1969, at the annual national meeting of the society. Two papers presented orally at the symposium are not included because the authors felt that their topics, deltaic sedimentation by J. M. Coleman and S. M. Gagliano, and reef environments by J. K. Rigby, were adequately published in other recent symposia of the S.E.P.M.

The problem of recognizing ancient sedimentary environments in the stratigraphic record is basic to essentially every appect of research in sedimentary rocks. The S.E.P.M. Publication Committee has approved the publication of this symposium which, it is hoped, will summarize much of what we currently know concerning environmental interpretation.

The editors are indebted to the National Science Foundation for a grant which permitted bringing J. J. Bigarella from Brazil, and H. E. Reineck from Germany to participate in the symposium. We also express our thanks to E. D. McKee for his advice and assistance in organizing the symposium and editing some of the papers. We are also indebted to our colleagues who reviewed the papers and helped with editorial discussions. The Department of Geology, Brigham Young University, provided funds to aid in the initial organization of the symposium and editorial costs.

J. K. RIGBY
W. K. HAMBLIN

CONTENTS

Preface		v
Environmental Indicators—A Key to the Stratigraphic Record	H. R. Gould	1
Classification of Sedimentary Environments	E. J. Crosby	4
Eolian Environments—their Characteristics, Recognition, and Importance	J. J. Bigarella	12
Recognition of Alluvial-Fan Deposits in the Stratigraphic Record	W. B. Bull	63
Physical Characteristics of Fluvial Deposits	G. S. Visher	84
Fluvial Paleochannels	S. A. Schumm	98
Criteria for Recognizing Lacustrine Rocks	M. D. Picard and L. R. High	108
Tidal Flats	H. E. Reineck	146
Recognition of Evaporite—Carbonate Shoreline Sedimentation	F. J. Lucia	160
Criteria for Recognizing Ancient Barrier Coastlines	K. A. Dickinson, H. L. Berryhill, Jr., and C. W. Holmes	192
Trace Fossils as Criteria for Recognizing Shorelines in the Stratigraphic Record	J. D. Howard	215
Recognition of Ancient Shallow Marine Environments	P. H. Heckel	226
Submarine Channel Deposits, Fluxoturbidites and other Indicators of Slope and Base-of-Slope Environments in Modern and Ancient Marine Basins	D. J. Stanley and R. Unrug	287

ENVIRONMENTAL INDICATORS—
A KEY TO THE STRATIGRAPHIC RECORD

H. R. GOULD
Esso Production Research Company, Houston, Texas

ABSTRACT

Since Leonardo da Vinci made his first environmental analysis in the fifteenth century, geologists have become increasingly concerned with sedimentary environments. Accordingly, their methods for recognizing environments of deposition have become increasingly more sophisticated, and their determinations have become increasingly more precise.

The major types of criteria now conventionally used in recognizing sedimentary environments are the physical, chemical, and biological characteristics preserved in the sediment. These are features that may be determined from a single small outcrop or subsurface core. Where larger or multiple outcrops are available, or where numerous subsurface cores are on hand, criteria of a much larger order of magnitude—namely, the lateral and vertical facies relationships and the three-dimensional geometric framework of the strata—can be employed and greatly strengthens and broadens the environmental interpretation.

The papers of this volume cover most of the major sedimentary environments and identify for each the criteria which permit its recognition. Such information is important, not only in interpretation of the stratigraphic record, but also in exploration and production of the bulk of our natural resources, including oil and gas, mineral deposits, and underground water supplies. Knowledge of sedimentary environments is essential also, in engineering geology studies of numerous and diverse types.

INTRODUCTION

In introducing this symposium volume on "Criteria for Recognizing Sedimentary Environments in the Stratigraphic Record," my purpose is to lay some of the groundwork for the papers to follow. Accordingly, I will (1) attempt to define what is meant by environments; (2) identify some of the major types of criteria used in recognizing various ancient, sedimentary environments; and (3) outline some of the important ways in which knowledge of sedimentary environments can be applied.

DEFINITION OF SEDIMENTARY ENVIRONMENT

By the term sedimentary environment we refer simply to the place of deposition and to the physical, chemical, and biological conditions which characterize the depositional setting. The frontispiece shows diagrammatically the great variety of depositional environments; these range from glaciated terrains through alluvial fan and dune environments of desert regions, alluvial belts and lakes of low-lying valleys, the numerous environments—bay, lagoon, beach, and barrier island—of coastal areas, and finally to the great array of marine environments grading from shelf and slope to deep ocean basin. Included within these latter settings are the clastic and carbonate reef environments of the shelf and the turbidite fan and abyssal plain environments of the deep sea. Each of these major environments is characterized by a definitive set of physical, chemical, and biological conditions which may leave an imprint on the deposited sediments. It is this record of past environmental conditions preserved in the sediments that provides the basic criteria for recognizing sedimentary environments in the stratigraphic column.

MAJOR TYPES OF CRITERIA

Possibly the earliest environmental analysis was made by Leonardo da Vinci in the fifteenth century, when he first recognized fossils in ancient rocks around the Mediterranean and concluded that they had been laid down in an ancient sea. Since that time, geologists have become increasingly concerned with sedimentary environments, and their criteria and ability to identify them have become increasingly more precise. The major types of criteria currently used in recognizing depositional environments are the physical, chemical, and biological characteristics of the sediments. These are features which may be determined from a single small outcrop or subsurface core. Where larger or multiple outcrops are available, or where numerous subsurface cores are on hand, criteria of a much larger order of magnitude—namely, the lateral and vertical facies relationships and the three-dimensional geometric framework of the strata in question—serve as invaluable additional environmental indicators.

Among the physical characteristics of principal value are such features as bedding and other large structures, nature of contacts between beds, sedimentary textures and related small-scale structures, and directional properties such

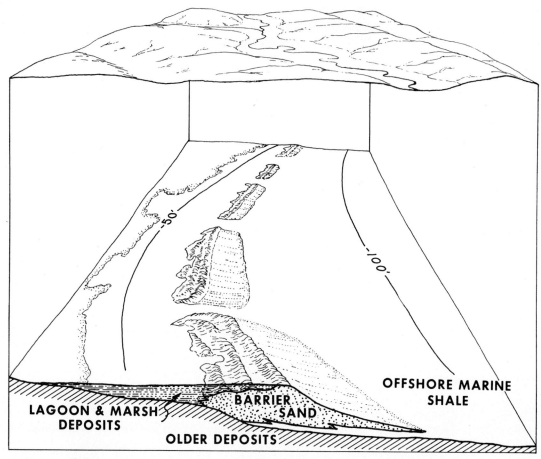

Fig. 1.—Ancient barrier-island sandstone facies illustrating use of lateral and vertical facies relationships and three-dimensional geometric framework in environmental analysis. The isopachous contours show thickness of finer-grained facies enclosing the barrier-island sandstone.

as flute casts, asymmetrical ripple marks, and preferred orientation of grains.

With regard to chemical characteristics, we are interested first, of course, in the gross composition of the rock—that is, whether it is a limestone, quartz sandstone, clay shale, evaporite, or other sedimentary type. Its major mineral constituents and particularly its authigenic minerals may be especially helpful as environmental indicators. For the biogenetical and chemical sediments, ratios of major elements, such as Ca-Mg ratios, and content of trace elements, such as strontium, boron, or others, may provide additional environmental clues. Isotopic analyses, including C^{13}/C^{12} and O^{18}/O^{16} ratios may yield other valuable information. Evidence of solution, occurring penecontemporaneously with or shortly after deposition such as that seen in the etched or partially dissolved tests of foraminifers in deep ocean environments, or the concentration of limestone nodules at the base of barrier-island quartz sand bodies, produced by precipitation of $CaCO_3$ leached from overlying shell-bearing sands, provides additional insight into the environmental setting. The type and composition of organic matter and the inclusion of soil zones within a depositional sequence can, likewise, furnish important environmental clues.

As to biological characteristics, the total faunal and floral assemblage, together with relative abundances and ratios of the various forms, is most useful. Careful attention to distinguishing endemic versus displaced biota can yield additional clues, particularly in recognition of turbidites, which commonly hold displaced faunas. The ecological characteristics of assemblages such as benthonic foraminifers that accurately record the depth of water in which they lived, together with trace fossils (the tracks and trails of organisms no longer in evidence) serve as additional valuable criteria for the recognition of sedimentary environments.

In addition to these features, the lateral and vertical facies relationships and the three-dimensional geometric framework, where conditions permit their observation, can greatly strengthen any environmental interpretation and at the same time add an important regional aspect to the determinations. Figure 1, which shows an ancient barrier-island sand body and related facies now deeply buried beneath the surface, demonstrates these points. In addition to definitive physical, chemical, and biological properties which characterize each of the facies, the lateral and vertical facies relationships, together with the three-dimensional geometry of the sand body, provide important confirming evidence of its environment of deposition. The lateral change in facies from offshore-marine shale on one side to lagoon and marsh deposits on the other indicates that this is a shoreline sand body and further that it was isolated or separated from the land either as an island or as a peninsula. The vertical facies sequence, which grades upward from offshore-marine shale at the base through the barrier sand mass into overlying marsh deposits, confirms this interpretation. The progradational sequence of facies is also typical of shoreline sands and demonstrates that they were formed through seaward outbuilding over finer-grained facies of the offshore-marine environment. The final confirming evidence that this is a barrier-island sand mass instead of a land-tied peninsula or spit is provided by its three-dimensional geometry, which demonstrates complete isolation from any land area. The overall shape of the sand body, including its smooth seaward margin and irregular back side, is in keeping with the barrier-island interpretation, as is the trend of the sand body, which parallels the regional depositional strike indicated by the trend of the isopachous contours of the enclosing beds. By these various types of criteria, it has been possible not only to define the depositional environment of this particular sand body, but also to establish a regional trend along which similar sand bodies may lie, as indicated by the predicted series of additional barriers shown in the figure.

APPLICATIONS OF ENVIRONMENTAL STUDIES

The following papers of this volume cover the major sedimentary environments and identify the most useful criteria for their recognition in the stratigraphic record. Such information is essential to any interpretation of geologic history and in that sense is fundamental to almost every other type of investigation undertaken by geologists. Moreover, knowledge of sedimentary environments—how to recognize them and how to predict them—has tremendous economic significance. Such knowledge is of utmost importance in the exploration and production of the bulk of our nonliving natural sources, including oil and gas, metallic minerals, and nonmetallic deposits such as sand, gravel, building stone, coal, and underground water supplies. Knowledge of sedimentary environments is essential, also, in engineering geologic studies of numerous and diverse types. Accordingly, this volume should be of great interest and value to all geologists concerned with sedimentary rocks, regardless of their specialized disciplines.

CLASSIFICATION OF SEDIMENTARY ENVIRONMENTS*

ELEANOR J. CROSBY
U.S. Geological Survey, Denver, Colorado

ABSTRACT

The ideal objective in classification of depositional environments is division of the entire depositional realm into useful discrete and consistently named categories. In practice, terms commonly used reflect place, medium, or process of deposition. Intergrading of environments makes necessary the recognition of environmental complexes, designated in terms of dominant aspect. Symposium papers are assigned to categories in an existing classification and in variant classifications for nonmarine and mixed categories. Marine classification needs extensive additional study.

Several uses of classification are self-evident: to bring order into a field of knowledge, to provide guidelines for those who must speak or write of such a field as a whole, and to establish a frame of reference for selective treatment of a subject. My first concern with an overall classification of environments of deposition arose from need for a filing system for published data on criteria for recognizing ancient environments.

The ideal objective of a classifier of depositional environments is to divide the entire depositional realm, without omission or overlap, into useful categories consistently named. In practice, utility of terms outweighs consistency and probably this is as it should be. The larger problem is in determining the rank and limits of categories.

In the introduction to this symposium, a depositional environment was defined as the place of deposition plus characteristic physical, chemical, and biological conditions; the choice is a practical one, especially at the level of classification at which a delta geologist communicates with an outer-continental-shelf or barrier-coast geologist. Environmental literature, however, yields few if any broad classifications that consistently use place terms, or any other single group of terms, at all levels. In addition to *place*, terms of *medium of deposition*, of *process*, and sometimes of *deposited material* seem to be required. A given term may imply two or more of these factors; most commonly an implication of place is included. Table 1 illustrates such a mixing of terms. Here, as in other illustrations, Roman numerals indicate classification levels.

The problem of defining fixed categories of classification arises from intergrading and from the enclosing of lesser by larger environments. Thus, a river environment may intergrade with a lacustrine environment of equal magnitude.

* Publication authorized by the Director, U.S. Geological Survey.

TABLE 1.—MIXING OF ENVIRONMENTAL TERMS AT SUCCESSIVE LEVELS OF CLASSIFICATION (I–IV)

I. Mixed nonmarine-marine (*medium* of deposition)
II. Tide-dominated (*process* affecting deposition)
III. Shore lagoon (*place* or *medium* of deposition)
IV. Oyster reef (*deposited material*)

Yet a major river and floodplain system may include enclaves of lake, marsh, and dune, representing environments that elsewhere may be independent and of magnitude comparable to that of the fluvial complex. For order and simplicity, it becomes necessary to designate an environmental complex in terms of its dominant aspect.

Distinguishing the dominant depositional character may present difficulties. Delaware Bay, in figure 1, is clearly an embayed coast environment, an estuary open to the sea. Pamlico Sound lies within a coastal barrier; is deposition within the sound influenced more by the barrier against the sea or by estuarine processes? Published comments suggest that the barrier influence is dominant, but they do not directly compare the two factors. Figure 2 shows areas of swamp and marsh in the Mississippi delta region and at the southern tip of Florida. At the first level of classification, both areas represent transitional or mixed nonmarine and marine environments; at the second level, both can be called swamp and marsh. Or, the Louisiana area may be said to represent a deltaic complex environment at the second level, while marsh and swamp are reduced to third-level components. This latter classification for the Louisiana area seems preferable because the environmental geologist's concern is less with a unit of deltaic sedimentation in the birdsfoot area and a unit of marsh sedimentation to the west than with the total product of deltaic and associated environments of deposition shifting through a span of geologic time. The conflict between time and en-

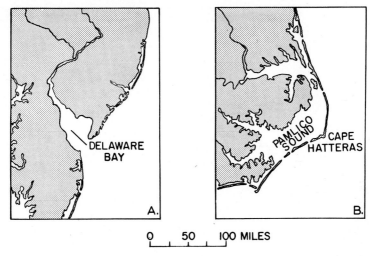

Fig. 1.—Embayed and barrier coasts. A. Delaware Bay, deposition dominated by estuarine processes; B. Pamlico Sound, deposition affected by both estuarine and barrier coast processes, without clear indication of dominant factor.

vironmental stability almost compels the geologist to think in terms of large environmental complexes, each in turn divided and subdivided to accommodate local and limited environmental studies.

To establish a framework of environmental classification for papers in this symposium, table 2 has been prepared from the running text of Dunbar and Rodger's Principles of Stratigraphy (1957). Many of the terms are the writer's condensations of the authors' more fully expressed concepts. All asterisks, queries and figures other than Roman numerals refer to number and placement of symposium papers as interpreted by the writer from abstracts.

In the nonmarine category desert and glacial environments of earlier classifications (e.g., Twenhofel, 1939, p. 51; Krumbein and Sloss, 1951, p. 196) are subdivided, and swamps and marshes are absorbed into other environments. Of the papers on nonmarine environments in this symposium, one on the subject of alluvial fans is assigned to a piedmont environment and possibly in part to desert; two on paleochannels and fluvial deposits are assigned to intermediate fluvial or floodplain fluvial environments; one to lake environment, and two to glacial environment, including ice deposits and perhaps ice-related fluvial and lacustrine deposits. The symposium paper on dunes may belong in part under desert, in part under coastal dune or in other nondesert eolian environments.

Under mixed or transitional environments, *process* categories, referred to as dominance by wave and by tidal action, are introduced, and these in turn are divided by *place*. Within this group, the symposium includes a paper on tidal flats; it also includes shore lagoon and wave-dominated low open shoreline under discussions of both barrier islands and evaporitic shorelines. Organic reefs and the subenvironments of marine deltas also are covered.

Symposium papers on shallow marine environments, shoreline or nearshore deposits, channeled marine slopes, and perhaps in part the paper on reefs, are in the division.

Figures 3, 4, and 5 are presented as another way of looking at the dividing of environmental categories. The recasting of the second level of nonmarine environments in figure 3 uses terms that, for the most part, have a strong connotation of *process* or *medium of deposition* as well as *place*. Most of them also are dominant-aspect terms for environmental complexes. Gravitational, representing the minor environments of talus and landslide formation, and eolian have been added to the list in table 2; desert has been omitted because it covers an assemblage of second-level environments, including most of the others shown here. Desert is a valid and useful term, but it has no strict counterpart applicable to humid regions. Therefore, in place of using desert as a category, it may be constructive to indicate that the nonmarine environments in figure 3 have both arid- and humid-region phases. In a very general way, the arrows represent relative development under arid and humid conditions.

If logic were to prevail, the total problem of environmental classification could be simplified at this point by including marine environments under a saline division of lacustrine, as both sea

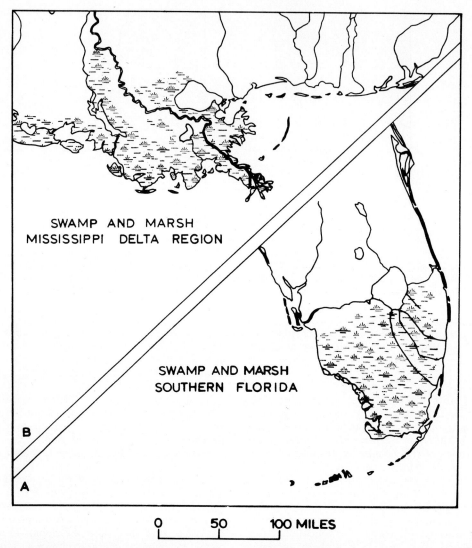

Fig. 2.—Classification of swamp and marsh coasts. A. Southern Florida coast, swamp and marsh at second level of classification; B. Mississippi Delta region, deltaic complex environment at second level of classification, swamp and marsh at third level.

and salt lake consist of undrained bodies of standing water. Such an arrangement would not—and should not—be tolerated long in formal classification. But acknowledging the relationship may underscore the difficulties in establishing second-level divisions of the single-medium marine environment that are comparable to the interacting but distinct media of nonmarine deposition.

In the mixed- or transitional-environments zone, between the clearly nonmarine and marine regions, we face the same problems that confront coastal morphologists (McGill, 1959). Classifications of convenient simplicity do not provide adequate place for complex environments, and a complete classification would be so long and cumbersome as to be unusable. The simplified second-level listing of mixed environments in figure 4 attempts to keep emphasis on geomorphic *place*, depends heavily on dominant-aspect terminology, and again allows for climatic influences. Use of coast in these designations is not wholly satisfactory, for coast is a vague term, usually implying a maximum seaward extent at low-tide line but providing no definition of the landward edge. The mixed environment zone has indefinite boundaries, both landward and seaward. Perhaps we can say only that the mixed zone ends where the invading influence, marine or nonmarine, is no longer significant—a highly subjective definition. Such a limit may be as close to the sea as the landward edge of supratidal storm-wave deposition above a beach; the seaward limit for a large delta may be well within the neritic marine zone defined by sea depth.

Of the second-level environmental complexes in figure 4, those of delta, barrier, and reef coasts are represented in this symposium and need little comment here. It should be noted, however, that in the previously shown classification (Table 2), the beach of a barrier island (low open shore) and the landward lagoon are separated under the second-level, wave-dominated and tide-dominated categories, respectively. In contrast, the barrier coast in figure 4 is defined as containing, at the third level, inner shore, lagoon, and barrier island or spit environments, all considered to be interdependent parts of a complex and each in turn subdivisible at a fourth level.

Other mixed-environment categories in figure 4 are not specifically discussed in the symposium. Under embayed coast are included estuarine environments and associated marshes, tidal flats, and beaches; fiords that function as estuaries; and other, nonestuarine and unbarred coastal indentations in which depositional conditions are modified by fresh water. The cliffed-coast environment is typified by much of the coast of

TABLE 2.—CLASSIFICATION OF SEDIMENTARY ENVIRONMENTS
[Adapted from Dunbar and Rodgers (1957)]

Level I	Level II	Level III
Non-marine	Fluvial*	Piedmont* Intermediate? Floodplain?
	Desert?	Enclosed basins Wide low-lying deserts
	Lake*	
	Glacial*	Ice deposition* Fluvial deposition? Eolian deposition Lacustrine deposition?
Mixed	Wave-dominated*	Cliffed shore* Low open shore*
	Tide-dominated	Tidal flat* Shore lagoon* Estuary
	Marine delta*	
	Organic reef*	Atoll Barrier reef Other reef
Marine	Neritic*	
	Bathyal-abyssal*	Open ocean bottom* Closed basin

Note: Asterisks and queries indicate inclusion or possible inclusion of category in symposium (determined on basis of prepresentation abstracts).

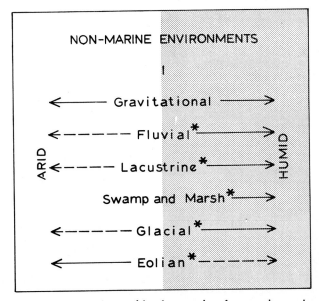

FIG. 3.—Alternative list of second-level categories of nonmarine environments.

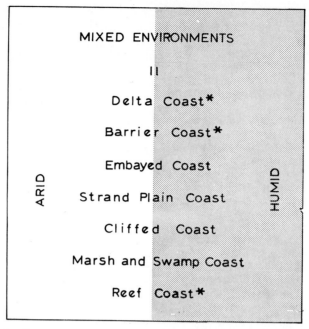

Fig. 4.—Alternative list of second-level categories of mixed nonmarine and marine environments.

central California. Marsh and swamp coasts, like other low coastal environments, run the risk of becoming barrier coasts, but enough maintain their independence to justify a second-level placement. The strand-plain coast environment would include the beach-ridge-and-marsh environment at the prograding edge of an area such as the Costa de Nayarit in Mexico or the chenier plain of Louisiana; some geologists might prefer to classify such environments as prograding barrier coast.

Marine depositional environments, including the major ocean basins and the marginal and mediterranean or epicontinental seas, are not easily confined within a fixed classification. Divisions of the comprehensive physiographic classification of the ocean floor developed by Heezen, Tharp, and Ewing (1959) are helpful, but do not closely correspond to depositional patterns and do not provide a direct basis for environmental classification. The physiographic and other factors that influence marine environments and their classification have been compressed by Sverdrup, Johnson, and Fleming (1942) into three groups: first, the general topography and depth of the site of deposition; second, the relation of the site of deposition to the source of inorganic material; and third, the physical and chemical conditions in the water overlying the site of deposition. Some investigators would add a biological factor. But these elements of environments do not, in themselves, set boundaries to divide the sea for depositional classification. Work in progress by H. E. Clifton, U.S. Geological Survey, and others may be of substantial assistance in dividing the nearshore and offshore marine zones on the basis of energy levels and resulting sedimentary structures.

Figure 5 is one of several possible worksheets of marine environmental factors within which specific depositional environments can be located with minimum descriptive designations. Except for the customary approximate limits of 100 and 1000 fathoms for neritic and bathyal depths, the factors shown are qualitative. On this version of the worksheet, some of the Sverdrup environmental factors are present by implication rather than by direct statement. For example, relative increase in distance from terrigenous sources accompanies depth changes. Or, in place of depth changes, one could substitute continental shelf and slope and deep ocean floor to represent general topography; "restricted" and "open circulation" imply differing physical and chemical factors and also the "closed basins" and "open ocean bottom" of Dunbar and Rodgers' bathyal-abyssal category.

The usefulness of such a chart as is shown in figure 5 may be in the questions raised, rather than in what is stated. (1) Can a boundary be defined between environments in which significant amounts of calcium carbonate accum-

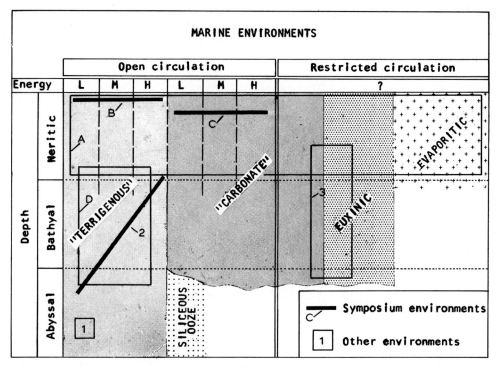

Fig. 5.—Experimental worksheet for showing relation between marine depositional environments in terms of depth, energy, circulation, and sediment character; e.g., environments in symposium: A. shallow marine; B. shoreline (nearshore); C. reefs (part); D. channelized slope and base of slope in marine basins; other environments: 1. red-clay deeps; 2. turbidity current; 3. barred basin. L, low energy; M, medium energy; H, high energy.

ulate and those in which it is not environmentally significant? (2) How should environmental energy levels be distinguished in deep water? Should maximum current energy be described as low because comparable energy is called low in the shallow neritic zone? The diagonal bar in figure 5 is one plotting of the depositional environment of a theoretical turbidity current, but it might be drawn otherwise, if the energy labels are shifted at depth.

In summary, the primary basis of classification for nonmarine environments may be depositional process or medium, and for mixed environments, geomorphic place; for marine environments, no single factor has emerged as dominant. Existing classifications are neither fixed nor final, and the value of an ultimate classification may be more theoretical than practical. Review of environmental literature suggests that, at this time, it is especially important for each writer to define his use of terms and to keep his discussions consistent with his own definitions.

For those concerned with devising classifications of environments of deposition, a sample of an intermediate stage in development of a classification follows this paragraph. The original listing was compiled by F. P. Shepard, Scripps Institute of Oceanography, and E. D. McKee, U.S. Geological Survey, in conference; additions and modifications were made by McKee and the writer, in part on the basis of available literature. Obviously, at this stage, the items listed have not been fully adjusted to the levels of classification shown in preceding tables and figures, nor have such adjustments been made as inclusion of beaches in all possible higher-level categories. Many of the terms would have to be modified to achieve the degree of consistency in terminology recommended in this paper. The listing does provide, however, terms for third- and fourth-level features not included elsewhere in this paper because such groups of terms are incomplete or are not available for many of the categories of more formal classification. A few references to helpful publications are included, but no attempt has been made to cite all of the material consulted.

ENVIRONMENTS OF DEPOSITION
(preliminary listing)

A. Terrestrial
   1. Landslide
   2. Talus
   3. Alluvial fans and plains (Blissenbach, 1954; Bull, 1964)
   4. River channels (Sundborg, 1956; Wolman and Leopold, 1957; Harms and others, 1963; Ore, 1964) Point bars, natural levees, ox bows
   5. Floodplains (see references under *River channels*)
   6. Glacial moraines
   7. Outwash plains
   8. Dunes (Hack, 1941; McKee and Tibbitts, 1964; McKee, 1966)
      a. Unidirectional wind types
         (1) Dome-shaped
         (2) Transverse
         (3) Barchan
         (4) Parabolic and blow-out
      b. Bidirectional wind types
         (1) Longitudinal (seif)
         (2) Reversing
      c. Multidirectional wind types: Star
B. Lacustrine
   1. Shallow lakes and playas
      a. Normal circulation
      b. Undrained (salt)
   2. Deep lakes
C. Delta and delta complex (Allen, 1965; Coleman and others, 1964; Fisk and others, 1954; Scruton, 1955; Shepard, 1956; van Straaten, 1959a)
   1. Types of deltas
      a. Embayed coast, lake
      b. Unprotected coast
         (1) Birds foot
         (2) Arcuate
         (3) Lobate
         (4) Cuspate
   2. Parts of deltas
      a. Subaerial or onshore
         (1) Distributary channel
         (2) Natural levee
         (3) Marsh and swamp
         (4) Interdistributary bay or trough
         (5) Beach
      b. Subaqueous or offshore
         (1) Channel and levee extensions
         (2) Distributary mouth bar
         (3) Delta front platform
         (4) Prodelta slope
         (5) Open shelf with delta influence
D. Beach (Inman, 1960; Shepard, 1963): Parts of beach
   1. Backshore
   2. Berm
   3. Foreshore (zone of swash)
E. Nearshore zone (approximately zone of single wave and breaker; see references, *Continental shelf*)
F. Offshore zone (see references under *Continental shelf*)
G. Barrier (Shepard and Moore, 1955; Fisk, 1959; Price, 1958; Phleger and Ewing, 1962): Parts of barrier
   1. Beach
   2. Dune field
   3. Barrier flat
   4. Washover fan
   5. Inlet
H. Bar (submerged)
   1. Longshore bar
   2. Bay bar
I. Tidal flat area (van Straaten, 1959b)
   1. Salt marsh
   2. Tidal flat
   3. Tidal channel
J. Lagoon (van Straaten, 1959b; Emery, Stevenson, and Hedgepeth, 1957; Phleger and Ewing, 1962; Fisk, 1959)
   1. Hypersaline (cut off from the sea)
   2. Brackish or normal (intermediate)
   3. Fresh (fed by streams)
K. Estuary (articles *in* Lauff, 1967)
   1. Shallow
   2. Deep
L. Continental shelf (Gorsline, 1963; Niino and Emery, 1961; Shepard and Moore, 1955; Greenman and LeBlanc, 1956; van Straaten, 1959b)
M. Epicontinental Sea
N. Deep intracontinental depression
   1. Trough
   2. Basin
      a. Elongate
      b. Subcircular
O. Continental borderland
   1. Basin
   2. Trough
P. Continental slope
Q. Deep sea
   1. Deep sea fan
   2. Abyssal plain: Areas of
      a. Organic debris: siliceous; calcareous
      b. Clay deposits
      c. Phillipsite and manganese
   3. Marine areas marginal to glaciers
R. Organic reef
   1. Linear reef, atoll, barrier
   2. Patch reef (pinnacle and knoll)
   3. Fringing reef

REFERENCES

Allen, J. R. L., 1965, Late Quaternary Niger delta, and adjacent areas: Sedimentary environments and lithofacies: Am. Assoc. Petroleum Geologists Bull., v. 49, p. 547–600.
Blissenbach, Eric, 1954, Geology of alluvial fans in semiarid regions: Geol. Soc. America Bull., v. 65, no. 2, p. 175–190.
Bull, W. B., 1964, Alluvial fans and near-surface subsidence in western Fresno County, California; U. S. Geol. Survey Prof. Paper 437-A, p. A1–A71.
Coleman, J. M., Gagliano, S. M., and Webb, J. E., 1964, Minor sedimentary structures in a prograding distributary: Marine Geol., v. 1, no. 3, p. 240–258.
Dunbar, C. O., and Rodgers, John, 1957, Principles of stratigraphy: New York, John Wiley and Sons, 356 p.
Emery, K. O., Stevenson, R. E., and Hedgepeth, J. W., 1957, Estuaries and lagoons, *in* Treatise on marine ecology and paleoecology: Geol. Soc. America Mem. 67, v. 1, p. 673–749.
Fisk, H. N., 1959, Padre Island and the Laguna Madre Flats, coastal south Texas, *in* Russell, R. J., chm., Coastal geography conf.: 2nd., April 1959, p. 103–151.
Fisk, H. N., McFarlan, Edward, Jr., Kolb, C. R., and Wilbert, L. J., Jr., 1954, Sedimentary framework of the modern Mississippi delta: Jour. Sed. Petrology, v. 24, no. 2, p. 76–99.
Gorsline, D. S., 1963, Bottom sediments of the Atlantic shelf and slope off the southern United States: Jour. Geology, v. 71, no. 4, p. 422–440.
Greenman, N. M., and LeBlanc, R. J., 1956, Recent marine sediments and environments of northwest Gulf of Mexico: Am. Assoc. Petroleum Geologists Bull., v. 40, no. 5, p. 813–847.
Hack, J. T., 1941, Dunes of the western Navajo Country: The Geographical Review, v. 31, no. 2, p. 240–263.
Harms, J. C., MacKenzie, D. B., McCubbin, D. G., 1963, Stratification in modern sands of the Red River, Louisiana: Jour. Geol., v. 71, no. 5, p. 566–580.
Heezen, B. C., Tharp, Marie, and Ewing, W. M., 1959, The North Atlantic—text to accompany the physiographic diagram of the North Atlantic, [Pt.] 1 *of* The floors of the oceans: Geol. Soc. America Spec. Paper 65, 122 p.
Inman, D. L., 1960, Shore processes, *in* McGraw-Hill Encyclopedia of Science and Technology: New York, McGraw-Hill, p. 299–306.
Krumbein, W. C., and Sloss, L. L., 1951, Stratigraphy and sedimentation: San Francisco, Calif., W. H. Freeman and Co., 497 p.
Lauff, G. H., ed., 1967, Estuaries [a symposium]: Am. Assoc. Adv. Sci. Pub. no. 83, 757 p.
McGill, J. T., 1959, Coastal classification maps—a review, *in* Russell, R. J., chm., 2d Coastal Geography Conf.: Coastal Studies Institute, Louisiana State University, April 6–9, 1959: Washington, D. C., Geography Br., [U.S.] Office Naval Research, p. 1–21.
McKee, E. D., 1966, Structures of dunes at White Sands National Monument, New Mexico (and a comparison with structures of dunes from other selected areas): Sedimentology, v. 7, no. 1, 69 p.
———, and Tibbitts, G. C., Jr., 1964, Primary structures of a seif dune and associated deposits in Libya: Jour. Sed. Petrology, v. 34, no. 1, p. 5–17.
Niino, Hiroski, and Emery, K. O., 1961, Sediments of shallow portions of East China Sea and South China Sea: Geol. Soc. America Bull., v. 72, no. 5, p. 731–762.
Ore, H. T., 1964, Some criteria for recognition of braided stream deposits: Wyoming Univ. Contr. Geol., v. 3, no. 1, p. 1–14.
Phleger, F. B., and Ewing, G. C., 1962, Sedimentology and oceanography of coastal lagoons in Baja California, Mexico: Geol. Soc. America Bull., v. 73, no. 2, p. 145–182.
Price, W. A., 1958, Sedimentology and Quaternary geomorphology of south Texas: Gulf Coast Assoc. Geol. Socs. Trans., v. 8, p. 41–75.
Scruton, P. C., 1955, Sediments of the eastern Mississippi delta, *in* Finding ancient shorelines: Soc. Econ. Paleontologists and Mineralogists Spec. Pub. 3, p. 21–51.
Shepard, F. P., 1956, Marginal sediments of Mississippi delta: Am. Assoc. Petroleum Geologists Bull., v. 40, no. 11, p. 2537–2623.
———, 1963, Submarine geology: New York, Harper and Row, 2d ed., 557 p.
———, and Moore, D. G., 1955, Central Texas coast sedimentation: Characteristics of sedimentary environment, recent history, and diagenesis: Am. Assoc. Petroleum Geologists Bull., v. 39, no. 8, p. 1463–1593.
Sundborg, Ake, 1956, The river Klarälven—a study of fluvial processes: Geog. Ann., Stockholm, arg, 38, h. 2–3, p. 125–316; Uppsala Univ., Geog. Inst., Medd. s. A., no. 115.
Sverdrup, H. U., Johnson, M. W., and Fleming, R. H., 1942, The oceans, their physics, chemistry, and general biology: New York, Prentice-Hall, 1087 p.
Twenhofel, W. H., 1939, Principles of sedimentation: New York, McGraw-Hill, 610 p.
van Straaten, L. M. J. U., 1959a, Littoral and submarine morphology of the Rhone delta: National Acad. Sci.—National Research Council, 2d Coastal Geograph. Conf., p. 233–264.
———, 1959b, Minor structures of some recent littoral and neritic sediments, *in* Symposium: Sedimentology of Recent and old sediments: Geologie en Mijnbouw (NW. ser.), 21e Jaargang, p. 197–216.
Wolman, M. G., and Leopold, L. B., 1957, River flood plains—some observations on their formation: U. S. Geol. Survey Prof. Paper 282-C, p. 87–109.

EOLIAN ENVIRONMENTS: THEIR CHARACTERISTICS, RECOGNITION, AND IMPORTANCE

JOÃO JOSÉ BIGARELLA

Instituto de Geologia, Universidade do Paraná, Curitiba, Paraná, Brasil

ABSTRACT

Eolian sandstone generally has large- to medium-scale tabular-planar and wedge-planar cross-beds. Trough cross-bedding is much less common. Cross-beds are composed mostly of steeply dipping laminae which normally are concave upward. In modern dunes the foreset beds near the top of the slip face have steep (29°–34°) dips, but in paleodunes this value is somewhat less (20°–29°), probably because erosion commonly removed the steeper, upper part of each lamina prior to deposition of the overlying set.

Dune cross-bedding is distinguished from other similar structures partly on the basis of more uniform grain size. The nature of the adjacent and/or intercalated beds may help to determine the environment of deposition. Attitude of the bounding surface also is a diagnostic feature.

In the absence of cross-bedding, other criteria are used to identify dune environments. Textural and mineralogical characteristics are not conclusive, and mean grain size seems to be of little value. Although much dune sand is slightly better sorted than other similar sediments, sorting is not distinctive. Positive skewness has been considered as an indication of dune environment; however, negative skewness also has been reported for dune sediments. Dune sand usually, but not always, is more rounded than beach sand.

Dune and beach sediments can be separated on the basis of the heavy-to-light mineral ratio, and the relation between the settling velocities of two or more minerals of different density values.

A combination of criteria, together with the stratigraphic relations of the deposit to adjacent beds, should be used in identifying a dune-sand envirionment.

INTRODUCTION

The purpose of this paper is to review sedimentary characteristics useful in identification of eolian environments. Emphasis is given to the interpretation of the ancient sandstones and of Quaternary structureless sand deposits with a review of the available literature. Results of this investigation include descriptions of the following: 1) Recent dune and beach structures; 2) cross-bedding as an indicator of prevailing winds; 3) paleowind patterns as derived from orientation of ancient dune cross-strata.

The bibliography on sand dunes is vast. Most references involve general problems such as geography, morphology, classification, genesis, transportation of sand by wind, and dune movement. Such topics are extensively treated in the scientific literature of many countries, especially in Europe and include classical papers by Sokolow (1894), Cornish (1897), Jentsch (1900), Solger (1910), Beadnell (1910), and Braun (1911).

The present work does not review all the early literature, but summarizes data and definitions from the more important studies, both classical and recent. Significant regional studies on coastal dunes are by Cressey (1928), Price (1958), and Cooper (1958, 1967) in the United States. Desert dune studies are by Madigan (1936) in Australia, Capot-Rey and Capot-Rey (1948) in the Sahara, and Holm (1960) in Arabia. Studies of structures shown in trenches cutting modern dunes have been made by McKee and Tibbitts (1964), McKee (1966), and Bigarella, Becker, and Duarte (1969). Many dune types still remain to be examined before geologists can be sure of correct interpretation of ancient eolian sandstones.

DUNE ENVIRONMENT

The characteristics of dunes depend to a great extent on environmental conditions. Dunes are formed (1) in all extremes of latitude, either as coastal features or in interior areas; and (2) under different types of climates ranging from extremely arid to humid.

Eolian topography, comprising deflation plains and concentrations of large sand bodies (dunes) is widely distributed, not only in dry regions, but also in humid regions. Eolian topography develops where a sufficient sand supply is available and where winds are strong enough to shift the sands, even in areas where the rainfall is relatively high, as along the coasts of northwestern United States and of southern Brazil.

The concentration of sand in relatively small areas, as shown by Kuhlman (1959c, p. 69), must be caused by sporadic changes in the capacity of the wind. The local accumulation of sand to form dunes depends on the nature of the deflation plain, or to obstacles such as grass and shrubs. A distinction may be made, according to Kuhlman, between a dune in which surface form controls its development and one caused by local changes in wind intensity.

In coastal regions, the landward transport of beach sands by the wind is an important process in the development of the coastal morphology. During the Quaternary, especially, a considerable amount of eolian sand was deposited on coastal plains. Active dune areas were much larger then than at present, and today these former dune fields are largely stabilized by vegetation. In many coastal areas, sand-binding plants above the shore are responsible for the development of a foredune zone consisting of hillocks or ridges.

Although the coastal dunes have several features in common with desert dunes, in some respects they are more complex, as their character depends to a great extent on the vegetation that grows on them (King, 1966, p. 136). In areas where the sand supply is not great, coastal dunes are limited to beach-dune ridges or simple foredune forms.

Large sand deserts are most common in areas of low relief, generally in long, narrow, low plains or in broad basins of low relief (Holm, 1960, p. 1370). Many of them are located in old alluvial plains (Fédorovich, 1957, p. 127). In desert regions, the sand is carried upslope, over topographic rises, and deposited on surfaces beyond where the angle of the slope is small (Bagnold, 1954; Holm, 1960, p. 1370).

Characteristics of desert dunes depend largely on the nature of the desert floor, on the supply of sand available, and on the wind regimen. The direction of the wind and its strength in relation to the sand supply are very important in determining the form of a dune (King, 1966, p. 134). The floor of the basin, adjacent to the dune field, may contain sand, usually trapped by pebbles (King, 1966, p. 134), which is immobile during periods of gentle winds but which becomes active during strong winds. At such times the moving sand tends to be deposited as sand patches. Strong winds have a tendency to increase the bulk of a dune, whereas the gentler winds tend to increase the length at the expense of the bulk (King, 1966, p. 134).

Recent studies on wind-driven sand have been made by Bagnold (1954), Ford (1957), Kuhlman (1959, a, b), Norris and Norris (1961), Sharp (1964), and McKee (1966), among others.

The relative movement of barchan dunes in southern Peru was determined by Finkel (1959, p. 628) from two sets of aerial photographs taken three years apart. The speed of movement of the barchans was greater for the smaller dunes and less for the larger ones. The following figures obtained by Finkel show an average movement of 75 barchans in relation to their crest heights.

Crest heights in meters:	1.0	2.0	3.0	4.0	5.0	6.0	7.0	Average 3.67
Annual movement in meters:	32.2	22.0	16.8	14.5	12.2	11.5	9.2	15.4

The forward movement affects the entire barchan without changing the size of the dune. It is made mostly by sliding on the slipface, caused by sand which is blown directly up the windward slope and over the crest. The sand leaves the dune between the horns, crossing the bare stretches between barchans and accumulating in the next dune which obstructs its path. The rate of sand loss from any one dune is roughly compensated by sand coming from the upwind barchan (Finkel, 1959, p. 630).

In coastal dunes of Lake Michigan the following figures for movement have been determined by Cressey (1928, p. 47): 5 feet in 4 months, 10 feet in 6 months, and 3 feet in 11 months.

At Guerrero Negro dune field, Mexico, the rate of movement in the winter has been determined to be 2.1 cm/day, whereas in the summer it is 8.4 cm/day (Inman, Ewing, and Corliss, 1966).

Parabolic dunes in Denmark are at present stabilized; however, they had a total movement of 350–750 meters during a period of about 90 years between 1795 and 1886, as can be demonstrated (Hansen, 1959, p. 90) by analysis of old maps (Videnskabernes Selskabs Kort and Geodetic Survey).

The rate of dune movement is not uniform. Barchan dunes of the Salton Sea area had average movement of 50 feet per year during the period 1941–1956, and 85 feet per year during the years 1956–1963 (Long and Sharp, 1964). The difference in the average movement was attributed by these authors to an increased sand supply.

Dune movement has been studied at White Sands by McKee (1966), where dunes close to the source area showed considerably greater movement than those downwind. Parabolic dunes farthest downwind not only moved less, but actually retreated on their windward sides because of accumulating sand (McKee, 1966, p. 10).

Measurements of the quantity of wind-transported sand under both dry and humid conditions were made in Skallingen, Denmark, by Kuhlman (1959a, b). The restraining effect of rain and moisture, surprisingly, was very small. Sand was able to drift again only a few minutes after a shower. Lenses of dry sand surrounded by wet sand developed close to the surface. These lenses were parallel to the surface, and did not become moist after rainy periods of hours or even days. Their presence seems important to sand drift in a moist climate for they permit sand transport even in rainy periods, provided the wind velocity is sufficiently great (Kuhlman, 1959a, p. 18).

In the large desert regions of the world, dunes dominate the topography and assume a variety of forms and sizes. In the Arabian desert, for example, most pyramidal dunes are more than 50 m in height but range up to 150 m, with diameters of 1 to 2 km (Holm 1960, p. 1373). Sigmoidal dunes in the same area range in size from tiny sharp ridges a few meters high up to large ridges 100 m high. In the southern Rub'al Khali desert a parallel system of complex linear dune ridges attains heights of 100 m and lengths of 20 to 200 km (Holm, 1960, p. 1373). Most of the ridges are 1 to 2 km wide.

In the barchan belts of southern Egypt and northern Sudan, the average height of dunes does not exceed 10 m (Bagnold, 1954, p. 219). The maximum height of a barchan dune in this area, according to Bagnold, is about 30 m.

The upwind portions of most dune fields are crowded with sand masses because the supply of sand is great. In barchan chains the crescentic forms tend to be irregular, whereas downwind portions are more isolated and generally develop more perfect forms (Bagnold, 1954, p. 218). The Algodones dunes in California are chains of giant barchans with slipfaces 60 to 100 feet high (Norris and Norris, 1961, p. 608). In the upwind area the dunes closely overlap, developing imperfect forms; downwind large, intradune, sand-free areas occur in which small barchans develop. These barchans have slipfaces from less than 3 feet to about 6 feet high (Norris, 1966, p. 292). The dune field ends rather abruptly.

Sand areas barren of dunes may be the result of rough ground surfaces over which the sand is bypassed owing to increased grain trajectories (Norris, 1966, p. 305). Sand tends to deposit on smooth or sandy, rather than rocky, surfaces. Mostly areas free of sand develop downwind from dunes during periods of strong cross-winds (Bagnold, 1954, p. 175). The sand-free depressions inside a dune field may be interrupted by small barchans or low sand ridges. In coastal areas such depressions may be covered by vegetation, especially grass, and may be filled with water during rainy seasons.

TEXTURAL AND COMPOSITIONAL FEATURES OF DUNES (EDITORIAL COMMENT)

In the preparation of this paper, original plans by the author called for a review of the pertinent literature in this field and an appraisal of its possible application and usefulness in determining criteria for the recognition of eolian depositional environments. Unfortunately, limitations of space in the present symposium make impossible the inclusion of the rather lengthy synthesis of textural and compositional studies of dune sands that were prepared. This synthesis included both a review of literature and contributions from original studies by the author and his associates. From it, the following broad conclusions that deal directly with the subject of this symposium are presented.

In general, textural and mineralogical characteristics, although helpful, cannot be considered diagnostic in the recognition of dune deposits. The mean grain size seems to be of little significance. Sorting usually is good and, although commonly a little better in dune sand than in various similarly sorted sediments, is not dependable. Positive skewness, considered by some an indicator of dune environment, is open to question because negative skewness has also been recorded in dune sands. Roundness of grains usually, but not always, is greater in dune than in beach sands, but at best is significant only in comparing adjacent sand environments. In such comparisons, separations on the basis of ratios of heavy to light minerals are, perhaps, most dependable.

SEDIMENTARY STRUCTURES
General

Probably the most important criteria for recognizing wind-laid deposits are distinctive features of sedimentary structure developed during deposition. These features are not only useful in distinguishing wind-laid sands from water-deposited sands but also in differentiating the main types of dunes. Information on structures is essential to the interpretation of dune characteristics preserved in the geologic record, as well as to the determination of wind trends. Measurement of true dips in cross-strata enables the determination of paleowind patterns responsible for the deposition both of recent and subrecent dunes and of ancient eolian sandstones. Furthermore, these data may be used for the determination of paleolatitudes and paleoclimatic conditions.

Description and Classification of Cross Strata

Sedimentary structures commonly are well exposed in outcrops of eolian sandstones. Most of the cross strata are composed of steeply-dipping laminae which normally are concave upward. Sets of cross strata are separated by bounding surfaces which may be either plane or curved. Plane bounding surfaces may be either parallel or convergent. These surfaces were developed as either erosional or nonerosional features. Most sets of cross strata with dips greater than 10° probably were deposited on the downwind sides of dunes. Strata dipping upwind are rarely preserved. Low-angle (<10°) cross-strata sets possibly represent deposits formed upwind or, more probably, near the base of downwind strata. The upwind strata of transverse dunes have been called topset (Shotton, 1937, p. 544) and backset (Smith, 1940, p. 161). Mostly they dip with angles less than 10°. Maximum dip angles up to 12° for the windward strata are recorded by several authors (McKee, 1945, p. 315; 1957, p. 1720; Cooper, 1958, p. 29; Laming, 1958, p. 180) but they may reach 16° in reversing dunes (Sharp, 1966, p. 1062). This last figure seems to be exceptional; the high dip is possibly caused by continuous reversals of piling up of sand under the action of winds with almost opposite directions.

In some places, large sand grains or even small pebbles cover bounding surfaces as a result of deflation.

The most steeply dipping cross strata are on the upper parts of slipfaces. Downslope, the angle of dip normally decreases. Near the base, most strata tend to become tangential with the underlying bounding surface and approach the horizontal. Measurements of modern dunes show the foreset strata near the top of the slipface with 29°–33° dips. In most paleodunes this value seems to be 20°–29°, probably because the surfaces preserved do not represent the original upper parts of slipfaces; such parts doubtless were eroded before deposition of the next overlying sets. The lower parts of the lee faces, therefore, constitute the main material represented in cross strata of eolian sandstone.

A great thickness of a set of cross strata has been considered as a criterion of the eolian origin of ancient sandstones (McKee, 1933; Reiche, 1938; Opdyke and Runcorn, 1960). This criterion applies especially to large dunes where a great part of the slip face has been preserved. In many occurrences of eolian cross-bedding, medium-sized cross strata also are characteristic. In the Botucatu paleodesert many outcrops contain large-scale cross strata with thicknesses of 10 m or even more; however, the cross strata commonly are less than 1 or 2 m in thickness.

Dips greater than 33° to 34° are difficult to explain. On Mustang Island, Texas, dips up to 42° occur. The suggestion is given that they result from salt spray along the beach giving to the sand a brine coating which acts as a cement and permits high angles of repose (McBride and Hayes, 1962, p. 550). In dunes at Rio Grande do Sul, Brazil, dips up to 46° have been measured, and at Praia de Leste in Paraná, some are recorded up to 39° and 42° (Bigarella, Becker, and Duarte, 1969). These latter dips are interpreted as the result of high moisture content of the sand. Moist grains stick together under action of surface tension and permit high angles of repose.

Classification of cross stratification is based on the character of lower boundary surfaces of the sets and on the shape of each set (McKee and Weir, 1953). In planar cross-stratification the lower bounding surface is a flat plane. The planar type is subdivided into (a) tabular planar, in which the bounding surfaces are practically parallel for long distances, and (b) wedge planar, in which the flat bounding surfaces are not parallel but convergent. In trough cross-stratification lower surfaces are curved. Where lower bounding surfaces are nonerosional, the term *simple* cross-stratification applies.

In dunes at White Sands, New Mexico, the tabular planar type of cross-stratification is the most common (McKee, 1966, p. 57). Wedge planar cross-strata sets are also present, but less common. The thin tabular sets generally end as wedges. Upwind, the bounding surfaces of the tabular sets are nearly horizontal or at very low angles. They change abruptly downwind into moderate- to high-angle (15°–25°), dipping surfaces. Trough cross-stratification, according to McKee (1966, p. 57), is represented in all the dune types of White Sands but is not a dominant structure in any of them. It is conspicuous in sections normal to the wind direction.

METHOD OF STUDY

The primary structures of recent dunes have been extensively studied in recent years. Seif dunes, formed by bidirectional winds in the Libyan desert, were examined by McKee and Tibbitts (1964). Four types of dunes developed by unidirectional wind in the White Sands National Monument, New Mexico, consisting of dome-shaped, transverse, barchan, and parabolic dunes, were analyzed by McKee (1966). The structures of coastal dune precipitation ridges at Praia de Leste (Paraná) have been studied by Bigarella, Becker, and Duarte (1969). In these studies, attempts were made to determine which structural features were common to or distinctive of each dune type.

In all studies, internal structures were plotted on graph paper at a convenient scale for illustrating significant features of the trench wall. For example, in the Praia de Leste precipitation ridges a grid was prepared using large nails and string (fig. 1). Several strata and the bounding surfaces were emphasized to show clearly in photographs.

Dune Types and Their Stratification

In dunes developed by unidirectional winds minor shifts in wind direction are not effective in the development of any distinctive features other than small and temporary irregularities. In dune fields where strong and persistent winds come from different directions, several types of dunes develop. The longitudinal or seif dune is attributed to the action of strong winds at approximately 90° to each other. Reversing dunes are formed when bidirectional winds blow in almost opposite directions. Multiple directional effective winds develop compound dunes like the star dunes of Saudi Arabia. The structures of many dune types are not yet known.

Dome-shaped dune.—Dome-shaped dunes, as illustrated at White Sands, New Mexico, are

FIG. 1.—Levelled grid pattern on wall of dune trench, using string and nail to indicate intervals either of 50 or 100 cm. The boundary surfaces were reinforced for drawing purposes. Jardim São Pedro precipitation ridge dune, Praia de Leste, Paraná (after Bigarella, Becker & Duarte, in press).

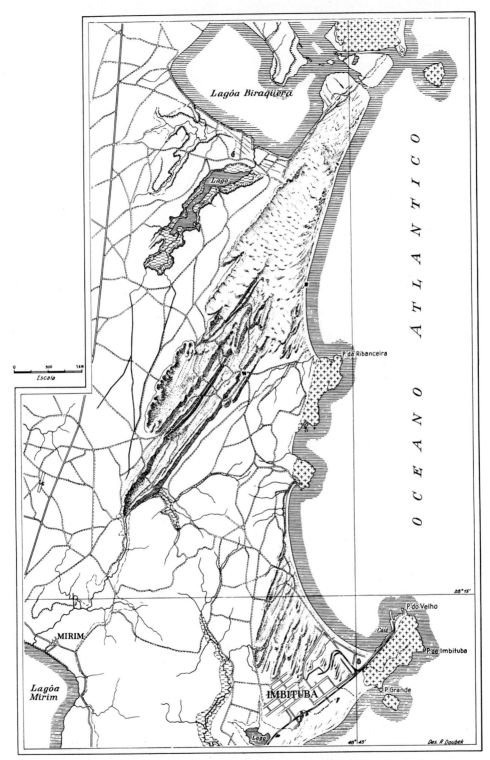

Fig. 2.—Typical transverse and parabolic dunes at the coast near Imbituba, State of Santa Catarina. The arms of the parabolic dunes attain great development originating a series of longitudinal ridges. The remaining area (with the exception of the crystalline hills) is made up mostly by a hilly sandy terrain constituted by stabilized, structureless and modified pleistocenic dunes. The longitudinal ridges are oriented parallel to the prevailing northeasterly winds. Drawing based on vertical air photos.

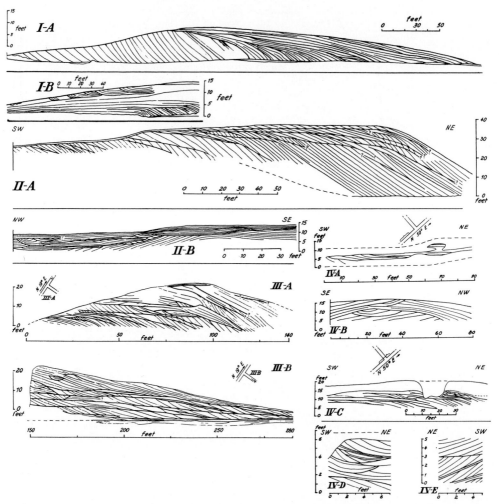

Fig. 3.—Dune structures at White Sands National Monument, New Mexico (based on McKee, 1966) I—cross-sections of dome-shaped dune; A—main trench oriented W-E; B—side trench, west wall, oriented S-N. II—Cross-sections of transverse dune; A—main trench, north wall, oriented SW-NE; B—side trench, east wall, oriented NW-SE. III—Cross-sections of barchan dune; A—main trench, northwest wall, oriented SW-NE; B—side trench, northeast wall, oriented NW-SE. IV—cross-sections of parabolic dune; A—main trench, northwest wall; B—side trench, southwest wall; C—main trench, northwest wall; D and E—details of the structures in the main trench. Lines queried(?) where the structure was obscure in the field.

located close to the source area (McKee, 1966, p. 26–27). They are low, circular, isolated mounds. A typical example was 450 feet (137 m) wide, 420 feet (128 m) across, in the direction of the predominant wind, and 18 feet (5.5 m) high.

Structures were exposed in two trenches approximately parallel and at right angles to the dominant wind direction. In the parallel trench, the structures showed two main phases in the evolution of the dome-shaped dune (fig. 3, 1A and 1B). The older phase possibly represented a former transverse-type dune with strata characterized by a continuous sequence of steeply dipping (28°–33°) foresets. Such strata occurred on the windward side of the dune up to its crest. Beyond this point both strata and bounding surface of sets were somewhat less steep, and on the lee part of the dune the dips were progressively lower.

Above the early structures a younger phase of deposition was developed with low-angle (3°–10°) strata. Some strata formed continuous planes extending to the lee base; others were topset planes leading into foresets (McKee, 1966, p. 27).

In the trench at right angles to the dominant wind direction, structures consisted largely of low-angle (1°–4°) strata dipping toward the

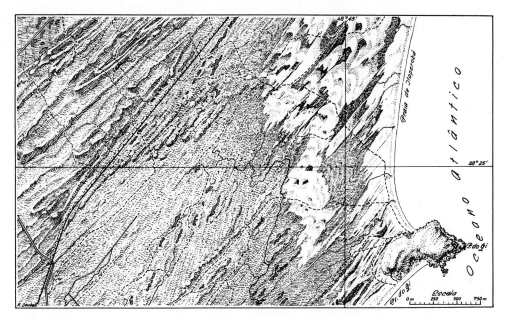

Fig. 4.—Barchan and parabolic dune at the coast of Santa Catarina, southern Brazil. The prevailing winds blowing from NE are the effective winds for sand transport at the Itaperobá Beach.

dune margin. Near the dune center, steep strata indicated that the dune crest had migrated laterally. In the upper part of the trench a series of cut-and-fill structures recorded erosion from fluctuating winds or velocity (McKee, 1966, p. 27).

Transverse dune.—A transverse dune is an almost straight sand ridge oriented at approximately right angles to the dominant, effective wind direction (fig. 2). The structure of this type of dune in the western part of the White Sands dune field, New Mexico, has been described by McKee (1966, p. 27, 31, 39). The dune was 400 feet (122 m) across in the direction of the dominant wind. The crest was nearly straight for 800 feet (244 m) and its maximum height was about 40 feet (12.2 m). Structures were exposed in two trenches (fig. 3, IIA and IIB).

In the main trench, oriented parallel to the dominant wind direction, two parts were recognized (fig. 3, IIA). The lower two-thirds of the dune sequence consisted of steeply dipping (30°–40°), large-scale, foreset strata. The upper part consisted of relatively thin, gently dipping to horizontal sets, which contained cross strata of moderate dip (11°–15°). Sets of cross strata in the upper part were progressively thinner and flatter upwards. In the uppermost part, upwind strata with windward dips of 5°–10° apparently were deposited and preserved (McKee, 1966, p. 31).

In the side trench, oriented parallel to the dune crest, both stratification and bounding planes of various sets were nearly horizontal or dipping at very low angles (fig. 3, IIB). Within the low-angle stratification was a series of troughs cut parallel to the wind direction and subsequently filled with symmetrical or asymmetrical curving laminae, which, when numerous, developed a festoon pattern (McKee, 1966, p. 29).

Barchan dune.—Structures of a barchan, a crescentic form with horns extending downwind (fig. 4), were studied in the White Sands dune field, New Mexico (McKee, 1966, p. 39–40). The dune was 170 feet (52 m) across its central part in the direction of the dominant wind, 290 feet (88 m) normal to this, from horn to horn, and 27 feet (8.2 m) high.

A main trench was dug normal to the crescent. Most strata showed leeward dips. The early stages of sand accumulation consisted of cross-strata dipping 26° to 34° and bounded by surfaces of low-angle downwind dip (2°–6°). Later deposits had both bounding surfaces and strata of steep dip. Among the leeward-dipping strata were a few that dipped (2°–5°) upwind. These probably represented either windward-side deposits or the result of a change in wind direction (fig. 3, IIIA).

A side trench, cut approximately normal to the main trench, showed structures in a curving barchan horn (fig. 3, IIIB). The cross-strata sets dipping 12°–23° were bounded by nearly flat-lying surfaces. Near the central part of the dunes, several trough structures indicated probable blowouts caused by fluctuations in the wind regime (McKee, 1966, p. 40).

Parabolic dune.—A parabolic dune is U-shaped or V-shaped, representing a type of blowout in which the middle part has moved forward with respect to the sides or arms (Hack, 1941, p. 242) (fig. 4). These arms commonly are anchored by vegetation and the entire dune is relatively stable. A typical V-shaped, parabolic dune in the downwind side of the White Sands dune field, New Mexico, has been described by McKee (1966, p. 50–51). The surveyed dune had a width of 265 feet (80.8 m) and arms about 900 feet (274.5 m) long.

A main trench was cut through the central part of the dune approximately parallel to the dominant wind direction (fig. 3, IVA and IVB). The lower part of the dune presented high-angle (20°–34°) cross-strata sets with bounding surfaces dipping 10° to 20°. The upper part, especially on the windward side, consisted of strata dipping (<12°) leeward and, near the dune crest, of shallow, asymmetrically filled troughs. In this type of dune, a conspicuous feature is the concave-downward curvature of high-angle foresets near the dune front (McKee, 1966, p. 50).

In the side trench, oriented normal to the main one, the structures were of several types. They included strata with apparent dips of low to moderate angle (fig. 3, IVC).

Precipitation ridge dune.—A precipitation ridge as defined by Cooper (1967, p. 22) is a linear sand body formed where dunes advance against a forest barrier. It is the ultimate stabilized form in such situations and may gradually be enveloped by the forest. At the coastal plain of Praia de Leste in the State of Paraná, Brazil, one or more dune ridges lie parallel to the coast line. The height of these ridges above the sand plain is between 5 and 10 m, but recent construction projects along the coast have removed most of the low dune ridges. Internal structures of these dune ridges have been studied by Bigarella, Becker, and Duarte (1969).

The Praia de Leste dune sediments rest on backshore deposits (fig. 5, sec. IV). The lower part of the dune structures is formed by medium-scale, sinuous layers which closely resemble the structures of produnes (Bigarella and Popp, 1966, p. 143–148). (See foredune structures, p. 24.) They actually represent dome-shaped deposits of eolian sand over which dunes have developed with characteristic structure. Low-angle foresets of former produnes dipping in all directions occur at the base of the dune and may represent a distinctive feature of this type of dune.

Cross-stratification types at Praia de Leste are essentially the same as those described in dunes at White Sands, New Mexico (McKee, 1966). They consist dominantly of tabular planar sets of two classes: those in which the bounding surfaces have low-angle dips and those in which they dip at moderate to high degree. Some sets of simple (nonerosional) tabular form are present and, uncommonly, those of trough type.

Rose diagrams, corresponding to every section studied, illustrate the dip directions of cross-bedding at different parts of the dune at the Jardim São Pedro (fig. 6, upper right hand cor-

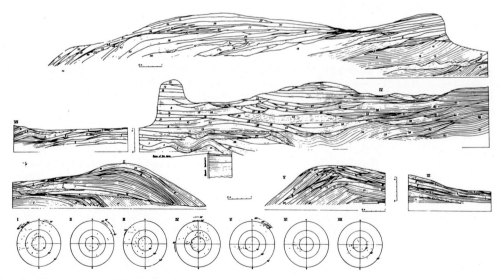

FIG. 5.—Cross-sections of the Jardim São Pedro precipitation ridge dune structures, Praia de Leste, Paraná. See figure 6 for tri-dimensional distribution of the sections. Stereonet refers to dip and dip direction for cross-strata from the several cross-sections (after Bigarella, Becker & Duarte, 1969).

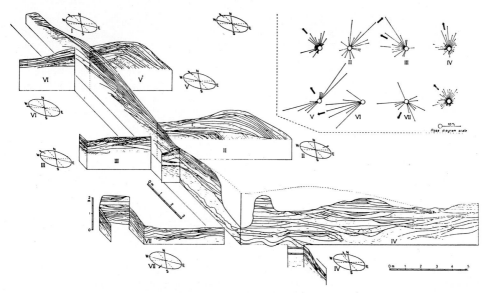

Fig. 6.—Diagram showing the tri-dimensional cross-section distribution of the cross-bedding pattern from the Jardim São Pedro precipitation ridge dune, Praia de Leste, Paraná. For every section, in a circumference the section dip direction resultant vector is indicated. In the upper right corner rose diagrams indicate dip directions and resultant dip vector for the several sections and for the whole dune (after Bigarella, Becker and Duarte, 1969).

ner). The average dip of the cross-bedding is 19.2°. Angles of repose between 34° and 39° have been measured: such unusually high dip values may result from humidity conditions at the time of deposition. The shape of the bounding surfaces is clearly indicated in figures 5 and 6. In sections normal to prevailing wind directions the bounding surfaces are roughly horizontal, whereas in profiles parallel to wind directions, they dip downwind. In the trenches cutting linguoid protuberances of the dune, boundary surfaces are strongly curving. The lower cross-bedding sets are larger than those above (some greater than 100 cm) and have higher dip angles; those above tend to be flatter and to follow in a general way the shape of the dune. The average thickness of these upper sets is mostly 10–30 cm (Bigarella, Becker, and Duarte, 1969).

About half of the cross-bedded sets of section VII and a few from sections II and IV (fig. 6) are located to windward. In the frontal part of section I and in sections V and VI (fig. 6) are sequences of cross-bedding with high angles of repose. Most of these high-angle surfaces face downwind to the northwest. Cross-bedding measurements indicate that, in addition to a concentration of high-dip angles downwind, many strata with high angles of repose ($>30°$) are spread over an arc of 170°. Possibly this spread decreases with the migration of the dune; however, no observations are available to show whether this is true or not (Bigarella, Becker, and Duarte, 1969).

Nearly horizontal layering in section IV (figs. 5 and 6), normal to the prevailing wind direction, is a feature also represented in transverse dunes (McKee, 1966, p. 59). It is characteristic only in a longitudinal profile, along the median part of the dune sand ridge, however; for outside this area, in the linguoid protuberance where crest and slipfaces are curved (section II, III, and V, fig. 6), the layering likewise is curved. Comparable structures are recorded in barchan and other dunes that have curving crest and slip-faces (McKee, 1966, p. 60).

In section V (fig. 6) a peculiar type of cross-stratification occurs, with cross-beds dipping in almost opposite directions. The same feature is less conspicuous in sections II and III (fig. 6). The shape of the linguoid protuberance probably is the main cause of this attitude in the strata. As the sand moved over the dune, it slipped off the linguoid body on all sides (Bigarella, Becker, and Duarte, 1969).

Longitudinal dune (seif).—A longitudinal or seif dune is a long, nearly straight sand ridge commonly parallel to other similar ridges and separated from them by wide, flat interdune surfaces. This type of dune normally is oriented along a vector resulting from two converging wind directions. It is common in parts of North Africa (Bagnold, 1954), in Australia (Madigan, 1936, p. 59), and in some other modern deserts.

Internal features of a seif dune located in the eastern margin of the Zallaf sand sea in southwestern Libya were studied by McKee and Tibbitts (1964). The structures were recorded from 9 test pits across the dune ridge. This dune was oriented with an east-west trend, being the product of bidirectional effective winds blowing in the morning from southeast and in the afternoon from northeast. Sand was deposited alternately on opposite sides of the dune crest. Cross strata did not dip in the wind direction but at right angles to the ridge. Cross strata within the dune mostly dipped 23° to 33°. However, near the base of the dune, the dip angle was considerable less.

Reversing dune.—Reversing dune, referred to by Merk (1960) as transverse dune, develops in an environment where the effect of the prevailing wind is opposed by strong, short-term, orographically controlled storm winds of almost opposite direction. In the Great Sand Dunes of Colorado, transverse ridges are oriented normal to the prevailing southwesterly winds (Merk, 1960). In the Kelso Dunes, California, both elongate and transverse ridges occur. From diagrams presented by Merk (1960) and by Sharp (1966), the conclusion is reached that the broad dune profiles and the small net movement of sand are actually controlled by the prevailing winds (fig. 8).

Many pits 4 to 5 feet deep were dug by Sharp (1966, p. 1062). The windward-slope bedding dip was as much as 16°, whereas most of the lee-slope strata were dipping considerably less than 30°. The pits, however, exposed only the upper 1 to 1.5 m of dune strata (fig. 8). The windward slope was composed of more than one set of cross strata, with the upper set dipping upwind. Unusually high dip for these strata (up to 16°) were caused mostly by storm winds. Below these strata, other sets dipped eastward (downwind) at higher angles (up to 32°). These strata possibly represented foresets deposited by the action of prevailing winds. Steeply dipping (>30°) slip-face deposits were rarely preserved (Sharp, 1966, p. 1062), suggesting that in those dunes the upper parts of lee slope deposits are subjected to severe erosion during reversals.

Star dune.—A star dune is a type of sand body, developed locally in Saudi Arabia and in parts of North Africa, that has a high central point from which three or more arms (ridges) radiate in various directions. The structure in two star dunes—a low one (35 feet) and a high one (100 feet)—near Zalim in Saudi Arabia has been determined on the basis of small trenches near the dune crest (McKee, 1966, p. 65–68).

Near the high dune at the time of the study, southerly or southeasterly winds maintained steep slipfaces on the lee sides of four ridges. High-angle (26°–34°) laminae in various sets showed three principal directions of dip, suggesting that three main wind directions—northerly, westerly, and southerly—were responsible for dune development.

The low star dune had two arms roughly

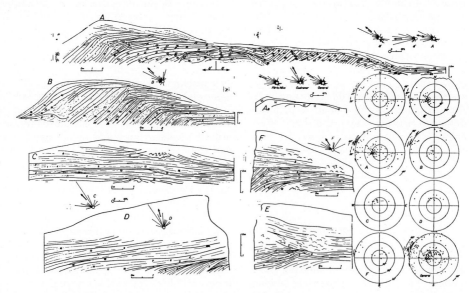

Fig. 7.—Precipitation ridge dune cross-sections showing the internal dune structures. A and B—Pôrto Nôvo dune; C, D, E and F—Guairamar dune, Praia de Leste, Paraná. The numbers given in tables refer to the attitude of cross-strata. Stereonets refer to dip and dip direction from several cross-sections. Rose diagrams indicate dip direction and resultant dip vector for the several sections surveyed in the dunes (after Bigarella, Becker & Duarte, 1969).

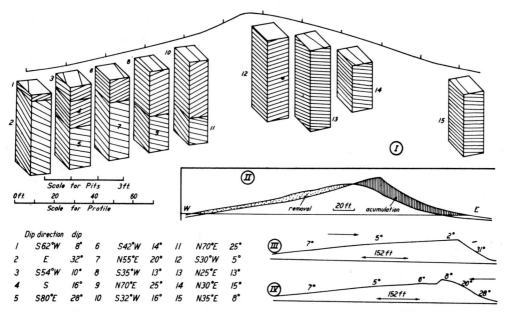

Fig. 8.—Cross-bedding attitudes at Kelso Dunes. I—Bedding attitudes in pits across dune (station 4, Kelso Dunes, November 20, 1960—based on Sharp, 1966, p. 1063). II—Summary of changes at poles of station 5, Kelso Dunes, from January 17, 1953 to February 15, 1964 (based on Sharp, 1966, p. 1057). III—Profile of a transverse dune taken through crest in direction of prevailing wind. Arrow indicates southwesterly wind direction. Great Sand Dunes, Colorado (after Merk, 1960, p. 128): IV—Profile of transverse dune, sowing the change produced by northeasterly storm wind. Great Sand Dunes, Colorado (after Merk, 1960, p. 128). The figures refer dip direction and dip values of the cross-strata.

parallel to and one almost at a right angle to the southerly winds. The flat, rounded surfaces of these arms were depositional features in which the strata dipped at low angles (2°–9°) northwestward and southwestward. High-angle (32°) strata dipping to the northeast were at about one meter below the surface.

Calcareous eolian deposits.—The term eolianite was proposed by Sayles (Bretz, 1960, p. 1732) to connote all wind-transported and wind-deposited material on Bermuda, irrespective of dune form or structure. Former dunes characteristically contain steep, parallel, lee-slope bedding, but much of the Bermuda sand rock, according to Bretz (1960, p. 1732), was accumulated in thin sheets with extremely variable cross-bedding attitude and without a characteristic dune profile.

Ninety-five percent of the exposed land of Bermuda, according to Mackenzie (1964, p. 1449), is composed of Pleistocene calcareous dunes, consisting predominantly of lobate sand bodies, which coalesced to form irregular transverse dune ridges. The lobate units are formed of high-angle (30°–35°) foresets and windward strata dipping 10° to 15°. The foreset cross strata are slightly convex upward.

In the Bahama Islands extensive, thick deposits of eolian carbonate sands include skeletal, pelletoidal, and oolitic grains (Ball, 1967, p. 571–573). These deposits form multiple seaward ridges. They differ from marine sediments of that area in greater thickness of cross-strata sets and in the presence of cross strata that are convex upward.

Preservation of the convex-upward cross-bedding is brought about by relatively rapid stabilization of the calcareous sand, where surface cementation by percolating rain water takes place, according to Mackenzie (1964). Where "stabilization was slow," he states, "the cross strata were eroded and appear in some sections like planar or festoon cross-bedding." Convex-upward cross-stratification is considered to be a characteristic feature of Bermuda and Bahama calcareous eolian deposits (Mackenzie, 1964, p. 1450); however, similar structures are recorded by Bigarella, Becker, and Duarte (1969) from the coastal precipitation ridge dune of Praia de Leste in southern Brazil, and by McKee (1966) from parabolic dunes at White Sands, New Mexico.

Features Common to Most Dunes

Features of cross-stratification that are common to the dome-shaped, transverse, barchan, and parabolic dunes in the White Sands field, according to McKee (1966, p. 59), are the following: a) numerous sets of medium to large-scale cross strata dipping (30°–34°) downwind; b) series of horizontal or low-angle bounding surfaces mostly developed in the upwind part of the dune; c) series of dipping bounding surfaces with moderately high angles (20°–28°) truncating higher-angle (28°–34°) foresets are generally developed in the downwind part of the dune; d) sets of cross strata that become progressively thinner from bottom to top of a dune and contain laminae that tend to flatten upward, especially in the upwind part of the dune; e) dipping foresets composed of progressively longer cross-laminae downwind; and f) local contorted bedding.

Structures Distinctive of Dune Type

Among the dunes of White Sands, New Mexico, some structural features are distinctive of particular dune forms (McKee, 1966, p. 58–60):

Dome-shaped dune.—Characterized by low-angle foresets dipping toward the lee side and toward both margins. High-angle foresets occur only in the middle part of the upwind side of the dune.

Transverse dune.—Nearly horizontal traces of stratification in sections at right angle to the wind direction. These strata represent relatively long slipface laminae dipping downwind. Low-angle (2°–5°) strata dipping upwind on the windward side of the dune seem to be also characteristic of transverse dunes, but they may also occur under some conditions in other dune forms. At White Sands they begin to form only after the dune reaches a height of 40 feet (12.2 m), according to McKee (1966, p. 60). These reverse strata were deposited on the low-angle, windward-dipping surface of the dune by winds that were not competent to carry the full load of sand up the dune slope.

Barchan dune.—Characterized in sections normal to the wind direction by moderate to high angles in the apparent dips of strata; these dips contrast with the usual low angles (1°–6°) of apparent dips in transverse dunes in similar situations.

Parabolic dune and precipitation ridge.—Have as their most distinctive feature many foreset surfaces that are convex, rather than concave, upward. In parabolic dunes, the windward sides are characterized by strata with low dips, whereas in the downwind side few foresets have high-angle dips. Precipitation ridges have strata that are convex upward with mostly high angles of dip.

Significance of the Cross-Stratification

The attitude of the bounding surface is used as a possible criterion for the determination of its relative position within the dune. The angles of dip, both of the bounding surface and of the cross strata, are, according to McKee (1966, p. 57), features indicative of the following situations: a) moderately to steeply dipping surfaces bounding tabular sets (planar and simple) of cross strata represent slipface deposits formed largely by avalanching of sand in the lower part of the dune. b) Horizontal or low-angle surfaces, bounding steeply dipping cross strata of the tabular planar type, are formed mostly high in a dune. They may or may not be the upwind equivalent of a steeply dipping unit (a, above). c) Low-angle dipping surfaces, bounding low-angle cross strata of the tabular planar type, probably indicate deposition of sand largely from suspension in very strong wind. This type of structure is common in dome-shaped dunes.

Structures recorded in the Praia de Leste dunes show features having the above-mentioned characteristics and confirm conclusions based on the White Sands dunes.

At the White Sands, wedge-planar cross-stratification is relatively uncommon and seems to represent changes in wind direction (McKee, 1966, p. 57). This type of structure, however, is rather common among cross-strata of various sandstones, inferred to be eolian, in the Colorado Plateau of the western USA and in the Botucatu Sandstone of Brazil.

Trough cross-stratification is not a common feature in the White Sands dunes of New Mexico. This structure is mostly restricted to the upper parts of dunes and apparently is formed by the development and subsequent filling of blowouts. In the Praia de Leste dunes and in the Botucatu Sandstone, in contrast, trough cross-stratification is a rather common feature. In these cases, the trough cross strata do not seem to result from blowouts but possibly indicate fluctuations in wind directions.

Foredune Structures

The dome-shaped dunes at the White Sands National Monument are essentially large-scale foredune features. Their structures were referred to in the preceding discussion. Likewise, beach-dune ridges of coastal areas are foredune features. The structures here discussed, however, refer to the pro-dune deposits (Portuguese equivalent: *ante-duna*) that consist of miniature dome-shaped eolian sand deposits located on the seaward side of a coastal precipitation ridge.

In a foredune area deposition of eolian sand starts around various obstacles, mostly grasses or small shrubs. The sand accumulates as successive shell-shaped strata (convex upward). The long axis of these dome-shaped deposits is

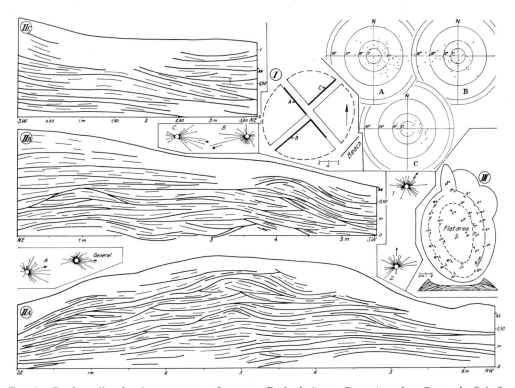

Fig. 9.—Produne (foredune) structures at Ipanema, Praia de Leste, Paraná, and at Barra do Sul, Santa Catarina. I—Location of the structures exposed in the walls of the trenches dug in the produne at Ipanema. II (A, B and C) cross-sections of the Ipanema produne; A, B and C are rose diagrams and stereonets representing the cross-strata dip directions and resultant vector for the cross-sections. III—Cross-strata attitude in a small produne at Barra do Sul, Santa Catarina. 1 & 2—rose diagram illustrating the dip direction and resultant vector of two groups of cross-strata measurements relative to the produne area.

parallel to the prevailing wind. Most of the dome-shaped produnes show blowout features. A blowout starts over the highest part of the dome, cutting both downward and laterally. It removes sand from the central portion of the dome, leaving an oval- to circular-rimmed pan, or small crater-shaped basin with openings upwind and downwind. Sands that were removed are then deposited somewhere downwind, starting development of a new dome-shaped pro-dune. This depositional and erosional process is continuously repeated.

The structures of an eolian complex at Barra do Sul, Santa Catarina, are briefly described by Bigarella and Popp (1966). The height of the pro-dunes commonly is less than one meter above the swale floor. One studied at Barra do Sul (fig. 9) measured 12 by 14 m in diameter and the rim was about 0.5 m high. The central part of this blowout was flat and slightly higher than the surrounding area. Cross-bedding measurements, taken at many points inside the pro-dune remnants, indicated superposition of a series of convex upward sets of strata. The attitude of these sets, obtained from cross-bedding measurements, differed greatly from those obtained in a precipitation ridge not far inland.

Because of the succession of convex-upward strata, the dip directions are distributed completely around the compass. Two sets of measurements, consisting of 49 and 44 readings, failed to give a resultant vector parallel to the prevailing wind (fig. 9). Moreover, the consistency ratio of both sets was very low, i.e. 0.13 and 0.24 (Bigarella and Popp, 1966, p. 146). In Barra do Sul (Santa Catarina), the average dip in the pro-dunes was determined as 14.3°, and the maximum dip was 31°.

In a produne, strata to windward with dips smaller than 10° frequently are preserved in the stratigraphic sequence, whereas in normal dunes such strata seldom remain.

Structures of the pro-dunes were investigated in considerable detail near Ipanema, Brazil (Praia de Leste Beach), directly behind the shoreline (fig. 9). The area showed blowout features associated with small-scale, dome-shaped, eolian deposits. The microtopography was ir-

TABLE 1.—CROSS-BEDDING MEASUREMENTS FROM THE PRO-DUNES OF PARANÁ AND SANTA CATARINA (SOUTHERN BRAZIL)

Locality	Number of measurements	Average dip direction	Maximum dip	Average dip	Consistency ratio
Ipanema (Paraná)					
A	56	N 76° E	34°	12.4°	0.56
B	39	S 71° W	33°	12.1°	0.44
C	23	S 82° E	19°	10.0°	0.50
General	118	E	34°	11.8°	0.22
Barra do Sul (Sta. Catarina)					
1	49	N 42° E	31°	15.0°	0.13
2	44	N 01° E	29°	13.4°	0.24
General	93	N 16° E	31°	14.3°	0.17

regular, being characterized by depressions (blowouts) and hillocks (pro-dunes). The dome-shaped deposits tended to coalesce.

Structure of a roughly circular pro-dune at Ipanema were exposed by digging three cuts, one approximately parallel and two normal to the prevailing wind direction (fig. 9, IIA, IIB, IIC). The general internal organization of cross-strata sets is illustrated but the resultant vector obtained from all measurements, here as at Barra do Sul, does not correspond to the prevailing wind direction. Thus, measurements made in pro-dunes and in the foredune area cannot be considered reliable for the deduction of paleo-wind directions. Moreover, dip directions do not indicate directions of sand transport, as many of the strata are upwind deposits. Consistency ratios are very low (0.22, Ipanema, and 0.17, Barra do Sul). In these localities prevailing winds blow from southeast and east respectively, so it is clear that the structures are not as completely developed and adjusted to prevailing winds as in most eolian environments.

The cross-stratification in pro-dunes consists dominantly of low- to moderate-angle, simple (non-erosional), tabular, and wavy forms. In some places steep foresets are present. Contorted bedding, although present, is not abundant as it is in a beach-dune ridge.

The average dip of pro-dune strata at Ipanema (11.8°) and Barra do Sul (14.3°) is higher than the average dip of beach-dune ridge strata (10.4°) and smaller than the average dip of cross-strata from precipitation dune ridges (19.5°). Dips between 6° and 14° are the most numerous, but a few high-angle strata occur in the pro-dunes, Table 1.

The shapes of the bounding surfaces are clearly indicated in figure 9 (IIA, B, and C). In a profile roughly parallel to the prevailing wind direction, the traces of the bounding surfaces dip seaward through much of the cross-section. The right lower part of the cross-section IIB (fig. 9) shows a convex upward succession of cross-strata. These sets of strata represent the former nucleus of the pro-dune, which has grown mostly by accretions on the seaward side (upwind). rather than on the downwind side. In the profile normal to the prevailing wind direction, the traces of the bounding surfaces dip toward the edges of the pro-dune (fig. 9 IIB and IIC).

Structures of Interdune Areas

Structures of flat, sandy interdune areas between longitudinal dunes near Sebhah, Libya, were recorded by McKee and Tibbitts (1964, p. 13). Far from the dune base, the structures exposed in a test pit 1.2 m deep consisted of well developed, flat-lying laminae but no cross-bedding. In two test pits near the dune base, low-angle strata dipping toward the dune were exposed. These strata did not dip normal to the crest, for they were not the result of avalanching.

The "serir," which is a plain of alluvial origin, and contains a great concentration of lag gravel from wind deflation, forms the margin of the sand sea. Deposits of the serir contain laminae that are horizontal or dip with low angles; also lenses and layers of structureless sediment. These poorly sorted sands and gravels suggest deposition by sheet floods (McKee and Tibbitts, 1964, p. 17).

At White Sands National Monument, New Mexico, beveled laminae of former dunes are exposed on interdune surfaces. The truncated layers show the crescentic shapes of former barchan and transverse dunes. Two trenches dug in an interdune area showed structures re-

Fig. 10.—Interdune areas, coastal dunes of Rio Grande do Sul, Brazil. Horizontal section of former dune shows bevelled structures in interdune area. The truncated dune structures indicate both shifting of wind direction and superposition of differentially oriented slipface deposits.

sembling those in nearby dome-shaped dunes (McKee, 1966, p. 55).

Interdune areas among the coastal dunes of Brazil show flat-lying laminae, as well as the patterns of truncated cross-strata like those of nearby active dunes (fig. 10). These structures are the remnants of eroded earlier dunes that migrated across the area. The overlapping pattern shows changes in orientation of the curved slipfaces, the result either of changes in wind direction or of irregular superposition of the curved foreset strata.

Beach-dune Ridge Structures

The beach-dune ridge deposits of Praia de Leste Beach are sediments intermediate and transitional between beach and dune sands. The beach-dune ridge is a very low, relatively broad and rounded ridge, parallel to the coast line (fig. 11). It is separated from the beach by the nip, but in many places there is no sharp separation. Where a transition zone occurs, it is characterized by the scarce marginal vetetation of the *Ipomoea* association.

Onshore winds carry the beach sands inland to the foredune or dune areas. In the first step, sand accumulates partly against the nip or on the transitional zone between beach and beach-dune ridge. Part of the sand is deposited on the beach-dune ridge (foredune in general) and part on the precipitation ridge dune. Typical structures are illustrated from two areas at Praia de Leste Beach in Paraná, Brazil (fig. 11).

Ripple Marks

Wind-formed ripple marks are infrequently preserved in ancient eolian sandstones. They have been noted in the Botucatu Sandstone of Brazil (Maack, 1966, pl. XVI) and in various Paleozoic and Mesozoic sandstones of the Colorado Plateau. They occur as parallel asymmetrical ripples with indices ranging from 20 to 50 (Poole, 1964, p. 398). Crests and troughs are subparallel to the dip direction of the cross-strata surfaces (McKee, 1945, p. 316; 1957; p. 1719). The high index and orientation relative to cross-stratification are distinctive features (Poole, 1964).

Most ripple marks are formed on the windward slopes of dunes. Their orientation depends on the part of the dune in which they develop, as crests and troughs commonly are normal to the local wind direction. Because the wind is constantly deflected by topography in a dune field, many different directions in ripple trend occur on the dune surfaces. Furthermore, the ripples are mostly formed by weak winds which may not correspond in direction to the prevailing wind.

Ripples of the windward side are nearly normal to the main direction of wind flow; however, these ripple marks are seldom preserved. Favorable conditions for burial and, therefore, preservation are almost entirely on the lee sides of the dunes. The orientation of ripples on the lee sides of dunes results from either the deflection of wind over the dune surface, or the temporary blowing of winds from other directions.

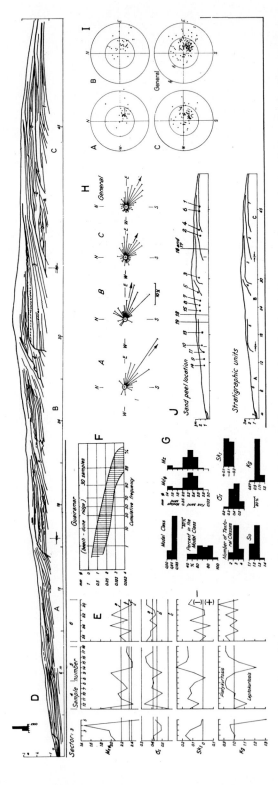

FIG. 11.—Beach-dune ridge structures at Guairamar, Praia de Leste, Paraná. A, B & C—detailed cross-sections of the beach-dune ridge oriented parallel to the prevailing wind (SE–NW). D—Whole beach-dune ridge indicating the main stratigraphic units. In cross-section A two nip scarps occur, one at the left is actually the present-day coast line limit. The one at the right side, as well as the one shown in cross-section B, are fossil nip scarps, documenting former coast-lines, and the retreat of the sea. The diagram E shows the variation of the parameters in the successive strata sets. In this diagram, the average values of Mz, σ_I for Guairamar, for the Praia de Leste dunes and for the Praia de Leste beach samples are represented as straight lines indicated respectively by the letters "a", "b" and "c". The number of the sector corresponds to the one of the stratigraphic unit. The diagram F illustrates the area occupied by the superposition of all the grain size analysis cumulative curves. The diagram G refers the parametric properties of the grain size distribution. H—rose diagram indicating the cross-strata dip directions and resultant vector for every cross-section and for the whole beach-dune ridge. I—Stereonets showing the attitude of the cross-strata in every cross-section and in the whole section. J—Location of the sand peels illustrated in figures 19 and 20. All the numbers referring dots correspond to sand peel plates parallel to the prevailing wind direction. The numbers 16, 17, 18 and 19 refer to sand peel plates normal to the prevailing wind direction. The lower cross-section refers to the location of the several stratigraphic units.

Two main types of wind ripple marks are recognized (Cressey, 1928, p. 32; Sharp, 1963, p. 617; Tanner, 1964, p. 432). A uniform-grained type (Cressey, 1928) or sand-ripple type (Sharp, 1963) is the most abundant and widespread. Granule ripple marks are not uncommon and are formed where local lag concentrations of coarse grains develop. The wave length in sand ripples of the Kelso dune area, California, commonly ranges between 3 and 6 inches (Sharp, 1963, p. 619) and in the Lake Michigan coastal dunes between 2 and 4 inches (Cressey, 1928, p. 32). The heights of sand ripples are mostly between 0.2 and 0.4 inches in the Kelso field and the ripple index mostly is larger than 18 (Sharp, 1963, p. 619).

Wind ripples commonly have larger indices than do water ripples (Kindle, 1917; Twenhofel, 1932; McKee, 1933; Sharp, 1963; and others). Indices below 15 are considered evidence of subaqueous deposition (McKee, 1933, p. 100).

Differences in orientation indicate variations in wind directions, and document the complexity of the wind currents near the ground (Sharp, 1963, p. 617). The currents along the sand surface may diverge at angles up to 90° from each other as a result of dune form. Therefore ripples are developed according to the direction of the wind current on the ground, and their orientation reflects, in part, the direct influence of the dune shape. "Strong currents have been observed to move longitudinally along the lee face of a transverse dune; at the same time, wind was blowing directly up the windward slope at velocities approaching 35 mph." Also, ripple marks may trend directly up and down the lee slope, at right angle to those ripples on the crest of the dune (Sharp, 1963, p. 619).

Small fluctuations in the wind direction do not affect the orientation of ripples (Sharp, 1963, p. 618); however, when wind direction changes greatly, a new set of ripples is initiated and superimposed on the old one, with a different trend. In a short time, the old ripple set commonly is destroyed under the new wind direction, but in many places the new set appears superimposed on the old one (figs. 12 and 13). In some places three superimposed sets of ripples of different trend and dimensions are preserved (Sharp, 1963, p. 618).

The ripple index depends mostly on grain size, composition, and wind velocity, according to Sharp (1963, p. 619). The influence of grain size on the shape and size of ripple marks has been discussed by Cornish (1897), Bucher (1919), Bagnold (1954), McKee (1945), Sharp (1963), and others. Experimental work (Bagnold, 1936) and field observations (Sharp, 1963) show that the wave length increases with the wind velocity. The height of the ripple increases with the coarseness of grains. Ripple marks in coarse sand mostly are larger than ripple marks in fine sand (Sokolow, 1894; Cornish, 1897; Sharp, 1963). Also, those in coarse sand have a lower index than those in fine sand. This index varies inversely with the grain size and inversely with the wind velocity.

The internal structures of ripple marks in the Kelso dunes of California were investigated by Sharp (1963, p. 621). The ripple marked sand, according to him, is relatively coarse and rests on a firm and generally smooth substratum of thinly bedded, finer sand. Most of the ripple marks are homogeneous without any visible internal structures; however, some show foreset bedding formed of fine grains, which, according to Sharp, must have been deposited in an interval of reduced wind velocity.

In wind ripple marks, the coarsest grains are concentrated on the crests (Cornish, 1914; Cressey, 1928; Twenhofel, 1932; Bagnold, 1954; Norris and Norris, 1961). "On the sand ripple of the Kelso Dunes coarser grains increase in number but not spectacularly in size up the windward slope to crest" (Sharp, 1963, p. 621). The leeward slope is composed of equally coarse material. The fine material in troughs is actually the exposed part of the substratum over which the ripple has moved, according to Sharp (1963).

Under conditions of relatively weak winds, ripple marks are abundant on the surface of dunes and they develop on surfaces undergoing neither marked erosion nor heavy deposition. They are more common on surfaces undergoing some accumulation than on surfaces of erosion (Sharp, 1963, p. 623). Ripple marks may start to form at a wind velocity of about 6.8 miles per hour but they will disappear when winds attain a speed of about 20 miles per hour (Cressey, 1928, p. 34).

Granule ripple marks are basically different from the sand type. They develop as concentrations of lag grains, mostly larger than 1 mm. In the Kelso dunes, California, they occur "in hollows, low on the windward flanks of dunes, or in scoured chutes where excessive deflation has produced a lag concentrate of grains between 1 and 3 mm in diameter" (Sharp, 1963, p. 632). Such ripple marks are more irregular than the sand type. They have a tendency to produce cuspate forms, "particularly in linear chains of crescent-shaped, barchan-like ripples" (Sharp, 1963). Granule ripples commonly contrast with sand ripples in having well-developed foreset laminae (Sharp, 1963, p. 633).

Shear-surface Structures

Scallop structures caused by lee-side avalan-

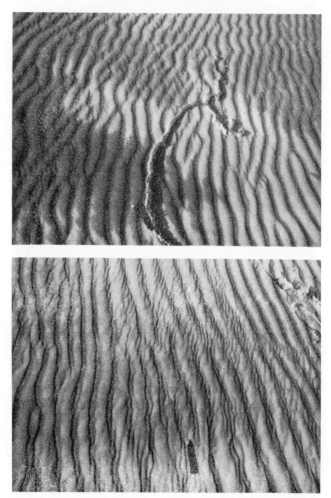

Fig. 12.—Typical wind ripples from the coastal dunes of Rio Grande do Sul, Brazil.

ching appear on essentially horizontal surfaces as repetitive segments of curves with wavelengths of 7 to 25 cm and amplitudes of 2 to 5 cm (fig. 14) (Sharp, 1966, p. 1064). Each scallop "consists of coarse, homogeneous sand without discernible structure, outlined by a thin layer of fine dark grains" (Sharp, 1966). The dark material is collected along the slip surface of the open U-shaped chute formed at the head of a sand avalanche. The bulk of the structure is composed of coarse material, which possibly represents the filling of the chute by wind-blown sand that accumulated on the upper part of the slip face (Sharp, 1966, p. 1064). The development of each new avalanche repeats the process. Scallops are not frequently seen because they develop in the uppermost part of the slip face and are, therefore, very seldom preserved (Sharp, 1966). This structure may be unique to dunes.

Contorted Bedding

General.—Penecontemporaneous deformation, caused by avalanching of foreset strata downslope, is a common phenomenon in coastal dune structures of southern Brazil (Bigarella, Becker, and Duarte, 1969) and occurs to some extent in most other dunes. In dunes at White Sands, New Mexico, it consists of wavy and contorted laminae among steeply dipping foresets near dune centers (McKee, 1966, p. 31).

Contorted beds occur in several forms, mostly irregular and with unsystematic folding patterns, with or without associated small faults. They form under the action of gravity at the time of, or shortly following, accumulation, and constitute one of the most characteristic features of strata on the upper slip face in some dunes.

Coastal dune contorted bedding.—Deformed beds occur between undeformed beds in coastal dunes at Praia de Leste (Paraná). Most of the

FIG. 13.—Wind ripples of coarse-grained sand in the interdune areas. Coastal dunes of Rio Grande do Sul, Brazil.

contortions affect more than one set of strata. The structures produced by the deformation seem to be chaotic. Folds and faults are very common. Some disrupted strata form breccias with a piling up of broken, rolled, and slumped masses of loose sand. Such deformational structures are numerous around the brinks of former dunes that have evolved into precipitation ridges. A single set of strata that forms topsets on the upwind side consists of foresets on the downwind side. Deformation is most frequent downwind from where the beds curve steeply downward. Contortions tend to be less prominent in lower downwind parts (Bigarella, Becker, and Duarte, 1969).

The arch formed by connected topset and foreset strata and the associated deformational structures seem to be distinctive characteristics of coastal dunes. They probably are preserved because the dunes cannot migrate easily against the vegetation line; upper parts of foreset strata, as well as corresponding topset strata, commonly remain. In most other types of dunes, however, the topsets normally are absent through removal by erosion, and the frequency of deformed structures seems to decrease downward on the slip face. The high angle of repose (up to 42°) on the lee side of dunes is the main cause of the great instability of the sediments accumulated there. They undergo frequent slipping which causes slump structures (Bigarella, Becker, and Duarte, 1969).

Deformational structures may occur in other parts of dunes but are less impressive than those in the upper part of slip-face strata. They occur in strata with gentler dip and may be different

 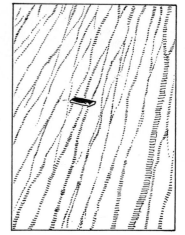

Fig. 14.—Cuspate structure developed by avalanching and exposed on wind-scoured surface. Drawing made after a photograph (after Sharp, 1966, pl. 3).

in origin. Although they have some resemblance to convolute lamination, they contain many features of avalanche structures, including faults. Characteristically, they consist of crumpling or intricate folding of strata between undeformed units. The deformation seems to have been produced before covering beds were deposited. The uppermost contact of deformed and normal beds is sharp and overlying undeformed strata lie directly on the disturbed beds without evidence of contemporary soft sediment thrusting (Bigarella, Becker, and Duarte, 1969).

Description of deformational structures.—Avalanching on the slip face of a dune occurs in a series of discontinuous movements with miniature landslides forming to leeward each time the equilibrium of the steep slope is disturbed beyond a critical value (McKee, 1945). Sliding of sand results mainly from overloading the brink of the dune with sediment, but it may also result from undermining of the basal part of the lee slope or by an accidental extra weight (such as an animal moving along the dune crest). Slump features of the lee slope show that, in plane view, borders of the slumped mass are marked by a series of variable and irregular lines (slump marks) roughly parallel to the direction of the slope (McKee, 1945).

Slump marks in dry-sand experiments show an infinite variety of form and detail (McKee, 1945, p. 320–322). Compound forms with nearly parallel lines commonly develop from a series of partially superimposed avalanches. When the sand is dampened by mist or dew, there is a crust of variable thickness according to moisture penetration. Slumping causes this crust to break up into a jumble or irregular, thin patches of sand curls and crusty fragments. Similar results are obtained after the sand has dried out (McKee, 1945, p. 322). When the dune is thoroughly wetted (either before or after drying out), a series of miniature steps develop by avalanching.

Sand peels of deformational structures from the Pôrto Nôvo, Brazil, precipitation ridge (Bigarella, Becker, and Duarte, 1969) were taken from vertical surfaces parallel to the prevailing wind (fig. 15), normal to the prevailing wind (fig. 16), and from a horizontal plane (figs. 17 and 18). In the several drawings, the complexity of structure patterns formed by lee-face avalanching is apparent. Peels that are parallel to the average wind direction show numerous folds and faults. In some sections, the folds have almost vertical axial planes as in convolute bedding; in other, recumbent folds are present.

The sand peels normal to the prevailing wind contain mostly wavy beds and "cut-and-fill" types of structure which indicate margins of an avalanching sand mass. Those peels representing a nearly horizontal plane view show complicated patterns of slip-face structure. Extensive folding is present (fig. 17) in many samples and faulting is evident in others (fig. 18).

Beach-dune ridge deformational structures.—Upper-beach and beach-dune sediments contain much irregular stratification associated with well-bedded deposits. The irregularities result largely from wind-blown sand deposited on an irregular surface. Contorted beds are very common in these deposits but are difficult to understand.

Along walls of cuts in beach-dune deposits at Guairamar (Praia de Leste, Paraná), a series of sand peels was made, both parallel (fig. 19) and normal (fig. 20) to the prevailing wind. Con-

Fig. 15.—Sand peel plates parallel to the prevailing wind direction. Deformational structures caused mostly by avalanching processes. Pôrto Nôvo precipitation ridge dune, Praia de Leste, Paraná (after Bigarella, Becker & Duarte, 1969).

torted bedding is in both low-angle and moderately inclined strata. It occurs in fine-grained and also in coarse-grained sediment.

The erosion surface and the structures developed at the foot of the nip, including contorted bedding, are illustrated in fig. 20b. Strata deposited against the nip attain moderate to high angles; this feature, together with undermining and disturbance of beds, as illustrated by the inclined crab burrow subsequently filled with sand (lower left corner of fig. 20b), explains the contorted bedding. Furthermore, some contorted beds are associated with paleosoil horizons (fig. 20a, upper part; fig. 19d). Distinct convolute lamination may develop in moderately inclined or even in almost horizontal strata. (fig. 19a–c).

Interpretation.—In spite of a lack of symmetry, slump structures may be used to determine the direction of slumping. A preferential orientation in axes of folding and corrugations result from slumping, roughly at right angles to

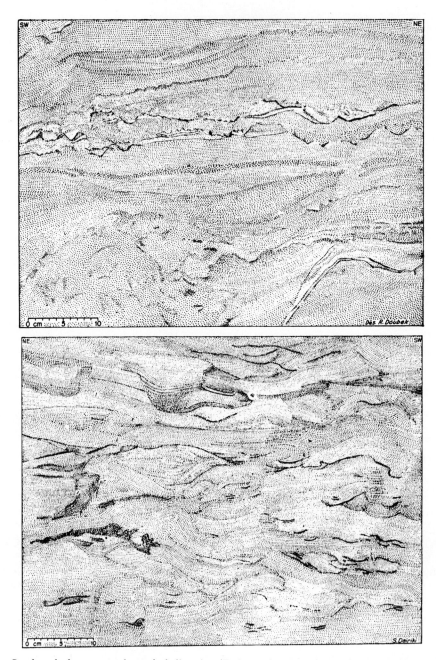

Fig. 16.—Sand peel plates normal to wind direction. Deformational structures in the slipface beds. Pôrto Nôvo precipitation dune ridge, Praia de Leste, Paraná (after Bigarella, Becker & Duarte, 1969).

the downslope (Jones, 1939; Murphy and Schlanger, 1962). From the attitude of structures, the slip face and, therefore, the prevailing wind direction can be inferred (Bigarella, Becker, and Duarte, 1969).

Oversteepening of the profile probably is the most important cause of contorted bedding. The deformation commonly is penecontemporaneous and seems to be developed under both dry and wet conditions. In a dune the windward face has closely packed sand, whereas in the slip face on the leeside, the sand is loosely packed and unstable like a dry quicksand (Bagnold, 1954). In such leeside deposits, a volume reduction of 7 percent might be expected to accompany disturbances of the slip-face deposits of the dunes

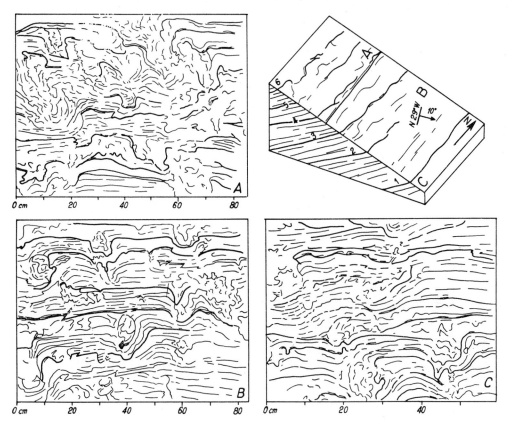

Fig. 17.—Deformational structures in coastal dunes at Aracaju, Sergipe. A, B and C—plan view of the deformations in the slipface deposits indicated in the block diagram.

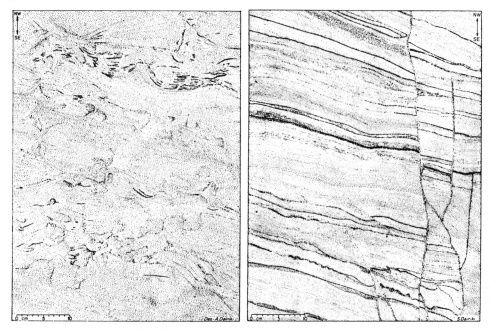

Fig. 18.—Sand peel plates, showing deformational structures in plan view (slipface beds). Pôrto Nôvo precipitation ridge dune, Praia de Leste, Paraná (after Bigarella, Becker & Duarte, 1969).

Fig. 19.—(A, B, C & D)—Sand peel plates of the beach-dune ridges stratifications and deformational structures. Guairamar, Praia de Leste, Paraná. Sand peel plates parallel to the prevailing wind direction. For location of the plates see figure 11.

Fig. 20.—(A, B, C & D)—Sand peel plates of the beach-dune ridges stratifications and deformational structures. Guairamar, Praia de Leste, Paraná. Sand peel plates parallel to the prevailing wind direction. For location of the plates see figure 11.

(Peacock, 1966, p. 161). Avalanching is one of the main causes of deformation; moreover, some slumps clearly did not originate under dry-sand conditions but were the product of rain-wetted sand. A crust of variable thickness may be formed by moisture penetration. Subsequent slumping causes this crust to break up following the movement of the sand and may result in folding or faulting of a particular group of strata.

Post-Depositional Changes in the Morphology and Texture of Dunes

After deposition, eolian sands undergo a series of changes which can preserve or destroy the morphology as well as modify the structural and textural characteristics of the dunes.

In the coastal dunes of Brazil, at least three periods of great dune development can be recognized in Pleistocene deposits by stratigraphic position and by differences in the color of sands. Among the stabilized dunes covered with grass, shrubs, or forest, both in northeastern and southern Brazil, different stages of oxidation coloration occur in the dunes, although internal structures may no longer be preserved and both grain size and mineralogical composition commonly are modified. Loss of internal structure probably was caused by heavy concentrated rainfall associated with change in climate. Textural and mineralogic changes are the result of weathering. The chemical decay of less stable minerals by intrastratal solutions conceivably is reflected in the grain-size parameter, as has been demonstrated for the Botucatu Sandstone of Brazil.

Ancient dunes commonly are reddish-brown, whereas more recent ones are brownish-yellow. Three stages of oxidation are recognized in the dune and barrier sands of the Gulf of Mexico coast (Price, 1962). The color changes correspond approximately with those observed in Brazil. Light reddish-brown color of the Algodones dunes in California is produced by a coating of ferric oxide on 25 to 60 percent of the grains and seems to be a measure of the age of the dunes (Norris and Norris, 1961, p. 611). Several ages of dunes in which the oldest dunes are darkest occur near Grants, New Mexico (Bryan and McCann, 1943, p. 282), and near Santa

Maria on the coast of Southern California and on San Nicolas Island, off the coast of Southern California (Norris and Norris, 1961, p. 611).

Eolian Structures Preserved in Sandstone

Eolian sandstones of the Colorado Plateau in the southwestern United States contain mostly large- and medium-scale cross strata of the wedge-planar and tabular-planar types. These structures are interpreted to have formed in transverse and barchan-type dunes. Cross-strata directions presented graphically show a spread of less than 120°, which is indicative of deposition by unidirectional winds. Frequency distribution plots of dip directions are mostly unimodal and show high consistency or small variation in direction (Poole, 1964, p. 400). No evidence is known of longitudinal dune structures, for in no one locality are dip directions of two distinct orientations or approximately 180° apart known to occur.

Trough cross-stratification is locally present in some eolian sandstone units. This type of structure is believed to represent blowouts and subsequent filling as a result of changes in wind velocity (Poole, 1964, p. 400). Simple cross-stratification also is locally present and is interpreted as having formed during constant deposition in which changes in wind strength and/or direction occurred.

One of the most characteristic features of the Botucatu and Sambaiba Sandstones in Brazil is the typical eolian type of cross-bedding, which consist mostly of large- to medium-scale, wedge-shaped sets, interpreted as products of transverse and barchan dunes. The size of the sets of cross strata ranges from small to large in the same geological section, which indicates that dune sizes differed extremely (Bigarella and Salamuni, 1961, p. 1092). In these sandstones, cross-strata are of the wedge-planar and tabular-planar types. (fig. 21). The attitudes of 3,520 cross-strata sets in the Botucatu Sandstone and those of 187 cross strata of the Sambaiba Sandstone in northern Brazil indicate that dip slopes did not exceed 35° (table 2). Dips of about 20° are common. The average dip for the Botucatu Sandstone cross strata is 20.3°, with the most common dips between 14° and 16° and between 20° and 22°. The average dip for the Sambaiba Sandstone is 19.5°.

Typical structures of the Botucatu Sandstone are illustrated in figure 21.

Criteria for distinguishing eolian structures in sandstone.—Sandstones developed in different environments may show similar structural features. Identification of the environment depends not only on the cross-bedding pattern, but also on associated features. Other structural characteristics, and some features of texture, are helpful in recognition of the environment.

In general, eolian sandstones are better sorted than subaqueous sandstones. Some structures and textures of sandstone in the Permian Rio do Rasto Formation of Brazil, believed to be of subaqueous origin, closely resemble those of eolian deposits. The non-eolian character of these strata, however, is demonstrated by associated and intercalated sequences of siltstone and claystone (fig. 21H, I, J, K, and L).

Some cross-bedded sandstones of the Pimenteira Formation, Devonian of Brazil, are composed of almost uniform, fine-grained sand, in which structures greatly resemble eolian types (fig. 21P, Q and R). The non-eolian nature of these is suggested by gravel zones containing well-rounded pebbles, though possibly such pebbles are lag deposits.

Convex-upward cross-strata are typical eolian features in some types of dunes like the parabolic

FIG. 21.—Botucatu Sandstone sedimentary structures. A comparison with the structures of other sandstone sequences. Botucatu Sandstone structures: A—section at 60° to the deduced prevailing wind direction. Highway Rio Claro—São Carlos, at the Serra de Santana (State of São Paulo); B—section approximately normal to the deduced prevailing wind direction at the quarry of "Asfalto Paulista Betumita S.A., São Paulo; C—section parallel to the prevailing paleowind direction, railway cut at the Serra de Botucatu, São Paulo; D—section approximately parallel to the prevailing paleowind direction, km 221 of the highway Rio Claro—São Carlos, São Paulo; E—section approximately normal to the locality paleowind direction, vicinities of Paula Freitas, Paraná; F—section approximately normal to the locality paleowind direction, km 293 of the highway Mafra—Lajes, Santa Catarina; G—section at 45° to the paleowind direction deduced for the locality of Santa Maria, Rio Grande do Sul (A to D—after Almeida, 1953; E to G—after Bigarella & Salamuni, 1967). Rio do Rasto Formation structures; H, I, J, K & L—cross-sections at the cuts of the highway between São Mateus do Sul and União da Vitória, Paraná (after Bigarella & Salamuni, 1967). M—Cross-bedding structures of the Pleistocene braided stream deposit of the Iguaçu River valley flat at Uberaba, Curitiba, Paraná. Section approximately transverse to the transportation trend (after Bigarella, Monsinho & Silva, 1965); N—Cross-section normal to the transportation trend relative to Pleistocene braided stream deposits at Uberaba, Curitiba (after Bigarella, Monsinho & Silva, 1965). Structures of the Devonian Sandstones of the Parnaiba Basin: O—cross-section of the

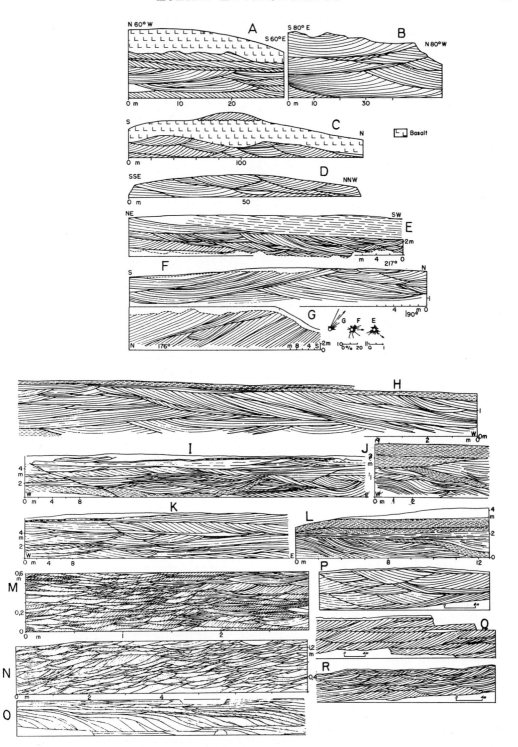

Serra Grande Formation Sandstone. The crossbedding pattern is parallel to the transportation trend, road Picos-Jaicós, Piaui, P and R—cross-bedding structures of the Pimenteira Formation sandstone in cross-sections parallel to the transport direction near Grossos, road Tianguá-Terezina, Piauí; Q—cross-bedding structures of the Cabeças Formation (overlying the Pimenteira Formation), section parallel to the transport direction, 5 km W of Picos, Piaui (O to R—after Bigarella, Mabesoone, Lins & Mota, 1965).

TABLE 2.—CROSS-BEDDING MEASUREMENTS: (A) FROM THE BOTUCATU SANDSTONE, CONCERNING LARGE GEOGRAPHIC AREAS, TWO NEW LOCALITIES IN PAULA FREITAS AND SANTA MARIA, AND PARAGUAY; (B) FROM THE SAMBAIBA FORMATION EOLIAN SANDSTONE OF PIAUI

Locality and geographic area	Number of measurements	Average dip direction	Consistency ratio	Maximum dip	Average dip
A—Botucatu Sandstone					
Mato Grosso Goiás	317	S 7° W	0.61	33°	20.2°
Minas Gerais	220	S 7° W	0.39	33°	19.6°
São Paulo	676	S 19° W	0.33	33°	20.3°
Northern Paraná	314	S 65° W	0.27	32°	19.6°
Southern Paraná	355	N 9° W	0.18	33°	19.5°
Paraná	669	N 73° W	0.15	33°	19.5°
Santa Catarina	391	N 80° E	0.45	33°	19.5°
Rio Grande do Sul	615	N 64° E	0.53	33°	20.8°
Uruguay	344	N 79° E	0.71	33°	21.2°
Vicinities of Paula Freitas—Paraná	94	S 54° E	0.24	30°	16.6°
New Road Santa Maria Julio de Castilho Santa Maria (RGS)	50	N 41° E	0.94	35°	26.3°
Paraguay					
a) Near Yhú	41	S 23° E	0.87	27°	17.7°
b) Independencia	28	N 48° W	0.57	25°	15.9°
c) Trinidad (Encarnación)	98	N 14° W	0.46	30°	20.2°
General (Paraguay)	167	N 6° W	0.52	30°	18.8°
B—Sambaiba Eolian Sandstone					
1) Regeneração, Natal and Barro Duro area	41	N 82° W	0.66	33°	18.0°
2) Between Regeneração and Amarante	19	N 62° W	0.98	26°	20.9°
3) Amarante	54	N 52° W	0.67	29°	16.5°
4) South of Amarante	73	N 70° W	0.81	30°	20.5°
General (Sambaiba Fm.)	187	N 67° W	0.74	33°	19.5°

and the precipitation ridges. Similar, convex-upward cross strata are known to occur also in strata interpreted as subaqueous, such as the Devonian of the Serra Grande Formation of Brazil (fig. 21-O). Recognition of the non-eolian nature of that sandstone is based on the poor sorting of its tabular-planar cross-strata and of its convex-upward strata.

WIND PATTERN DEDUCED FROM DUNE MORPHOLOGY AND INTERNAL STRUCTURES

General

The morphology and internal structure of dunes commonly indicate the wind direction responsible for them. It is important to determine whether the dune-forming winds represent prevailing winds or merely storm winds caused by passing cyclones. The wind direction interpreted from dunes may or may not correspond to the more persistent prevailing wind. Exceptional storm winds may be far more effective than moderate winds of long duration in transporting sand.

On the basis of the predominance of relatively small grains and the uniformity of particle size, central European dunes are considered by Poser (quoted by H. E. Wright, Jr., 1961, p. 954) to have been formed by relatively constant winds. The presence of coarse-grained laminae in some of the otherwise fine-grained sets indicates that strong winds blew periodically and brought coarse sand into the dune field. He believes dunes in northeastern Germany and Poland, however, were deposited by changeable winds associated not with passing cyclonic depressions but rather with a wind system around an anticyclone.

Present day dune-forming winds in Europe have the same mean directions as does the prevailing wind pattern deduced for the late-glacial dunes, i.e., southwesterly winds in Belgium, westerly in eastern Germany northwesterly in Poland and northerly in Hungary (Poser, cited by H. E. Wright, Jr., 1961, p. 954). This pattern agrees with that on weather maps which show a high-pressure area over Europe as an eastward extension from the Azores high.

In order to test the validity of paleowind trends, determined on the basis of internal structures, an extensive area of coastal dunes in South America from the mouth of the Amazon River to the La Plata Estuary was examined. The resultant vector, obtained from cross-bedding measurements, was compared with present-day wind-rose diagrams and with the prevailing-wind pattern. These investigations were based on the general assumption that lee-side strata are deposited by prevailing winds, which are related to major wind belts and to geographic pole positions.

Wind Belts

Trade winds blow toward the equator from regions of high pressure at latitudes of about 30°. They are deflected by the Coriolis force, which imparts a westerly component. The resulting wind regime is present over both the Pacific and Atlantic Oceans, in Africa north of the Sahara, and in the south Indian Ocean (Irving, 1964, p. 234). At the poleward limit of the trade winds, change to a westerly direction, usually between the latitudes 30°–40°, occurs. The latitude which marks the mean boundary between these westerly winds and the trade winds is called the "wheel-round latitude."

In the Sahara and other trade-wind deserts of the world, the critical wheel-round latitude is marked by a distinct change in the alignment of sand dunes and dune movement. In the northern hemisphere, north of this critical latitude, dunes are aligned northwest to southeast; as this latitude is approached the dunes become aligned north-south, and below this latitude they are aligned northeast-southwest (Opdyke, 1958, Ph.D. thesis, Univ. Durham, England). In the southern hemisphere, the "wheel-round" is established in an opposite direction, i.e., southwest-northeast, south-north, and southeast-northwest approaching the equator.

In the present dune field of northern Africa, the wheel-round latitude is about 25°N., whereas for the Pleistocene dune field, this latitude was located southward at about 15°N. (Fairbridge, 1964, p. 38). This indicates a climatic shift of more than 1,000 km. The present southern limit of loose sand is in the Libyan desert and is about 15°N., whereas during Pleistocene (glacial) time it was about 5° N. in the Nile-Congo watershed (fig. 22). A comparable situation is in southern Africa with regard to Recent and Pleistocene Kalahari sand dunes in relation to the equator (Fairbridge, 1964).

Paleowinds

Although the internal structure of a dune differs according to its type, in the great majority of dunes valid determinations of the paleowind directions can be obtained. Eolian sandstones lately have received considerable attention in connection with studies of ancient wind patterns. A number of formations considered to have been formed as ancient dunes are both thick and widespread. They have been analyzed in determining paleowind directions, in making paleographic and paleoclimatic interpretations, and in checking directional trends based on paleomagnetic determinations (Opdyke and Runcorn, 1960; Creer, 1958; Poole, 1962; Bigarella and Salamuni, 1961).

In the interpretation of eolian cross-bedding dip directions, the assumption is made that the general planetary circulation of the earth's atmosphere throughout the past consisted of 1) low latitude trade-wind zones; 2) middle latitude westerly wind zones, and 3) polar calm zones. This wind pattern applied to both northern and southern hemispheres and should have been symmetrical around the axis of rotation. The extention in latitude of the trade-wind belt would be different in the past because of 1) different temperature gradients between equatorial and polar regions, and 2) changes in angular velocity of the earth during geological time. Local and regional changes in this idealized circulation pattern, therefore, would have been produced mainly by relief of continents and by daily and seasonal changes (Bigarella and Salamuni, 1961, p. 1100).

Depending on the topography, present-day trade winds blow more or less constantly between about 20° N. Lat. and 20° S. Lat., except in the monsoon belt of India. The trade winds are very constant over the oceans, but are far less consistent and uniform over the continents (Opdyke and Runcorn, 1960, p. 960). Modern low-latitude deserts show seasonal variations in the trade-wind pattern because continents and oceans alter the distribution of temperature and hence alter the atmospheric pressure. Paleodesert areas must have been subjected to these same deviations from the general trade wind pattern.

In certain respects, the Botucatu paleodesert, Brazil, the late Paleozoic and early Mesozoic deserts of the Colorado Plateau, USA, and the

FIG. 22.—African desert characteristics. A—Non-glacial stage; B—glacial state. Sketch maps of Africa to illustrate hypothetical distribution of subtropical fronts and mean July air streams during a non glacial (interglacial) and a glacial stage. Thick arrows indicate moist monsoon and tropical air stream; thin arrows indicate westerlies and trade-winds. Rows of dots indicate sand dunes; those of the glacial stage that lie out on the continental shelf are related to the prevailing low sea levels; those in subequatorial latitudes are today vegetated. Dunes of the last interglacial and of the climatic optimum are mainly incorporated in present dune deserts, though in Mediterranean latitudes they are partially vegetated owing to the return of "cool pluvial" conditions during the last few millenia. WRL—wheel round latitude marks the westerly/trade-wind transition and thus the mean dune belt, at about 30° N or S in nonglacial and migrating down to 15°, 10° or even 5° in glacial phases,

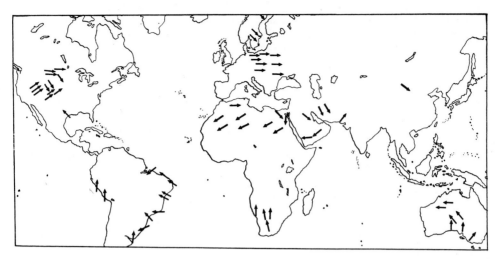

Fig. 23.—Sand dune movement indicated by orientation and internal structures of recent and pleistocenic dunes in the world (based in Opdyke, Ph.D. Thesis Univ. of Durham; Price, 1958; Bigarella, this paper.

Permian desert of England are similar to the modern trade-wind deserts, of which the Sahara is an example. Paleowind directions determined for Permo-Pennsylvanian eolian sandstones across more than 1000 miles in the western United States, and those from the Botucatu paleodesert of Brazil, more than 2,500 km wide, are very constant over wide areas and the few minor variations probably resulted from features of local geography. A paleowind pattern so constant over large areas "is most easily explained as arising from planetary causes" (Opdyke and Runcorn, 1960, p. 968).

Dune Orientation and Cross-bedding Analysis in Areas Other Than South America

Meteorological data are meager for large parts of the sandy deserts, and, are lacking for some entire regions; a similar lack of data exists in numerous areas of coastal dune development. For many dune studies investigators have necessarily used data from meteorological stations located far from the dunes in question. Notwithstanding, most results have been in agreement with regional circulation patterns. Only in restricted areas where the effects of the prevailing winds are counteracted by strong, short-term storm winds are results not clear.

A. *Europe*.—On the basis of the morphology and structure of dunes, westerly winds are inferred for the formation of the northern and central European sand dunes (fig. 23) (Högbom, 1923). In Denmark the axis of parabolic dunes is oriented parallel to the strong northwesterly gales (Hansen, 1959, p. 90).

B. *United States*

1) Kelso dunes, Calif.—The Kelso dune area is located in the zone of prevailing westerlies. This area is rimmed on the northern, eastern, and southern sides by mountains. In addition to the dominant winds, other strong winds, caused by storms, blow from different, sometimes almost opposite directions, greatly reducing the effect of the westerlies (Sharp, 1966). The most frequent sand-moving winds blow from the westerly quadrant, but winds from northerly, easterly, and southerly quadrants are also very active in sand transport, although less frequent and of shorter duration. These winds nearly balance the effects of prevailing westerlies.

Studies by Sharp (1966) of the Kelso dunes furnish significant information on sand bodies developed in an environment controlled by alternating winds with almost opposite directions. Measurements of cross-bedding dip orientation did not yield any strong indication of the dominant westerly wind.

Lee-slope bedding in some ancient eolian

---

a total climatic boundary shift of 2,000–3,000 km. Note the relatively poor development of dunes during the non-glacial (i.e. "normal") geological periods, but the broad development of the "laterite zone" (after Fairbridge, 1964, p. 357). C—Lineation of the Kalahari Dunes (after Opdyke, Ph.D. Thesis). D, E and F—dune patterns in Libya reproduced from aerial photographs, compiled by I.C. Conant (after McKee, 1964). G—Orientation of the dunes in Sahara and Arabia (based on Price, 1958). (1)—longitudinal dunes with clockwise migration direction; (2) transverse dunes; (3)—rhourds and couchets, large eolian form of the Sahara; (4)—Ergs with unknown dune types.

sandstones may be the result of occasional strong winds rather than of prevailing winds, according to Sharp (1966, p. 1067). Although this conclusion probably is correct for some local areas, it does not represent the general pattern obtained elsewhere from extensive cross-bedding measurements. These indicate clearly the importance of the prevailing winds both in recent dunes and in eolian sandstones (e.g., Botucatu paleodesert).

The internal structures of the Kelso dunes are not yet well known in detail. Whether or not the attitude of cross strata deep inside the dune (i.e., those which represent the main periods of dune movement) show any relationship to prevailing winds has not been determined.

2. *Mustang Island, Texas.*—On Mustang Island, Texas, the attitudes of 130 cross strata at 7 stations along an 18 km stretch of beach have been determined (McBride and Hayes, 1962, p. 547). The resultant vector is WNW (N. 68°W.), the consistency ratio 0.35, and the average dip of the cross strata 23.7°. For this region, a similar direction of sand transport was determined by Price (1958, p. 53). This northwesterly-trend is in agreement with present-day wind conditions (fig. 24).

3. *White Sands National Monument, New*

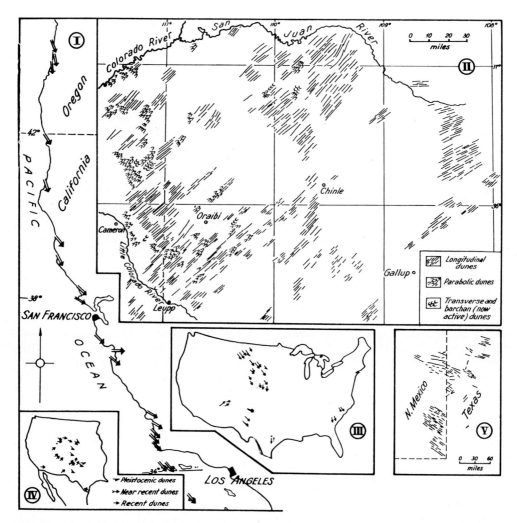

FIG. 24.—Orientation of the dunes in United States. I—Orientation of the coastal dunes in California and Oregon (based on Cooper, 1958, 1967); II—Distribution of the major dune types in Navajo County (based on Hack, 1941, p. 244); III—Clockwise wheel round lineations of ancient dunes in U.S. (based on Price, 1958); IV—Direction of Recent and Pleistocene dune movements in U.S. (based on Opdyke—Ph.D. Thesis, Univ. of Durham); V—Lineaments of former longitudinal dune fields of Llano Estacado, New Mexico & Texas (based on Price, 1958).

TABLE 3.—CROSS-BEDDING MEASUREMENTS FROM THE WHITE SAND DUNES (NEW MEXICO, U.S.A.)

Locality	Number of measurements	Average dip direction	Maximum dip	Average dip	Consistency ratio
A	58	N 84° E	36°	21.3°	0.83
B	52	N 70° E	34°	21.2°	0.77
C	73	N 53° E	39°	18.2°	0.72
D	161	N 88° E	32°	14.6°	0.68
General	344	N 77° E	39°	17.5°	0.71

Obs.: A—2 miles north from the pyramidal dunes over the alkali flat; B—Parabolic dune; C—Parabolic dune; D—Parabolic dune.

Mexico.—Measurements of the attitude of cross-bedding dips at several areas inside the White Sands dune field, New Mexico, as illustrated in figure 25 and presented in table 3, show an average dip direction that does not differ greatly from one area to the next (McKee, 1966, p. 58).

4. *Western coast of United States.*—The general orientation of dunes along the western coast of the United States is shown in figure 24. The "resultants of the winds exerting control over dune orientation are closely similar regionally," according to Cooper (1967, p. 16). The extreme values are N. 22°W. and N. 60°W. The average is N. 42°W., which indicates northwesterly winds are responsible for sand movement.

C. Africa.—The whole dune system of northeastern Africa constitutes a vast wheel-round from a NNW-SSE axis in the north, through N–S to an axis running ENE-WSW in the south (fig. 22). Dune chains lie on arcs of a circle centered roughly in the neighborhood of Lat. 23°N., Long. 25°E. (Bagnold, 1954, p. 235). The axial directions of seif dunes correspond more or less to that of the prevailing winds, according to Bagnold (1954, p. 234) and Fairbridge (1964, p. 356). Seif dunes in Libya are oriented east-west, being formed by the alternation of strong winds, which blow in the morning from SE and in the afternoon from NE (McKee and Tibbitts, 1964).

Pleistocene dunes in Africa extend far south to the equator. The orientation of ancient seifs is not compatible with the present wind patterns (Fairbridge, 1964, p. 356).

D. Australia.—Two prevailing winds affect

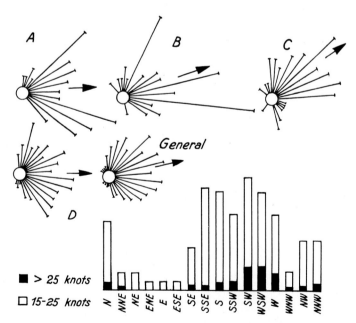

FIG. 25.—Rose diagrams representing dip directions of the cross-strata sets from the dunes of White Sands National Monument, New Mexico. A—2 miles north from the pyramidal dunes over the alkali flat· B, C & D—Parabolic dune. The histogram (after McKee, 1966, p. 11) refers the surface wind directions at Holloman Air Force Base, which is adjacent to White Sands dune field.

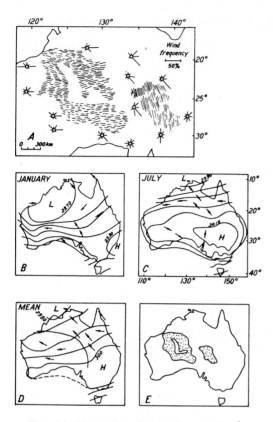

Australia, the westerlies of the southern coast and the southeast trades to the north (Madigan, 1936), modified by a monsoon effect. The distribution and alignments of dunes, as shown in both low and high pressure areas, and also wind-rose diagrams for Australia, indicate that a general circulation pattern remains essentially the same throughout the year (fig. 26). The southeast trades blow continuously in the central and northeastern areas; easterly winds blow in the western interior (Madigan, 1936, p. 207). Trade winds are dominant in autumn and winter. A wind rose shows clearly the swing to the left of the winter southerlies (fig. 26). Alignment of dune ridges is parallel to the known prevailing winds (Madigan, 1936, p. 208).

E. *Brazil-Uruguay Coast.*—The general wind circulation pattern and the resultant vectors obtained from cross-bedding measurements in recent and near-recent dunes along the Brazilian and Uruguayan coasts are summarized from statements by Serra (1956), Andrade (1964), and Santos (1965).

The basic elements of the present-day general circulation pattern include 1) air masses originating in both South Atlantic and North Atlantic high-pressure cells; 2) periodic advances of air masses of polar origin.

The Southern Atlantic, semimobile, anticyclone cell constitutes the main center of circulation (fig. 27). During the southern winter (July), the high-pressure cell expands toward and over the continent, whereas in the southern summer it retreats into the middle of the ocean.

In the eastern and northern parts of the South Atlantic, high-pressure cells start trade winds blowing toward the equator. They constitute the

Fig. 26.—Australian desert characteristics. A—Sand ridge desert showing the trend of the sand ridges and the annual wind roses for the meteorological station nearest to the sand ridge area. B and C—Isobars for Australia and predominant winds for summer (January) and winter (July). D—Isobar and predominant winds for the year. E—Sandy deserts of Australia (after Madigan, 1936).

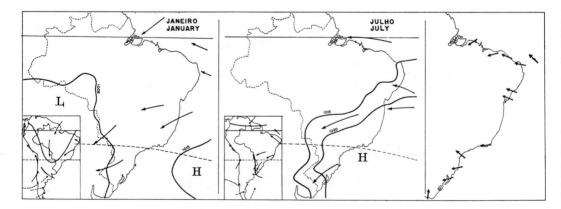

Fig. 27.—Schematic diagrams of the present-day wind circulation in Brazil (after Andrade, 1964). The right side diagram indicates the resultant vectors from the dune cross-bedding measurements, which are in agreement with the general wind pattern circulation at the Brazilian coast.

southeasterly and easterly winds that blow on the coast of northeastern Brazil. In the northern part of the high-pressure cell the "return" trade winds begin and they blow over the eastern Brazilian coast as northeasterlies or easterly winds. The westerly winds begin along the western and southern margins of the high-pressure cell.

The zone where the trade winds and the "return" trade winds diverge, is called the zone of divergence. At the coast, this zone migrates toward the equator in the winter and southwards in the summer (southern seasons). It is located north of Salvador, Bahia, during January and south of this city during July (fig. 27). North of the divergence zone, along the coast, winds blow mostly from southeast to east, whereas south of this zone the winds blow mostly from northeast to east.

The zone of contact (doldrums) between the trade winds from the southern and the northern Atlantic is called the intertropical front. During the winter (July), all the northern coast of Brazil is under the influence of the southeasterly trade winds, the intertropical front being in the northern hemisphere. During the summer (Jan-

FIG. 28.—Weather forecast charts for Brazil indicating the position of: A—high pressure areas· B—low pressure area· cold fronts; intertropical front. These charts indicate the frequent presence of polar air masses in southeastern and southern Brazil, not only in the winter, but also frequently in the summer (based on the weather forecast published by the newspaper "O Estado de São Paulo").

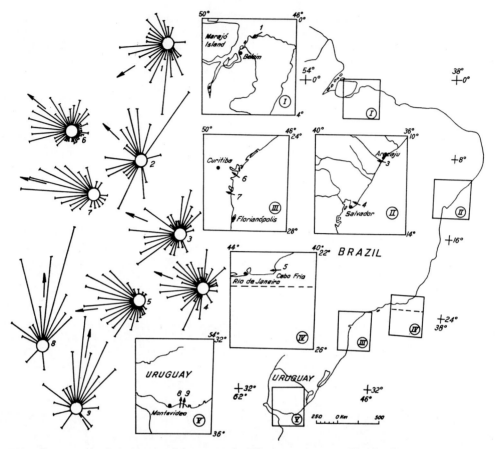

FIG. 29.—Recent and subrecent coastal dune cross-bedding measurements. The dip direction resultant vector is indicated in the map relatively to northern, northeastern, eastern, southeastern Brazil, and to Uruguay (data complementary to figures 30 and 31). The rose diagrams indicate the dip direction and results of vector for the coastal dunes: 1—Salinópolis (Pará); 2—Territory of Fernando de Noronha; 3—Aracaju (Sergipe); 4—Salvador (Bahia); 5—Cabo Frio (Rio de Janeiro); 6—Praia de Leste (Paraná); 7—Barra do Sul (Santa Catarina); 8 & 9—Uruguay.

uary), the northern coast of Brazil is under the influence of the northeasterly trade winds from the North Atlantic high-pressure cell. These winds then penetrate deep into the continent, toward the Chaco region (fig. 27). Cross-bedding measurements made in coastal dunes in eastern Brazil are nearly parallel with prevailing winds (fig. 27).

The southern Atlantic polar air mass, reinforced by the Pacific polar air mass, advances periodically northward along the coast and throughout the central plains of South America. During the winter (July), along the coast, the polar air masses reach as far north as northeastern Brazil, while in summer (January), the northern limit of periodic advances is somewhere in Bahia. The advance of the polar air masses is rather complex and is accompanied by cyclonic depressions moving northeastward. Behind the polar front, anticyclonic cells are developed.

Figure 28, based on forecasts published daily by the newspaper "O Estado de São Paulo," shows the frequent presence of anti-cyclonic polar air masses along the southern and southeastern Brazilian coasts.

Northern Brazil.—Cross-bedding measurements made at Salinópolis near the mouth of the Amazon, Pará, in a small dune field indicate the predominance of winds blowing from northeast (fig. 29). They are actually trade winds derived from the North Atlantic high-pressure cell. The rose diagrams of figure 30 for Belem, Pará, indicate for the region that the more frequent winds occur in July and October (winter-autumn season), corresponding to the less rainy period of the year (fig. 31). However, up to now no information is available on the period in which the sand transport is effective.

Northeastern Brazil.—The cross-bedding measurements made in the coastal dunes of

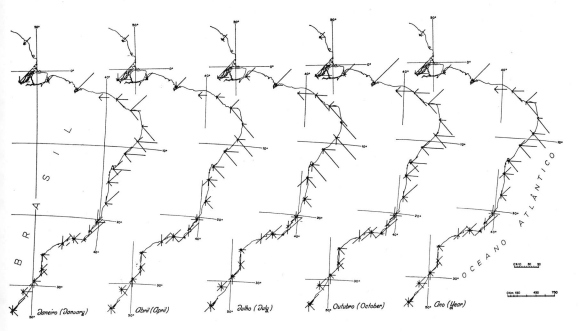

Fig. 30.—Wind rose diagrams representing the present wind frequency (January, April, July, October and yearly) for the meteorological stations along the Brazilian coast (after Serra, 1955, 1956).

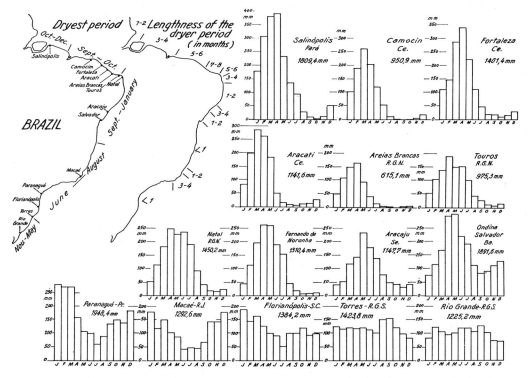

Fig. 31.—Rainfall diagrams relative to the meteorological stations close to the dunes surveyed along the Brazilian coast. In the two maps are indicated the dryest period and the length of the dryest period. The amount of rainfall is referred in mm.

TABLE 4.—CROSS-BEDDING MEASUREMENTS FROM THE COASTAL DUNES OF PARÁ, PIAUI, CEARÁ, RIO GRANDE DO NORTE, FERNANDO DE NORONHA, SERGIPE, BAHIA, RIO DE JANIERO AND PARANÁ

Locality	Number of measurements	Average dip direction	Consistency ratio	Maximum dip	Average dip
State of Pará Salinópolis	60	S 58° W	0.54	38°	22 °
State of Piaui Luiz Correia	145	S 76° W	0.69	38°	21.2°
State of Ceará Paracuru Fortaleza Majorlândia	109 47 67	N 77° W N 81° W N 51° W	0.78 0.76 0.80	40° 34° 34°	26.1° 20 ° 18.8°
Ceará—General result	223	N 70° W	0.76	40°	22.6°
State of R.Grande do Norte Tibau São Bento do Norte Touros Genipabu Natal	48 70 37 41 83	N 87° W S 83° W N 76° W N 29° W S 84° W	0.38 0.72 0.48 0.62 0.31	32° 34° 37° 28° 36°	20.5° 18.3° 19.7° 14.8° 20 °
R.G.N.–General result	279	N 80° W	0.45	37°	18.8°
Territory of Fernando de Noronha	35	N 47° W	0.34	36°	20.6°
State of Sergipe Aracaju	69	N 70° W	0.41	39°	26.1°
State of Bahia Itapoã	70	N 72° W	0.44	—	—
State of Rio de Janeiro Cabo Frio a) b)	107 83	S 77° W S 88° W	0.71 0.76	34° 33°	21.1° 22.7°
R.J.–General result	190	S 82° W	0.73	34°	21.8°
State of Paraná Jardim São Pedro Guairamar-Pôrto Novo	233 218	N 43° W N 59° W	0.38 0.59	39° 42°	19.2° 19.8°
Pr.–General result	451	N 53° W	0.47	42°	19.5°

northeastern Brazil, from Salvador, Bahia, up to the mouth of the Parnaiba River, Piaui, as well as in the dunes of the Fernando de Noronha Island, indicate the predominance of winds blowing from southeast and east (fig. 32 and 30, table 4). These winds are trade winds derived from the eastern and northern flanks of the South Atlantic high-pressure cell. South of Recife, Pernambuco, towards Salvador, these winds are frequent throughout the year (fig. 30). In Aracaju, the driest period of the year occurs between August and March (fig. 31). Dune sands in this area probably are moved during this period of less rain fall.

North of Recife, at the coast of Rio Grande do Norte, Ceará and Piaui, southeasterly-trade-winds are dominant during the winter (July). On the northern coast of Rio Grande do Norte and Ceará, the climate is dryer than on the eastern coast of northeastern Brazil (Paraíba, Pernambuco, Alagoas, Sergipe, and Bahia). Rainfall in Natal, Rio Grande do Norte, is 1450 mm; in Areias Brancas, 615 mm (fig. 29). In Natal, the driest months are from August to February; in Areias Brancas, from June to February.

On the basis of cross-bedding studies, the winter season (July) seems to be the most effective for sand movement (fig. 30). In spite of an abundance of northeasterly winds in the summer

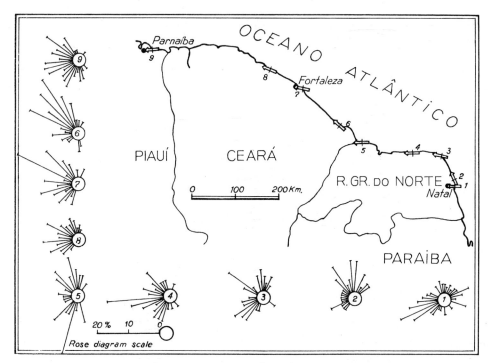

FIG. 32.—Coastal dune cross bedding resultant vector and rose diagrams indicating cross-strata dip directions, Northeastern Brazil. The arrows indicate the predominance of the southeasterly to easterly trade-winds.

season (January), no evidence exists in the dune cross-bedding pattern (rose diagrams, fig. 32) that this wind was effective in moving sands. Up to now, no observations are available concerning the period of movement of sands on the coastal dunes of northeastern Brazil.

Eastern Brazil.—South of the zone of divergence of the trade-winds, at Cabo Frio in the state of Rio de Janeiro, the north-easterly "return" trade-winds are dominant throughout the year. The cross-bedding resultant vector indicates winds blowing from east-northeast. In the Cabo Frio coastal dune field, northeasterly winds are slightly deflected by local relief. Notwithstanding, a good agreement occurs between the resultant vector and the prevailing winds (figs. 27 and 29).

Southern Brazil.—Cross-bedding measurements made in the coastal dunes of Santa Catarina Island, on the southern coastal plain of the State of Santa Catarina, and along the coast of Rio Grande do Sul (fig. 33), indicate the dominance in these areas of northeasterly- to easterly-winds ("return" trade-winds).

In Santa Catarina, the driest months are June to August, while in Rio Grande do Sul the driest months are between November and May (fig. 31).

On the Island of Santa Catarina, attitudes of cross-strata were measured in four localities (Praia dos Inglêses, Praia do Santinho, Lagôa and Praia do Pântano do Sul, fig. 33, nos. 1 to 4; table 5.) Results are not in agreement with the general circulation pattern. In the first two localities, the main sand movement is toward NNW because the dune field is protected from prevailing winds by local relief. Both in Praia dos Inglêses and Praia do Santinho, occasional and strong southerly winds are responsible for dune structures and movement (consistency ratios: 0.72 and 0.83).

At Lagôa and Praia do Pântano do Sul (fig. 33, nos. 3 and 4) dune fields also are largely affected by local relief. At Lagôa (fig. 33, no. 3) the resultant vector indicates wind blowing from east-northeast (prevailing wind quadrant), but the consistency ratio is very low (0.12) and the rose diagram (fig. 33, no. 3) shows dips all around the quadrant. Nevertheless, dune morphology indicates that occasional strong southerly winds result from advances of polar air masses over the region. Short term observations in the field (not accurate determinations) indicate that the northeasterly prevailing winds are deflected by the topography, changing to northerly winds. This situation favors the development of an environment of reversing dunes.

At Praia do Pântano do Sul (fig. 33, no. 4),

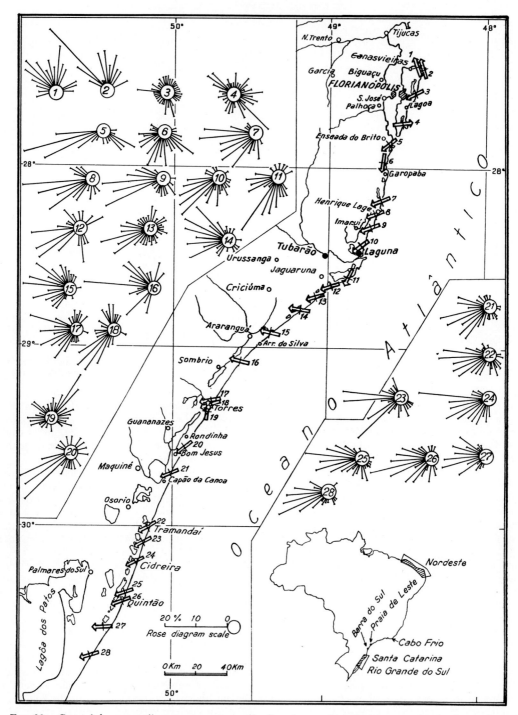

Fig. 33.—Coastal dune rose diagram cross-strata dip directions and resultant vector for the dunes along the southern Brazilian coast from Florianópolis to Lagôa dos Patos. The arrows indicate usually the prevailing northeasterly winds, exceptions occur at the localities no. 1, 2, 4 & 19 where the dune field is protected from the prevailing winds but exposed to the short term strong storm southerly winds.

TABLE 5.—CROSS-BEDDING MEASUREMENTS FROM THE COASTAL DUNES OF SANTA CATARINA BRAZIL

Locality	Number of measurements	Average dip direction	Consistency ratio	Maximum dip	Average dip
Barra do Sul	78	N 76° W	0.61	32°	12.3°
Praia dos Inglêses	78	N 22° W	0.72	35°	17 °
Praia do Santinho	44	N 17° W	0.83	34°	20.2°
Lagôa	158	S 65° W	0.12	32°	17.7°
Praia do Pântano do Sul	75	N 84° E	0.12	30°	14.6°
Pinheira	70	S 52° W	0.59	33°	18.7°
Zareia, Garopaba	96	S 8° W	0.64	36°	22.1°
Praia N da Ponta da Careca do Velho	125	S 67° W	0.85	37°	24.2°
5 km south of Henrique Lage	52	S 71° W	0.84	35°	22.7°
Itaperuba	92	S 70° W	0.79	37°	24.8°
Laguna	150	S 51° W	0.79	41°	22.7°
Camacho	57	S 38° W	0.68	33°	21.1°
11 km NE of Balneário Jaguaruna	36	S 70° W	0.54	33°	19.4°
Balneário Jaguaruna	143	S 72° W	0.53	34°	20.3°
Balneário Rincão	100	N 81° W	0.78	35°	18.6°
Môrro dos Conventos	136	N 76° W	0.75	37°	21.9°
24 km SW of Môrro dos Conventos	34	N 75° W	0.70	36°	23.6°
Passo de Tôrres	76	S 80° W	0.61	32°	18.7°
S.C.–General results	1600	S 77° W	0.50	41°	20.2°

cross-bedding measurements show a very low consistency ratio (0.12). The resultant vector is not in agreement with the wind pattern. Here, also, dune morphology indicates the action of strong and occasional southerly winds.

The influence of topography is clearly indicated in cross-bedding measurements from the Tôrres (south) locality shown in figure 33, no. 19. This locality is protected from prevailing winds by hills. The resultant vector (table 6) indicates the result of occasional strong southerly winds. At Tôrres (north) and Passo de Tôrres (fig. 33, no. 17 and 18), just northward of and close to the locality of Tôrres (south), however, no relief protects the dune field from the prevailing winds and at these localities the resultant vector coincides with the prevailing wind direction.

The localities on the Island of Santa Catarina and at Tôrres (south) document the influence of the local relief on cross-bedding pattern. Twenty-three other localities (fig. 33; tables 5 and 6) along the southern coast of Santa Catarina and along the coast of Rio Grande do Sul, confirm the validity of the principle used for determination of the prevailing wind direction from cross-bedding measurements and rose diagrams (fig. 33) show no influence of the strong and occasional southerly winds responsible for dune structures in fields protected from the prevailing wind by topographic relief. Those data furnish excellent evidence of the importance of prevailing winds in accumulating sand and the ephemeral nature of the short-term strong winds.

Southeastern Brazil.—In the northern coast of Santa Catarina and at the coast of Paraná, the cross-bedding resultant vector indicates a dominance of winds blowing from southeast and from east (figs. 29 and 27). These results are in agreement with the direction of regional prevailing winds. The southeasterly- and easterly-winds may be from the South Atlantic high-pressure cell (Bigarella, Becker, and Duarte, 1969); however, they also may have originated in the temporary but frequent anticyclonic cells of advanced polar air masses established on and off the southern Brazilian coast, mostly during the winter and less frequently during the summer (fig. 28). Rose diagrams (fig. 30) show that such winds from the southern and eastern quadrants are frequent, together with the "return" trade winds.

On the southern Santa Catarina and Rio Grande do Sul coasts possibly only the "return" trade winds are effective in sand transport, but in the northern Santa Catarina and Paraná coasts the former are important. The transport of sand in these areas is most effective in the winter season, which coincides with the dry months and the frequent presence of polar air masses (Bigarella, Becker, and Duarte, 1969).

Uruguay.—Two localities, 48 km apart on the southern coast of Uruguay between Montevideo and Punta del Este, indicate the

Table 6.—Cross-bedding measurements from the coastal dunes of Rio Grande do Sul (Brazil) and Uruguay

Locality	Number of measurements	Average dip direction	Consistency ratio	Maximum dip	Average dip
State of Rio Grande do Sul					
Tôrres (north)	52	S 72° W	0.56	36°	22 °
Tôrres (south)	115	N 7° E	0.47	35°	21.3°
Praia da Figuerinha	39	S 53° W	0.60	34	20.1°
6,5 km NE Capão da Canôa	90	S 65° W	0.72	34°	19.3°
Tramandai	125	S 63° W	0.69	35°	18.3°
Barro Preto	51	S 64° W	0.68	34°	18.7°
Cidreira	66	S 66° W	0.86	31°	18.3°
Pinhal (north)	130	S 72° W	0.74	32°	21.2°
Pinhal (south)	105	S 70° W	0.79	37°	19.3°
18 km south of Pinhal	69	S 87° W	0.87	46°	24.5°
40 km south of Pinhal	84	S 72° W	0.86	36°	22 °
R.G.S.–General result	926	S 74° W	0.63	46°	20.4°
Uruguay					
El Pinar (km 31 Carretera Interbalnearia)	33	N 01° E	0.79	36°	16.5°
Km 79 (Carretera Interbalnearia)	32	N 10° E	0.65	35°	16.4°
Uruguay–General result	65	N 05° E	0.51	36°	16.5°

dominance of southerly winds in sand movement among the dunes (fig. 29, table 6). These winds are closely related with the advance of the polar air masses.

Dip Direction Rose Diagrams

Rose diagrams representing attitudes of dip in dune cross-strata at Mustang Island, Texas, show a bimodal distribution which, according to McBride and Hayes (1962, p. 548), could be the result of: 1) insufficient number of measurements; 2) variations in wind direction; and 3) migration of the dune with slip faces dipping at oblique angles to the major wind direction. On Mustang Island sand moves in the form of an "asymmetrical pyramidal dune," having two faces dipping in the directions of cross-bedding modes (McBride and Hayes, 1962, p. 549).

The standard deviation of cross-strata dip azimuths is low in fluvial deposits and high in marine (Pryor, 1960, table 4). The mean standard deviation (75°) of Mustang Island cross-strata is intermediate between fluvial and marine (McBride and Hayes, 1962, p. 551). It results from bimodal distribution. The standard deviation seems to be influenced both by the type of cross-bedding and by the degree of variation in the direction of sand transport. The planar type of cross-strata tends to have unimodal distribution; the trough type mostly has bimodal distribution.

Cross-bedding measurements normally are taken at random in the field from strata dipping more than 10°. The cross-strata show mostly curving surfaces, deposited as foresets, either concave upward, as in transverse and barchan dunes, or convex upward, as in precipitation ridges and in the front part of parabolic dunes. Rarely, measurements made in the field coincide with average dip directions (prevailing wind), which correspond to axes of cross-strata sets. More frequently the measured attitude of the cross-strata is to one side of the axis of a cross-strata set. This situation results in bimodal character of a rose diagram but should not necessarily be interpreted as the result of wind variation or bidirectional wind pattern.

In well developed transverse dune fields, in which the lee sides of dunes do not show pronounced concave fronts, cross-bedding measurements tend to have unimodal distribution.

The bimodal character of dip directions in cross-bedding may be of two types. In one, the spread of the two modes is less than 120°, as in most transverse, barchan, and precipitation-ridge dunes. In a second type, the spread of the two modes is more than 120°, as illustrated by longitudinal dunes, the arms of parabolic dunes that form long parallel ridges, and other blowout types of dunes. The widely spread type of bimodal distribution may be the result of either a bidirectional wind pattern or moderate shifting of a unidirectional wind.

PALEODESERTS

Late Paleozoic to Middle Mesozoic Paleodeserts in the United States

Extensive outcrops of eolian sandstones, believed to have been deposited in paleodesert environments, occur in the Colorado Plateau. They are chiefly in the states of Wyoming, Utah, Colorado, Arizona, and New Mexico, and range in age from Late Pennsylvanian to Late Jurassic.

Numerous investigators have studied the local stratigraphy, but specialized studies of the nature, structure, and paleowind patterns of these sandstones are mostly by Reiche (1938), McKee (1945), Poole (1962, 1964), Opdyke and Runcorn (1960), and Irving (1964). Eolian sandstones of the western United States occur in Pennsylvanian to Jurassic rocks (Poole, 1962, p. 148).

These sandstones, postulated as eolian, are mostly light colored and are predominantly fine grained but with coarse to medium-sized grains concentrated along planes of cross-stratification. The sandstones are moderately to well sorted and grains range from subangular to well rounded. Pitted and frosted grains are common.

Most of these sandstones are considered dominantly eolian, although they contain strata of fluvial and/or marine origin, especially near their depositional margins.

1. *Pennsylvanian and Permian time.*—Regional stratigraphic relations indicate that in general the supposed eolian sandstones tongue into marine sequences westward and into other types of continental deposits eastward. The eolian sandstones occur as a belt extending from southwestern Wyoming through eastern Utah, and southward into central Arizona. They are believed to represent in the western part a coastal-plain dune area and in the eastern part an inland desert. The sand probably was derived from beach deposits and regressive marine sands.

Measurements of cross strata dip directions indicate "that the prevailing winds during Pennsylvanian and Permian time were from the north and northeast during deposition of the upper part of the Weber Sandstone; from the northwest, north, and northeast for the White Rim and Coconino Sandstone; from the northwest for the Cedar Mesa Sandstone Member; and from the north and northeast for the De Chelly" (Poole, 1962, p. 148) (fig. 34).

2. *Triassic time.*—Local eolian sandstone units in the Early and Middle (?) Triassic Moenkopi Formation indicate that paleowinds were from the northwest at that time (McKee, 1945). The Late Triassic Wingate Sandstone intertongues with the upper part of the Chinle Formation and is considered to represent a

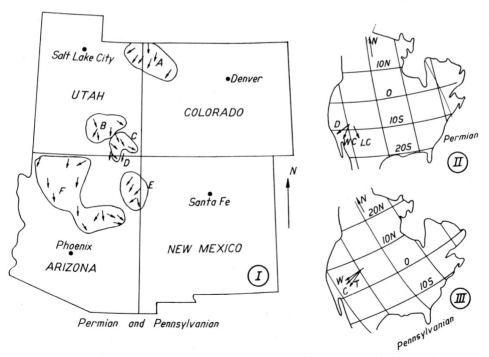

Fig. 34.—Wind directions during Permian and Pennsylvanian times, Colorado Plateau. I—data after Poole, 1962: A—Upper part of Weber Ss.; B—White Rim Ss., Member of Cutler Fm.; C—Cedar Mesa Ss., Member of Cutler Fm.; D & E—De Chelly Ss., Member of Cutler Fm. (E—data from Reiche, 1938); F—Coconino Ss. (data from Reiche, 1938). II & III—data after Irving (1964): D—De Chelly Ss.; WC—Western Coconino Ss.; LC—Little Colorado Coconino Ss.; C—Casper Ss.; T—Tensleep Ss.

humid-tropical or subtropical climate with distinct dry seasons (Poole, 1964, p. 403).

3. *Late Triassic (?) and Early Jurassic time.*—The eolian sandstones of this age are parts of continental sequences and are inferred to have been deposited in an inland-desert environment; they extended southwestward from central Wyoming through Utah, northern Arizona, southeastern Nevada, and into southern California.

Cross strata measurements for the Nugget and Aztec sandstones indicate that paleowinds were from the north and northeast (Poole, 1962, p. 148–150). The Navajo Sandstone seems to have been deposited by winds more from the northwest, except in southwest Utah and northwest

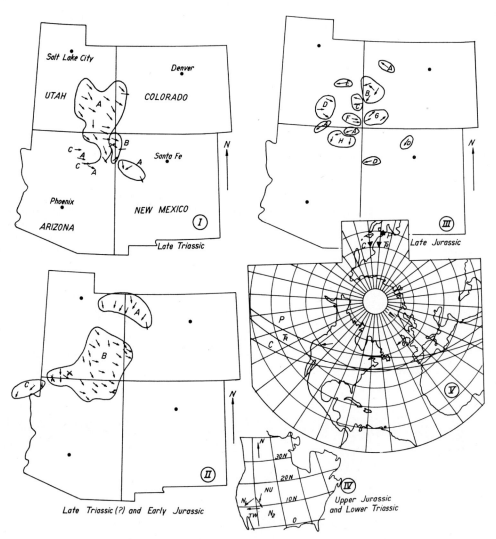

FIG. 35.—Wind directions during Mesozoic time, Colorado Plateau (after Poole, 1962). I—Late Triassic: A—Wingate Ss.; B—Tongues of Wingate Ss. in upper part of Chinle F.; C—Dinosaur Canyon Ss. Member of Moenave F. II—Late Triassic (?) and Early Jurassic: A—Nugget Ss. and equivalent rocks; B—Navajo Ss.; C—Aztec Ss. III—Late Jurassic: A to D—Entrada Sandstone. A—undifferentiated; B—Slick Rock Member; C—Moab Member; D—Upper Sandy Member; E—basal sandstone bed of Carmel Formation; F—Bluff Ss.; G—Junction Creek Ss.; H—Cow Spring Ss. (data from Poole, 1962; Reiche, 1938; Wilson, 1959; Cadingan, 1952; Harshbarger, 1949; Harshbarger, Repenning & Irving, 1957. IV—Paleowind direction, paleomeridian and paleolatitude for the Upper Triassic and Lower Jurassic eolian sandstones of the Colorado Plateau: NU—Nugget Ss., Ni—Upper Navajo Ss. (Kiersh); Nz—Middle Navajo Ss. (Kiersh); TW—type Wingate Ss. (Irving, 1964, p. 235). V—Position of the Equator and Poles in Carboniferous, Permian and Triassic Times (after Collinson & Runcorn, 1960, p. 969).

Arizona, where northerly paleowinds are indicated (fig. 35).

4. *Late Jurassic time.*—In the western part of the region most of the inferred eolian sandstones of this age are associated with marine units, whereas those in the eastern part probably were deposited in an inland-desert environment (Poole, 1964, p. 403). The supposed coastal dune deposits are associated with evaporites, suggesting arid climate conditions. Some of the younger sandstones are associated with fluvial deposits.

During Late Jurassic time, the paleowind regime apparently was more complex than before, as dip direction vectors indicate two almost opposite wind directions. The Entrada Sandstone was deposited by northeasterly and easterly paleowinds. An exception is its upper member, in south-central Utah, which probably was deposited by northwesterly and westerly winds. During Cow Springs Sandstone deposition, in Arizona, northerly and northeasterly winds prevailed. The Bluff and Junction Creek sandstones represent paleowinds blowing from west and southwest. Prevailing easterly winds probably were responsible for the deposition of the basal sandstone of the Carmel Formation (fig. 35).

5. *Conclusions.*—In western United States, the late Paleozoic and early Mesozoic sandstones, believed to be eolian, are considered to have been deposited under a trade wind circulation regime. The present area of distribution of these sandstones is in a region far north of the present trade-wind belt. The dune beds indicate, on the average, paleowinds blowing from the north to the south. This direction is entirely different from that of present-day westerly winds, characteristic of these latitudes (figs. 34 and 35).

A northerly extension of the trade wind belt at the time of deposition of these sandstones is suggested (Poole, 1957). The possibility that the region was formerly in a lower latitude and has since been rotated clockwise relative to the meridian is also suggested (Irving, 1964, p. 234). Because the trade wind belt must always have been in existence, its range in latitude apparently changed. Thus, continental drift and polar wandering probably are responsible for a change in position of the belt relative to the continent (Opdyke and Runcorn, 1960, p. 960).

Paleowind directions determined by Shotton (1956) for the British Permian and those for the late Paleozoic and early Mesozoic of the western United States represent northeasterly winds within a belt from the equator to 20° north at that time (Opdyke and Runcorn, 1960, p. 968). Paleomagnetic measurements indicate that

Fig. 36.—Paleowind directions for the Permian eolian sandstones of Great Britain (after Shotton, 1956).

North America and Europe moved 24° by the rotation of one continent with respect to the other about the present pole.

The position of the equator, based on paleomagnetic data, offers the best explanation for the paleowind determinations, both in western United States and in the Botucatu paleodesert of Brazil. Not until Late Mesozoic time did North America and Europe move out of the trade wind belt into the westerly wind belt (Opdyke and Runcorn, 1960, p. 970).

Wind directions obtained from a study of the cross-stratification in eolian sandstones of the Colorado Plateau could have been produced by storm winds, rather than by the prevailing winds, according to Sharp (1966, p. 1067). From the great number of measurements made on recent and near recent dunes along the eastern coast of South America, from the Amazon River to the La Plata Estuary, however, as well as from the great number of measurements made in the Mesozoic eolian sandstones of South America, one can be relatively sure that the resultant vector obtained from cross-bedding measurements indicates, in a majority of cases, the direction of prevailing winds. Only where topographic features interfere are the prevailing winds not registered in the cross-strata patterns of dunes. On the other hand, it seems clear that in areas subjected to almost opposite effective wind directions, cross-bedding measurements are not decisive in determining the prevailing wind direction. For the most part, the dip directions recorded in rose diagrams will indicate localities that have been subjected to such complex wind circulation patterns.

New Red Paleodesert, Great Britain

Cross-bedding in the eolian Lower Bunter Sandstone of North Worcestershire and East Shropshire was studied by Shotton (1937). The average dip direction of the Lower Bunter Sandstone in the outcrops west and northwest of Birmingham ranges between S. 54° W. and N. 59° W., with a general average of S. 86° W., which indicates that winds were blowing directly from the east. This sandstone is believed to represent barchan dunes (Shotton, 1937).

The paleowind directions for the British Permian eolian-type sandstones are shown in figure 36. These data indicate that the poles were in a different position from those of today (Shotton, 1956). The easterly paleowind directions seem to represent trade winds when corrected for paleomagnetic declination (Runcorn, 1961). Permian dunes of Britain probably were deposited in a hot desert (Shotton, 1956) in the trade wind belt of the northern hemisphere.

Botucatu Paleodesert, South America

The large-scale, cross-stratified facies of the Botucatu Sandstone, considered eolian, are in places associated with fluvial sediments (Pacheco, 1913). The Botucatu Sandstone is now considered to be Late Jurassic to Early Cretaceous in age, based on recent radiometric dates of intercalated basalts.

Paleowind directions of the Botucatu Sandstone.—Paleowind circulation in the Botucatu

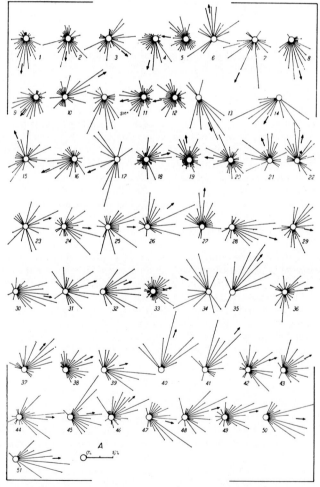

Fig. 37.—Rose diagrams showing cross-bedding dip trends for the Botucatu Sandstone. For the location of the sites see fig. 38 (after Bigarella & Salamuni, 1967, p. 269).

paleodesert has been investigated by Almeida (1953, 1954), Bigarella and Salamuni (1959a, 1959b, 1961), Bigarella and Oliveira (1966). Measurements in the correlative Sambaiba Formation of northern Brazil were made by Bigarella, Montenegro, and Coutinho (unpublished data).

Cross-bedding dip directions in the southern region of the Botucatu paleodesert (Santa Catarina, Rio Grande do Sul, and Uruguay) indicate that paleowinds blew from west and west-southwest (fig. 37 and 38). In southern Paraguay, these winds were deflected northward (fig. 38 II). Cross-bedding measurements in the northern region of the Botucatu paleodesert (Mato Grosso, Goiás, Minas Gerais, São Paulo, Paraná, and part of Paraguay) indicate mean paleowind trends from the north or north-northeast, which are "return" trade winds (fig. 38). These northerly paleowinds were deflected toward the west in Paraná.

On the Botucatu paleodesert, southern air masses moved northward, leaving their traces mainly in Paraná and São Paulo and, less frequently, the northern air masses penetrated the southern areas. Because of the conflict of cold air masses with warmer ones, the possibility of rainfall was greater north of Santa Catarina; this is indicated also by the presence of the subaqueous-type deposits of the Piramboia and Santana facies. These deposits, at the base of and interbedded with inferred dune deposits, are

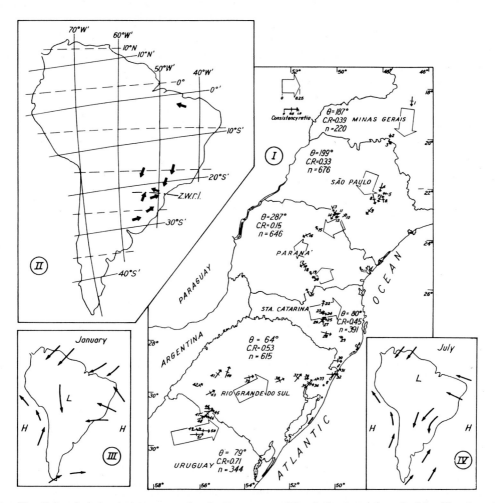

FIG. 38.—Paleowind circulation pattern for the Botucatu and Sambaiba formations. I—Map Showing wind circulation during Upper Jurassic—Lower Cretaceous time, as inferred from cross-bedding measurements in the eastern outcrops of the Botucatu Sandstone (θ—average dip direction; C.R.—consistency ratio; n—total number of measurements). II—Average paleowind directions for the Botucatu and Sambaiba formations. Paleomeridian and paleolatitudes according to Irving (1964, p. 238); z.w.r.l.—zone of wheel round latitude. III and IV—Present-day general circulation pattern for South America during January and July (after Monteiro, 1958).

rather common in Paraná and São Paulo, decreasing northward. Waterlaid deposits are rare in the southern part of the Botucatu paleodesert (Bigarella and Salamuni, 1961, p. 1101). Cross-bedding measurements in Paraná have a very low consistency ratio, presumably mostly the result of the conflicting action of two almost opposite wind directions represented by westerly and northerly winds. The area of conflict between these winds extends westward into Paraguay (fig. 38 II).

Cross-bedding measurements of the Sambaiba Formation furnish a clue, important in interpreting the environment of the Late Jurassic-Early Cretaceous South American paleodesert (Botucatu and Sambaiba Formations). The average dip direction of N. 67° W. in the Sambaiba Sandstone indicates that prevailing winds responsible for sand movement were the southeasterly trade winds (fig. 38; table 2).

In conclusion, the South American Mesozoic eolian sands were deposited in a low-latitude paleodesert resulting from 1) southeasterly paleotrade-winds (Sambaiba Formation); 2) northerly and north-northeasterly "return" paleotrade-winds (northern Botucatu paleodesert region); and 3) westerly paleowinds (southern Botucatu paleodesert region). The Botucatu paleodesert is similar to present trade wind deserts, of which the Sahara is an example. The trade-wind deserts have a distinctive pattern called the "wheel-round," indicated by changes in the alignment of the sand dunes, and of dune movement, at an apparently critical latitude. Cross-bedding in the Botucatu eolian sandstone has shown the existence of a critical latitude, near the border between Paraná and Santa Catarina (25°–27° Lat. south).

Westerly winds entering the southern part of the Botucatu paleodesert probably were derived from a high-pressure cell situated over the former Jurassic-Cretaceous sea to the west. The Andean Range did not exist at that time to prevent penetration of westerlies into the area of deposition of the Botucatu Sandstone. The east-southeasterly winds (Sambaiba Formation) and northerly winds (northern Botucatu paleodesert) were derived from the former South Atlantic high-pressure cell.

The general circulation pattern obtained from cross-bedding measurements in the Botucatu and Sambaiba Sandstones indicates that the position of South America relative to the equator in Late Jurassic-Early Cretaceous time was not very different from what it is now; this conclusion is consistent with paleomagnetic studies of Creer (1958).

REFERENCES

ALMEIDA, F. F. M., 1953, Botucatú, a Triassic desert of South America: XIX Congres Géologique International Comp. Rendus de la 19e Session p. 9–24.

———, 1954, Botucatú, um deserto triássico da América do Sul: DNPM Div. Geol. Min., Notas Prelim. e Estudos, no. 86, 21 p. Rio de Janeiro.

ANDRADE, G. O. DE, 1964, Os climas: in A. de Azevedo (Ed.), Brasil a terra e o homem. p. 397–457. Companhia Editora Nacional, São Paulo.

BAGNOLD, R. A., 1936, The movement of desert sand: Proc. Roy. Soc. A. 157, London.

———, 1954, The physics of blown sand and desert dunes: 265 p., Methuen & Co. Ltd., London.

BALL, M. M., 1967, Carbonate sand bodies of Florida and the Bahamas: Jour. Sed. Petrology, v. 37, no. 2, p. 556–591.

BEADNELL, H. J. L., 1910, Sand dunes of the Libyan desert: Geogr. Jour. v. 35, p. 379–395.

BIGARELLA, J. J., BECKER, R. D. AND DUARTE, G. M., 1969, Coastal dune structures from Paraná (Brazil): Marine Geology, v. 7, no. 1, p. 5–55.

———, MOUSINHO, M. R. AND SILVA, J. X., 1965, Processes and environments of the Brazilian Quaternary: Imprensa Univ. Fed. Paraná. Curitiba 71 p.

———, AND OLIVEIRA, M. A. M., 1966, Nota preliminar sôbre as direções de transporte dos arenitos Furnas e Botucatú na parte setentrional da Bacia do Paraná: Bol. Par. Geogr. v. 18/20, p. 247–256, Curitiba.

———, AND POPP, J. H., 1966, Contribuição ao estudo dos sedimentos praiais recentes, IV: Praia e dunas de Barra do Sul (S.C.), Bol. Par. Geogr., v. 18/20, p. 133–149.

——— AND SALAMUNI, R., 1959a, Contribuição ao estudo da estratificação cruzada nos arenitos mesozóicos do Brasil meridional e Uruguai: Dusenia, v. 8 (2), p. 45–60, Curitiba.

———, AND ———, 1959b, Nota sôbre a estratificação cruzada do Arenito Botucatú nos Estados de Minas Gerais, São Paulo e Paraná: Inst. Biol. Pesq. Tecn., Notas Prelim. e Estudos no. 3, 4 p., Curitiba.

———, AND ———, 1961, Early Mesozoic wind patterns as suggested by dune bedding in the Botucatú Sandstone of Brazil and Uruguay: Geol. Soc. Amer. Bull., v. 72, p. 1089–1106.

———, AND ———, 1967, Some palaeogeographic and palaeotectonic features of the Paraná Basin: in Bigarella, Becker & Pinto (Eds.), Problems in Brazilian Gondwana Geology: pp. 235–301. Curitiba.

BRAUN, G., 1911, Entwicklungsgeschichtliche Studien an europäischen Flachlandküsten und ihren Dünen. Veröff. des Inst. für Meereskunde und des Geog. Inst. an der Univ. Berlin, Heft 15, p. 1–174.

BRETZ, J. H., 1960, Bermuda: A partially drowned, late mature, Pleistocene karst: Geol. Soc. Amer. Bull., v. 1729–1754, 71, p.

BRYAN, KIRK, AND MCCANN, F. T., 1943, Sand dunes and alluvium near Grants, New Mexico: Amer. Antiquity, v. 8, no. 3, p. 281–295.

BUCHER, W. H., 1919, The origin of ripples and related sedimentary surface forms: Amer. Jour. Sci. 3 d series, XLVII, p. 149–210.
CADIGAN, R. A., 1952, The correlation of the Jurassic Bluff and Junction Creek Sandstones in Southeastern Utah and Southwestern Colorado; Pennsylvania State College Master's thesis, 163 p.
CAPOT-REY, R. AND CAPOT-REY, F., 1948, Le déplacement des sábles eoliens et la formation des dunes désertiques, d'après R. A. Bagnold: Trav. de l'Inst. de Recherches Sahariennes, Tome 5, p. 47–80.
COLLINSON, D. W. AND RUNCORN, S. K., 1960, Polar wandering and continental drift: new evidence from paleomagnetic observations in the United States: Geol. Soc. Amer. Bull. v. 71, p. 915–958.
COOPER, W. S., 1958, Coastal sand dunes of Oregon and Washington: Geol. Soc. Amer. Memoir 72, 169 p.
———, 1967, Coastal dunes of California: Geol. Soc. Amer. Memoir 104, 131 p.
CORNISH, V., 1897, On the formation of sand dunes: Geogr. Jour. IX, p. 278–309.
———, 1914, Waves of sand and snow: London, T. F. Unwin, 383 p.
CREER, K. M., 1958, Preliminary palaeomagnetic measurements from South America: Ann. Geophys., v. 14, no. 3, p. 373–390.
CRESSEY, G. B., 1928, The Indiana sand dunes and shore lines of the Lake Michigan Basin: The Geogr. Soc. of Chicago Bull., no. 8, 80 p., 20 plates, Univ. Chicago Press.
FAIRBRIDGE, R. W., 1964, African Ice-Age aridity: *in* Nairn (Ed.)—Problems in Palaeoclimatology, p. 356–360.
FÉDOROVITCH, B. A., 1957, L'origine du relief des déserts du sable actuels: Moscou, Inst. Géogr. de l'Acad. Sci del'U.R.S.S., Essays de Géographie, p. 117–129.
FINKEL, H. J., 1959, The barchans of southern Peru: Jour. Geol., v. 67 (6), p. 614–647.
FORD, E. W., 1957, The transport of sand by wind: Trans. Am. Geophys. Union, v. 38, no. 2, p. 171–174.
HACK, J. T., 1941, Dunes of western Navajo Country: Geogr. Rev. 31, p. 240–263, New York.
HANSEN, V., 1959, Sandflugten i Thyog dens indflydelse pa Kulturlandskabet: Meddelelser fra Skalling—Laboratoriet, v. 16, p. 69–91, København.
HARSHBARGER, J. W., 1949, Petrology and stratigraphy of Upper Jurassic rocks of central Navajo Reservation, Arizona: Arizona Univ. Ph.D. Thesis.
———, REPENNING, C. A., AND IRVING, J. H., 1957, Stratigraphy of the uppermost Triassic and Jurassic rocks of the Navajo Country: U.S. Geol. Surv. Prof. Paper 291, 74 p.
HÖGBOM, I., 1923, Ancient inland dunes of Northern and Middle Europa: Geografiska Annaler.
HOLM, D. A., 1960, Desert geomorphology in the Arabian Peninsula: Science, v. 132, no. 3437, p. 1369–1379.
INMAN, D. L., EWING, G. C., AND CORLISS, J. B., 1966, Coastal sand dunes of Guerrero Negro, Baja California, Mexico: Geol. Soc. Amer. Bull., v. 77, p. 787–802.
IRVING, E., 1964, Paleomagnetism and its application to geological and geophysical problems: John Wiley & Sons, 399 p, New York.
JENTSCH, K. A., 1900, Die Geologie der Dünen, in Gerhardt, Handbuch des deutschen Dünenbaues: Pt. 1, p. 1–124.
JONES, O. T., 1939, The geology of the Colwyn Bay district; a study of submarine slumping during the Salopian period: Quat. Jour. Geol. Soc., London, v. 95, p. 335–376.
KINDLE, E. M., 1917, Recent and fossil ripple marks: Canadian Geol. Surv., Museum Bull. 25.
KING, C. A. M., 1966, Techniques in Geomorphology: E. Arnold, London.
KUHLMAN, H., 1959a, Sandflugt og Klitdannelse Meddelelser fra Skalling: Laboratoriet, v. 16, p. 1–19, København.
———, 1959b, Kornstörrelser i Klit-og Strandsand: Meddelelser Fra Skalling—Laboratoriet, v. 16, p. 20–56, København.
———, 1959c, Quantitative measurements of aeolian sand transport: Meddelelser Fra Skalling—Laboratoriet, v. 16, p. 51–74, København.
LAMING, D. J. C., 1958, Fossil winds: Jour. Alberta Soc. Petroleum Geologists, v. 5, p. 179–183.
LONG, J. T. AND SHARP, R. P., 1964, Barchan-dune movement in Imperial Valley, California: Geol. Soc. Amer. Bull., v. 75, p. 149–156.
MAACK, R., 1966, Os problemas da Terra de Gondwana relacionados ao movimento tangencial de migração da crosta terrestre: Bol. Par. Geogr., v. 18/20, p. 25–70, Curitiba.
MACKENZIE, F. T., 1964, Geometry of Bermuda Calcareous dune cross-bedding: Science, v. 144, no. 3625, p. 1449–1450.
MADIGAN, C. T., 1936, The Australian sand ridge deserts: Geogr. Rev., v. 26, p. 205–227.
MCBRIDE, E. F. AND HAYES, M. O., 1962, Dune cross-bedding on Mustang Island, Texas: Amer. Assoc. Petroleum Geologists Bull., v. 46, p. 546–551.
MCKEE, E. D., 1933, The Coconino Sandstone—its history and origin: Carnegie Inst. Washington, Pub. 440, p. 78–115.
———, 1945, Small-scale structures in the Coconino Sandstone of northern Arizona: Jour. Geol., v. 53, no. 5, p. 313–325.
———, 1957, Primary structures in some recent sediments: Bull. Amer. Assoc. Petrol. Geologists, v. 41, no. 8, p. 1704–1747.
———, 1966, Structures of dunes at White Sands National Monument, New Mexico (and a comparison with structures of dunes from other selected areas): Sedimentology, v. 7, no. 1, p. 1–70, Amsterdam.
——— AND TIBBITTS, G. C., 1964, Primary structures of a seif dune and associated deposits in Libya: Jour. Sed. Petrology, v. 34, no. 1, p. 5–17.
———, AND WEIR, G. W., 1953, Terminology for stratification and cross-stratification in sedimentary rocks: Bull. Geol. Soc. Amer., v. 64, p. 381–390.
MERK, G. P., 1960, Great sand dunes of Colorado: Guide to the Geology of Colorado, Rocky Mt. Assoc. Geologists, p. 127–129.
MONTEIRO, C. A. F., 1958, A circulação atmosférica e os tipos de tempo; *in* Atlas Geográfico de Santa Catarina; Depto. Estadual de Geografia e Cartografia—Florianópolis.

MURPHY, M. A. AND SCHLANGER, S. O., 1962, Sedimentary structures in Ilhas and São Sebastião formations (Cretaceous), Recôncavo Basin, Brazil: Bull. Amer. Assoc. Petrol. Geologists, v. 46, p. 457–477.
NORRIS, R. M., 1966, Barchan dunes of Imperial Valley, California: Jour. Geol., v. 74, no. 3, p. 292–306.
———, AND NORRIS, K. S., 1961, Algodones dunes of southeastern California: Geol. Soc. Amer. Bull., v. 72, p. 605–620.
OPDYKE, N. D., 1958, Palaeoclimatology and palaeomagnetism in relation to polar wandering and continental drift: Ph.D. thesis, Univ. Durham, England, 235 p.
———, AND RUNCORN, S. K., 1960, Wind direction in the western United States in the Late Paleozoic: Geol. Soc. Amer. Bull., v. 71, no. 7, p. 959–972.
PACHECO, J., 1913, Notas sôbre a geologia do vale do Rio Grande, etc: São Paulo, Com. Geog. e Geol, p. 33–38, São Paulo.
PEACOCK, J. D., 1966, Contorted beds in the Permo-Triassic aeolian sandstones of Morayshire: Bull. Geol. Surv., Gr. Brit., v. 24, p. 157–162.
POOLE, F. G., 1957, Paleowind directions in late Paleozoic and early Mesozoic time on the Colorado Plateau as determined by cross-strata: Geol. Soc. Amer. Bull., v. 68, p. 1870.
———, 1962, Wind directions in Late Paleozoic to Middle Mesozoic time on the Colorado Plateau: U. S. Geol. Surv. Prof. Paper 450-D, art. 163, p. 147–151.
———, 1964, Palaeowinds in the western United States: *in* Nairn, A.E.M., Problems in Palaeoclimatology, p. 394–405, John Wiley & Sons, London.
PRICE, W. A., 1958, Sedimentology and Quaternary Geomorphology of South Texas: Gulf Coast Assoc. Geol. Soc. Trans., v. 8, p. 41–75.
———, 1962, Stages of oxidation coloration in dune and barrier sands with age: Geol. Soc. Amer. Bull., v. 73, p. 1281–1284.
PRYOR, W. A., 1960, Cretaceous sedimentation in Upper Mississippi Embayment: Amer. Assoc. Petroleum Geologists Bull., v. 44, no. 9, p. 1473–1504.
REICHE, P., 1938, An analysis of cross-lamination: the Coconino Sandstone: Jour. Geol., v. 46, p. 905–932.
RUNCORN, S. K., 1961, Climatic change through geological time in the light of the paleomagnetic evidence for polar wandering and continental drift: Quat. J. Roy. Meteorol. Soc., v. 87, no. 373, p. 282–313.
SANTOS, E. O., 1965, Características climáticas: *in* A. de Azevedo (Ed.) A baixada santista, aspectos geográficos, v. 1, p. 95–150, Editôra da Univ. de São Paulo.
SERRA. A., 1956, Atlas climatológico do Brasil: Conselho Nac. de Geogr., 433 p., Rio de Janeiro.
SHARP, R. P., 1963, Wind ripples: Jour. Geol., v. 71, p. 617–636.
———, 1964, Wind-driven sand in Coachella Valley, California: Geol. Soc. Amer. v. 75, p. 785–804.
———, 1966, Kelso Dunes, Mojave Desert, California: Geol. Soc. Amer. Bull., v. 77, p. 1045–1074.
SHOTTON, F. W., 1937, The lower Bunter sandstones of north Worcestshire and east Shropshire: Geol. Mag., v. 74, p. 534–553.
———, 1956, Some aspects of the New Red Desert in Britain, Liverpool Manchester: Geol. J., v. 1, p. 450–466.
SMITH, H. T. U., 1940, Geologic studies in southwestern Kansas: Kansas State Geol. Surv. Bull. 34, 212 p.
SOKOLOW, N. A., 1894, Die Dünen; Bildung, Entwicklung und innerer Bau (German translation from Russian 4th ed., assisted by A. Arzruni): Berlin, Springer, 298 p.
SOLGER, F., 1910, Geologie der Dünen: *in* Dünenbuch, Stuttgart, F. Enke.
TANNER, W. F., 1964, Eolian ripple marks in sandstone: Jour. Sed. Petrology, v. 34, no. 2, p. 432–433.
TWENHOFEL, W. C., 1932, A treatise on sedimentation: 2nd Ed., Baltimore.
WILSON, R. F., 1959, The stratigraphy and sedimentology of the Kayenta and Moenave formations, Vermilion Cliffs region, Utah and Arizona: Stanford Univ. Ph.D. thesis, 337 p.
WRIGHT, H. E., JR., 1961, Late Pleistocene climate of Europe: A review, Geol. Soc. Amer. Bull., v. 72, p. 933–984.

RECOGNITION OF ALLUVIAL-FAN DEPOSITS IN THE STRATIGRAPHIC RECORD

WILLIAM B. BULL
University of Arizona

ABSTRACT

Alluvial fans, commonly are thick, oxidized, orogenic deposits whose geometry is influenced by the rate and duration of uplift of the adjacent mountains and by climatic factors.

Fans consist of water-laid sediments, debris-flow deposits, or both. Water-laid sediments occur as channel, sheetflood, or sieve deposits. Entrenched stream channels commonly are backfilled with gravel that may be imbricated, massive, or thick bedded. Braided sheets of finer-grained sediments deposited downslope from the channel may be cross-bedded, massive, laminated, or thick bedded. Sieve deposits are overlapping lobes of permeable gravel.

Debris-flow deposits generally consist of cobbles and boulders in a poorly sorted matrix. Mudflows are fine-grained debris flows. Low viscosity debris flows have graded bedding and horizontal orientation of tabular particles. Viscous flows have uniform particle distribution and vertical preferred orientation that may be normal to the flow direction.

Logarithmic plots of the coarsest one-percentile versus median particle size may make patterns distinctive of depositional environments. Sinuous patterns indicate shallow ephemeral-stream environments. Rectilinear patterns indicate debris-flow environments.

Fans consist of lenticular sheets of debris (length width ratio generally 5 to 20) and abundant channel fills near the apex. Adjacent beds commonly vary greatly in particle size, sorting, and thickness. Beds extend for long distances along radial sections and channel deposits are rare. Cross-fan sections reveal beds of limited extent that are interrupted by cut-and-fill structures.

Three longitudinal shapes are common in cross section. A fan may be lenticular, or a wedge that is either thickest, or thinnest, near the mountains.

INTRODUCTION

Alluvial fans are distinctive terrestrial stratigraphic units. A fan is a deposit whose surface forms a segment of a cone that radiates downslope from the point where the stream leaves the mountains. Although the surface is not preserved in the stratigraphic record, distinctive suites of alluvial-fan deposits are preserved in many parts of the world.

Alluvial fans are important economically. The deposits form the principal ground-water reservoir in many areas, and the recharge of many groundwater basins is through the alluvial fans that fringe the basin. The surfaces of many fans are highly desirable for agricultural, urban, and industrial uses.

Modern alluvial fans have the following definitive characteristics that aid in interpretation of alluvial-fan deposits preserved in the stratigraphic record. Each fan is derived from a source area with a drainage network that transports the erosional products of the source area to the fan in a single trunk stream. The result of fluvial deposition by the stream is a cone-shaped deposit. The plan view of the deposit commonly is fan-shaped, and the contours bow downslope from the fan apex. Overall radial profiles are commonly concave, and cross-fan profiles are convex. Bedrock knobs, such as are commonly associated with pediments, rarely protrude through the thick alluvium. If present, they are most common near the fanhead.

A vertical aerial view of an alluvial fan having the above characteristics is shown in Figure 1. The Copper Canyon fan forms a 180° cone because of the absence of large fans that would restrict lateral expansion of the fan. Generally, adjacent fans restrict the lateral extent of the individual deposits. Thus, most fan deposits occur as a series of coalescing alluvial cones that form a piedmont slope that is sometimes called a bajada. Small coalescing fans are shown in the upper part of Figure 1.

Alluvial fans are most widespread in the drier parts of the world, but also occur in humid regions such as Japan (Murata, 1966); the Himalayan Mountains (Drew, 1873); Canada (Winder, 1965); and in the Arctic regions of Scandanavia (Hoppe and Ekman, 1964) and Canada (Legget, and others, 1966).

The effect of source-area characteristics on fan size has lead to the development of equations that relate the two. For example variations of the coefficient in the equation

$$A_f = cA_d^n$$

(where A_f and A_d are the fan and drainage-basin areas) reflects the erodibility of the source rocks and the tectonic environment affecting the erosional-depositional system (Bull, 1962a, 1968,

Fig. 1.—Vertical view of the Copper Canyon fan, Death Valley, California. Photograph by Howard Chapman, U.S. Geological Survey.

p. 104–105; Denny, 1965, p. 38; Melton, 1965; Hawley and Wilson, 1965; Hooke, 1968, p. 609–621).

Purpose and Scope

This article discusses the characteristics of alluvial-fan deposits that may be useful in the recognition of alluvial-fan deposits in the geologic record.

Characteristics that can be determined in small outcrops or core samples will be described first, to provide criteria for the identification of five types of fluvial processes that construct fans. Then the stratigraphy of alluvial fans will be discussed to provide the three-dimensional framework needed to complete the identification of the alluvial-fan depositional environment.

ACKNOWLEDGMENTS

The author is indebted to Roger LeB. Hooke of the University of Minnesota and W. K. Hamblin of Brigham Young University for critical review of the manuscript.

ALLUVIAL-FAN DEPOSITS

Although some depositional environments can be identified by a combination of physical, chemical, and biological data, alluvial fans are identified mainly by a distinctive suite of physical properties. A few general chemical and biological characteristics will be noted at this point, then the remainder of the article will be devoted to the physical properties of fan deposits.

Two types of chemical data pertain to alluvial fans—the degree of oxidation and the salts that have been deposited in, or have accumulated in, alluvial-fan deposits. Most fan deposits are oxidized. Therefore, lack of reduced deposits is a typical characteristic of fans that have accumulated in the arid and semiarid parts of the world. Fine-grained oxidized fan deposits that contain visible organic matter have been described by Miller, Green, and Davis (1971). Meade (1967, p. 6) notes that the oxidized color of fan deposits can be retained even after burial to more than 1,500 feet below the water table.

Alluvial-fan deposits commonly contain salts such as gypsum and calcite that have been deposited with the sediments or have accumulated as a result of weathering of the surficial materials. Because a fan is the product of a single stream, marked variations of soluble salts may occur in adjacent fans if the lithologies of the source areas are markedly different. Chemical characteristics can be used to distinguish between fans from different source areas. For example, Bull, (1964a, p. 61) lists mean gypsum contents of 2.2 and 0.03 percent for adjacent fans. Burial below the groundwater table permits circulation of groundwater to change the chemical character of the contained salts by solution and deposition. Maximum salt contents commonly occur near the downslope edge of fans, because this may be an area of ground-water discharge or of flow of saline water from adjacent playa deposits. The concentric banding near the downslope edge of the fan shown in Figure 1 is the result of larger salt and moisture contents and more intense chemical weathering, than the upslope parts of the fan.

The solution of calcium carbonate in the upper part and deposition in the lower part of the soil profile is characteristic of fan deposits that are not receiving additional increments of sediment and have insufficient water infiltrating through the soil to remove the authigenic carbonate. Detailed descriptions of carbonate accumulations in desert soils, such as can occur on alluvial fans, are given by Gile, Peterson, and Grossman (1966). Alternating periods of deposition and soil-profile formation are characteristic of most alluvial fans as the depositional area shifts from one part of the fan to another during the process of constructing the cone-shaped deposit. If sufficient time is available for weathering to occur between periods of deposition, a series of soil profiles will result, each profile representing a time of no deposition. An example of multiple buried soil profiles is shown in Figure 2. Three zones of carbonate accumulation are exposed in the 35-foot bank where the main stream channel has trenched the fanhead. The fan deposits consist entirely of water-laid sediments derived from a carbonate-rock source area.

Fossils not only are rare in alluvial-fan deposits but are of little use in differentiating the fan depositional environment from other continental depositional environments such as flood plains. Plant fragments are rare in gravelly alluvial-fan deposits. Fragments of organic matter have not been noted in many fan deposits, but

Fig. 2.—Caliche layers in the Wheeler Wash fan, south side of the Spring Mountains, Nevada.

the presence of organic material should not be used as an indicator of the lack of an alluvial-fan environment. Legget and others (1966, p. 20) describe alluvial-fan deposits in northern Canada containing decomposed and undecomposed organic matter in silts having a reduced color. Meade (1967, p. 7) describes core-hole samples from alluvial fans in the western San Joaquin Valley, California, that contain disseminated small fragments of organic material in sand, silt, and clay that has an oxidized color. The present climate of the area consists of hot, dry summers and a mean annual rainfall of 6–8 inches that occurs during the winter months.

It is the physical characteristics that provide the surest means of identifying alluvial-fan deposits in the stratigraphic record. Some modes of deposition on alluvial fans are common in other depositional environments. Stream-channel deposits are found in several depositional environments, but sheetflood and sieve deposits are most likely to be preserved if deposited on an alluvial fan. Debris flows also can occur in several different environments. They are found on hillslopes (an erosional environment) and in river valleys. However, debris flows generally are stored only temporarily in river valleys, because they are associated with terrains that produce rapid runoff and hence are likely to be eroded by subsequent water flows. Little erosion occurs on an actively aggrading alluvial fan; thus, the fan environment is ideal for preservation of debris flows in the stratigraphic record.

The tendency of most modes of fan deposition to occur as sheets results in a diagnostic geometry of bedding that helps identify modes of deposition—such as mudflows—as occuring on alluvial fans rather than in other depositional environments. This important aspect is discussed in the sections on bedding and stratigraphic geometry.

Alluvial fans have greatly differing lithologies. Some fans consist of organic silt (Legget and others, 1966, p. 21), others consist largely of pebble- to boulder-size material without fines (Hooke, 1967, p. 456). More than one mode of deposition occurs on most fans, and the proportions of different types of deposits may vary both vertically and in the downslope direction from the fan apex. Each bed of a fan represents a single depositional event that has resulted from one of a wide spectrum of precipitation and erosion events within the source area. The runoff that is supplied to the main stream channel leading to the fan may be the result of rainfall over the entire basin, or of snowmelt runoff from all or part of the basin. Thus, differences in runoff characteristics, source and amount of sediment load, mode of transport, and other factors vary greatly and are reflected in the individual beds preserved in the fan.

The distribution of maximum and mean particle sizes in modern arid-region alluvial fans shows that particle size decreases downslope with increasing distance from the fan apex (Sharp, 1948; Blissenbach, 1954; Beaty, 1963; Drewes, 1963; Bull, 1964a; Ruhe, 1964; Bluck, 1964; and Hooke, 1967). The uniformity of particle size decrease is greatly affected by the amount of temporary channel entrenchment, which causes the loci of deposition to occur in different areas along a given radial line (Buwalda, 1951).

Water-laid Deposits

Two types of water-laid deposits can be found in most alluvial fans, and a third type occurs on fans with certain source-area conditions. Most of the water-laid sediments consist of sheets of sand, silt, and gravel deposited by a network of braided distributary channels. The second type of sediment consists of fillings in stream channels that were temporarily entrenched into the fan. The third type is sieve deposits (Hooke, 1967, p. 453–456) which are formed when surficial fan material is so coarse and permeable that even large discharges infiltrate into the fan completely before reaching the toe. Gravel deposited in the area where the flow infiltrates forms lobate deposits which further decrease the slope and promote additional deposition.

Sheetflood Sediments

Sheets of sediments are deposited by surges of sediment-laden water that spread out from the end of the stream channel on a fan. Deposition is caused by a widening of the flow into shallow bands or sheets and concurrent decrease in depth and velocity of flow, rather than a change in gradient at the end of the stream channel. Depths of water generally are less than a foot. The shallow distributary channels rapidly fill with sediment and then shift a short distance to another location. The resulting deposit commonly is a sheetlike deposit of sand, or gravel, that is traversed by shallow channels that repeatedly divide and rejoin.

The deposits commonly consist of gravel, sand, or silt that contains little visible clay. In general, they are well sorted and may be cross-bedded, laminated, or massive. The characteristics of sediments deposited by braided streams are described in detail by Doeglas (1962).

The low bars and anastomosing channels characteristic of water-laid sheets of sediments on fans are shown in Figure 3A. The photograph is a view of part of an area of water-laid deposition that is 1 mile long. A map of a similar depositional area on the same fan and a summary

FIG. 3.—Braided-stream sediments, Western Fresno County, California. A. Braided water-laid sediment on the Arroyo Hondo fan. B. Outcrop of braided water-laid sediments in the ancestral fan of Panoche Creek.

Fig. 4.—Cross-section of deposits filling a small stream channel in the Tumey Gulch fan, Western Fresno County, California.

of particle-size distribution of 14 samples are given by Bull (1964a, p. 27).

Lenses caused by deposition as low bars and islands of braided-stream sediments are readily recognized in exposures of older fans that are normal to the direction of flow. The sands, silts, and gravels shown in Figure 3B were deposited near the apex of a Pliocene and Pleistocene alluvial fan that had an area of more than 200 square miles.

Stream-channel Sediments

The deposits that backfill stream channels temporarily entrenched into a fan generally are coarser grained and more poorly sorted than the sheets of water-laid sediments deposited by networks of braided distributary channels. Bedding of the channel-fill sediments may not be as well defined as for the sheetlike deposits. The thickness of individual beds ranges from less than one inch to more than five feet, but bed thicknesses of two inches to two feet are the most common.

A cut-and-fill structure in a canal bank along the fan contour is shown in Figure 4. The adjacent deposits consist mainly of sheets of clayey sand deposited by mudflows, and the channel filling consists of beds of gravel and clayey sand.

Comparison of the sorting of the sediments of braided stream and stream-channel deposits for 36 samples is summarized by Bull (1964a, p. 28). The samples were derived from a source terrain that consists of sandstone and mudstone, and the mean clay content of the samples was 6 percent. The sorting[1] was as follows:

Sieve Deposits

If the source area supplies little sand, silt, and

[1] The three sorting indices used are the Trask sorting coefficient, $S_0 = \sqrt{Q_{75}/Q_{25}}$; the phi quartile deviation,

$$QD_\phi = \frac{\phi 25 - \phi 75}{2}$$

and the phi standard deviation,

$$\sigma_\phi = \frac{\phi 16 - \phi 84}{2}$$

clay to the fan, the deposits may be sufficiently permeable to allow water from a flood discharge to infiltrate entirely before reaching the toe of the fan. Such conditions promote deposition of lobes of gravel. Hooke (1967, p. 453–456) has studied these lobate gravel deposits in detail and has named them "sieve deposits." He stated that because "water passes through rather than over such deposits, they act as strainers or sieves by permitting water to pass while holding back the coarse material in transport."

Thus, unique source-area conditions are responsible for fans composed of sieve deposits. Source areas for sieve deposits are underlain by rocks such as jointed quartzite; and the clasts supplied to the fans characteristically are subangular blocks instead of well-rounded gravel. The excellent sorting of sieve deposits results in a massive bed and poorly defined contacts between beds.

Sieve deposits are much less common than other types of water-laid deposits on alluvial fans but are among the most distinctive.

Sieve deposits on a small, steeply sloping fan are shown in Figure 5. Most of the surface has been darkened by desert varnish coating the particles. The man is standing on a finer-grained, recently deposited sieve lobe.

Debris-flow Deposits

Water flows can selectively deposit part of their sediment load as a result of decrease in velocity or depth of flow. When a flow incorporates sufficient sediment, sediment entrainment becomes irreversible, and the flow behaves more like a plastic mass than a Newtonian fluid. Debris flows have a high density and viscosity compared to stream flows. Because of these traits, debris-flow deposits are poorly sorted, have lobate tongues extending from sheetlike deposits, have well-defined margins, and are capable of transporting boulders weighing many tons. Factors that promote debris flows are abundant water (usually intense rainfall) over short periods of time at irregular intervals, steep slopes having insufficient vegetative cover to prevent rapid erosion, and a source material that provides a matrix of mud. Deposits of high viscosity flows are most common near the fan apexes (Hooke, 1967, p. 452; Crowell, 1954).

The proportions of water-laid and debris-flow deposits vary greatly from fan to fan and may change during the history of accumulation of the deposits of a single fan. Where source-area conditions are not conducive to the production of debris flows, the fan deposits consist entirely of water-laid sediments. Other fans consist mainly of debris flows. Most fans whose source areas produce debris flows also have flood events that result in the deposition of water-laid deposits. Thus, the deposits of many fans consist of inter-

FIG. 5. Sieve deposits on a small fan in Death Valley, California.

Fig. 6. Debris flow on the Sparkplug Canyon fan, west side of the White Mountains, California.

bedded deposits of debris and water flows in varying proportions.

Part of a thick viscous debris flow is shown in Figure 6. Abrupt margins of lobate tongues of debris in the foreground indicate that the flow was highly viscous. The round protuberances further upslope in the flow are 3- to 8-foot, mud-covered boulders. The smooth surface of most of the flow is characteristic of fresh debris flows deposited on alluvial fans. Old debris flows commonly consist of surficial lobes and levees of cobbles and boulders, because rainwash has removed much of their mud matrix. Later water flows may cut through the flow. Debris-flow levees occur adjacent to the channels of some flows (Sharp, 1942). The maximum thickness of the flow shown in Figure 6 is 4 to 6 feet (Beaty, 1968, p. 18).

A mudflow is a type of debris flow that consists mainly of sand-size and finer sediment. Many workers use the term "mud-flow" in a genetic sense for all types of debris flows, because a matrix of mud is the distinguishing feature of this type of flow, as contrasted with water flows.

The mudflow shown in Figure 7 consists of a poorly sorted clayey sand. The viscosity of the flow was not great, as is indicated by the thin margins of the lobate tongue. The flow thickens rapidly away from the margins and is 3 to 4 inches thick at the location of the shovel. Although the flow contained virtually no material larger than 4 millimeters, the cobble train shown at the right side of the photograph was part of a lobate tongue of an earlier debris flow that transported larger particles. Polygonal dessication cracks are characteristic of clayrich mudflows that contain little gravel.

The bedding of debris-flow sequences commonly is not well defined, but upon close examination bedding planes between flows can be discerned in outcrops (Beaty, 1970, fig 4). Where interbedded with water-laid sediments, debris-flow beds are readily apparent.

Several features aid in the recognition of debris flows in the geologic record. The uniform thickness of the central parts of the sheet-like deposits produces beds that are remarkably consistent in thickness when observed in outcrop. An indication of the viscosity of a given flow can be obtained by study of the position and orientation of the larger clasts. A fluid debris flow will have graded bedding and a horizontal or imbricated orientation of the tabular gravel fragments. The more viscous flows have the larger clasts distributed uniformly throughout the thickness of the flow. The most viscous flows not only have uniform distribution of the larger clasts, but the tabular particles commonly have

Fig. 7.—Mudflow on the Santiago Creek fan, north side of the San Emigdio Mountains, California.

a vertical prefered orientation normal to the direction of flow.

Poor sorting is characteristic of debris flows. Many debris-flow deposits are so coarse grained that it is difficult to obtain a representative sample for determining the particle-size distribution of the material. As a result, few particle-size analyses have been made of debris flows (Crawford and Thackwell, 1931; Sharp and Nobles, 1953 p. 556), and even fewer comparisons have been made of the sorting of debris flows and water-laid sediments from the same source areas. Bull (1964a p. 24, 25, 65, 66) made 50 particle-size analyses of mudflows, and the mean sorting characteristics of the suite of samples are summarized in Table 1. The samples are from the same source terrain as noted for the water-laid samples in Table 1. The mean clay content of the mudflow samples was 31 percent.

CM Patterns

Logarithmic plots of the coarsest 1-percentile particle size (C) and the median particle size (M) may make patterns distinctive of the depositional environment according to Passega (1957, 1960). Bull (1962b), using the parameters C and M from the particle-size analyses of 102 surficial and 50 corehole samples of alluvial fan deposits, found distinctive patterns that could be related to the modes of deposition on the alluvial fans.

The mode of deposition was determined in the field for the surface samples, and the CM patterns for these samples are shown in Figure 8. Passega found that a sinuous pattern was definitive of the tractive-current mode of deposition associated with water-laid deposits of perennial rivers, and that a rectilinear pattern roughly parallel to the limit C=M was characteristic of turbidity-current deposits. These modes of deposition apparently have analogs on arid-region alluvial fans.

Three of the four segments of Passega's tractive-current pattern are represented by the wa-

TABLE 1.—SORTING OF ALLUVIAL-FAN DEPOSITS DERIVED FROM THE DIABLO RANGE, CALIFORNIA

	So	σ_ϕ	QD_ϕ
Braided-stream deposits			
Range	1.1–2.7	0.48–2.4	0.15–1.4
Mean	1.5	1.0	.56
Stream-channel deposits			
Range	1.3–4.8	0.82–3.4	0.42–23
Mean	2.1	2.0	1.1
Mudflow deposits			
Range	5.0–25.	4.1–6.2	2.3–4.7
Mean	9.7	4.7	3.1

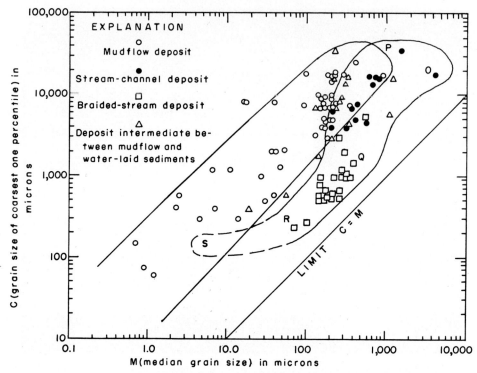

Fig. 8.—CM patterns of surficial alluvial-fan deposits, western Fresno County, California. Letters are discussed in the text.

ter-laid deposits shown in Figure 8. The segment representing the deep protected channel of Passega (RS) is not present, suggesting that this type of deposition is not characteristic of alluvial-fan sediments deposited by shallow ephemeral streams.

The CM pattern for the mudflow deposits is the same type that Passega shows for turbidity currents, which suggests that both turbidity currents and mudflows entrain fine-grained material. The chief difference between the patterns for mudflows and turbidity currents appears to be the sorting of the coarsest half of the deposits. Passega mentions that C is 2.3 to 4.2 times M for points along the axes of his turbidity current patterns. C ranges from about 40 to 80 times M for points along the axis of the mudflow CM pattern. This poorer sorting suggests that the mudflows had a much higher density and viscosity then did the turbidity currents studied by Passega.

CM patterns can be used to ascertain the mode of deposition of alluvial-fan deposits from subsurface samples. Figure 9 shows CM patterns for core-hole samples from two fans adjacent to the Diablo Range in California. About 68 percent of the source area for the Arroyo Ciervo fan is underlain by soft clayey rocks such as mudstone and shale, but only 34 percent of the source area of the Martinez Creek fan is underlain by clayey rocks.

Ten samples from a 70-foot core hole in the Martinez Creek fan are represented by the points with pattern T in Figure 9. Most of the samples are moderately well sorted sand and form a pattern suggestive of deposition by tractive currents.

Forty samples from five core holes in the Arroyo Ciervo fan are represented by the points within pattern M in Figure 9. The wide range of sorting between the 50th and 99th percentile suggests that a mixed depositional environment may be represented. However, all but four of the points can be included in the rectilinear mudflow-type pattern. Points in the vicinity of A probably indicate mudflow deposition, and the four points near C may represent water-laid sediments. C is about 45 times M at points along the axis of the "mudflow" pattern.

STRATIGRAPHY OF ALLUVIAL FANS

Bedding

Bedding of deposits is one of the best methods of identifying the alluvial-fan environment of deposition. Within a single outcrop a variety of strata generally can be observed, each bed representing a particular set of hydraulic conditions that determined the thickness, particle-size and distribution, particle orientation, and type of contact with the underlying bed. Even in fans composed entirely of water-laid sediments, differences in flow result in marked differences in the sedimentological characteristics of the beds (Figure 3B).

Bedding differences are even more striking in the many fans composed of both water-laid and debris-flow deposits. The poorly sorted, massive beds of debris-flow deposits stand out in marked contrast to the beds of water-laid sediments. Thus, one distinctive feature of alluvial-fan deposits is the variety of depositional types that can be observed.

Because the bulk of the fan deposits are deposited as sheets and lobes, uniform thickness for a given bed is common in most outcrops, particularly for debris-flow deposits. The thickness of the bedding of water-laid deposits usually is a function of the amount of relief between the bars and the braided stream channels in the area at the time of deposition, and the degree of erosional modification by post-depositional flows.

An example of bedding variety is shown in Figure 10. The massive thick bed of uniform thickness above the hat consists of clayey gravel and was deposited as a viscous debris-flow. Beneath the debris flow are beds of well-sorted, water-laid sand. A ½-inch bed of water-laid clay immediately above the sands is typical of the waning phase of ephemeral water flooding on a fan, when the competence is sufficient to transport only silt and clay. Beds of poorly sorted, silty gravel occur above the debris-flow bed and beneath the water-laid beds. These beds may be interpreted as low viscosity, debris-flow deposits, or as poorly-sorted, water-laid deposits.

When individual beds can be identified at

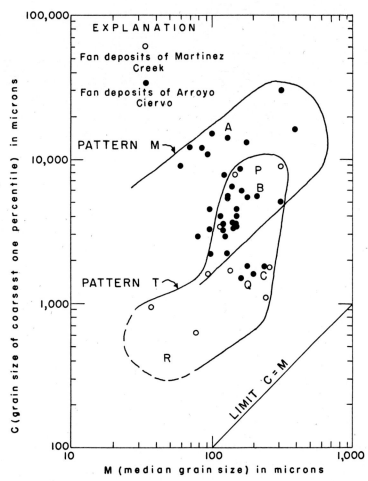

Fig. 9.—CM patterns of alluvial-fan deposits from core holes, western Fresno County, California, Letters are discussed in the text.

Fig. 10.—Bedding of Late Tertiary alluvial-fan deposits, south side of the Santa Catalina Mountains, Arizona.

more than one outcrop, another diagnostic stratigraphic indicator becomes readily apparent—the extensive sheet-like aspect of the individual beds. Sheets of fan deposits commonly are 10 to more than 100 times the width of the channel that transported the material to the fan. Thus, one of the identifying characteristics is the ratio of the trunk channel widths, as shown by cut-and-fill structures, to the width of the former areas of deposition represented by individual beds. This relation is best observed in exposures parallel to the former fan contour.

The sheetlike character of the bedding commonly is not apparent in those fans consisting mainly of gravel. The braided-stream mode of deposition that is characteristic of these gravel deposits results in small-scale, cut-and-fill structures being common in those exposures parallel to the fan contour. Thick, sheetlike beds of water-laid gravel, if present, can be attributed to large floods that reworked large volumes of previously deposited material.

The sheets and tongues of water-laid and debris-flow deposits on modern alluvial fans are generally 5 to 20 times as long as they are wide. The length of the sheets ranges to a few tens of feet to many miles. Exposures usually are not available to indicate the extent of the larger sheets comprising fans. However, by examining the beds that represent flows of small extent, one finds that these commonly are narrow compared to the length of the same bed.

Although local variation of flow direction on a given fan may exceed 30 degrees (Bull, 1971, fig. 52), the lack of meandering channels on most fans results in a high consistency of current flow directions. Statistical studies by Howard (1966, p. 152) and Nilsen (1969, p. 50) indicate that a high consistency of flow directions is characteristic of ancient alluvial-fan environments.

Some of the structures examined by Nilsen have up-current dips that indicate a flow direction that is opposite that of the majority of the orientations. Instead of concluding that the up-current results are in error, Nilsen concludes that the up-current-dipping strata resulted from the preservation of antidune bed forms, such as have been described by Harms and Fahnestock (1965).

Stratigraphic Geometry of a Fan

The overall geometry of an alluvial fan reflects the accumulation of vast numbers of beds of differing extent and thickness and the changes in loci of deposition caused by entrenchment and backfilling of the trunk stream channel. The typical situation is shown diagrammatically in Figure 11. The area portrayed is one where recent uplift along a boundary fault has induced rapid accumulation of alluvial-fan deposits adjacent to the mountain front. The surface of the fan is not entrenched and is traversed by a network of braided distributary streams, one of which is associated with the most recent episode of deposition.

Considerable variety exists in the radial and cross-fan stratigraphic relations. Along the radial sections of a fan, individual beds may be traced for long distances, and channel-fill deposits are rare.

In contrast, the cross-fan sections reveal overlapping beds of limited extent that are interrupted by cut-and-fill structures. Because some channels were entrenched only a short distance downslope from the fan apex and others were entrenched into a midfan area, cut-and-fill structures are most common near the fan apex and rare or absent near the toe of the fan.

Three basic shapes of fans can be preserved in the stratigraphic record. Cross sections of these shapes are shown in Figures 12, 13, and 14. The examples shown in Figures 12 and 13 are actual field examples, where the shape of the body of fan deposits is discussed relative to a time line indicated by the top of a lacustrine clay.

Figure 12 shows a wedge of deposits that is thickest near the mountains. In this area uplift

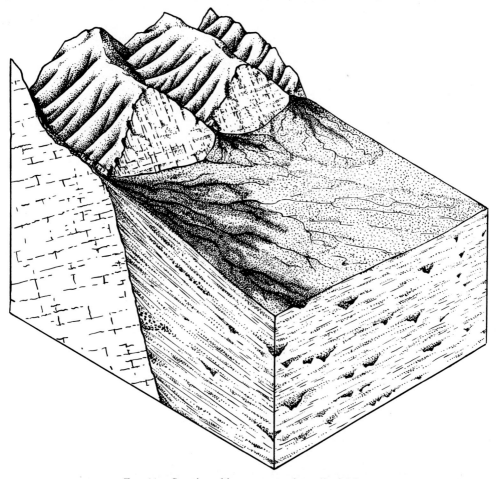

FIG. 11.—Stratigraphic geometry of an alluvial fan.

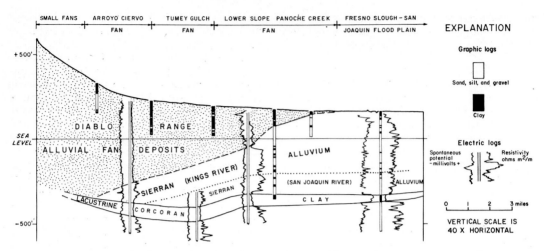

Fig. 12.—Longitudinal section of fan deposits that are thickest adjacent to the mountain front. Modified from Magleby and Klein (1965, Plate 5).

of the mountains occurred mainly before deposition of the Corcoran lake clay, which is only slightly deformed. Uplift of the mountains has increased the sediment yield of the source area and the fans that were only 2½ miles long at the end of the lacustrine period have expanded to 14 miles.

A lense-shaped mass of fan deposits is shown in Figure 13. In this case, the Pleistocene lake clay has been folded along its western extent, as has an undetermined amount of the overlying fan deposits. The largest thicknesses of post-pluvial fan deposition have occurred not adjacent to the mountain front, but in the midfan area. The increase of sediment yield as a result of tectonic uplift has increased the fan areas, thus allowing clayey sands from the sedimentary rocks of the Diablo Range to be deposited on top of the arkosic sands derived from the granitic rocks of the Sierra Nevada.

The relations shown in Figures 12 and 13 reveal the influence of an active orogenic environment—an environment typical of thick accumulations of fan deposits. After tectonic activity

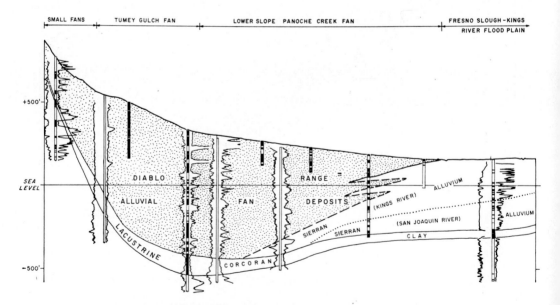

Fig. 13.—Longitudinal section of fan deposits that are lenticular. Modified from Magleby and Klein (1965, Plate 4). See figure 12 for explanation of symbols.

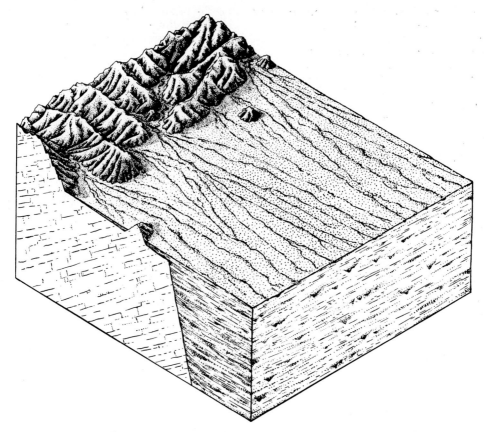

Fig. 14.—Longitudinal section of fan deposits that thicken away from the mountains.

ceases, the streams will continue to downcut within the mountains and eventually will cut below the altitude of the fan apex (Eckis, 1928, p. 237–238). The resulting entrenching of the upslope part of the fan can be regarded as a permanent type of entrenchment that removes the trenched part of the fan as a potential area of deposition. Erosion of the mountain front and the fan deposits above the stream channel continue, and as a result the surface loses much of its aspect of a segment of a cone.

A change from a depositional to an erosional environment may remove tens or even hundreds of feet of surficial fan deposits, and if accompanied by erosional retreat of the mountain front will result in the formation of a pediment upslope from the former mountain front (Figure 14). The stratigraphic relations and depositional characteristics described earlier will still identify the deposits downslope from the former mountain front as being part of a former alluvial fan. However the shape of the remaining deposit will be that of a wedge that thickens away from the former mountain front.

The shape of the overall sequence of deposits that is preserved in the stratigraphic record will in part be a function of the thickness of the deposits that accumulated before erosion of the uppermost deposits occurred, and the thickness of deposits removed by erosion. Part of the body of deposits portrayed in Figure 14 may be the result of redistribution of deposits from the upper to the lower parts of the fan.

Relation of Alluvial-fan Deposits to Adjacent Depositional Environments

Alluvial-fan deposits commonly are in contact with deposits of adjacent alluvial fans or deposits of flood-plain and lacustrine environments.

The deposits of individual alluvial fans interfinger in zones of coalescence. Where core hole and electric-log data are available, the distinctive lithologies between fans derived from differing source-area lithologies are readily discernible. An example is the situation shown in Figure 13, which shows the relations between the "small fans," the Tumey Gulch fan, and the Panoche Creek fan, each of which can be identified by the logs.

Examination of many logs in an area, such as shown in Figures 12 and 13, suggests that the lateral expansion of fans that occurs during the history of accumulation of deposits is minor. Rates of accumulation of deposits on a given fan may increase or decrease with time, but the rates of sediment-yield change in adjacent drainage basins are likely to be similar; hence, little change in the boundaries of adjacent fans occurs with time.

If the downslope edge of the fan is adjacent to a through-flowing stream, the fan deposits are likely to be in contact with floodplain deposits (Figures 12 and 13). If the rate of fan deposition exceeds the rate of floodplain deposition, the fans will expand their area by encroachment over the floodplain depositional area until the rates of deposition in the two areas are approximately equal. In the Western Fresno County area, accelerating rates of sediment yield during the last 600,000 years (the age of the top of the lake clay: Janda, 1965, p. 131) and the tendency toward an equilibrium defined by equal rates of accumulation have resulted in overlap of the fan deposits on the flood-plain deposits.

The tendency toward equal rates of deposition among coalescing alluvial fans and between fans and the playa in a closed basin has been described by Hooke (1968, p. 614–616). Hooke's steady-state model suggests that the areas of accumulation of fan and playa deposits are directly proportional to the volumes of material being supplied to each fan and to the playa, per unit time. Using Hooke's model, the playa-fan contact will be an intertonguing one, as will the contacts between coalescing fans if sediment inputs fluctuate.

In closed basins, pluvial lakes may have formed and inundated parts of the alluvial fans. Lake beds deposited during pluvial intervals form extensive blanket-like deposits that occur as layers in the sequence of fan deposits. Where the areal extent of a lake bed can be defined in the subsurface, the upslope extent of the lake beds will depict not only the shore line, but also the fan contour at that point in the history of accumulation of the fan deposits.

Some workers have described fan deposits that are associated with talus or sedimentary breccias. Blissenbach (1954) notes the interfingering of fan deposits toward the source area with talus deposits, and Drewes (1963, p. 34, 52) describes coarse unbedded sedimentary breccias adjacent to a former mountain front. The lack of bedding, the resistant rock types, and the general lack of fine-grained material noted by these workers suggests that the types of coarse-grained deposits noted by them may be water-laid sieve deposits, such as described by Hooke (1967). Talus can be distinguished in the field from sieve deposits because the particle size increases in the downslope direction in talus, and talus is skewed toward the coarser particle sizes.

Comparison of Alluvial-fan Deposits with Other Coarse-grained Depositional Environments

Some workers regard coarse clastic beds as fan deposits if they appear to be terrestrial deposits. Fans can be readily distinguished from other environments associated with coarse clastics, such as stream-channel gravels, pediment gravels, and marine conglomerates.

Marine conglomerates are extensive in the stratigraphic record, and the transgressive or regressive shorelines associated with their deposition tend to produce sheets of gravel whose interstices are filled with finer-grained material. However, these deposits differ from most fan deposits in the following aspects. (1) The littoral gravels are gray, green, or blue, indicating a reducing environment. (2) The interstices are filled with well-sorted sand instead of a poorly sorted matrix. Gravels filled with sand are found in fan deposits also, but are characteristic of stream-channel deposits and thus generally do not have the lateral extent associated with sheets of littoral gravels. (3) The degree of rounding of both the large and small clasts in marine coarse-grained clastics is much better than found in the alluvial-fan environment.

Most of the differences that can be noted between the marine and alluvial-fan environments also pertain to the littoral facies of lacustrine environments.

Stream-channel and flood-plain deposits, although terrestrial, differ markedly from the deposits of the alluvial-fan environment. Pointbar deposits and channel fills constitute the bulk of the deposits associated with an aggrading river. In contrast alluvial-fan deposits have only a minor amount of stream-channel deposits, the majority of the deposits consisting of lenticular sheets of water-laid sediments and debris flows. Debris flows are not characteristic of most floodplain and stream-channel deposits and floods erode debris-flow deposits deposited within the mountains. The change from an erosional to a depositional environment that occurs at the downslope end of the stream channel terminating on the fan results in conditions that are ideal for the accumulation of hundreds, or thousands, of feet of deposits that consist in part of debris flows.

The fluvial deposits that are found mantling some pediments differ from fan deposits in that they are thin sequences of water-laid sediments that represent a zone of transportation rather than a depositional area. Of course this does not preclude the possibility of the pediment being buried by thick fan deposits (Williams, 1969, p. 191–200). Even some of the temporary accumulations of alluvium on pediments have cross-slope profiles indicative of small alluvial fans (Mabbutt, 1966, fig. 5).

ALLUVIAL FANS IN THE STRATIGRAPHIC RECORD

Thick alluvial fans are orogenic deposits, not only because uplift creates mountainous areas that provide debris and increase stream competence, but also because the loci of deposition on alluvial fans is controlled largely by the rate and magnitude of uplift of the adjacent mountains (Bull, 1964b, 1968). Optimum conditions for accumulation of thick sequences of fan deposits occur where the rate of uplift exceeds the rate of downcutting of the trunk stream channel at the mountain front. The orogenic interpretation of fan deposits applies particularly to the thick sequences of fan deposits found in the stratigraphic record.

Small alluvial fans are found also in areas of nontectonic, base-level change, such as where a river or glacier has eroded its floor at a more rapid rate than have the tributary streams (Suggate, 1963; Carryer, 1966). Changes in the mode and loci of fan deposition ascribed to climatic changes are discussed by Lustig (1965, p. 183–186).

Deposition of thick sequences of fan deposits have occurred during many orogenic periods. Examples are the Late Pre-Cambrian Keweenawan red beds of the Canadian shield, the Devonian fan deposits of Norway that are associated with the post-Caledonian Svalbardian disturbance (Nilsen, 1968a, 1968b, 1969), the Pennsylvanian Fountain Formation of Colorado (Tieje, 1923; Hubert, 1960; Howard, 1966), and the Triassic Newark Group associated with the Appalachian Revolution (Dunbar, 1949, p. 311–316; Krynine, 1950; Reinemund, 1955; and Klein, 1962). It is not the purpose of this article to review the various alluvial-fan sequences that have been described in the stratigraphic record. However, some of the characteristic features of fans can be illustrated by using two suits of fan deposits as examples—the late Pre-Cambrian fans of Scotland and some Pliocene fans of Southern California.

Torridonian Alluvial-fan Deposits of Scotland

Proterozoic fan deposits have been described as overlying Archean gneiss along the northwest coast of Scotland (Maycock, 1962; Williams, 1966, 1969). Maycock (1962, p. 124) describes the Ardheslaig facies as consisting of "thick bedded, very rudely stratified, very coarse, poorly sorted breccia-conglomerate. Interbedded with this material occur varying percentages of thin to thick bedded, medium- to coarse-grained, poorly sorted feldspathic sandstone."

Rare interbedded shale sequences are interpreted as being playa deposits. In Figure 15, playa deposits are preserved as red shales and fine-grained sandstone beds. Overlying the playa deposits are crudely stratified red fanglomerates. Maycock (1962, p. 129) points out that the presumed lack of extensive terrestrial vegetation in Pre-Cambrian times favored the production of debris flows.

Williams (1966, p. 1305) concludes that sheetfloods deposited the tabular conglomerates and that the water-laid sediments consist of (1) braided-stream deposits, and (2) point-bar deposits indicating a channel environment.

By using paleocurrent directions of Maycock and other workers, Williams (1966, fig. 1) concluded that two sets of radiating current directions indicated the existence of two large coalescing Torridonian alluvial fans. Similar paleocurrent studies made by Howard (1966) of the Fountain Formation of Colorado indicated the general positions of fan apexes for a bajada of Pennsylvanian fans.

Sedimentary structures indicative of fluvial deposition have been described in the alluvial fans preserved in several areas. The common sedimentary structures noted in the Precambrian and Paleozoic fan deposits of Norway, Scotland, and Colorado are summarized in Table 2.

TABLE 2—Common sedimentary structures described in alluvial fans in the stratigraphic record

Sedimentary structure	Stream-channel deposit	Sheetflood deposit
Trough cross-stratification	N. S. C.	
Planar cross-stratification	N. S. C.	
Graded bedding	N. S. C.	
Flat-stratification	N.	N. S.
Primary current lineations	N.	N. C.
Flow casts		C.
Ripple-drift bedding		N. S.
Micro-cross lamination		C.

N. Devonian fans of Norway (Nilsen, 1968)
S. Precambrian fans of Scotland (Williams, 1969)
C. Pennsylvanian fans of Colorado (Howard, 1966)

Fig. 15.—Pre-Cambrian Torridonian fanglomerates overlying playa deposits, Scotland. Photo by Ian Maycock, Continental Oil Company.

Ridge Basin Alluvial-fan Deposits, California

Crowell (1954) uses the Ridge Basin area in Southern California to illustrate a closed intermontane basin that was filled with more than 25,000 feet of clastic deposits during Late Miocene and Pliocene time. The deposits of the Ridge Basin Group reveal features that are characteristic of many fan deposits in the stratigraphic record. The deposits exhibit a radial inhomogeneity downslope from the former fan apexes, great thicknesses, and a close association with areas of active tectonic movement.

A diagrammatic section of the Ridge Basin area is shown in Figure 16. The deposits adjacent to the San Gabriel fault zone consist of a poorly bedded cobble and boulder conglomerate with a mudstone matrix that is suggestive of debris-flow deposition. The debris-flow unit grades rapidly away from the fault zone into water-laid conglomerates and sandstone. This is consistent with the laboratory and field studies made by Hooke (1967, p. 452–453) that show that the debris-flow deposition predominates on the upper parts of alluvial fans, and that water-laid deposits predominate on the lower parts of fans because of the ability of water floods to transport debris further downslope.

The conglomerate and sandstone adjacent to the San Gabriel fault scarp has a statigraphic thickness of 27,000 feet (Crowell 1954). Thicknesses of alluvial-fan deposits in the stratigraphic record that exceed 10,000 feet are common. The Newark group in the Appalachians exceeds 20,000 feet, and the Devonian fan deposits in Norway have a maximum thickness of 17,000 feet. Even the remnants of the Pre-Cambrian Torridonian fans in Scotland are 8,000 feet thick.

The great thickness of the Ridge Basin conglomerates, as well as the time span represented by the sequence, provides clear evidence of repeated uplift of the source areas (Crowell, 1954).

The overall sequence of the Ridge Basin deposits from the San Gabriel fault scarp to the center of the depositional basin is from breccia and conglomerate to sandstone to shale, which can be presumed to represent playa deposition in the center of the basin.

SUMMARY OF DIAGNOSTIC CRITERIA OF ALLUVIAL-FAN DEPOSITS

Many alluvial fans have the following diagnostic features that identify them as the products of a distinctive terrestrial depositional environment.

1. Alluvial fans are oxidized deposits that

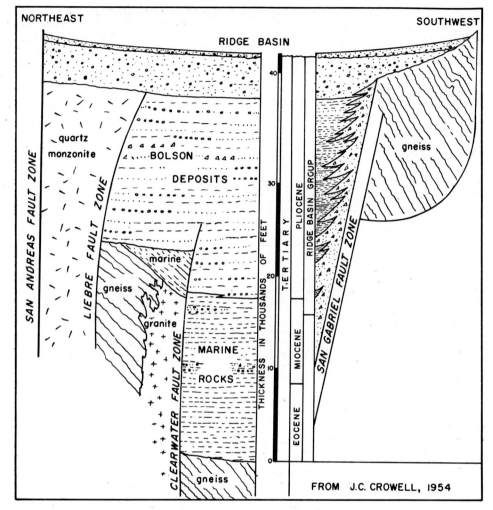

Fig. 16.—Diagrammatic section of the Ridge Basin bolson deposits, California. Modified from Crowell (1954).

rarely contain well-preserved organic material.

2. Alluvial fans commonly consist of thick sequences of water-laid sediments deposited by braided distributary streams and by stream-channel deposition; of mudflow and coarse-grained, debris-flow deposits; or of both water-laid and debris-flow deposits.

3. The bulk of the deposits consist of sheets that have length/width ratios of roughly 5 to 20. Channel-fill deposits comprise a minor proportion of most fans.

4. The proportion of debris-flow deposits decreases downfan from the apex in those fans where both water-laid and debris-flow deposits are present.

5. Particle size decreases downfan from the apex, but the uniformity of particle-size decrease with distance is affected greatly by the amount of temporary channel entrenchment during the history of the fan. Fanhead trenches result in coarse debris being deposited on the middle and lower parts of a fan, instead of being deposited on the upslope part of a fan.

6. Cut-and-fill structures are common near the fan apexes and are rare near the toes of most fans.

7. When compared with other depositional environments, the hydraulics of transport and deposition are greatly different for the individual beds within a sequence of fan deposits. The result is a sequence of beds that vary greatly in particle size, sorting, and thickness.

8. Logarithmic plots of the coarsest 1-percentile particle size (C) and the median particle size (M) of alluvial-fan deposits provide two distinctive patterns. A sinuous pattern representing a tractive-current type of ephemeral-stream deposition is common, and a rectilinear type of pattern that roughly parallels the limit $C = M$ is typical of debris flows. The rectilinear pattern differs from that for turbidity-current deposits in that values of C along the axis of the pattern range from 40 to 80 times M—values that are more than 10 times the values for turbidity-current deposits.
9. Alluvial fans commonly have transgressive or intertonguing relations with the deposits of other depositional environments, such as flood plains or lakes.
10. Depositional structures of fans reflect a radial flow direction from the apex. Individual beds continue for long distances in exposures that are parallel to the radial lines of a fan, but cross-fan exposures reveal overlapping beds of limited extent that are interrupted by cut-and-fill structures. Paleocurrent direction data—such as stream-trough data on horizontal bedding surfaces—produces directions that are suggestive of radial-flow directions on former alluvial-fan surfaces.

REFERENCES

BEATY, C. B., 1963, Origin of alluvial fans, White Mountains, California and Nevada: Ann. Assoc. Amer. Geographers, v. 53, p. 516–535.
———, 1968, Sequential study of desert flooding in the White Mountains of California and Nevada: U.S. Army Natic Laboratories, Earth Science Laboratory, Tech. Rept. 68-31-ES, 96 p.
———, 1970, Age and estimated rate of accumulation of an alluvial fan, White Mountains California, U.S.A.: Amer. Jour. Sci., v. 268, p. 50–77.
BLISSENBACH, ERICH, 1954, Geology of alluvial fans in semiarid regions: Geol. Soc. America Bull., v. 65, p. 175–190.
BLUCK, B. J., 1964, Sedimentation of an alluvial fan in southern Nevada: Jour. Sed. Petrology, v. 34, p. 395–400.
BULL, W. B., 1962a, Relations of alluvial-fan size and slope to drainage-basin size and lithology in western Fresno County, California: U.S. Geol. Survey, Prof. Paper 450-B, p. 51–53.
———, 1962b, Relation of textural (CM) patterns to depositional environment of alluvial-fan deposits: Jour. Sed. Petrology, v. 32, p. 211–216.
———, 1964a, Alluvial fans and near-surface subsidence in western Fresno County, California: U.S. Geol. Survey Prof. Paper 437-A, 70 p.
———, 1964b, Geomorphology of segmented alluvial fans in western Fresno County, California: U.S. Geol. Survey Prof. Paper 352-E, p. 89–129.
———, 1968, Alluvial fans: Jour. Geol. Ed., v. 16, p. 101–106.
———, 1971, Prehistoric near-surface subsidence cracks in western Fresno County, California: U.S. Geol. Survey Prof. Paper 437-C, (in press).
BUWALDA, J. F., 1951, Transportation of coarse material on alluvial fans: Geol. Soc. America Bull., v. 62, p. 1497.
CARRYER, S. J., 1966, A note on the formation of alluvial fans: New Zealand Jour. Geol. and Geophys., v. 9, p. 91–94.
CRAWFORD, A. C., AND THACKWELL, F. E. 1931, Some aspects of the mudflows north of Salt Lake City, Utah: Utah Acad. Sci. Proc., v. 8, p. 97–105.
CROWELL, J. C., 1954, Geology of the Ridge Basin Area: Calif. Div. Mines, Bull. 170, map sheet no. 7.
DENNY, C. S., 1965, Alluvial fans in the Death Valley region, California and Nevada: U.S. Geol. Survey. Prof. Paper, 466, 62 p.
DOEGLAS, D. J., 1962, The structure of sedimentary deposits of braided rivers: Sedimentology, v. 1 p. 167–190.
DREW, FREDERICK, 1873, Alluvial and lacustrine deposits and glacial records of the upper Indus basin: Geol. Soc. London Quart. Jour., v. 29, p. 441–471.
DREWES, HARALD, 1963, Geology of the Funeral Peak Quadrangle, California, on the east flank of Death Valley: U.S. Geological Survey Prof. Paper 413, 78 p.
DUNBAR, C. O., 1949, Historical Geology: New York, John Wiley and Sons; 573 p.
ECKIS, ROLLIN, 1928, Alluvial fans in the Cucamonga district, Southern California: Jour. Geol., v. 36, p. 111–141.
GILE, L. H., PETERSON, F. F. AND GROSSMAN, R. B. 1966, Morphological and genetic sequences of carbonate accumulation in desert soils: Soil Science, v. 101, p. 347–360.
HARMS, J. C. AND FAHNESTOCK, R. K. 1965, Stratification, bed forms, and flow phenomena (with an example from the Rio Grande): In primary sedimentary structures and their hydrodynamic interpretation, Soc. Econ. Paleontologists and Mineralogists Spec. Publ. No. 12, p. 84–115
HAWLEY, J. W., AND WILSON, W. E. 1965, Quaternary geology of the Winnemucca area, Nevada: Nevada Univ. Desert Research Inst. Tech. Rept. No. 5.
HOOKE, R. L. B., 1967, Processes on arid-region alluvial fans: Jour. Geol., vol. 75, p. 438–460.
———, 1968, p. 456, 609–621.
———, 1967, p. 453–456.
———, 1968, Steady-state relationships on arid-region alluvial fans in closed basins: Amer. Jour. Sci., v. 266, p. 609–629.
HOPPE, GUNNAR, AND EKMAN, STIG-RUNE, 1964, A note on the alluvial fans of Ladtjovagge, Swedish Lapland: Geografiska Annaler, v. 46, p. 338–342.

HOWARD, J. D., 1966, Patterns of sediment dispersal in the Fountain Formation of Colorado: Mountain Geologist, v. 3, P. 147–153.

HUBERT, J. F., 1960, Petrology of the Fountain and Lyons Formations, Front Range, Colorado: Colorado School of Mines Quart., v. 55, no. 1, p. 1–242.

JANDA, R. J., 1965, Quaternary alluvium near Friant, California: Internat. Assoc. Quaternary Research, 8th Cong., U.S.A., 1965, Guidebook for field conf. 1, p. 128–133.

KLEIN, G. DeV., 1962, Triassic sedimentation, Maritime Provinces, Canada: Geol. Soc. America Bull., v. 73, p. 1127–1146.

KRYNINE, P. D., 1950, Petrology, stratigraphy, and origin of Triassic sedimentary rocks of Connecticut: Conn. Geol. and Nat. Hist. Survey Bull. 73. 247 p.

LEGGET, R. F., BROWN R. J. E., AND JOHNSTON G. H., 1966, Alluvial-fan formation near Aklavik, Northwest Territories, Canada: Geol. Soc. America Bull., v. 77, p. 15–30.

LUSTIG, L. K., 1965, Clastic sedimentation in Deep Springs Valley, California: U.S. Geol. Survey Prof. Paper 352-F, p. 131–192.

MABBUTT, J. A., 1966, Mantle-controlled planation of pediments: Amer. Jour. Sci., v. 264, p. 78–91.

MAGLEBY, D. C., AND KLEIN, I. E., 1965, Ground-water conditions and potential pumping resources above the Corcoran Clay—an addendum to the ground-water geology and resources definite plan appendix, 1963: U.S. Bur. Reclamation open-file report, 21 Plates.

MAYCOCK, IAN, 1962, The Torridonian Sandstone, Round Loch, Torridon, Wester Ross: Unpub. Ph.D. thesis, Univ. Reading, England, 305 p.

MCKEE, E. D., 1957, Primary structures of some recent sediments: Amer. Assoc. Petroleum Geologists Bull. v. 41, p. 1704–1747.

MEADE, R. H., 1967, Petrology of sediments underlying areas of land subsidence in central California: U.S. Geol. Survey Prof. Paper 497-C, 83 p.

MELTON, M. A., 1965, The geomorphic and paleoclimatic significance of alluvial deposits in southern Arizona: Jour. Geol. v. 73, p. 1–38.

MILLER, R. E., GREEN J. H., AND DAVIS G. H., 1971, Geology of the compacting sediments in the Los Banos-Kettleman City subsidence area, California: U.S. Geol. Survey Prof. Paper 497-E (in press).

MURATA, TEIZO, 1966, A theoretical study of the forms of alluvial fans: Geographical Rept., Tokyo Metropolitan Univ., v. 1, p. 33–43.

NILSEN, T. H., 1968a, Old red sedimentation in the Solund District, western Norway: International Symposium on the Devonian System, Calgary, Canada. Sept. 1967, v. 2, p. 1101–1115.

———, 1968b, The relationship of sedimentation to tectonics in the Solund Devonian district of southwestern Norway. Universitetsforlaget, Olso Norges Geolgiske Undersokelse No. 359, 108 p.

———, 1969, Old Red sedimentation in the Buelandet-Vaerlandet Devonian district, western Norway: Sedimentary Geology, v. 3, p. 35–57.

PASSEGA, RENATO, 1957, Texture as characteristic of clastic deposition: Amer. Assoc. Petroleum Geologists Bull., v. 41, p. 1952–1984.

———, 1960, Sedimentologie et recherche de petrole: Inst. Francais petrole Rev. et Annales combustibles liquides: v. 15, p. 1731–1740.

REINEMUND, J. A., 1955, Geology of the Deep River Coal Field, North Carolina: U. S. Geol. Survey Prof. Paper 246, 159 p.

RUHE, R. V., 1964, Landscape morphology and alluvial deposits in southern New Mexico: Annals. Assoc. Amer. Geographers, v. 54, p. 147–159.

SHARP, R. P., 1942. Mudflow levees: Jour. Geomorphology, v. 5, p. 222–227.

———, 1948, Early Tertiary fanglomerate, Big Horn Mountains, Wyoming: Jour. Geol., v. 56, p. 1–15.

———, AND NOBLES L. H., 1953, Mudflow of 1941 at Wrightwood, Southern California: Geol. Soc. America Bull. v. 64, p. 547–560.

SUGGATE, R. P., 1963, The fan surfaces of the central Canterbury Plain: New Zealand Jour. Geol. and Geophys., v. 6, p. 281–287.

TIEJE, A. J., 1923, The red beds of the Front Range of Colorado: a study in sedimentation: Jour. Geol., v. 31, p. 192–207.

WILLIAMS, G. E., 1966, Paleogeography of the Torridonian Applecross Group: Nature, v. 209, no. 5030, p. 1303–1306.

———, 1969, Characteristics and origin of a Pre-Cambrian pediment: Jour. Geol., v. 77, p. 183–207.

WINDER, C. G., 1965, Alluvial cone construction by alpine mudflow in a humid temperate region: Canadian Jour. Earth Sciences, v. 2, p. 270–277.

PHYSICAL CHARACTERISTICS OF FLUVIAL DEPOSITS

GLENN S. VISHER
University of Tulsa

INTRODUCTION

Fluvial processes and stream deposits have been subjects of study for many centuries, but modern studies started with the fundamental work of Gilbert over fifty years ago. This was the beginning of a period of detailed measurements and collection of data for the purpose of mathematically predicting stream behavior. Field observations were tested and checked by laboratory flume studies to arrive at more precise mathematical descriptions of the variables. This work led to the development of a new discipline called "fluvial hydrology." Centers for the study of hydrology developed in many of the engineering schools across this country and throughout the world. The construction of navigable water-ways, dams, canals, and spillways provided the impetus for research into the physical processes controlling erosion, sediment discharge, deposition, floods, and channel stability.

Parallel to this development and also following the fundamental work of Gilbert, the geomorphologist and sedimentologists studied meandering, sediment transport, texture, and sedimentary structures. The engineering and geological studies were rarely correlated, and little attempt was made to obtain a synthesis of the two approaches. To my knowledge the first attempt was by Sundborg (1956), in which he showed the relation of hydrology to the resulting sedimentary deposits of the Klaralven River. Other workers, including Leopold, Wolman, and Mackin, extensively studied channel geometry, flood plain development, and sediment transport, considerably advancing the understanding of fluvial processes.

Two fluvial hydrologists, Kennedy (1961) and Simons et al. (1961), published, within a few months of each other, similar studies on the relation of fluid flow to movable beds of sand. These studies provided the missing relationship between the parallel lines of study carried on by the fluvial hydrologists and the sedimentologists. In a short time it was apparent to a number of workers that stream hydraulics, bedforms, sedimentary structures and textures were all interrelated. From this it was an easy step to add the information obtained by the geomorphologists to provide a rational model of fluvial sedimentation, and during the period of 1963 to 1965, five workers presented models. Each worker, however, had a slightly different approach, but together they provided sufficient information to construct a coherent model of fluvial sedimentation.

FLUVIAL MODELS
Point Bar Model

The first physical model of fluvial sedimentation was presented by Bernard and Major (1963). They provided extensive field and laboratory work on the sequence of structures and textures within a point bar of the Brazos River (Table 1). In addition they reported on the mechanisms of flood-plain development and the relation of the point bar to the valley fill.

The recognition of a standard sequence of structures and an overall upward decrease in grain size provided the criteria for recognizing fluvial deposits in outcrops and cores. They recognized the relation of the size of the bedforms to the scale of the cross-bedding, the presence of horizontal lamination, and the importance of the basal bedload zone. Little quantitative hydraulic data was provided with the model, but the genetic association of specific bedforms and structures was emphasized.

Flow Regime Model

Harms and Fahnestock (1965) provided additional details on the distribution of specific sedimentary structures within a channel and their relation to the hydraulics of sediment transport and fluid flow (Fig. 1). In a previous paper

TABLE I

Downward Gradation of Grain Size from Fine to Coarse
Four Classes of Sedimentary Structures (Sequence from Top to Bottom)
- Small Ripple Cross-Bedding
- Horizontal Lamination
- Giant Ripple or Medium Scale Cross-Bedding
- Poor Bedding

The Section Is on Offlap Sequence Deposited on the Convex Bank
Thickness of Section Equal to Average Maximum Flood Depth
Valley Fill Sequence May Be 18–20 Times Stream Width

(After H. A. Bernard and C. F. Major, 1963)

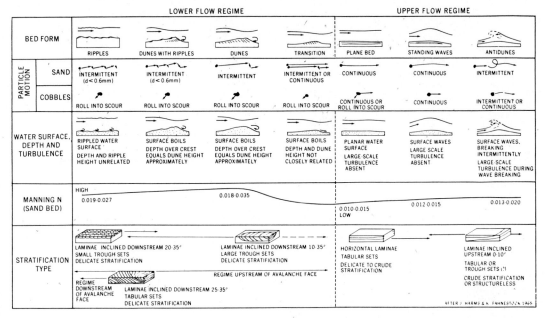

Fig. 1.—Relation of hydraulic parameters to bedforms, sedimentary structures and particle transport. Data based on observation of the Rio Grande.

Harms, MacKenzie, and McCubbin (1963) described sedimentary structures in a point bar of the Red River.

The specific association of bed forms, sedimentary structures, and flow regime had not previously been demonstrated in a natural river. The significance of their work lies in the assignment of sedimentary units and sedimentary structures to flow regime categories, therefore allowing the generalized interpretation of depositional flow environment. For example, horizontal stratification was shown to be a product of plane bed transport produced in the upper flow regime.

Geomorphic Model

The most complete summary of data concerning fluvial deposits was published by Allen (1965). This summary article is still the best available on the morphology of fluvial deposits. He presents seven hypothetical geometric models produced under varying conditions. Examples of four of these are illustrated in Figure 2. These four are thought to be the most important types of fluvial deposits preserved in the ancient record. These models emphasize the possible geometric variability of fluvial sand bodies, and they also offer the possibility of interpreting the origin of specific sands when more information is known about their geometry and internal characteristics. Allen's approach adds the dimension of geomorphology to the interpretation of fluvial deposits.

Paleocurrent Model

One of the most thorough surface and subsurface studies of a group of sandstones was carried out by Potter (1963). Most of the sands he describes are attributed to channel processes. The fluvial origin of these sands was demonstrated by their geometry, sedimentary structures, and most importantly by the presence of unidirectional currents. A summary of the criteria used for interpreting the origin of fluvial sandstones was presented in a later paper (Potter, 1967). The model he uses is taken directly from an outcrop section of a Pennsylvanian sandstone and illustrates those characteristics he believed to be important (Fig. 3). In this model paleocurrent directions, bedding thicknesses, vertical grain-size variations, and sedimentary structures are indicated.

The contribution of this approach is that the paleocurrents, paleoslopes, and sedimentary structures are related to the geometry of the sand body. The resulting sedimentary model and environmental interpretations are related to the sedimentary basin and provide additional information to interpret the genesis of a particular sand.

Vertical Profile Model

The model developed illustrates the relation of structures and textures and how they may be distributed throughout a valley fill (Fig. 4). Of special importance is that similar vertical sections occur throughout the valley, and they are

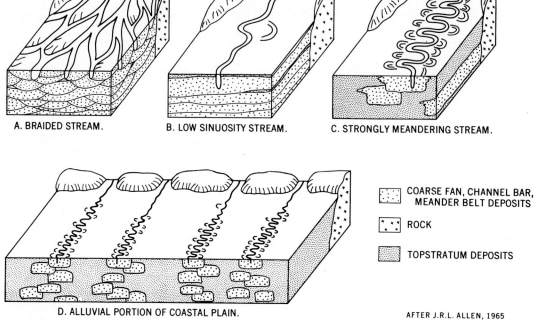

Fig. 2.—Conceptual models of fluvial channel geometry.

formed by the same processes. For a sedimentary model to be of value, readily observable characteristics must be associated in a way which permits only one interpretation. A study made of possible modes of deposition and direct comparisons to other vertical profiles formed by differing processes indicated that the fluvial model is unique, and no other sedimentary process produces the same pattern of structures and textures (Visher, 1963; 1965a).

Other aspects of the fluvial model were later developed (Visher, 1965b). The approach was primarily to define in more detail the possible bedding, sedimentary structure, and textural variations that were present within single vertical sections from the same channel complex (Fig. 5). Also, more detail on the textures of fluvial sands and how they related to the sedimentary structures was added. From these studies it was possible to show the vertical and lateral variability that could be developed within a channel complex.

PHYSICAL ASPECTS OF FLUVIAL DEPOSITS

The depositional models that have been presented are based upon observable physical aspect of the sediments. The generalized model has been accepted by most sedimentologists, but the details of the physical characteristics and their interpretation are still areas of active research.

Much work is now in progress on the origin and significance of textural variations, mineralogy of the fresh-water sediments, the origin of specific sedimentary structures, and the importance of morphology and surface textures of detrital sands. These are the areas where advancement in the interpretation of fluvial deposits is possible.

Sedimentary Structures

The sequence of specific sedimentary structures in fluvial deposits is still the primary basis for recognizing deposits formed by unidirectional confined flow. The four most common types are illustrated in Figure 6. All of these structures may be formed in other depositional environments, but the only place where the specific combination of structures and their characteristic vertical distribution is developed is within channels.

Controversy, however, still exists concerning the distribution of festoon or trough cross-bedding. Recent work by McKee (1966) has shown that these forms are rare in dunes. Also there is no evidence that they can develop without the presence of confined unidirectional flow. Consequently, they would commonly be developed only in tidal, distributary, and fluvial channels. If this generalization is correct, a powerful indicator of the depositional process is available.

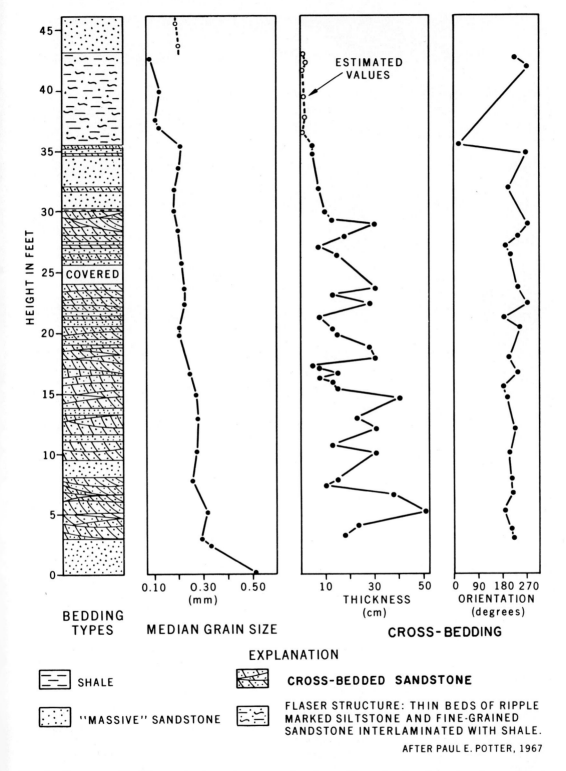

FIG. 3.—Summary of bedding, grain size and transport data from a Pennsylvanian channel sandstone, Illinois basin. Data suggests the composite nature of thick channel sequences.

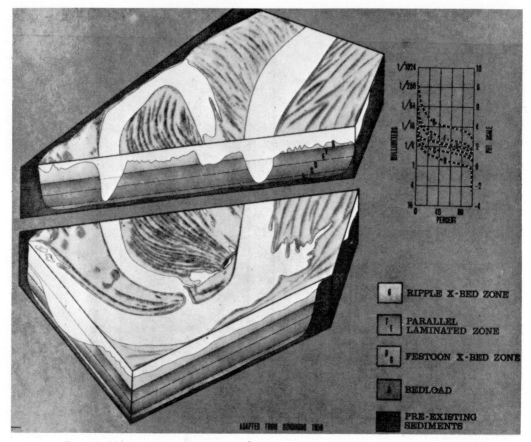

FIG. 4.—Characteristics of textures and structures from the Klaralven River alluvial channel. Data provides the basis of the fluvial model.

Other types of structures are more ambiguous. Current and wave ripples are developed both in marine and fluvial environments. Planar, tangential, and tabular cross-bedding has been identified in all sedimentary deposits and therefore does not appear to be diagnostic. Horizontal or current lamination appears to be uniquely associated with the upper flow regime, but not restricted to confined unidirectional flow. Beaches, shallow marine bars, flood plains, and areas of tidal surge all may show this sedimentary feature.

The vertical arrangements of the sedimentary structures within the fluvial channel appear to be primarily depth controlled, as illustrated by Figure 5 (Visher, 1965b). The amplitude of the structures decreases upward with large-scale sand wave or trough cross-bedding near the bottom, overlain by small-scale festoon cross-bedding, followed by either current lamination or ripple cross-bedding (Allen, 1967). Each type may be repeated between scour surfaces, but the amplitude in each successive depositional unit is slightly less.

Fabric and Texture

Some new studies are available concerning the fabric and texture of fluvial sands. For many years attempts have been made to find definitive environmental characteristics based on particular aspects of the detrital fraction. Many textural studies were conducted in an attempt to show particular grain size distributions or statistical measures. New work in this area suggests that this may indeed be possible. Other studies on surface texture have proven successful in the identification of a number of environments. Also grain shapes, especially of the coarse fraction, provide insight into the dynamics of transport. A number of studies have been made of grain orientation, and a strong correlation has been demonstrated between grain orientation, the sedimentary structures, and transport directions. Each of these four types of measurements provide independent lines of evidence for determining depositional processes and in turn depositional environments.

Surface Textures

Research by Krinsley et al. (1962; 1968) has shown that processes related to beach, glacial, and aeolian sands have characteristic surface textures. Preliminary data showed also that grain surfaces of fluvial sands tend to be smooth, with no characteristic fracture or etch patterns. However, the absence of the other diagnostic characteristics may be an important supporting line of evidence to interpret fluvial processes. Additional work in this area may, however, provide more direct evidence of sedimentary dynamics.

Grain Morphology

Many studies have been made concerning roundness and sphericity of clastic particles. Present evidence is that sphericity is primarily an inherited trait and is not an environmental indicator (Krumbein, 1941). Roundness, however, does indicate transportational history. Little quantitative data is available directly comparing the roundness of detrital grains from differing environments. This may well be a fruitful area of research.

The coarse fraction may be more diagnostic of transportational processes. Dobkins and Folk (1968) have shown that pebbles of the same size tend to be equant and possibly rod-shaped in fluvial channels, and discoidal on beaches. They attribute this to differences in hydraulic behavior and transport dynamics. Equant particles would be preferentially removed during fluvial transport because of their high relative settling velocities, and the discoidal pebbles would be transported in greater abundance to the deltas, beaches, and marine environments.

VERTICAL SIZE CHANGES

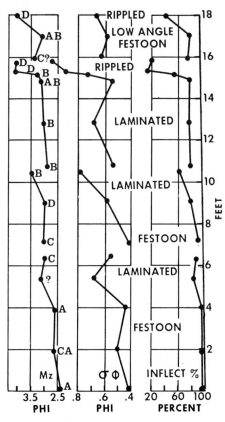

FIG. 5.—Textures, bedding and sedimentary stuctures from an ancient channel sandstone. This ancient channel unit is comparable to modern fluvial sands.

RIPPLE X-BEDDING & LINEATION

CURRENT RIPPLES

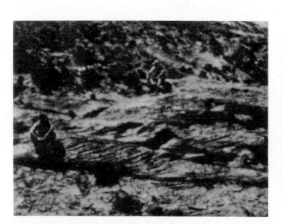
TROUGH X-BEDDING

CURRENT LINEATION & RIPPLE X-BEDDING

Fig. 6.—Typical sedimentary structures developed by fluvial processes.

Grain Orientation

The only studies available on this aspect show that the long axes of quartz grains tend to be parallel to the direction of transport (Potter and Mast, 1963). Grains in current ripples, festoon or trough cross-beds, and especially in current laminated units all show a strong long axis orientation parallel to the current. The graining or parting lineation on the surface of current laminated units reflects this preferred orientation and may indicate deposition by a strong unidirectional current (Allen, 1964).

Size Distributions

Size distributions of a number of environments have been reported by Friedman (1967) and Folk (1966). Criteria have been established to distinguish environments on the basis of statistical measures of the size distribution. These criteria have been moderately successful on modern deposits, but consistent results have not been obtained on ancient sands. Many other workers have studied textures of sands, usually obtaining similar conclusions. From the data published it is apparent that statistical measures of fluvial size distributions do not provide a diagnostic criterion, and other aspects of the grain-size distribution must be used if an environmental indicator is to be developed.

Transportation of detritus by running water has been an area of study for many years, and from the basic data provided by this research a new approach to interpreting sedimentary processes from textural data has been suggested (Moss, 1962; 1963). The suggestion is that sediment is transported by surface creep, saltation, and suspension, and that each of these transport mechanisms is reflected as separate modes in a single grain-size distribution. Also it is suggested that the number, amount, size-range, mixing, and sorting of these populations varies systematically in relation to provenance, sedimentary process, and sedimentation dynamics (Visher, 1969). Data from the Mississippi River (U.S. Waterways Experiment Station, 1939) and from Jopling (1966) illustrate the relation between sediment transport and the deposited sediment laminae (Fig. 7). The presence of three populations (Pop. A, B, and C) is indicated by the log probability plot of the Mississippi River data. Also it shows that the suspension load (Pop. C) increases rapidly above the depositional interface. Note also that the data from Jopling show that the grain-size distribution calculated by Enstein's bedload formula, and a sample from a delta foreset, is directly comparable to the Mississippi River bedload data. The three curves represent approximately the same transport velocities, and grain size was similar, thus allowing a direct comparison to be made. Of particular importance is that the deposited laminae and the sediment in transport at the depositional interface are strikingly similar.

Comparisons between grain-size distributions from differing environments have been reported (Visher, 1967; 1969), and they show that systematic differences are present in the log probability plots. Differences appear to be related to the depositional processes, thereby providing a basis for determining the origin of individual sands. Data collected suggests that most sands deposited by strong unidirectional currents contain two fundamental populations (Fig. 8 and 9). The saltation and suspension populations are predominant unless the sand sample is particularly coarse. The size interval for the saltation populations is from 1.75 to 3.0 phi, and there is a break between the saltation population and the suspension population between 2.75 and 3.5 phi. The only material that would be transported by surface creep and not be included in these two populations would usually be coarser than .5 to 0.0 phi. These are characteris-

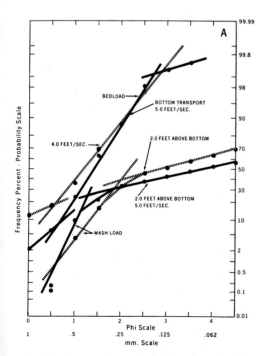 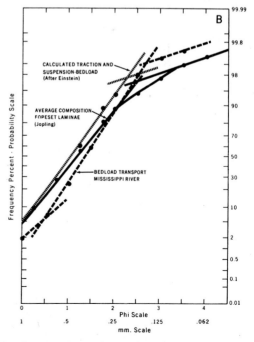

FIG. 7.—Comparison of bedload and washload grain-size distributions (Figure 7-A). The similarity of modern theoretical and ancient grain-size distributions produced by a current depositional process is illustrated (Figure 7-B). Data supports a process interpretation of log-probability plots of grain-size distributions.

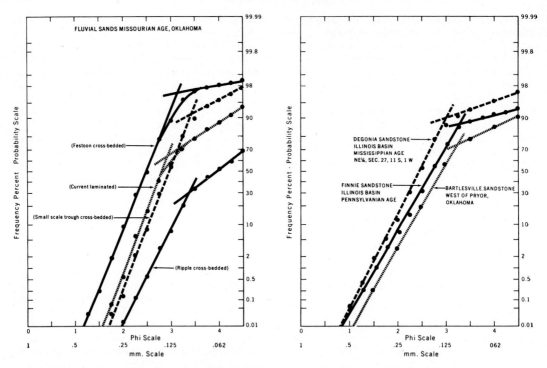

FIG. 8.—Examples of log-probability plots of textural data from ancient fluvial sandstones. Similarity of the plots suggests a similar depositional process.

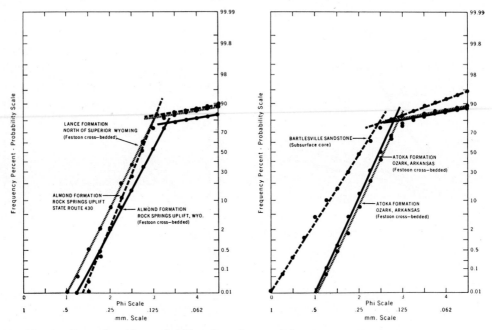

FIG. 9.—Examples of log-probability plots of textural data from ancient fluvial sandstones.

tic aspects of fluvial log probability plots, and when developed appear to be diagnostic. Other types of log probability plots are developed in fluvial regimes under special conditions, but these can easily be recognized if a number of samples are compared. From the data currently available no examples have been obtained from other environments that can be confused with the fluvial type (Visher, 1969).

Mineralogy

The literature on fluvial deposits does not provide reference to any diagnostic minerals or mineral assemblages. The only information that has been reported indicates an absence of certain minerals in fluvial environments. The most widely used mineral is glauconite. It has been found not to persist in a fresh-water, oxidizing environment. Other detrital ferrous iron compounds—for example, chlorite, siderite, and magnetite—are not common in fluvial deposits. Most stream waters are slightly oxidizing with nearly neutral to slightly acidic (Garrels and Christ, 1965, p. 381), which would account for these mineral deficiencies. The lack of a mineral is not an environmental criterion, but it does provide some supporting evidence when correlated with other physical criteria.

Clay minerals may provide a more direct indication of the depositional environment. A systematic x-ray diffraction study of more than 70 core samples from the Bluejacket-Bartlesville Sandstone from various environmental units indicates that there is an inverse relation between kaolinite and a probable diagenetic ferrous chlorite. This relation also is environmentally sensitive, with kaolinite more abundant in fluvial sands and chlorite more abundant in marine sands. The amount of illite in these sands also shows the same relation, with abundant illite in marine sands and very minor amounts in fluvial sands. Figure 10 illustrates these mineralogical relationships. The disappearance of kaolinite and the development of illite and chlorite in a single sand unit is possibly a significant environmental indicator. The origin of the illite and chlorite is a particularly interesting problem, and the petrographic evidence suggests that both are diagenetic. If a diagenetic origin can be confirmed, these minerals may be directly related to the changes in water chemistry and possibly provide a salinity indicator.

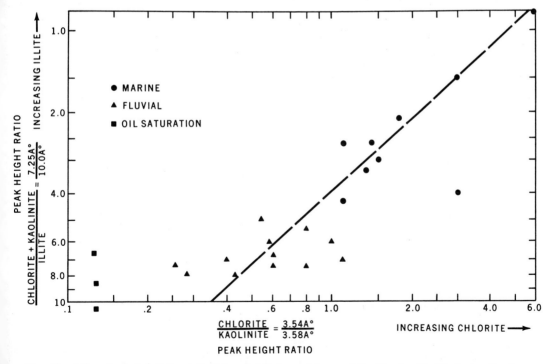

Fig. 10.—Mineralogical data from fluvial and deltaic sandstones of the Bartlesville-Bluejacket Sandstone. Data suggests that illite and chlorite may be associated with marine sandstones, and kaolinite with fresh-water sandstones.

In one example, the Bluejacket-Bartlesville Sandstone, a direct measure of the salinity of the water may be indicated.

Summary of Physical Aspects

As indicated under each topic, these are the areas where research appears to be needed. Additional criteria to be used in identifying depositional environments are often required in individual cases, and the more independent criteria available the more certainty can be placed in the final interpretation. Even more important, however, is the determination of the details of a channel system. The volume of sediment in transport, characteristics of the river, the valley, seasonal discharge, salinity, and channel patterns are all questions that may be answerable if more precise understanding can be obtained of the physical characteristics of a fluvial system.

GENETIC IMPLICATIONS OF FLUVIAL PROCESSES

There are two ways of examining the fluvial model: (1) from a geomorphological approach; and (2) from a process-response approach. Each of these provides new insights into the various aspects of fluvial sedimentation and the physical characteristics of the resulting deposits. Geomorphology deals with tectonics, channel slope, sediment and water discharge, and climatic cycles. These provide the framework of the fluvial system. Only some of these parameters have been quantified and their relationships between the variables examined. The other aspect would be the detailed process-response relationships characteristic of a single reach, or vertical profile. These are more closely related to the physics of open channel flow and can be more easily quantified and interpreted.

Geomorphology of the Fluvial Model

Several of the fluvial models presented in the first part of this paper are fundamentally geomorphic in character. Little new information is available summarizing all of the possible variations that can be exhibited in fluvial channels and valley fill deposits. As a way of summarizing some of the characteristics of meandering streams with seasonal floods, Figure 11 is presented. This illustrates the relation of sheet sands, channel fill, point bars, and crevasse deposits. It also shows some of the internal geometry of the valley fill sequence and the dominant textural and sedimentary structure patterns. Bedforms, flow regimes, and detailed vertical profiles cannot be included in this type of model. The reasons for and significance of local variations have not been ascertained, and for the purpose of this model they must be omitted. As more information is obtained on various types of fluvial deposits, and as details are gathered on the areal variations within one channel, possibly more insight can be achieved and more sophisticated models developed.

Process-Response Model

There are certain processes that are universal to fluvial sedimentation and rarely associated with other depositional environments. These processes and the physical responses that are tied to them should be the bases for interpreting the genesis of a sand body. Table II is presented to summarize some of these characteristic fluvial processes and the physical aspects that are necessarily produced.

For a system to be fluvial it must show unidirectional flow; typically channeling is developed, and most streams show meandering in their lower reaches. Other aspects, such as seasonal discharge, velocity pulsations, and a specific

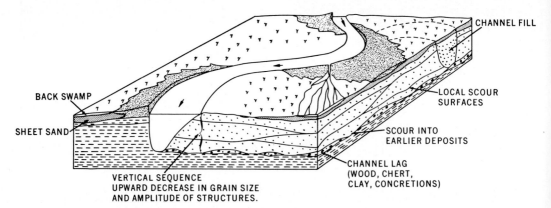

FIG. 11.—Geomorphology of the fluvial depositional model. It is suggested that each depositional unit may be related to specific depositional processes and positions within a fluvial channel complex.

water chemistry, are also very much a part of most rivers. These six processes produce the characteristics that we associate with fluvial deposits.

Channeling

Channels are a product of scour, and the repeated floods within a stream valley produce repeated scour surfaces. Associated with these surfaces are lag deposits concentrated by the removal of portions of the preexisting sediments. One of the first observations to be made in studying an outcrop is to determine the location and distribution of scour surfaces. These surfaces form the base of individual sedimentation units and are the logical starting points for reconstructing the history of sedimentation.

Meandering

Meandering produces many of the aspects of a fluvial deposit. Meander bends provide a sloping surface on which sedimentation must take place. This produces a depth variation and consequently a vertical change in velocity, grain size, and amplitude of the sedimentary structures. The basic fluvial model recognized a number of years ago was based on this process. The lateral extension and down-valley progress of the meander loops produce a valley fill complex of point bar deposits.

Depth, velocity, flow regime, and sediment concentration are all related to stream power and shear stress (Simons et al., 1965, p. 47). The geometry of channel cross-section of a meandering stream requires that specific sequences of sedimentary structures and textures be developed. These relationships are the primary control for the fluvial model that has been developed. This particular model may not be applied to channels or fluvial regimes that do not meander. Other criteria must be used to determine the significant criteria for their interpretation.

Unidirectional Flow

Of all processes related to fluvial sedimentation, the single unvarying aspect is unidirectional currents. The only environmental realm that has this as the predominate sedimentation process is fluvial, but other sedimentary deposits—for example, fluxoturbidites, subaqueous fans, and possibly some distributary channels—may also be formed by unidirectional currents. However, characteristics produced by this process must be present before a fluvial interpretation can be considered.

Three types of bedforms have been specifically associated with unidirectional currents: (1) dunes, (2) lingoid ripples, and (3) climbing

TABLE II

Processes	Responses
Channeling	Scour Surfaces Lag Deposits—Clay Chips, Wood, Coarse Detritus Concretions, Reworked Fauna
Meandering	Point Bars Depth Variation—Depositional Surface Vertical Changes in Grain Size Vertical Changes in Amplitude of Structures Valley Fill

Unidirectional Flow	Bedforms
	  Sand Waves and Bars—Planar Cross-beds
	  Dunes—Trough Cross-beds
	  Ripples—Lingoid and Climbing
	Entrainment of Detritus
	  "Saltation" 100 to 750μ
	  Intense Turbulence—Suspension
	  Positive Skewness—Fair Sorting
	Fabric and Surface Textures
	  Alignment of Particles
	  Few Impact Marks
	  Equant to Rod Shaped Particles

Seasonal Discharge Peaks	Sand Unit as Thick as Channel Depth Rapid Deposition after Peak Flow Separation of Channel and Floodplain Deposits
Velocity Pulsations	Lamination Sediment Accretion
Chemical—Water pH $\sim$ 6.8 Eh $\sim$ +0.4 T.S. <5000 ppm	Dispersion of Clays Elimination of Detrital Ferrous Iron Compounds Hydrolysis of Potassium and Magnesium Elimination of Illite and Chlorite Colloids of Silica, Alumina, Iron and Organic Organisms Restricted

ripples. Other types of features including sand waves, bars, and plane bed are typically developed by unidirectional currents but have also been observed in other environmental realms. The structures associated with these bedforms have been identified in deposits of fluvial origin.

The mechanisms of transport of clastic detritus also control the textural characteristics of the sediment. The current produces a moving grain layer in response to the bed shear, and this layer represents a single population of the size distribution. The sorting and size range of the moving grain layer or saltation population is variable but usually occurs in the size interval from 100 to 750 microns, and typically represents from 70 to 98 percent of the total distribution. The range of grain size in transport in this layer governs to a large measure the sorting of the deposited sand. In addition, all currents develop turbulence, and some portion of the sediment load is carried in suspension. Most fluvial sediments contain a separate fine population related to the suspension mode of particle transport. This population is usually intermixed in the interstices of the coarser saltation population. The amount of this population is reflected by the degree of positive skewness of the distribution.

Nonreversing currents and the linear relation of particle transport to river length combine to produce characteristic fabrics, grain shapes, and surface textures. The most significant aspect is the high degree of preferred orientation of sand-size particles. This produces a directional permeability, which often results in directional cementation patterns. Equant and rod-shaped grains are preferentially deposited in fluvial channels, due to their higher fall velocity. Finally, the smooth surface texture reflects the lower incidence of particle collision due to the distance of particle transport and the lower impact energy developed in the moving grain layer.

Seasonal Discharge Peaks

Periodic floods provide for the formation of successive depositional units within the channel. The thickness of a fluvial sand will be equal to the depth of the average flood channel. Major floods will produce overbank deposits and will form a flat flood-plain surface of wide areal extent. The separation of deposits into channel and flood plain is a typical aspect of the fluvial regime. Finally, the rate of deposition of individual sand units within the channel is rapid. After peaks in discharge, thick sedimentary sections are accumulated in lenses of local areal distribution. The lensing of depositional units within the fluvial regime is one of the most characteristic aspects of these deposits. Velocity pulsations are the rule in all current regimes.

Velocity Pulsations

Velocity pulsations produce changes in the grain-size distribution of the moving grain layer, and particles not in equilibrium will either be eroded or deposited. This produces interlamination of coarse and fine sediment populations characteristically developed within fluvial sediments. During periods of declining flow velocity, grains of a particular size will be deposited until either that size is exhausted or the velocity changes. An abrupt switch to another grain size will then occur and a sediment laminae will develop. Once a laminae is developed a major velocity change is necessary before it is eroded; consequently sediment accretion will take place during stable or decreasing velocity conditions.

Water Chemistry

The water chemistry relates to mineral stability, biologic activity, and to physical aspects of fine-particle transport. Slightly acid, oxidizing, low-salinity water is characteristic of fluvial conditions. These characteristics cause the oxidation of ferrous iron compounds; the hydrolysis of magnesium and potassium within chlorite and illite, resulting in their elimination; dispersion of clay minerals; and, the solution of fine-grained carbonates. Other aspects include the absence of organisms requiring sea water to support life, and because of the stability of silicon, alumina, iron, and organic colloids in fresh water they would not be deposited with the fluvial sediments. These colloids would be removed in waters of higher salinity and may result in contrasting diagenetic mineral suites in marine and fluvial sediments.

SUMMARY

An attempt has been made to relate the characteristics of fluvial deposits to the geomorphic and hydraulic processes that are responsible for their formation. Without a physical rationale for specific characteristics of fluvial deposits, their recognition must be subjective, and observers with the most experience would be most likely to make proper interpretations. For unambiguous interpretations something more than empirical data is required. and this means we need to develop physical models and relate them to the dynamics of sedimentation. The progress that has been made in identifying fluvial deposits must be attributed directly to the insights of the fluvial hydrologists and the quantitative geomorphologists. The outline of the process-response characteristics presented here may provide the framework for the interpretation of fluvial deposits.

REFERENCES

ALLEN, J. R. L., 1964, Primary current lineation in the Lower Old Red Sandstone (Devonian), Anglo-Welsh basin. Sedimentology, v. 3, no. 2, p. 89–108.

———, 1965, A review of the origin and characteristics of recent alluvial sediments. Sedimentology, v. 5, no. 2, p. 89–191.

———, 1967, Depth indicators of clastic sequences: Marine Geology, v. 5, nos. 5–6. p. 429–446.
BERNARD, H. A., AND MAJOR, C. F., Jr., 1963, Recent meander belt deposits of the Brazos River: An alluvial "sand" model (Abst.): Am. Assoc. Petroleum Geol. Bull., v. 47, no. 2. p, 350–1.
DOBKINS, J. E., JR., AND FOLK, R. L. 1968, Pebble shape development on Tahiti—Nui (Abst.). Am. Assoc. Petroleum Geol. Bull., v. 52, no. 3, p. 525.
FOLK, R. L., 1966, A review of grain-size parameters: Sedimentology, v. 6, no. 2, p. 73–93.
FRIEDMAN, G. M., 1967, Dynamic processes and statistical parameters compared for size frequency distributions of beach and river sands: Jour. Sedimentary Petrology, v. 37, no. 2, p. 327–54.
GARRELS, R. M., AND CHRIST, C. L., 1965, Solutions, Minerals, and Equilibria: New York: Harper and Row, 450 p.
HARMS, J. C., MACKENZIE, D. B., AND MCCUBBIN, D. G., 1963, Stratification in modern sands of the Red River, Louisiana: Jour. Geol., v. 71, no 5, p. 566–80.
———, AND FAHNESTOCK, R. K., 1965, Stratification, bed forms, and flow phenomena (with an example for the Rio Grande): in Primary Sedimentary Structures and their Hydrodynamic Interpretation, ed. G. V. Middleton, Soc. Econ. Paleontologists and Mineralogists, Spec. Pub. No. 12, p. 84–115.
JOPLING, A. V., 1966, Some principles and techniques used in reconstructing the hydraulic parameters of a paleo-flow regime: Jour. Sedimentary Petrology, v. 36, no. 1, p. 5–49.
KENNEDY, J. F., 1961, Stationary waves and antidunes in alluvial channels: California Inst. of Tech., W. M. Keck Lab. of Hydraulics & Water Resources, Rept. KH-R-2; 146 p.
KRINSLEY, D. H., AND TAKAHASHI, T., 1962, Applications of electron microscopy to geology: New York Acad. of Science Trans., ser. 2, v. 25, no. 1, p. 3–22.
———, AND DONAHUE, J., 1968, Environmental interpretation of sand grain surface textures by electron microscopy: Geol. Soc. America Bull., v. 79, no. 6, 743–8.
KRUMBEIN, W. C., 1941, Measurement and geologic significance of shape and roundness of sedimentary particles: Jour. Sedimentary Petrology, v. 11, p. 64–72.
MCKEE, E. D., 1966, Structures of dunes at White Sands National Monument, New Mexico (and a comparison with structures of dunes from other selected areas): Sedimentology, v. 7, no. 1, p. 1–69.
MOSS, A. J., 1962, The physical nature of common sandy and pebbly deposits. Part I: Am. Jour. Science, v. 260, no. 5, p. 337–73.
———, 1963, The physical nature of common sandy and pebbly deposits. Part II: Am. Jour. Science, v. 261, no. 4, p. 297–343.
POTTER, P. E., 1963, Late Paleozoic sandstones of the Illinois basin: Ill. Geol. Survey, Rept. Invest., no. 217, 92 p.
———, 1967, Sand bodies and sedimentary environments: a review: Am. Assoc. Petroleum Geol. Bull., v. 51, no. 3, p. 337–65.
———, AND R. F. MAST, 1963, Sedimentary structures, sand shape fabrics, and permeability, Part I: Jour. Geol., v. 71, no. 4, p. 441–71.
SIMONS, D. B., AND RICHARDSON, E. V. 1961, Forms of bed roughness in alluvial channels: Am. Soc. Civil Engineers, Proc., v. 88, no. HY 3, p. 87–105.
———, RICHARDSON, E. V. AND NORDIN, C. F. Jr., 1965, Sedimentary structures generated by flow in alluvial channels: in Primary Sedimentary Structures and their Hydrodynamic Interpretation, ed. G. V. Middleton, Soc. Econ. Paleontologists and Mineralogists, Spec. Pub. no. 12, p. 34–52.
SUNDBORG, A., 1956, The River Klaralven, a study of fluvial processes: Geografiska Annaler, v. 38, p. 127–316.
U.S. Waterways Experiment Station, 1939, Study of materials in suspension, Mississippi River: Tech. Memo. 122-1, Vicksburg, Miss., 27 p.
VISHER, G. S., 1963, Use of the vertical profile in environmental reconstruction (Abst.): Am. Assoc. Petroleum Geol. Bull., v. 47, no. 2, p. 374.
———, 1965a, Use of the vertical profile in environmental reconstruction: Am. Assoc. Petroleum Geol. Bull., v. 49, no. 1, p. 41–61.
———, 1965b, Fluvial processes as interpreted from ancient and recent fluvial deposits: in Primary Sedimentary Structures and their Hydrodynamic Interpretation, ed. G. V. Middleton, Soc. Econ. Paleontologists and Mineralogists, Spec. Pub. No. 12, p. 116–132.
———, 1967, Grain size distributions and depositional processes: Pre-print VII International Sedimentologic Congress, Reading and Edinburg, England, 4 p.
———, 1969, Grain size distributions and depositional processes: Jour. Sedimentary Petrology, v. 39, p. 1074–1106.

FLUVIAL PALEOCHANNELS

S. A. SCHUMM
Dept. of Geology, Colorado State University

ABSTRACT

The width (w), depth (d), meander wavelength (l), gradient (s), shape (w/d), and sinuosity (P) of stable alluvial river channels are dependent on the volume of water moving through the channel (Q_w) and the type of sediment load conveyed through the channel (Q_s).

$$Q_w \propto \frac{w, d, l}{s}$$

and

$$Q_s \propto \frac{w, l, s}{dP}$$

Empirical equations developed from data collected along modern alluvial rivers permit calculation of the effects of changes of hydrologic regimen (Q_w, Q_s) on channel morphology. Conversely, these relations permit estimation of paleochannel gradient, meander wavelength, sinuosity, and discharge from the dimensions of the paleochannel as exposed in cross section.

The recognition of paleochannels within valley-fill or other complex fluvial deposits is a major problem, but criteria for the delineation of paleochannel cross-sectional shape and dimensions have been developed from studies of shapes and sediment characteristics of Australian paleochannels.

INTRODUCTION

Numerous detailed studies of modern river systems have been made by geomorphologists, civil engineers, and sedimentologists, and one would assume that a means of predicting both the third-dimensional character as well as the hydrology of a paleochannel from information obtained at an exposure of the channel cross section in outcrop is possible. This sanguine expectation, however, has not been realized due primarily to two factors. First, the statistically significant relations between channel morphology and water discharge are not sufficiently accurate for predictive purposes. As an example, a range of 50 to 100,000 cubic feet per second was obtained when Moody-Stuart (1966, p. 1113) used channel depth to estimate discharge in paleochannels of Devonian age in Spitsbergen. Secondly, the student of modern river systems frequently finds that the paleochannels of interest to his geological colleagues are an order of magnitude larger than anything that is familiar to him. Most modern alluvial rivers flow in wide valleys and on valley-fill deposits that may be hundreds of feet thick; therefore, an explanation would seem to be that in one case we deal with a stream channel and in the other with a valley-fill deposit. For example, a comparison of river and valley meanders reveals that valley meander wavelength is about ten times that of the river which occupies the valley (Dury, 1964).

The research of the fluvial geomorphologist is aided by the complete exposure of the object of his study. He can obtain complete data on channel morphology, but it is a rare occasion when he has the opportunity to observe the sedimentary deposits that comprise a flood plain and valley-fill. Conversely, paleochannels are commonly observed only in cross section or they can be studied in even less detail from a series of well logs. Hence, a stream channel may be difficult to distinguish within a larger valley-fill deposit, and therefore, quantitative geomorphic relations may be of limited use.

In spite of the pessimistic statements made above, procedures for the estimation of the third dimensional character of paleochannels (gradient, meander wavelength, sinuosity) and the hydrology of the system (mean annual discharge, mean annual flood) will be developed and evaluated.

ACKNOWLEDGMENTS

The multiple regression analyses were performed by Ronald Corbett. Data on dimensions of the Finke River were obtained in the field with J. A. Mabbutt of the University of New South Wales. Figure 6 is based on data provided by Mark Stannard of the New South Wales Water Conservation and Irrigation Commission. Dr. John A. Campbell of Colorado State University offered suggestions leading to revision of the manuscript.

FLUVIAL MORPHOLOGY

The morphology of river channels and their adjustment to changing hydrologic and hydraulic conditions have long been of interest to man. Most of the relevant literature on this subject is reviewed in Leopold, Wolman and Miller (1959) and Morisawa (1968). All studies confirm the fact that the greater the quantity of water moving through a river channel the larger will be the size of that channel. In general, as water discharge (Q_w)[1] increases, width, depth, and meander wavelength increase and gradient decreases. Usually with increasing discharge downstream channel width and depth increase (Leopold and Maddock, 1953). However, if discharge did not increase in a downstream direction then neither would channel width and depth. The Finke River of Central Australia provides an example of this reversal of downstream trends. Near its headwaters west of Alice Springs, the river is 380 feet wide and 6 feet deep, but about 250 miles downstream at Finke it is 250 feet wide and 4 feet deep. Further downstream a progressive decrease in size occurs until the channel eventually disappears in the sands of the Simpson Desert. Thus, size changes in a downstream direction are, indeed, dependent on discharge. If discharge decreases, when a river flows into a progressively more arid region, the dimensions of the channel will decrease.

The nature of the sediment load moved through the channel also has a significant effect on channel morphology, and an attempt has been made to classify alluvial channels on the basis of the type of sediment load that is moved through the channel (Schumm, 1968). On Table 1 three types of stream channels are distinguished as follows: relatively narrow, deep and sinuous suspended-load channels that on the average transport less than three percent of sand size or larger sediment; relatively wide, shallow, and straight bedload channels that on the average transport greater than eleven percent of sand size or larger sediment; and mixed-load channels which have morphologic and sediment characteristics intermediate between these two end members. There apparently is a continuous series of channel types both single and multiple, depending on the nature of the sediment load moved through the channel.

Analysis of data collected at 36 river cross sections in Australia and on the Great Plains of the western United States supports this classification. For example, it has been demonstrated that channel shape expressed as a width-depth ratio (F) and channel sinuosity (P) are controlled primarily by the type of sediment load, such that a bedload channel has a high width-depth ratio but low sinuosity (Schumm, 1963). Numerous other studies have demonstrated that channel width (w), depth (d), gradient (s), and meander wavelength (l) are related to water discharge (Q_w) as follows:

$$Q_w \propto \frac{w, d, l}{s} \qquad (1)$$

The above relation indicates that channel width, depth, and meander wavelength will increase with an increase in discharge, but channel gradient will decrease. However, for a constant discharge a change in the average quantity of bedload (Q_s) moved by the stream, that is the percentage of the average total load that is sand size or larger (percentage of bedload), is related

[1] See appendix for definition of symbols.

TABLE 1.—CLASSIFICATION OF STABLE ALLUVIAL CHANNELS

Type of sediment transport	Channel sediment (percentage of silt and clay)	Bedload (percent of total load)	Type of River	
			Single Channel	Multiple Channel
Suspended load and dissolved load	20	<3	Suspended-load channel. Width-depth ratio less than 10; sinuosity greater than 2.0; gradient relatively gentle.	Anastomosing system
Mixed load	5–20	3–11	Mixed-load channel. Width-depth ratio greater than 10, less than 40; sinuosity, less than 2.0, greater than 1.3; gradient moderate. Can be braided.	Delta distributaries Alluvial plain distributaries
Bedload	5	>11	Bedload channel. Width-depth ratio greater than 40; sinuosity, less than 1.3; gradient relatively steep. Can be braided.	Alluvial fan distributaries

to channel morphology as follows:

$$Q_s \propto \frac{w, l, s}{d, P} \quad (2)$$

Equation 2 shows that as the sand load or bed-load (Q_s) increases, channel width (w), meander wavelength (l), and gradient (s) increases, whereas channel depth (d) and sinuosity (P) decrease. Width-depth ratio will increase under these conditions.

The above generalizations, which are based on the statistical analysis of data from very different types of rivers (Schumm, 1968), suggest that a change in discharge and type of sediment load will induce major adjustments of river morphology. Except under controlled laboratory conditions, a change in discharge will be accompanied by a change in the type of sediment load. The channel response to these changes will be complex. For example, if both discharge and percentage of bedload increase (direction of change of each variable is indicated by a plus or minus exponent), channel morphology will be affected as follows:

$$Q_w^+ Q_s^+ \propto \frac{w^+ l^+ F^+}{P^-} s\pm d\pm \quad (3)$$

The effects of increased discharge and percentage of bedload combine to increase channel width, meander wavelength, and width-depth ratio and to decrease sinuosity. However, their effects are opposed so that an increase in discharge will act to decrease gradient and increase depth, but an increase of bedload will increase gradient and decrease depth. The ultimate changes depend on the magnitude of the changes of discharge and type of sediment load, but in the above case if sinuosity decreases then gradient should increase, and if both width and width-depth ratio increase then depth should decrease or remain unchanged. Of course, if both discharge and percentage of bedload decrease, their effects would be the opposite of these shown by equation 3. For a discussion of the effects of other changes of water and sediment discharge on channel morphology, see Schumm (1969).

The influence of water discharge and type of sediment load on alluvial channels has been demonstrated by an analysis of data collected from rivers on the Great Plains of the United States and along the Murrumbidgee River of New South Wales, Australia. In these studies (Schumm 1968, 1969) the percentage of silt and clay (sediment finer than 0.074 mm) in the perimeter of the channels (M) was used as an index of type of sediment load (Q_s). The relation is inverse with high silt-clay content being characteristic of low percentage of bedload (Table 1).

Empirical equations were developed which are statistically very significant and which can be used to predict channel morphology from the discharge and the percentage of silt and clay in the bed and bank sediments. The empirical equations are based on data obtained from rivers that transport only sands, silts and clays in a stable (no progressive aggradation or degradation) alluvial channel composed of sediments that are moving through the channel. In addition, the channels are not located near the coast, and all are in subhumid or semiarid environments.

ESTIMATE OF PALEOCHANNEL MORPHOLOGY AND HYDROLOGY

The independent variables that influence the morphology of both modern and ancient channels are water (Q_w) and type of sediment load (Q_s). Obviously data on water discharge and type of sediment load are not available for paleochannels, and it is necessary, therefore, to attempt to estimate these variables from information that can be obtained from cross sections of the paleochannel. A possible solution to these difficulties is at hand. The width-depth ratio (F) of modern channels is closely related to channel silt-clay content (M) as follows:

$$F = 225\ M^{-1.08} \quad (4)$$

Width-depth ratio, therefore, can be substituted for M and will reflect the nature of the sediment in the paleochannel as well as the type of sediment load. Channel cross-sectional area should be a useful index of discharge when combined with width-depth ratio. For the rivers from which the basic data were obtained, channel width is significantly related to channel cross section and to discharge, as other investigators have demonstrated (e.g., Leopold and Maddock, 1953). Although not independent variables, channel width and depth are significantly related to other morphological characteristics of a channel and to its hydrology. Therefore, an estimation of paleochannel gradient, meander characteristics, and hydrology can be obtained by using width-depth ratio as an index of type of sediment load and width as an index of discharge.

Multiple regression analyses of width-depth ratio (F) and channel bankfull width (w) on meander wavelength (l) and gradient (s) were performed. The equations obtained are as follows:

$$l = 18\ (F^{.53}\ w^{.69}) \quad (5)$$

or in log form

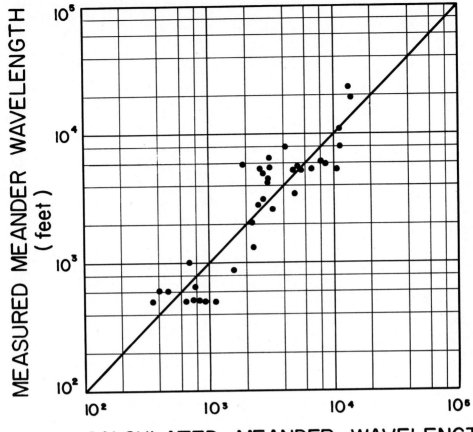

Fig. 1.—Relation between measured meander wavelength and calculated meander wavelength. Meander wavelength was calculated by use of equation 5.

$$\log l = 1.27809 + 0.52822 \log F + 0.68774 \log w$$
$$r = .91$$
$$Se = .21 \text{ log unit}$$

$$s = 30 \left(\frac{F^{.95}}{w^{.98}}\right) \quad (6)$$

or in log form

$$\log s = 1.48085 + 0.94774 \log F - 0.87937 \log w$$
$$r = .84$$
$$Se = .16 \text{ log unit}$$

In the above equations width-depth ratio is directly related to meander wavelength and gradient. Width is directly related to meander wavelength and is inversely related to gradient, and these are the expected relations. That is, an increase in width-depth ratio reflects a greater percentage of bedload, which is associated with a larger wavelength and gradient. Channel width increases with discharge, which is associated with a larger wavelength but smaller gradient.

In order to show graphically the accuracy of the predicting equations, calculated and measured values of meander wavelength and gradient are plotted on figs. 1 and 2. The plotted points fall within a half log cycle, and it is doubtful that a better relation can be obtained with the present state of knowledge concerning river morphology.

Multiple regression equations of width-depth ratio (F) and channel bankfull width (w) on mean annual flood (Qma) and mean annual discharge (Qm) were performed. The equations obtained are as follows:

$$Qma = 16 \left(\frac{w^{1.56}}{F^{.66}}\right) \quad (7)$$

or in log form

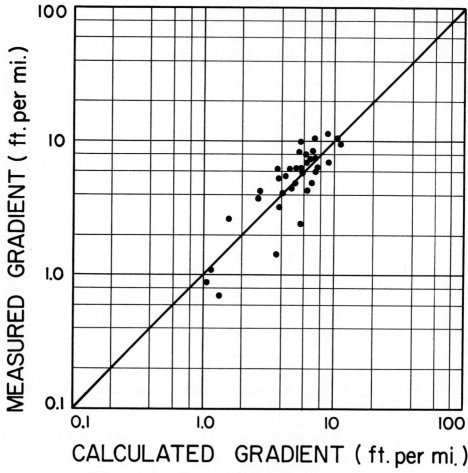

Fig. 2.—Relation between measured channel gradient and calculated gradient. Gradient was calculated by use of equation 6.

$\log Q_{ma} = 1.19416 - 0.66227 \log F + 1.55584 \log w$

$r = .90$

$S_e = .20$ log unit

$$Q_m = \frac{w^{2.43}}{18\, F^{1.13}} \quad (8)$$

or in log form

$\log Q_m = -1.24661 - 1.13327 \log F + 2.42853 \log w$

$r = .90$

$S_e = .31$ log unit

As expected, discharge is directly related to width in equations 7 and 8. Width-depth ratio previously has not been found to be dependent to a major extent on discharge; therefore, the inverse relation of width-depth ratio to discharge in equations 7 and 8 requires explanation. Both equations reduce to an approximation of channel area. Channel area is, of course, closely related to discharge.

The plot of calculated versus measured mean annual flood is truly impressive (fig. 3). It is generally agreed that the channel-forming discharge is a discharge that approaches bankfull; therefore, mean annual flood should bear a close relation to channel dimensions. This is demonstrated by the plot of Figure 3, except for one point located at the lower end of the plot. In this case the calculated mean annual flood is 5 times that of measured mean annual flood. The explanation seems to be that this river drains from the Sand Hills of Nebraska, and it is fed primarily by ground water. The discharge of this channel is very regular and flood peaks are not high. Nevertheless the channel is relatively wide and has a high width-depth ratio as a result of a high sand load. This combination yields an erroneously high estimate of mean annual flood. Such an error probably will occur only in the unusual situation of a stream that is dominated by ground water discharge from a drainage basin of high permeability.

The plot of calculated versus measured mean annual discharge (fig. 4) is less impressive and

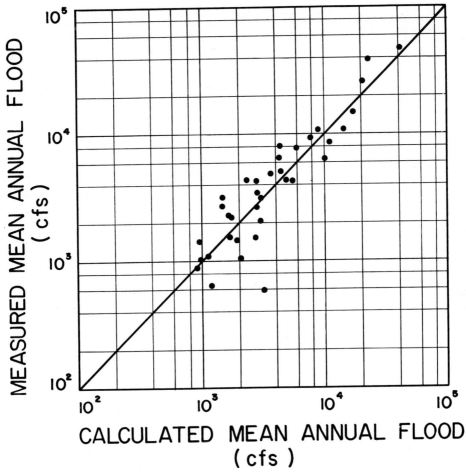

Fig. 3.—Relation between measured mean annual flood and calculated mean annual flood (recurrence interval of 2.33 years). Mean annual flood was calculated by use of equation 7.

appears to reflect a climatic influence. If as stated above the dimensions of a channel are established primarily by a flood event, then mean annual discharge will be less closely related to channel morphology than will the mean annual flood. This occurs because two channels, one carrying infrequent ephemeral flow and the other carrying perennial discharge, can both have similar dimensions dependent on a near bankfull discharge. At the lower end of the plot of Figure 4 scatter is greatest. All of the points in the upper half of the plot cluster closely about the line, and these points represent perennial streams. Those that fall well below the line in the lower half of the plot are ephemeral streams draining semiarid regions. The equation yields a much higher estimate of mean annual discharge than is appropriate for ephemeral-stream channels. Therefore, if evidence exists that a paleochannel is associated with a dry climate, an estimate of mean annual discharge based on equation 8 should be reduced by one-half.

The statistical relations are very good and suggest that if paleochannel width and depth can be obtained at an outcrop or from well data, then an estimate of both paleochannel gradient, meander characteristics, and hydrology can be obtained.

It is possible to obtain additional information concerning paleochannel hydraulics. For example, if mean annual flood (Qma) is considered to be approximately equal to bankfull discharge, then when it is divided by channel area (A) an estimate of the velocity (V) of flood discharge is obtained as follows:

$$V = \frac{Qma}{A} \tag{9}$$

Information can also be obtained concerning the valley through which the paleochannel flowed. Sinuosity (P) is the ratio of valley

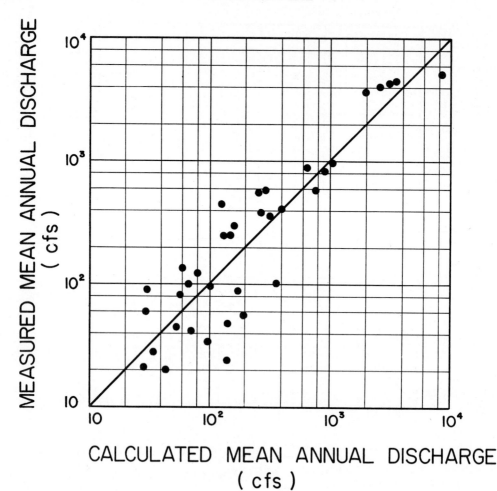

Fig. 4.—Relation between measured mean annual discharge and calculated mean annual discharge. Mean annual discharge was calculated by use of equation 8.

gradient (Sv) to channel gradient (s). An estimate of sinuosity from width-depth ratio alone can be obtained as follows:

$$P = 3.5 \; F^{-.27} \qquad (10)$$

Depending on the history of the river system any estimate of sinuosity can involve major errors (Schumm, 1963). However, when gradient (s) is calculated from equation 6, the slope of the surface over which the paleochannel flowed, the valley gradient, (Sv), can be obtained as follows:

$$Sv = s \cdot P \qquad (11)$$

Further, an estimate of meander wavelength from equation 5 provides the spacing between meanders. Combined with an estimate of sinuosity a fair picture of the pattern of a paleochannel can be developed.

Determination of Paleochannel Dimensions

The problem of distinguishing between a paleochannel and valley-fill deposits is a major one. Unless the dimensions of a paleochannel can be obtained, the quantitative relations developed for modern rivers cannot be applied to paleochannels.

One means by which it may be possible to set limits to the dimensions of fluvial paleochannels is based on a comparison of channel size. The largest river in the world today is the Amazon. It drains an area of 2.3 million square miles with an average annual precipitation of 80 inches. The width of the Amazon ranges from 1 to 3 miles and the depth from 20 to 300 feet along the lower 900 miles of its course. At Obidos, where the U. S. Geological Survey has made measurement of discharge, the channel is 200 feet deep and about 7500 feet wide (Oltman, et al., 1964).

In contrast, the Mississippi River, the world's fourth largest, (Morisawa, 1968), is about 2000 feet wide and 60 feet deep at Vicksburg. These channels are enormous, but they need to be in order to transport an average flow of about 5.5 million cfs and 0.75 million cfs respectively. The Amazon, in fact, transports 15 percent of all the fresh water running off to the oceans. It is unlikely that there are many paleochannels that have accommodated discharges of this magnitude, and most would probably be very much smaller, for example, on the order of a few hundred feet wide and up to 20 feet deep. Information on the dimensions of some paleovalley deposits (Table 2) reveals that most are smaller than the Amazon and Mississippi Rivers, but if they are assumed to be paleochannels they must have drained enormous areas of high rainfall. In many instances such huge source areas did not exist.

There can be little doubt that many paleochannel deposits are, in fact, valley-fill deposits composed of sediments deposited in a shifting and/or aggrading channel (see for example Rubey and Bass, 1925; Friedman, 1960; Howard and Showe, 1965; Siever, 1951; Charles, 1941; Lins, 1950). John Harms (1966) has made very clear that the "J" sand which is about 2400 feet wide and 60 feet deep is a valley-fill deposit. How then is it possible to identify and relate the dimensions of a paleo-river channel to the paleohydrology of the system? The answer lies in the fact that, although the relations between climate and runoff may have changed through geologic time, the laws of hydraulics did not. A flood was and is capable of moving only certain volumes of sediment of a given size range. Colby's (1964a, 1964b) work provides a basis for the estimation of sand load and bed scour during floods. Of major importance here is his conclusion that the scour of a channel during floods is not great. Certainly the evidence from paleochannels, where crossbedded units are on the order of a foot thick, indicates that the channel was not capable of setting in motion the previously deposited sediments to any great depth. Therefore, a channel must have occupied several positions in a valley. At one stage it must have flowed near the floor of the valley while eroding the bedrock floor, but when the upper few feet of fluvial sediment was being deposited, the channel was flowing across the upper surface of the valley-fill deposit just as the rivers of today commonly are flowing on valley-fill sediments that may be in excess of 100 feet. Nevertheless, the problem of identification and delineation of a paleochannel cross section in a paleovalley deposit remains.

TABLE 2.—PALEOVALLEY DIMENSIONS

Channel	width	depth (feet)
J channel (Harms, 1966)	1500 ft.	50 ft.
Rocktown channel (Rubey and Bass, 1925)	0.5 mile	25 (range 15–100 ft.)
Indiana channels (Friedman, 1960)		
New Goshen	0.5 mile	40
Terre Haute	0.25 mile	40
Winslow	2600 ft.	50
Englevale channel (Howard and Schowe, 1965)	0.5 mile	60
Pre-Pennsylvanian channel (Siever, 1951)	3.75 mile	200+
Bush City channel (Charles, 1941)	1000 to 2000 ft.	55
Tonganoxie channel (Lins, 1950)	14 miles	1000

An example will be used to demonstrate that estimates of paleochannel width and depth can be made. On the aerial photograph of Figure 5, a Holocene paleochannel on the Riverine Plain of New South Wales, Australia is visible. Borings into this channel reveal a lens-shaped sand deposit much like those described by stratigraphers investigating more ancient paleochannels (fig. 6). Although the maximum width of this deposit is about 1500 feet, the width of the channel that flowed on the surface of the deposit and that is still visible on the surface of the plain (fig. 5) was about 500 feet.

Several lines of evidence indicate that the Riverine Plain paleochannel was not deeper than about 10 feet, during at least the latter part of its existence (Schumm, 1968, p. 30–36). For example, on bends lateral migration of the channel occurred, planing away the heavy clay of the alluvial plain at a depth considerably less than the maximum thickness of the sand deposit. The sand body is composed of crossbedded sedimentary units which average about one foot thick. This implies that when the stream flowed near the surface of the plain its depth was only a few feet below the contact between overlying fine sediments and the fluvial sand (8 to 12 feet below the surface of the plain, Figure 6). Further, distributary systems related to the major stream are about 10 feet deep (Pels, 1964).

The Australian paleochannels were wide, shallow streams that transported relatively

Fig. 5.—Aerial photograph of paleochannel on the Riverine Plain of New South Wales, Australia. The width of the channel at the surface of the plain is about 500 feet. See figure 6 for cross section of this fluvial deposit. (Photograph courtesy of New South Wales Lands Dept.)

large quantities of sand. They were, therefore, bedload or mixed load channels, according to the classification of Table 1. Quantitative studies of channel morphology reveal that the Australian paleochannels conform to the relations developed between type of sediment load, hydrology, and morphology of rivers of the Great Plains of the western United States. These relations indicate that only a part of the width and a fraction of the depth of the valley-fill deposit can be related to the stream that transported the sediment (fig. 6). Therefore, in most cases the cross-sectional area, width, and depth of a lens-shaped sand body is greater than that of the stream that transported the sediment. If the surface of the paleochannel deposit is flat, then channel width could be that of the sand body, but if the upper surface is convex then the

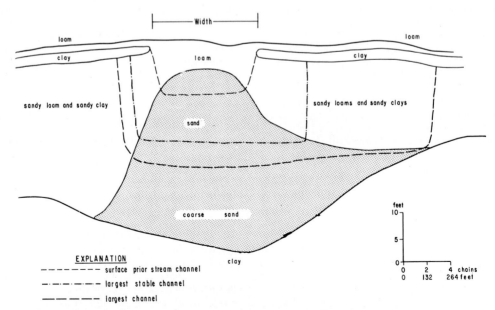

Fig. 6.—Generalized cross section of Australian paleochannel on Riverine Plain of New South Wales from bore data obtained by M. Stannard (from Schumm 1969). Location is just off of eastern edge (right side) of figure 5. Dashed lines show outline of three possible channel cross sections. The smallest is that associated with the channel visible on the surface of the plain. The intermediate and largest cross sections shown are the largest stable cross section and the largest possible cross section that could be associated with an alluvial channel. In each case the depth of the cross sections is considerably less than the depth of the sand body.

width of the channel is probably less than half that of the sand body (fig. 6). As the width-depth ratio of a bedload channel will usually be greater than 40, one may search for evidence of a channel floor at a depth about 1/40 the width of the sand body.

Perhaps other evidence may be found to assist in the delineation of a paleochannel. For example, a pebble layer may be evidence of a commonly achieved depth of scour (Livesey, 1963), or an abrupt change in the texture of the sediment at the top of the channel sand may be evidence of the onset of a final period of paleochannel aggradation (Schumm, 1968, p. 36).

It remains for the stratigrapher to recognize within a fluvial deposit the lateral and vertical extent of the channel. Once this is done then the equations can be used to estimate paleochannel hydrology, gradient, and meander characteristics.

DISCUSSION

Several problems associated with determining the third dimensional character and hydrology of paleochannels have been considered but an important aspect of the problem deserves mention. The apparent exactitude of equations 5 through 8 should not mislead an investigator into assuming that they provide anything but improved estimates of paleochannel character.

Geologists have frequently been intimidated when the apparently conclusive results of other disciplines have been applied to geologic problems with disconcerting results. A reasonable approach to the problems discussed here would be to use the equations to estimate paleochannel discharge, gradient and pattern, but if the estimates are at variance with other geologic evidence, then the geologic evidence should be considered to be of a higher order of significance.

Appendix 1—Definition of Symbols

A channel cross-sectional area in square feet
d channel maximum depth in feet
F width-depth ratio
l meander wavelength in feet
M percentage of silt and clay in channel perimeter (sediment finer than 0.074 mm)
P sinuosity, ratio of channel length to valley length
Qw water discharge, Qm or Qma, in cubic feet per second
Qs percentage of total mean sediment load that is sand size or larger (bedload)
Qma mean annual flood in cubic feet per second
Qm mean annual discharge in cubic feet per second
s channel gradient in feet per mile
Sv valley gradient in feet per mile
V velocity in feet per second

REFERENCES

CHARLES, H. H., 1941, Bush City Oil Field, Anderson County, Kansas: in Stratigraphic Type of Oil Fields: American Assoc. Petroleum Geol., Tulsa, Okla., p. 43–56.
COLBY, B. R., 1964a, Discharge of sands and mean-velocity relationships in sand-bed streams: U. S. Geol. Survey Prof. Paper 462-A, 47 p.
———, 1964b, Scour and fill in sand-bed streams: U. S. Geol. Survey Prof. Paper 462-D, 32p.
DURY, G. H., 1964, Principles of underfit streams: U. S. Geol. Survey Prof. Paper 452-A, 67p.
FRIEDMAN, S. A., 1960, Channel-fill sandstones in the Middle Pennsylvanian rocks of Indiana: Indiana Geol. Survey, Report of Progress 23, 59p.
HARMS, J. C., 1966, Stratigraphic traps in a valley fill, western Nebraska: Am. Assoc. Petroleum Geol. Bull., v. 50, p. 2119–2149.
HOWARD, L. W. AND SCHOEWE, W. H., 1965, The Englevale channel sandstone: Kansas Acad. Science Trans., v. 68, p. 88–106.
LEOPOLD, L. B. AND THOMAS MADDOCK, JR., 1953, The hydraulic geometry of stream channels and some physiographic implications: U.S. Geol. Survey Prof. Paper 252, 57p.
LEOPOLD, L. B., WOLMAN, M. G., AND MILLER, J. P., 1964. Fluvial Processes in Geomorphology: W. H. Freeman and Co. 511 p.
LINS, T. W., 1950, Origin and environment of the Tonganoxie sandstone in northeastern Kansas: Kansas Geol. Survey Bull. 86, p. 105–140.
LIVESEY, R. H., 1963, Channel armoring below Fort Randall Dam: U.S. Dept. Agriculture Misc. Pub. 970, p. 461–469.
MOODY-STUART, M., 1966. High and low sinuosity stream deposits with examples from Devonian of Spitsbergan: Jour. Sed. Pet. v. 36, p. 1101–1117.
MORISAWA, MARIE, 1968. Streams—their dynamics and morphology: McGraw-Hill Book Co., New York, 175 p.
OLTMAN, R. E., H. O'R. STERNBERG, F. C. AMES, AND L. C. DAVIS, JR., 1964, Amazon River investigations reconnaissance measurements of July 1963: U.S. Geol. Survey Circular 486, 15p.
PELS, SIMON, 1964, Quaternary sedimentation by prior streams on the Riverine Plain, southwest of Griffith, New South Wales: Royal Soc. New South Wales, Jour. and Proc., v. 97, p. 107–115.
RUBEY, W. W. AND N. W. BASS, 1925, Geology of Russell County, Kansas: Kansas Geol. Survey Kansas, Bull. 10, p. 1–86.
SCHUMM, S. A., 1963, Sinuosity of alluvial channels on the Great Plains: Geol. Soc. America Bull., v. 74, p. 1089–1100.
———, 1968, River adjustment to altered hydrologic regimen—Murrumbidgee River and paleochannels, Australia: U.S. Geol. Survey Prof. Paper 598, 65p.
———, 1969, River metamorphosis: Jour. Hydraulics Div., Proc. American Soc. Civil Eng. HY 1, p. 255–273.
SIEVER, RAYMOND, 1951, The Mississippian Pennsylvanian unconformity in southern Illinois: Am. Assoc. Petroleum Geol., v. 35, p. 542–581.

CRITERIA FOR RECOGNIZING LACUSTRINE ROCKS[1]

M. DANE PICARD[2] AND LEE R. HIGH, JR.[3]

ABSTRACT

Lacustrine deposits can be recognized through the use of a variety of criteria that include physical, chemical and biologic aspects of sedimentary rocks. Because large lakes closely resemble shallow, epicontinental seas in physical properties, large differences in lithologic characteristics, sequences, facies relations, sedimentary structures, paleocurrent patterns and other physical parameters are not to be expected. Comparisons of these features in lacustrine and epicontinental rocks indicate the scarcity of significant diagnostic differences. In contrast, fluvial and lacustrine environments differ considerably in physical aspects, and the two environments can be differentiated on the basis of the physical characteristics of the rocks.

Large lakes and epicontinental seas differ mainly in size, chemistry and biota. Size differences are evident: few lakes exceed 10,000 sq. mi. in area. Regional stratigraphic and paleogeographic relations are useful therefore in distinguishing former lakes and seas. Geochemical differences are more definitive, however. Normal sea water has been relatively constant in composition for most of geologic history and changes related to evaporation, precipitation or dilution are predictable. In contrast, the chemistry of lacustrine water is not uniform, but is determined almost wholly by the lithology and climate of the hydrographic basin. The geochemical character of lakes therefore differs widely from one area to another and rarely approximates that of sea water. Accordingly, lakes and seas can differ in authigenic and diagenetic minerals. The products of evaporite cycles are especially useful for environmental interpretation.

Marine and nonmarine environments commonly are distinguished by their faunas. Paleontologic differentiation of nonmarine environments is uncertain, however, and requires further study.

Despite the qualifications mentioned here, reconstructions of lacustrine environments can be made with confidence. Lacustrine or marine environments can be differentiated from fluvial environments on the basis of physical properties of the deposits. Lakes and seas can then be differentiated by geochemical and biologic criteria.

INTRODUCTION

Lakes are present in a variety of geological settings. They range from small ephemeral ponds to large lakes in structural basins that cover thousands of square miles. Because lakes are especially sensitive to local conditions such as climate and source areas, their water chemistry ranges widely. Similarly, deposition can be clastic, organic or evaporitic. Lake bottoms can be continuously agitated by currents or they can be stagnant. Biologic activity ranges from intense to slight. In short, lacustrine deposits are characterized by a seemingly bewildering array of physical and biological parameters that superficially defy generalization.

Lakes have an astonishing variety of origins. They originate through a variety of processes that produce depressions bounded on all sides by a rim. The actual processes involved in the formation of a lake are classified as *constructive* when the rim is actively built, *destructive* when the lake is excavated, and *obstructive* when a pre-existing valley is dammed (Davis, 1882; Hutchinson, 1957, p. 156). For convenience, Hutchinson (1957, p. 157–163) suggested a classification of lakes according to the general nature of the processes responsible for building, excavation and damming. The major groups of Hutchinsons' classification are tectonic basins, lakes associated with volcanic activity, lakes formed by landslides, lakes formed by glacial activity, solution lakes, lakes due to fluvial activity, lake basins formed by wind, lakes associated with shorelines, lakes formed by organic accumulation, lakes produced by higher organisms, and lakes produced by meteorite impact. Seventy-six separate types are included in his classification, but most of them would not exist long enough for a thick sequence of rocks to accumulate and be preserved.

To leave a significant stratigraphic record, a lake must accumulate sediment over a long period of time (table 1). With few exceptions, only such long-lived lakes have been recognized in the pre-Pleistocene record. Accordingly, we restrict our consideration of lacustrine criteria to these lakes. Other types are more properly considered as subenvironments of fluvial, deltaic, or glacial settings.

A reliable estimate of the volume of lacustrine rocks in the stratigraphic record is not possible on present information. The amount, however, is low, perhaps no more than a few percent. The scarcity of known ancient lakes reflects, in part, the inability of geologists to recognize rocks of this environment, despite the excellent early studies of Lake Bonneville (Gilbert, 1891)

---

[1] Manuscript received April 15, 1969.

[2] Department of Geological and Geophysical Sciences, University of Utah, Salt Lake City, Utah.

[3] Department of Geology, Oberlin College, Oberlin, Ohio.

TABLE 1.—AREAS AND APPROXIMATE DURATIONS OF TYPICAL LAKES
(Data from: Van Houten, 1962; Bradley, 1963; Picard, 1957c; 1963; this paper)

Rock Unit, Lake	Location	Age	Area (in mi²)	Duration (in m.y.)
Lake Bonneville	Utah, Nevada, Idaho	Pleistocene	19,750 (max.)	1.0
Searles Lake	California	Pleistocene	385	0.17 (?)
Lake San Augustin	New Mexico	Pleistocene	225	0.65
Lake Uinta (Green River, lower Uinta fms.)	Utah	Eocene	7,700	13.3
Gosiute Lake (Green River Fm.)	Wyoming, Colo.	Eocene	17,000 (max.)	4.0
Flagstaff Limestone	Utah	Paleocene-Eocene	7,000 (max.)	2.75
Todilto Limestone	New Mexico, Colo., Ariz.	Late Jurassic	34,600	0.02
Popo Agie Formation	Wyo., Utah	Late Triassic	50,000 (+)	3.0 (?)
Lockatong Formation	New Jersey, Pa.	Late Triassic	2,250	5.1

and Lake Lahontan (Russell, 1885). During the Pleistocene and Tertiary, when lakes were most abundant and extensive (figs. 1–3), they covered only a small percent of the land surface. Ancient lakes also existed during pre-Tertiary time and, in places, dominated the deposition of large regions (fig. 4).

The geological significance of lakes is out of

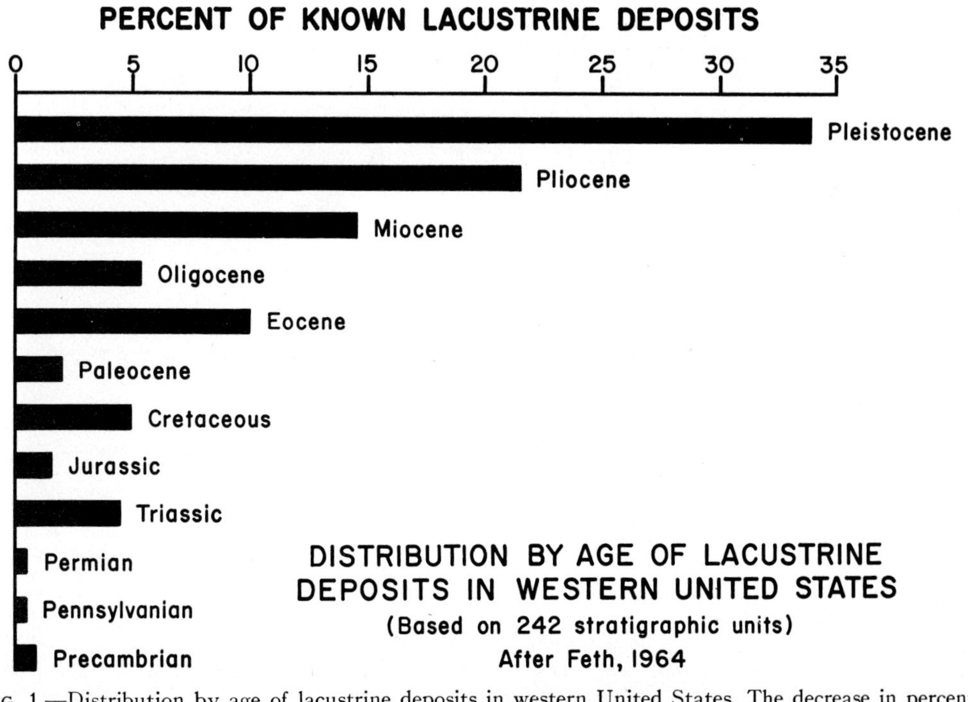

PERCENT OF KNOWN LACUSTRINE DEPOSITS

[Bar chart showing: Pleistocene ~35%, Pliocene ~25%, Miocene ~22%, Oligocene ~8%, Eocene ~15%, Paleocene ~5%, Cretaceous ~7%, Jurassic ~4%, Triassic ~5%, Permian ~2%, Pennsylvanian ~2%, Precambrian ~2%]

DISTRIBUTION BY AGE OF LACUSTRINE DEPOSITS IN WESTERN UNITED STATES
(Based on 242 stratigraphic units)
After Feth, 1964

FIG. 1.—Distribution by age of lacustrine deposits in western United States. The decrease in percent of lacustrine rock units with increasing age suggests that many ancient lakes have not been recognized. Some of the stratigraphic units assigned to the Eocene are probably partly Paleocene in age, reflecting the Paleocene-Eocene boundary problem. Difficulties of age assignment also exist at the Cretaceous-Paleocene boundary. Triassic lacustrine deposits have been distinguished and studied much more extensively than Jurassic lacustrine deposits.

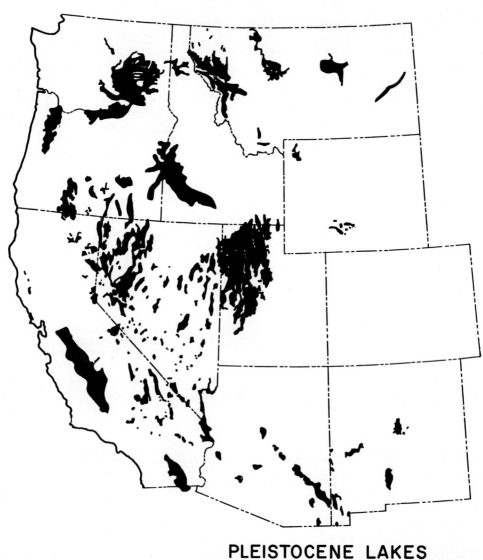

FIG. 2.—Distribution of maximum known or inferred extent of Pleistocene lakes reported in the literature for western United States. Pertinent references are given in Feth (1961). The Pleistocene probably is characterized by the maximum concentration of lacustrine environments. However, most of the Pleistocene lakes were too short-lived to accumulate significant deposits and, if older, would be unrecognizable.

all proportion to their limited occurrence. They constitute local base levels, fluctuations of which control the erosion or deposition of sediment in surrounding areas. Documented examples of local base level control are the Triassic lake in New Jersey (Van Houten, 1962, p. 575–576) and Lake Uinta (Eocene) in Utah (Picard and High, 1968b). Inasmuch as lakes are sensitive to small changes in climate, they are valuable for paleoclimatic reconstructions. In addition, some ancient lakes have yielded remarkable fossil assemblages. In these and other ways, lake deposits provide definitive means of paleoenvironmental reconstruction.

TYPES OF LACUSTRINE CRITERIA

It is not possible to review here all of the criteria for reconstruction of lacustrine environments. Therefore, we have selected for evaluation criteria that are generally applicable.

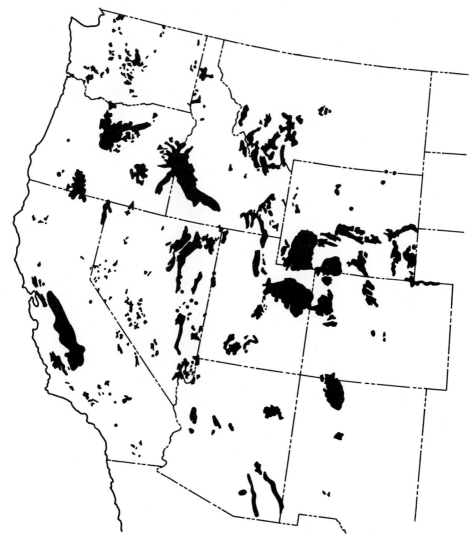

FIG. 3.—Distribution of reported Tertiary lacustrine deposits in the western United States. References for each epoch are given in Feth (1963). During the Tertiary, uplift of the Cordilleran geosyncline and associated block faulting created numerous intermountain basins with interior drainage. Large lakes then developed within some of these basins.

For the interpretation of depositional environment, a wide selection of information usually is available. However, different types of information frequently are contradictory. Petrographic evidence may suggest one climatic interpretation, but the fauna another. In short, the evidence used for paleogeographic reconstructions is diverse, scattered, incomplete and not always diagnostic.

Rocks are the sum of physical, chemical and biologic parameters. Most of the evidence that is useful for paleoenvironmental reconstruction is physical, and includes stratigraphic, petrographic and sedimentologic information. Specific areas of investigation include sequence, facies relationships, lithology, texture, bedding, sedimentary structures and paleocurrent patterns. Chemical evidence, which is obtained mainly from the fields of mineralogy and geochemistry, includes authigenic minerals, mineral

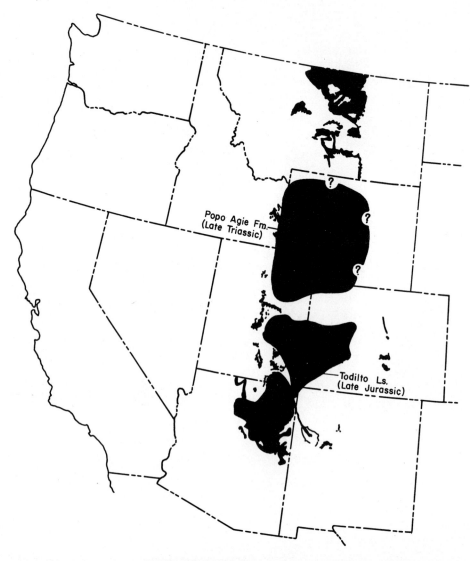

FIG. 4.—Distribution of pre-Tertiary lacustrine deposits in western United States. The causes of the remarkable north-trending alignment are unknown.

varieties, cation and isotope ratios, and the reconstructed chemistry of the water body. Biologic information consists of the preserved fauna and flora. For most depositional environments, biologic information is generally the most diagnostic but frequently is insufficient in quantity. Consequently the interpretation ultimately rests on the less diagnostic but more abundant physical parameters. Lacustrine criteria that have been cited most frequently are summarized in figure 5.

STRATIGRAPHIC RELATIONS

Size and shape.—Twenhofel (1932, p. 824) considered most lakes to be small and generally round or elliptical. However, lakes that originate from structural movements tend to be significantly larger than other types of lakes and

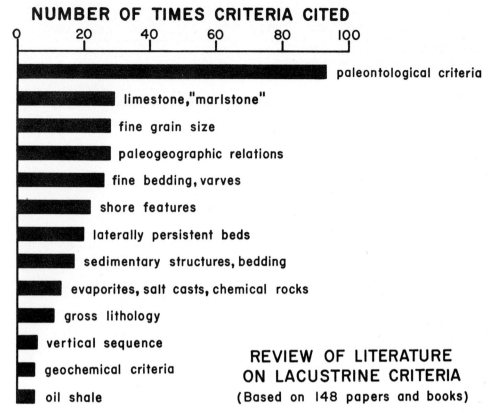

Fig. 5.—Cited criteria for the recognition of lacustrine deposits. All criteria cited for a lacustrine origin in each reference were noted.

may be far from round (e.g., Rift valley lakes of Africa, Lake Baikal, and so forth). Compared with epicontinental seas, even the largest lakes are small. Recent large lakes have areas on the order of 10,000 sq. mi. Larger lakes are known from the past but none of them exceed in area about 100,000 sq. mi. For example, one of the largest lakes yet recognized covered at least 50,000 sq. mi. (High and Picard 1965, 1969a). Even though this extent is comparable with marine embayments and lagoons, it is an order of magnitude less than that of epicontinental seas. Although lacustrine deposits are limited in extent, the mean thickness of known lacustrine rocks is significant (figs. 6–7).

Large lakes are irregular in shape. In detail, the outline of a lake can be as indented as marine coastlines. Precise shoreline positions are difficult to establish locally, however, because ancient depositional features and local irregularities are smoothed or are lost from the record by erosion. Because of differences in size, lacustrine deposits on a regional scale frequently are roundish "blobs," in contrast to marine deposits which tend to form linear map patterns.

Facies pattern.—Large lakes closely resemble shallow seas in processes of deposition, but there are several differences that are useful. Lakes show only minor tidal effects and, because of limited fetch and wave amplitude, have a shallower wave base than do seas. Shore and nearshore deposits therefore are less extensively developed in lakes than in seas. Although the general facies progression of fluvial-deltaic to nearshore to offshore is the same in both cases, the fluvial to offshore transition is more abrupt for lakes than it is for epicontinental seas.

There are exceptions, however, to both the general facies pattern and the abruptness of the transition. Twenhofel (1932, p. 824) considered the ideal lacustrine pattern to consist of an outer belt of shore gravel, an intermediate belt of sand, an inner belt of sandy marly mud, and mud in the center (figure 8). If wave action is limited, the nearshore and shore phases can consist of organic peat and muck. In contrast, Longwell (1928, p. 82, 95) and Feth (1964, p. 21–22) cited examples of fine-grained lacustrine mudstone that abuts directly against bedrock of the enclosing basin walls, and coarser grained

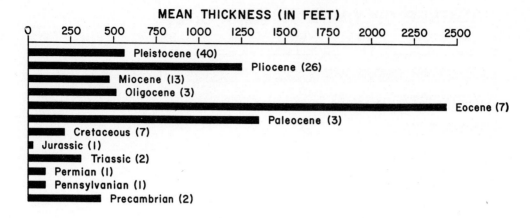

FIG. 6.—Mean thickness of lacustrine rock units, by age, in western United States. Some lacustrine units attain considerable thickness, but most of them are generally thin.

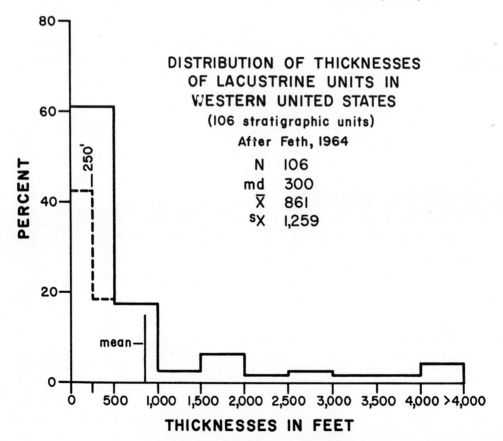

FIG. 7.—Histogram of thicknesses of lacustrine units in western United States. Thin rock units are dominant.

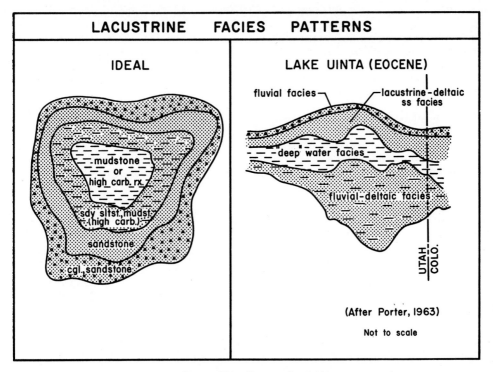

Fig. 8.—Comparison of "ideal" lacustrine facies pattern and simplified facies pattern of Lake Uinta (Eocene).

nearshore and shore deposits are absent. Similarly, Visher (1965, p. 57–58) suggested that because lakes are filled largely by deposition of fine material over the entire basin floor, rather than by peripheral accretion, the sediment deposited below wave base will be extensive and that deposited in shallow lacustrine and shore environments will be restricted and irregular. If the lake is stable, this relationship is valid. Twenhofel (1932, p. 824) noted, however, that fluctuations in water level can produce extensive shore deposits. For example, the size of Lake Uinta repeatedly fluctuated throughout its history. On the northern side where the regional slope was steep, the offshore-fluvial transition is abrupt (Picard and High, 1968b, p. 378), in contrast to the southern side where the regional slope of the basin was gentle and an extensive deltaic facies was deposited (fig. 8).

The facies patterns of saline lakes differ from those of fresh-water lakes. The lack of inflow leads to reduced amounts of clastic material, and shore phases can be muddy. As the lake dries, salts are deposited, the least soluble around the margins and the most soluble in the center of the basin. This pattern differs from that of coastal lagoons and restricted marine basins in its concentricity.

In summary, lakes resemble shallow seas in their physical processes of deposition, and the resulting facies patterns are similar. Shore phases of lakes tend to be poorly developed, and fluvial-offshore transitions can be abrupt. The presence of markedly narrow shore and nearshore deposits is suggestive of lacustrine deposits. In contrast, the presence of extensive shore phases does not eliminate a lacustrine interpretation because fluctuations in lake level can produce widespread, nearshore, lacustrine and deltaic complexes.

Sequence.—Stratigraphic sequence, inasmuch as it represents the orderly succession of events, is a valuable means of interpreting depositional environments. Walther's Law (1893–94) enables one to predict the sequence if the lateral facies and history are known (High and Picard, 1969b, p. 723). Inasmuch as all lakes are ultimately filled, regression dominates the history of a lake, and many workers believe that the ideal lacustrine sequence grades upward from fine-grained rocks deposited below wave base into coarser shore and fluvial deposits (Twenhofel, 1932, p. 827; Lahee, 1941, p. 72–73; Visher, 1965, p. 57–58).

Because the facies patterns of lakes and shallow seas are similar, the ideal lacustrine sequence resembles the regressive marine sequence. Visher (1965, p. 57–58) suggested that the two environments can be differentiated on the relative abundance of units within the sequence. In the lacustrine sequence, offshore deposits should be more important quantitatively than in regressive marine sequences. However, local exceptions to this generalization are common and stratigraphic sequence is only suggestive of lacustrine origin.

Figure 9 is a comparison of generalized stratigraphic sequences of three large ancient lakes with the ideal lacustrine sequence. The deposits of Lake Uinta approximate the ideal cycle, and fine-grained lacustrine mudstone and carbonate of the Green River and lower Uinta formations (both Eocene) are succeeded by coarser fluvial clastics of the upper Uinta and Duchesne River formations (Eocene?). The sequence of the Late Triassic lake in New Jersey is less obvious. Although the lacustrine deposits of the Lockatong Formation are succeeded by the fluvial deposits of the Brunswick Formation (Late Triassic), a simple textural trend is not apparent. The Lockatong Formation contains sandstone, mudstone and chemical rocks; the Brunswick Shale is dominantly mudstone, except along basin margins where conglomeratic

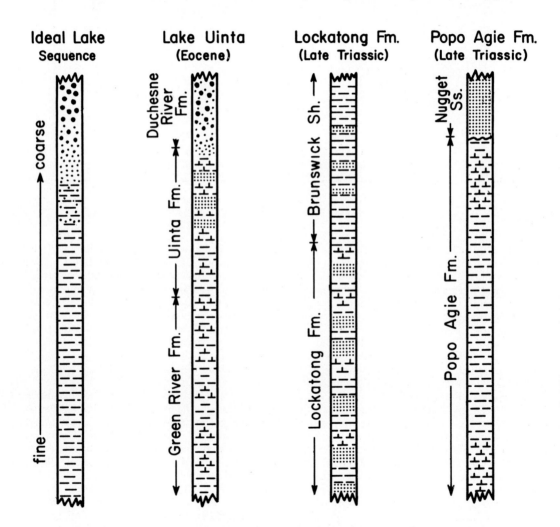

FIG. 9.—Comparison of "ideal" lacustrine stratigraphic sequence with those of several lacustrine rock units.

sandstone is common. A final example is that of the Late Triassic lake in Wyoming. Here the sequence is incomplete, and lacustrine rocks of the Popo Agie Formation are truncated by an unconformity at the base of the Nugget Sandstone (Jurassic?).

Regional relations.—Lakes are continental features and are associated with other continental deposits. Recognition of fluvial deposits surrounding a suspected lacustrine sequence is suggestive of lacustrine origin.

The association with fluvial rocks is useful for differentiating coastal lagoon and lacustrine deposits. Lakes are enclosed completely by fluvial settings and fluvial rocks must occur between a possible lacustrine unit and known marine rocks. If the unknown unit is in close proximity with marine rocks, and no intervening fluvial rocks are present, a lagoonal setting that opens to the sea is indicated.

PALEONTOLOGY

Feth (1964, p. 17–20) stated that fossils furnish definitive criteria of lacustrine environments. The evidence presented, however, serves mainly to distinguish fresh-water from marine environments. The means of recognition of saline lakes and their organisms is not considered, thereby limiting greatly the utility of fossils as indicators of lacustrine environments. Fossil lacustrine faunas are too poorly studied to be useful generally. It is possible to distinguish fresh-water from saline faunas (table 2), but it can be difficult to distinguish fresh-water lake from stream faunas because the two environments frequently are closely associated. For example, in the living molluscan fauna of North America, some species are surprisingly adaptable and can accommodate with equal ease to running water and stagnant backwaters (La Rocque, 1969, pers. commun.). Interpretation is difficult, even at the generic level. The genus *Goniobasis* has living representatives in the high-energy environments of the shores of the Great Lakes and also in the low-energy environments of small lakes and ponds.

Lakes are characterized by their diversity. Although fluvial and marine subenvironments generally can be considered uniform in terms of water chemistry, sediment, hydrography and so forth, lakes range from fresh-water to supersaturated conditions and the faunas reflect this diversity. In addition, a single lake can evolve, or even fluctuate, from fresh to saline during its history. The resulting ecologic stresses induced by rapid fluctuations should be reflected in the fauna. However, stress conditions are also encountered in other environments and all low diversity faunas are not lacustrine. In marine and fluvial settings, stress conditions are generally local and are associated with areas of more normal conditions. In contrast, stress conditions are likely to exist throughout an entire lacustrine basin at a given time, and "normal" conditions may be ephemeral or entirely lacking. This principle, then, together with regional stratigraphic information, can be useful in determining lacustrine environments.

TABLE 2.—Fossil fresh-water and marine organisms (X, common in habitat; O, occurs; —, absent; R, fossils rare; [1], distinctions at lower taxonomic levels)

	Marine	Fresh-Water
Myxophyceae (blue-green algae)	X	X
Chlorophyceae (green algae)	X	X,R
Rhodophyceae (red algae)	X	O,R
Phaecophyceae (brown algae)	X	—
Charales	—	X
Bacillariophyceae (diatoms)	X	X
Foraminifera	X	O,R
Radiolaria	X	—
"Thecamoebians"	X	O
Porifera	X	O
Scaphopoda	X,R	—
Amphineura	X,R	—
Monoplacophora	X,R	—
Cephalopoda	X	—
Pelecypoda[1]	X	X
Gastropoda[1]	X	X
Coelenterata	X	—
Bryozoa	X	—
Brachiopoda	X	—
Protoarthropoda	O,R	O,R
Trilobita	X	—
Branchiopoda	O	X,R
Ostracoda[1]	X	X
Cirripedia	X	—
Malacostraca	X,R	X,R
Xiphosura	O,R	O?,R
Eurypterida	O?	X?,R
Arachnida	—	X,R
Insecta	O,R	X,R
Echinodermata	X	—
Pisces[1]	X	X
Amphibia	O,R	X
Reptilia[1]	O	X
Aves	X,R	X,R
Mammalia	O	X

Plants.—Although large plants are characteristic of terrestrial settings, the occurrence of plant fossils is not diagnostic of continental environments. Indeed, fossil plants may yield little information concerning the depositional environment if they have been transported to the environment following death (Schopf, 1957a, p. 703; Brown, 1957, p. 729). Wilson (1957, p. 719) suggested that terrestrial spores and pollen that are transported into the sea decrease in both diversity and abundance at increasing distances from the shoreline. The resulting gradients can be diagnostic of transported floras.

Spore and pollen present taxonomic problems, especially for older deposits, and their use as environmental indicators is restricted. Frequently, the identification of a particular spore as marine or nonmarine is based on associated fossils or lithology, which precludes an environmental interpretation. For spore and pollen that originate from known terrestrial plants, the problem of transportation remains. However, an assemblage that is exclusively terrestrial suggests that the associated beds are nonmarine. Differentiation of fluvial and swamp from lacustrine environments is still a problem, although the presence of a diversity and quantity gradient is suggestive of a large lake. Annotated bibliographies of spore and pollen occurrences are given in Schopf (1957a, 1957b) and in Wilson (1957).

Diatoms have adapted widely to virtually all types of water. Their extreme specific differentiation and wide distribution makes them excellent ecologic indicators (Lohman, 1957, p. 731). However, the usefulness of diatoms in ancient rocks is limited by their short range (restricted to post-Paleozoic rocks) and the difficulty of interpreting the environment if there are no living species in the assemblage.

The preservation of leaves can be significant. Flattened and whole leaves suggest quiet water deposition, and macerated material is suggestive of swamp, deltaic or fluvial environments (Pardee and Bryan, 1926, p. 8; Twenhofel, 1932, p. 162–163).

Algae are the most widely used plants for environmental reconstructions. Although marine-nonmarine distinctions can be made, detailed subdivision of nonmarine environments is uncertain. Blue-green algae are the most widely adapted but are not significant for geological use. The suggestions of Rezak (1957, p. 145–146) concerning growth forms are potentially useful, but have not been confirmed. He noted that the dominant forms in marine or saline lake environments are stromatolitic, and arborescent or free concretionary masses are characteristic of fresh-water streams and lakes. Green algae also are both marine and nonmarine. However, the families Codiaceae and Dasycladaceae are the only green algae that have calcareous skeletons and are entirely marine. Thus, fresh-water forms are rare in the fossil record. Red algae are almost exclusively marine, in contrast to charophytes, which are fresh-water algae. The latter group is used widely for dating nonmarine beds. Despite their restriction to nonmarine environments, charophytes are not sufficiently studied to permit separation of fluvial, pond and lacustrine deposits. The charophytes were reviewed by Peck (1953, 1957). Important fossil algae were discussed by Johnson (1961).

Invertebrates.—Among the invertebrates, many forms are exclusively, or dominantly, marine (table 2). They include foraminifera, radiolaria, coelenterates, bryozoa, brachiopods, scaphopods, amphineura, monoplacophora, cephalopods, trilobites, cirripeds and echinoderms. In addition, fresh-water forms of other groups are rarely found as fossils. These include sponges, malacostracans, arachnids and insects. There are exceptions, however. The Green River Formation, the Florissant lake beds, the Wellington Formation (Tasch, 1964) and other units in Europe contain large numbers and varieties of insect remains that indicate lacustrine environments, especially the larvae. The scarcity of fresh-water sponges may merely reflect neglect of this group (La Rocque, 1969, pers. commun.). Fresh-water sponges have minute needle-like spicules of silica that are easily preserved, but few people have studied or even bothered to collect them, probably because they are almost impossible to identify generically or specifically and their environmental significance is not known. Pelecypods, gastropods and ostracods are probably the most common nonmarine invertebrate fossils (table 2). Fresh-water pelecypods and gastropods are listed in table 3.

Although the distinction between fresh-water and marine mollusks usually can be made, the recognition of specific nonmarine environments

TABLE 3.—FRESH-WATER MOLLUSKS
(After Yen, 1951)

Pelecypoda
Unionidae
Sphaeriidae
Anodontidae
Corbiculidae
Gastropoda
Viviparidae
Pleuroceratidae
Amnicolidae
Ampullariidae
Neritidae
Assiminaeidae
Thiaridae
Valvatidae
Lymnaeidae
Planorbidae
Physidae
Ellobiidae
Orygoceratidae
Lancidae
Ancylidae

is uncertain. Baker (1928) noted that the fauna of numerous small lakes in Wisconsin generally is less diverse than that of nearby rivers. Individuals also tend to be smaller in lakes according to Baker. The distribution of mollusks can also be suggestive of lakes. Because most mollusks only inhabit shallow water, large lakes can be rimmed by fossiliferous bands while the center is relatively fossil-free, except for sparse Sphaeriidae. Yen (1951, p. 1379) suggested that some assemblages of pelecypods and gastropods are indicative of fluvial or lacustrine environments. Although mixing of forms is possible, a deposit yielding an assemblage of Unionidae, Sphaeriidae, Viviparidae and Pleuroceratidae and a smaller number of pulmonate gastropods can be considered fluvial. Lacustrine environments are indicated by an assemblage of aquatic pulmonates.

Ostracods are potentially useful (Swain, 1956a, 1964a,b) and numerous fresh-water forms are recognized. However, there is not sufficient detail known to distinguish fluvial and lacustrine environments. Feth (1964, p. 18) and Bradley (1969, pers. commun.) consider the presence of numerous bedding planes covered with ostracod tests to be indicative of a lacustrine environment.

Vertebrates.—Except for fish, vertebrates are dominantly terrestrial; marine forms are readily recognized by skeletal characters. Feth (1964, p. 19) stated that some varieties of turtles and crocodilians are apparently restricted to fluvial deposits, but details were not given. He also suggested that it is likely that articulated skeletons will occur in lacustrine units, and that disarticulated skeletons suggest the stronger currents of fluvial environments. However, in the quiet waters of lakes, many skeletons are disarticulated by scavengers.

Footprints of shore birds in a terrestrial setting are suggestive of lacustrine environments (Curry, 1957; Moussa, 1968). However, bird prints also are present in coastal environments or on flood plains.

GEOCHEMISTRY

General.—One of the most significant differences between lacustrine and other settings is in the geochemistry of the water. In contrast to oceans and streams, lakes are characterized by their variability. The salinity of closed lakes ranges from less than 1 percent to over 25 percent by weight of salts (Langbein, 1961, p. 1). Lakes collect water from local streams and reflect the chemistry of source areas. Accordingly, the geochemistry of lakes varies widely (Livingstone, 1963, p. G16-G18). In addition, individual lakes show rapid temporal fluctuations in response to changes in the climate of the basin or changes in the source areas (Belt, 1968, p. 166-167). Nevertheless, lakes, including those that fluctuate widely in concentration, tend strongly to reflect the lithology and geochemistry of their hydrographic basins. Lakes and their sediments tend to be throughout their histories Na-rich, Cl-rich, SO_4-rich, borate-rich and so forth. The Green River lakes generally were poor in SO_4 and Cl and rich in Na and CO_3. In Wyoming, the Green River Formation does contain halite, but the amount is not large compared with other evaporite minerals, Searles Lake has much more SO_4 than the Green River lakes and so do its sediments. Lakes that are poor in Cl and SO_4 tend to be CO_3-rich but almost any combination of the three anion radicals is possible. There is enough CO_3 in Great Salt Lake, for example, to form significant algal reefs in association with SO_4 salts, and the largest part of the Ca^+ is used in forming $CaCO_3$ and $CaMg(CO_3)_2$ and not $CaSO_4$. In lakes it is possible, therefore, to have minerals, mineral assemblages and sequences that are unusual for other environments and that are characteristic of particular regions.

Evaporites.—During their history, many large lakes become saline and evaporites are deposited. Because the lake is restricted and complete evaporation is possible, soluble salts (partly a function of the geochemistry of the lake water) are formed that only rarely are deposited from marine water. In addition, the variable chemistry of lake waters results in unusual salts, some of which are unique to a particular lake. Table 4 lists some of the minerals that occur in the deposits of Searles Lake, Lake Gosiute and Lake Uinta. Deposition of saline minerals in lacustrine environments was discussed briefly by Borchert and Muir (1964).

Feth (1964, p. 20, 21) suggested that tabular or broadly lenticular deposits of gypsum, borates, celestite, trona and other evaporites are indicative of lakes. Only borates and trona are restricted to lakes, but the other evaporites, if found in a probable nonmarine setting, are strongly suggestive of lacustrine deposition.

Evaporite sequences of saline lakes are more complicated than those of sea water, 1 liter of which yields, in order, calcite and dolomite (0.1g), gypsum (1.7g), halite (29.7g) and bittern salts (6.9g; Clarke, 1924). At the present time, it is not possible to predict the sequences or amounts of salts resulting from the evaporation of complex saline lakes. McKee (1954, p. 51) suggested that the absence of limestone immediately below bedded gypsum may reflect local source areas bounding a lake. Marine

TABLE 4.—LACUSTRINE AUTHIGENIC MINERALS
(Data from: Smith and Haines, 1964; Milton and Eugster, 1959; Hunt and others, 1954)

Searles Lake (late Quaternary)	Green River Formation (Eocene)	
Carbonates	*Carbonates*	*Zeolites*
Aragonite $CaCO_3$	"Alstonite-bromlite" (?)	Analcime $NaAlSi_2O_6 \cdot H_2O$
Burkeite $2Na_2SO_4 \cdot Na_2CO_3$	$CaBa(CO_3)_2$	Clinoptilolite $NaAlSi_{4.2-5}$
Calcite $CaCO_3$	Barytocalcite $CaBa(CO_3)_2$	$O_{10.4-12} \cdot 3.5-4H_2O$
Dolomite $CaMg(CO_3)_2$	Brayleyite $Na_3PO_4 \cdot MgCO_3$	Mordenite $NaAlSi_{4.5-5}O_{11-12}$
Gaylussite $Na_2Ca(CO_3)_2 \cdot 5H_2O$	(carbonate-phosphate)	$\cdot 3.2-3.5H_2O$
Nahcolite $NaHCO_3$	Burbankite $Na_2(Ca,Sr,Ba,Ce)_4-$	
Pirssonite $Na_2Ca(CO_3)_2 \cdot 2H_2O$	$(CO_3)_5$	*Sulfides*
Trona $Na_2CO_3 \cdot NaHCO_3 \cdot 2H_2O$	Calcite $CaCO_3$	Marcasite FeS_2
Tychite $2Na_2CO_3 \cdot 2MgCO_3 \cdot Na_2SO_4$	Dawsonite $Na_3Al(CO_3)_3 \cdot 2Al(OH)_3$	Pyrite FeS_2
	Dolomite $CaMg(CO_3)_2$	Pyrrhotite $Fe_{1-x}S$
Silicates	Eitelite $Na_2Mg(CO_3)_2$	Wurtzite ZnS
Adularia $KAlSi_3O_8$	Gaylussite $Na_2Ca(CO_3)_2 \cdot 5H_2O$	
Searlesite $NaBSi_2O_6 \cdot H_2O$	Magnesite $MgCO_3$	*Sulfates*
	Nahcolite $NaHCO_3$	Anhydrite $CaSO_4$
Zeolites	Pirssonite $Na_2Ca(CO_3)_2 \cdot 2H_2O$	Barite $BaSO_4$
Analcime $NaAlSi_2O_6 \cdot H_2O$	Shortite $Na_2Ca_2(CO_3)_3$	(?) Bassanite $CaSO_4 \cdot 1/2 H_2O$
Phillipsite $KCa(Al_3Si_5O_{16}) \cdot 6H_2O$	Siderite $FeCO_3$	Gypsum $CaSO_4 \cdot 2H_2O$
	Thermonatrite $Na_2CO_3 \cdot H_2O$	
Sulfates	Trona $Na_2CO_3 \cdot 2H_2O$	*Chlorides*
Aphthitalite $K_3Na(SO_4)_2$	Witherite $BaCO_3$	Halite $NaCl$
Mirabilite $Na_2SO_4 \cdot 10H_2O$		Northupite $Na_2CO_3 \cdot MgCO_3 \cdot NaCl$
Thenardite Na_2SO_4	*Silicates*	
	Acmite $NaFe^{3+}Si_2O_6$	*Phosphates*
Chlorides, Fluorides	Albite $NaAlSi_3O_8$	Collophane $Ca_{10}(PO_4)_6CO_3 \cdot H_2O$
Galeite $Na_2SO_4 \cdot Na(F,Cl)$	Clay Minerals	Fluorapatite $Ca_{10}(PO_4)_6F_2$
Halite $NaCl$	Elpidite $Na_2ZrSi_6O_{15} \cdot 3H_2O$	
Hanksite $9Na_2SO_4 \cdot 2Na_2CO_3 \cdot KCl$	Garrelsite $(Ba,Ca,Mg)B_3SiO_6$	*Hydrocarbons*
Northupite $Na_2CO_3 \cdot MgCO_3 \cdot NaCl$	(OH)	albertite
Schairerite $Na_2SO_4 \cdot Na(F,Cl)$	Labuntsovite (K,Ba,Na,Ca,Mn)	coal
Sulfohalite $2Na_2SO_4 \cdot NaCl \cdot NaF$	$(Ti,Nb)(Si,Al)_2-(O,OH)_7H_2O$	gilsonite
Teepleite $Na_2B_2O_4 \cdot 2NaCl \cdot 4H_2O$	Leucosphenite $CaBaNa_3BTi_3Si_9-$	ingramite
	O_{29}	ozokerite
Borates	Loughlinite $(Na_2,Mg)_2Si_3O_6(OH)_4$	tabbyite
Borax $Na_2B_4O_7 \cdot 10H_2O$	Quartz SiO_2	uintahite
Tincalconite $Na_2B_4O_7 \cdot 5H_2O$	Reedmergnerite $NaBSi_3O_8$	utahite
	Riebeckite-magnesioriebeckite	wurtzilite
	$(Na_2(Mg,Fe^{2+})_3(Fe^{3+},Al)_2Si_8O_{22}-$	
	$(OH)_2$	
	Searlesite $NaBSi_2O_6 \cdot H_2O$	
	Sepiolite $Mg_2Si_3O_6(OH)_4$	

deposition is not excluded, however. Despite the lack of models of lacustrine evaporite deposits similar to those for marine deposition, known saline lakes can be used as examples of sequences to be expected. Table 5 compares the evaporite sequence of several lacustrine deposits with the normal marine sequence.

Authigenic minerals.—In addition to evaporites precipitated from solutions, saline lake deposits contain a number of other minerals, the presence of which is useful in interpreting environment. These include salts, feldspars, zeolites, other silicates and iron minerals.

More than one-half of known carbonate min-

TABLE 5.—GENERALIZED EVAPORITE SEQUENCES
(Data from: Krauskopf, 1967; Smith and Haines, 1964; Fahey, 1962; Van Houten, 1965)

	Marine	Searles Lake (California)	Gosiute Lake (Wyo., Colo.)	Triassic Lake (N.J.)
Order of Precipitation ↑	variable salts	hanksite, aphthitalite		
	halite, $MgSO_4$, $MgCl_2$	mirabilite, thenardite, burkeite, halite	halite	
	gypsum, halite	trona, nahcolite	trona	glauberite (?), gypsum
	gypsum	gaylussite, pirssonite	shortite	analcime
	calcite-dolomite	calcite-aragonite	calcite-dolomite	calcite-dolomite

erals are present in lacustrine beds of the Green River Formation (Milton, 1961, p. 564). Most of them are alkali (sodium), alkali-earth (calcium, magnesium, barium), or alkali-alkali earth carbonate, but a few contain rare earths, aluminum, chlorine or phosphate (Milton and Fahey, 1960, p. 242).

Authigenic feldspars, zeolites and other silicates found in the Green River Formation constitute a remarkable assemblage of minerals (table 4). Although the Green River is not intruded by igneous rocks or regionally metamorphosed, it contains authigenic silicates that are found elsewhere only in alkalic pegmatites or other magmatic environments. These minerals probably formed in the Green River at moderate pressures and low temperatures ($<200°C$; Milton and others, 1960). Several of the silicate species (reedmergnerite, garrelsite and loughlinite) are unique to the Green River; other silicate species of the Green River originated in conditions unlike those of their formation elsewhere in the world (Milton and others, 1960, p. 171–172).

Authigenic zeolites and feldspars commonly are found in deposits of saline lakes. They are not valid criteria, however, for environmental interpretation, because these minerals also form in fresh-water, marine and soil environments. High silica zeolites (clinoptilolite, mordenite) are rare in marine deposits. The suggestion of Crowley (1939) that secondary growths of potassium feldspar originate only in marine environments is not valid; Milton and others (1960, p. 179–180) found authigenic microcline in "tuff-beds" of the Green River. Hay (1966) recently reviewed current knowledge of sedimentary zeolites and feldspars.

Iron minerals.—Sedimentary iron ore, particularly chert-banded deposits, generally is considered to be lacustrine in origin (Sakamoto, 1950; Backlund, 1952; Hough, 1958; Govett, 1966). This interpretation is based on rather precise knowledge of iron-carbonate-silica geochemistry (Garrels, 1960, p. 204–206). Modern marine environments are not capable of concentrating iron to the degree observed in the ore. In contrast, modern lakes do concentrate iron and provide a mechanism, the annual overturn of water, that could produce iron-silica banding. Govett (1966, p. 1201), who reviewed the chemical factors of both marine and lacustrine environments, concluded that iron is precipitated in lakes during the autumn and silica is precipitated during the summer. Borchert (1964, p. 179–180) disagreed and suggested that sedimentary iron ore is marine. He cited the similarity of young and old deposits, despite the drastic changes in continental environments that occurred when land plants appeared in the Devonian. The origin of sedimentary iron deposits remains controversial.

Iron minerals also are found as accessory minerals in many lacustrine deposits. Disseminated pyrite is common in the Green River and Lockatong Formations and goethite in the Popo Agie Formation. Although these minerals do not, by themselves, indicate the depositional environment, they may, in association with other authigenic minerals, indicate pH-Eh conditions that are suggestive of nonmarine waters. Together with regional relations, these mineral associations can constitute strong evidence of lacustrine deposition. Krumbein and Sloss (1951, p. 183) noted that iron oxide (mainly "limonite") forms in some fresh-water lakes, and Bradley (1931, p. 31) suggested that aggregates of microgranular or microscopic pyrite are a distinctive feature of lacustrine deposits in the Green River Formation.

Hydrocarbons.—The general dogma that oil and gas in substantial quantities originate only in marine environments has been slow to expire. Recent work indicates, however, that continental rocks contain adequate petroleum source beds. Furthermore, some evidence indicates that nonmarine petroleum differs from marine petroleum. Hunt, Stewart and Dickey (1954) demonstrated that there is a correlation between certain solid hydrocarbons of the Uinta Basin and extracts from Tertiary beds believed to be their source. As environment of deposition and salinity of Lake Uinta changed, the hyrocarbons, ozocerite, albertite, gilsonite and wurtzilite formed successively (Hunt, Stewart and Dickey, 1954, p. 1690–1692).

Swain, Blumentals and Prokopovich (1958, p. 184) concluded that the high content of saturated hydrocarbons and low asphaltenes and tars in the Precambrian Thompson Slate and Rove Formation graywacke is suggestive of deposition in oligotrophic lakes. In contrast, bitumens in the argillite of the Rove Formation and the Cuyuna ("Biwabik") Formation resemble those of modern Gulf of Mexico sediment, suggesting a marine environment.

Picard (1960) compared physical properties of oils from nonmarine fields in Utah and Neveda. The overall similarities were interpreted to reflect similar lacustrine origins. Similarly, Bass (1964, p. 204) compared the curves of correlation indices of crude oils of different reservoirs and concluded that the composition of oil in the lacustrine Green River Formation was controlled primarily by the depositional environments of the reservoir and source rocks.

A partial list of papers that discuss the nonmarine origin of oil and gas follows: Degens,

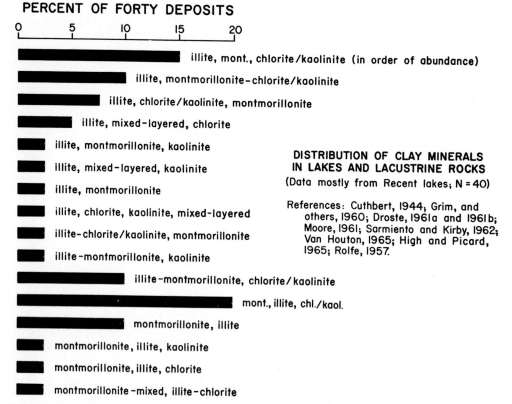

FIG. 10.—Associations of clay minerals in Recent lakes and lacustrine rocks. Apparently, there is not a single association of clay minerals that would be considered typical of lakes. Rather, lacustrine deposits are characterized by diverse clay minerals that reflect source materials and climate.

Chilingar and Pierce, 1963; Felts, 1954; Judson and Murray, 1956; Nightingale, 1930 and 1935; Picard, 1956, 1959 and 1962; Smith, 1954; Swain, 1956b; and Swain and Prokopovich, 1954.

Other geochemical evidence.—Trace elements have been used to differentiate depositional environments. Degens, Williams and Keith (1957, p. 2441–2452) found that B, Ga and Rb are useful in differentiating fresh-water and marine shale. Similarly, Keith and Degens (1959, p. 42–44) suggested that B, Li, F and Sr are concentrated in marine shale and Ga and Cr in fresh-water shale. In partial contrast, Potter, Shimp and Witters (1963, p. 682–683) suggested that B, Cr, Cu, Ga, Ni and V are each significantly more abundant in marine than in fresh-water argillaceous sediments. More recently, Lonka (1967) found that B, Ga, Rb, Ba and Be are salinity sensitive. The use of boron in predicting paleosalinity trends also was suggested by Walker (1968), but anomalous results were reported recently by Cody (1970). Despite the apparent success of some of these studies, trace element distributions in lacustrine rocks are not established. The problems of differentiating saline lakes from marine settings and fresh-water lakes from streams are unresolved.

The use of clay mineralogy in environmental interpretation is uncertain. Degens, Williams and Keith (1957, p. 2452–2453) suggested that authigenic illite is indicative of marine environments and that degraded illite and kaolinite are detrital minerals. Keller (1956, p. 2730) also considered illite to be characteristic of marine deposits. It was recently suggested, however, that authigenic illite originated in the Green River Formation of Wyoming (Tank, 1969). Grim (1958) outlined expected changes in clay minerals that are dependent on the water chemistry. In contrast, Weaver (1958) suggested that clay minerals have scant value as environmental indicators because they mainly reflect source materials rather than environment. The strong influence of source areas on clay mineral suites is shown by the distribution of clay minerals in the Gulf of Mexico (Griffin, 1962).

The distribution of clay minerals in some Recent lakes is summarized in figure 10. These distributions suggest that the clay minerals are related to the nature of the source areas. In the absence of more extensive studies, we conclude that the clay mineralogy of a rock unit is not indicative of whether or not the unit is lacustrine.

Studies of stable isotopic ratios may furnish useful criteria for recognizing lacustrine rocks. However, at the present time, the results are suggestive but not compelling. Fractionation of oxygen isotopes is predominantly temperature dependent, and only poorly and indirectly reflects salinity (Epstein, 1959, p. 220–221). In contrast, carbon isotopes may be salinity sensitive and marine water is relatively enriched in C^{13} (Keith and Degens, 1959, p. 54–55). Silverman and Epstein (1958, p. 1000–1001) suggested that C^{13}/C^{12} ratios of petroleum and related organic deposits are related to the environmental conditions in which the source materials accumulated. They found that petroleum and organic extracts derived from marine samples have δ-values that range from -22.2 to -29.4 permil, but that oils and bituminous materials derived from the Green River Formation range from -29.9 to -32.5 permil. Similarly, plants growing in marine environments have higher C^{13}/C^{12} ratios than do terrestrial plants (Wickman 1952; Craig, 1953). Silverman and Epstein (1958, p. 1003) concluded, however, that observed differences in C^{13}/C^{12} ratios between marine and nonmarine oils can be inherited from initial differences in source materials, but they may be more directly controlled by the degree of water circulation than the salinity of the water. Since lake water can range from fresh to hypersaline and circulation varies greatly, it may not be possible to discriminate between lacustrine and marine environments through the use of C^{13}/C^{12} ratios.

Studies of sulfur isotopes are more promising. Thode and others (1958) found that hydrocarbons of "nonmarine" origin, or hydrocarbons formed in the bottom sediments of lakes or seas cut off from oceans, are considerably enriched in S^{34} compared to hydrocarbons of "marine" origin. They attributed the enrichment in S^{34} to bacterial action. Further, lacustrine hydrocarbons apparently are characterized by large variations in sulfur isotope ratios (S^{34}/S^{32}), in contrast to nearly constant ratios in marine hydrocarbons. These relations suggest that the sulfur isotope ratio of petroleum is related closely to the sulfur environment in which the hydrocarbons formed (Harrison and Thode, 1958, p. 2648). Although it has not been demonstrated that the sulfur environment always coincides with the depositional environment, Jensen (1963, p. 284) suggested that hydrocarbons, pyrite and SO_4 incorporated in rocks of the Green River Formations essentially originated syngenetically with the enclosing rock. Jensen concluded that marine hydrocarbons show a narrow range in S^{34}/S^{32} ratios because oceans have varied little in sulfur isotopic composition during Phanerozoic time. Lacustrine hydrocarbons in the Uinta Basin, on the other hand, became enriched in S^{34} by loss of S^{32} from the lake in the form of H_2S that either reacted with iron to form syngenetic sulfides or escaped to the atmosphere.

PETROGRAPHY

Composition.—Two elements of composition, geochemistry and authigenic mineralogy, have been discussed. Two additional elements, detrital mineralogy and gross lithology, are considered here.

Detrital mineralogy requires little discussion. Allogenic grains are inherited from source areas and, although modified during transportation and burial, are more indicative of provenance than of depositional environment. Although it might be expected that rapid lateral variations are indicative of the local and restricted source areas of streams entering lakes, similar changes also occur in marine environments.

In general, lithologic criteria are insufficient to differentiate between lacustrine and other environments. Because many physical properties of marine and lacustrine environments resemble each other, the resulting clastic rocks also resemble each other. However, within a single area fluvial and lacustrine rocks differ (Picard, 1957a). Although the distinctions summarized in table 6 may not be valid for rock units other than those studied, the trends illustrated may be characteristic of fluvial-lacustrine gradients.

The problem of recognition and interpretation of lacustrine limestone has been given special attention by several workers; their views are summarized by Feth (1964, p. 28–30). Characteristics that have been suggested to be indicative of nonmarine limestone include dense nodular limestone that is brecciated because of shrinkage or syneresis (Pettijohn, 1957, p. 411; Krumbein and Sloss, 1951; Feth, 1964, p. 29–30), and powdery marl (Krumbein and Sloss, 1951; Feth, 1964, p. 29). Krumbein and Sloss also suggested that fresh-water limestone is lacustrine in origin. Picard (1957a, p. 376–377) found that limestone in the Green River and Uinta Formations is dominantly of lacustrine origin, and that only thin limestone beds are present in fluvial sequences that are associated with lacustrine rocks. Brief descriptions of several ancient lacustrine limestone deposits were summarized by Sanders and Friedman (1967, p. 203–213) but petrographic criteria were not given.

In the Great Salt Lake only unicellular algae

TABLE 6.—CHARACTERISTICS OF FLUVIAL AND
LACUSTRINE ROCKS IN TERTIARY BEDS
OF UINTA BASIN, UTAH
(After Picard, 1957a)

Lacustrine	Fluvial
Sandstone	
Quartz arenite, subarkose, sublitharenite	Quartz arenite, subarkose, sublitharenite
Subrounded	Subangular
Very fine to fine	Very fine to medium
Good to excellent sorting	Poor to good sorting
Calcite, dolomite, silica cement	Calcite cement
Accessory oolites	Accessory rock fragments
Light colors	Darker colors, red
Sheet	Lenticular
Mudstone	
Brown, gray, green	Red or green
Indurated	Friable
Subwaxy-resinous	Earthy
Iron sulphides, chert, silica, saline minerals	Rock fragments
Calcite, dolomite, silica cement	Calcite cement
Varved, shaley	Poorly bedded
Carbonate	
Extensive and varied limestone and dolomite	Thin, spotty limestone; no dolomite

are significant biostrome builders; the blue-green colonial alga *Aphanothece packardii* is dominant (Carozzi, 1962, p. 246). Typically, these biostromes show a distinct morphological zonation, from the lake to the land, of subparallel festooned ridges, tongue-like festooned ridges, composite rings and flat-topped mounds, and small isolated mounds. The morphological zones reproduce and frequently exaggerate the underlying topography which is eroded in firm argillaceous and oölitic sand (Carozzi, 1962, p. 247–248).

The carbonate petrology of Recent deposits in Lake Constance (Germany) was described recently by Schöttle and Müller (1968). Detrital and biogenic calcite, detrital dolomite and biogenic aragonite are all deposited in Lake Constance. Calcite is precipitated by abundant carbonate-producing plants (Schizophyceae and blue-green algae).

Calcitic onkolites and aragonitic shells of Lake Constance have unusually high Sr/Ca ratios, most of which are in the range of marine environments (Müller, 1968, p. 116). The chemistry of lake waters can vary greatly, however, and Lake Constance water is characterized by high Sr/Ca ratios (Müller, 1968, p. 126).

Studies of both Recent and ancient lacustrine carbonate, although few in number, indicate that lacustrine carbonate cannot be distinguished from marine carbonate on the basis of petrographic relationships. Unless the carbonate contains organisms that are characteristic of the environment, other criteria must be sought.

Texture.—Grain size and sorting are widely used indicators of relative "energy" or agitation of the depositional site. Although modified by source characteristics and rate of deposition, texture largely reflects depositional processes. It may be possible, therefore, to distinguish beach, dune and river sand (Moiola and Weiser, 1968). However, lakes and shallow seas are closely similar in depositional processes, and lacustrine textures may not be diagnostic (Solohub and Klovan, 1970).

BEDDING, SEDIMENTARY STRUCTURES AND
PALEOCURRENT PATTERNS

General.—Because bedding, sedimentary structures and paleocurrent patterns are almost exclusively the result of depositional processes, they are powerful means for interpreting environment. Unfortunately, it is difficult to differentiate between lacustrine and shallow marine rocks on the basis of these attributes. The processes in both environments are similar and the products are similar. In contrast, the differences between fluvial processes and those of lacustrine (or marine) environments are sufficiently great so that the products can be distinguished (Picard and High, 1970).

In evaluating bedding and sedimentary structures, several approaches are possible. Firstly, features that are unique to a particular environment are sought. However, the existence of such "keys" is doubtful in most deposits. Secondly, preferred associations can be noted and possibly correlated with particular environments. This latter approach has been used by most workers, and table 7 summarizes lacustrine associations of other workers. Finally, if the actual abundance of features can be determined, relatively minor changes may be indicative of different environments (Picard, 1967a, p. 49).

Bedding.—Lacustrine rocks are considered to be both finely laminated and massive (non-bedded). Because of their limited extent, lakes have small fetches and shallow wave-bases. Wide areas in the central part of a lake may be below wave-base. Sediment deposited below wave-base can be undisturbed and thin and parallel lamination can be preserved. The presence of fine lamination, formerly considered to be diagnostic of lakes (Darton, 1901, p. 558), is now used with caution. Wilson (1958, p. 1749) noted that such "quiet" water conditions can also exist on floodplains and in lagoons and seas. In contrast to Darton's views on lamination, Twenhofel (1932,

TABLE 7.—BEDDING TYPES AND SEDIMENTARY STRUCTURES BELIEVED TO BE CHARACTERISTIC OF LACUSTRINE ENVIRONMENTS
(X, characteristic; r, rare)

Bedding Types, Sedimentary Structures	(Klein 1962a)			Van Houten (1964)	Visher (1965)	Greiner (1962)	Bradley (1931)
	Keuper Marl (below wave base)	Keuper Marl (above wave base)	general lacustrine	Lockatong Fm.	general lacustrine	Albert Fm.	Green River Fm.
thin bedding, rhythmic bedd., varves	X	X	X	X	X	X	X
even, horizontal bedding	X	X	X	X	X	X	X
cross-stratification		X	X	X	X	r	r
trough		X	X	X			
planar		X					
ripple-stratification	X	X	X	X			r
disturbed stratification	X		X	X	X	X	X
graded bedding	X		X	X	X		
asymmetrical ripple mark			X		r	r	r
symmetrical ripple mark	X	X	X		r	X	r
shrinkage cracks		X		X		X	X
parting lineation		X					
rib-and-furrow		X					
groove cast		X					
load cast	X					X	
pull-apart structure	X			X			X
raindrop impression		X					
burrowed structure, worm trail	X	X	X	X			

p. 824) suggested that deposition in the center of large lakes is continuous and bedding may not develop, or it can be destroyed by organisms (Twenhofel, Carter and McKelvey, 1942, p. 536–537; Twenhofel and McKelvey, 1941 p. 847).

The Popo Agie and Green River formations illustrate the wide variety of bedding and sedimentary structure associations that can develop in lakes. Bedding in the lacustrine deposits of the Popo Agie is poorly developed and many units are massive (table 8). Burrowing and disturbed bedding are common, suggesting that original bedding was destroyed. Horizontal lamination and cross-stratification are rare. In contrast to the Popo Agie, the Green River contains an abundant suite of bedding types and sedimentary structures (figs. 11–18). Fine lamination and disturbed bedding are dominant, but wavy, cross, ripple and algal stratification are common. The variety of bedding types present in the Green River (table 9) compares closely with that found in shallow marine deposits, and it would be difficult to distinguish the Green River from a shallow marine deposit on the basis of bedding.

Graded deposits, which are rare in the Green River Formation, have been studied in some Recent lakes. Turbidites in Lake Mead (Gould, 1951) consist of fine material discharged from river mouths into shallow basins. In contrast, turbidite deposits in Crater Lake (max. depth, 1932 ft.) resemble those found in continental borderland basins of the ocean (Nelson, 1967, p. 847).

Varving, which is characteristic of Pleistocene glacial lakes, is a suggestive criterion of lacustrine environments. Definite varves and possible varves have been found in ancient lacustrine deposits (Bradley, 1929a; Anderson and Kirkland, 1960; Rayner, 1963, p. 132; Van Houten, 1964, p. 509; McLeroy and Anderson, 1966, p. 605; Dineley and Williams, 1968, p. 251). It is necessary in considering varves to distinguish between glacial varves, which are composed of clastic sediment or rock, and nonglacial varves of the lacustrine type, which are characterized by carbonate and organic material that is of biologic and chemical origin. There could be nonglacial clastic varves, but it would be difficult to prove that they were annual deposits.

The general utility of varves in environmental interpretation is limited by the following factors: few lacustrine deposits are varved, varving can be partially destroyed by compaction and diagenetic changes, and some marine deposits contain varve-like stratification. Further, "varve-like" alternations of light and dark layers can reflect other local conditions of deposition (Smith, 1959, p. 452; Hansen, 1940).

TABLE 8.—BEDDING TYPES AND SEDIMENTARY STRUCTURES IN SHALLOW MARINE AND LACUSTRINE UNITS. DATA FROM PICARD, 1967a; HIGH AND PICARD, 1965; THIS PAPER.

(X, characteristic; r, rare)

	Marine	Lacustrine	
	Red Peak Fm. (Triassic), Wyo.	Green River Fm. (Eocene), Utah	Popo Agie Fm. (Triassic), Wyo.
Bedding:			
varve		X	
horizontal	X	X	r
wavy	X	X	
graded	r	r	
ripple	X	r	
cross	X	r	r
convolute	r	r	
disturbed	X	X	X
massive	r		r
intra. fm. cgl.	X	r	X
Structures:			
asymmetric ripple mark	X	r	
symmetric ripple mark	r	r	
interference ripple mark	r	r	
flat-topped ripple mark	r		
mega ripple	r		
parting lineation	X	r	
core-and-shell	r		
shrinkage polygon	X	r	
linear-shrinkage crack	r	X	
burrow, trail	X	r	X
psuedo rib-and-furrow	X		
rill mark	r		
rain drop impression	r	r	
salt cast	r	X	
channel	r	r	
sedimentary dike	r	X	r
patterned cone	r		
groove cast	r		
load cast	r		
flute cast	r		

Sedimentary structures.—With the possible exception of fine lamination, no lacustrine criterion has been so widely misapplied as ripple marks. The familiar, and largely erroneous, classification of ripple marks into symmetrical (wave-formed) and asymmetrical (current-formed) types has led to serious misinterpretation of this structure. The symmetry of a ripple mark is determined by its external form which is not necessarily related to mode of formation. All wave-formed ripple marks are not symmetrical and all current-formed ripple marks are not asymmetrical. The mode of formation, which is determined by the presence or absence of foreset laminae, is not always reflected in the external shape of the ripple mark. This conclusion is supported by experiments (McKee, 1965, p. 72–74), observations of modern ripple marks (Evans, 1941), and studies of ancient marine (Picard and High, 1968a, p. 412–413) and lacustrine deposits (this paper). Therefore, the conclusion of several workers that the absence of currents in lakes will result in dominantly symmetrical ripple marks is without foundation.

Morphometric studies of ripple marks in which "wave" and "current" ripples are differentiated on the basis of symmetry (Tanner, 1967; Harms, 1969) do not explain the occurrence of nearly symmetrical ripple marks that have definite foreset laminae. Although the symmetry can be indicative of formation by waves, the foresets indicate that transport by currents has taken place. In our studies of shallow marine (Picard and High, 1968a) and lacustrine units, we found that asymmetric ripple marks with definite foreset laminae are dominant. Figures 19–21 show relevant measurements for shallow marine and lacustrine ripple marks. All the ripple marks, even those with perfect symmetries (RSI = 1.0), display foreset laminae. Nearly all of these ripple marks were formed by waves in shallow water. Ideally, waves are non-translatory. However, in shallow water, wave motion is distorted and there is a net forward movement. Thus, in coastal areas, circulation is established by onshore (wave-drift) and offshore (rip) currents. The resulting ripple marks indicate wave formation by their symmetry but they also show sediment transport by the inclined foresets. We have not seen any symmetric ripple marks that have a chevron-like internal structure (symmetrical foreset and topset laminae).

In comparing shallow-water marine and lacustrine ripple marks, significant distinctions are not evident (figs. 19–21). The persistent view that lakes are characterized by ripple marks that are more symmetrical than those of other environments is not valid. Fluvial ripple marks are generally more asymmetrical than those of shallow marine and lacustrine settings. However, exceptions to this generalization are common, especially in backwaters along the margins of channels.

Bradley (1926, p. 125) suggested that fluvial cross-stratification should be larger than nearshore cross-stratification. This relationship is confirmed in the Green River Formation where lacustrine cross-stratification is only half as thick as equivalent fluvial cross-stratification (fig. 22). However, we see no reasons for shallow-water marine and lacustrine cross-stratification to differ in scale.

Earlier suggestions that marine shrinkage

TABLE 9.—CHARACTERISTIC FEATURES OF LACUSTRINE ENVIRONMENTS REPRESENTED IN PARACHUTE CREEK MEMBER, GREEN RIVER FORMATION (EOCENE), RAVEN RIDGE AREA, UTAH AND COLORADO
(Major features capitalized; secondary features in small letters; minor but useful elements in parentheses)

Nearshore, Shallow, Open Water
    Horizontal Stratification
    Wavy-S ratification
    Siltstone

    small-scale cross-stratification
    disturbed bedding
    oriented, linear-shrinkage cracks
    asymmetric ripple marks

    (ripple-stratification)

Nearshore Shoal
    Horizontal Stratification
    Small- and Medium-Scale Cross-Stratification
    Oolite
    Algal Mats

    asymmetric ripple marks
    polygonal shrinkage cracks (hairline)
    pisolite
    chert pebble conglomerate

    (ripple-stratification)
    (bone fragments)

Beach-Shoreface
    Horizontal Stratification
    Small- and Medium-Scale Cross-Stratification
    Sandstone

    bifurcating ripple marks
    oriented, linear-shrinkage cracks
    ripple-stratification
    burrowed structure
    chert pebble conglomerate

    (large-scale cross-stratification)
    (disturbed bedding)
    (small channels)

Lagoon
    Horizontal Stratification
    Claystone

    fine lamination
    clastic lenses
    oolitic sandstone
    algal mats
    carbonate pebble conglomerate

Offshore
    Horizontal Stratification
    Oil Shale
    Claystone

    incomplete shrinkage cracks
    thin oolite beds

    (algal beds)

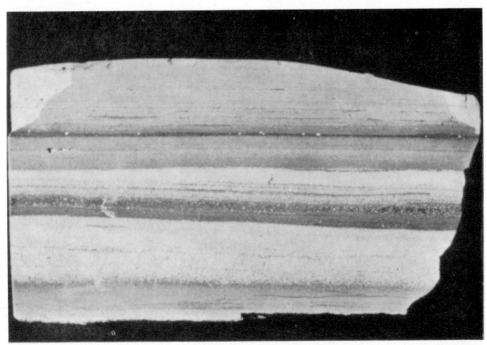

FIG. 11.—Fine horizontal lamination and evaporite bedding in dolomite from saline facies, lower Uinta Formation (Eocene Lake Uinta). Length of specimen is 14 cm.

Fig. 12.—Wavy, ripple and small-scale cross-stratification in Parachute Creek Member of Green River Formation.

Fig. 13.—Small-scale cross-stratification (micro cross-lamination) in Parachute Creek Member. Dominantly trough cross-stratification. Length of scale is 15 cm.

FIG. 14.—Syneresis cracks in low-grade oil shale of Parachute Creek Member.

cracks curl downward, in contrast to those of fresh-water environments (Kindle, 1917), were discounted by Bradley (1933). A similar suggestion of Kindle that marine shrinkage cracks are smaller than fresh-water cracks has not been substantiated.

Unusual forms of mud cracks are also not diagnostic of lacustrine environments. In the Green River Formation, incomplete linear cracks (fig. 23) are abundant in nearshore beds (Picard, 1966). Similar features have also been found in marine units (Picard, 1969) and in modern stream deposits (Picard and High, 1969), indicating once again that sedimentary structures are controlled by physical processes that may be active in several unrelated environments.

Paleocurrent patterns.—There are few studies of lacustrine paleocurrents. Paleocurrent directions of inferred lacustrine deposits in the upper Triassic of Connecticut were determined recently by Sanders (1968, p. 284), but the measurements are few and he was unable to reconstruct the current system of the lake. Picard (1967b) found an unequal, bimodal pattern in the Green River Formation that he interpreted to be the result of onshore and offshore directions (1967b, p. 387–392; 1967c, p. 2472–2474). In thin lacustrine deposits of the lower part of the Morrison Formation (Late Jurassic) of northern New Mexico, Tanner (1968, p. 191) found that the strikes of the crests of 69 percent of the ripple marks are approximately parallel with the interpreted shoreline and that 78 percent of the asymmetric ripple marks indicate onshore di-

FIG. 15.—Disturbed bedding in Douglas Creek Member of Green River Formation.

Fig. 16.—Algal stromatolites in Parachute Creek Member. Hammer for scale.

Fig. 17.—Parting lineation in Parachute Creek Member. Quarter for scale.

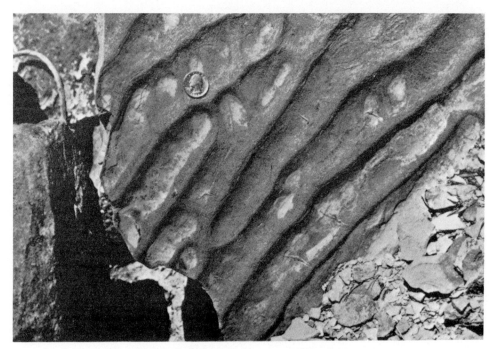

Fig. 18.—Linear asymmetric ripple marks in Douglas Creek Member. Interference ripple marks near quarter.

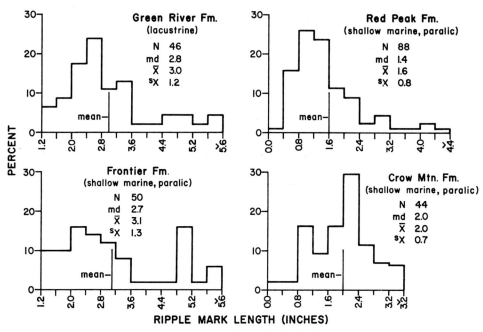

Fig. 19.—Comparison of lengths of linear asymmetric ripple marks in shallow marine and lacustrine deposits.

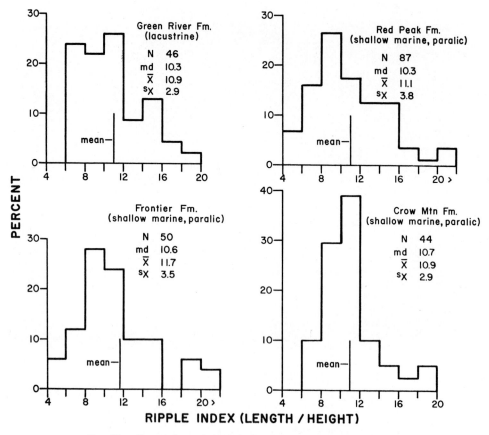

Fig. 20.—Comparison of ripple indices of linear asymmetric ripple marks in shallow marine and lacustrine deposits.

rections. Similarly, the orientation (strike of crests) of ripple marks in modern lakes is dominantly parallel with the shoreline, although oblique and long-shore orientations also are present (Wulf, 1963, p. 693–694; Davis, 1965, p. 863). However, similar patterns are found in shallow marine environments (fig. 24), and without additional studies distinctions cannot be made.

Our recent comparison of fluvial and lacustrine paleocurrents in the Wasatch-Green River transition zone indicates that these environments may not always be differentiated on the basis of paleocurrent patterns (Picard and High, 1970). Both fluvial and lacustrine rock units yielded unimodal patterns and the amount of scatter was the same. The unimodal fluvial pattern was expected. The unimodal lacustrine pattern may reflect only the dominant onshore mode and the secondary offshore and longshore modes may not have been defined by the small number of measurements available at each stratigraphic section.

MISCELLANEOUS CRITERIA

Geomorphology.—Numerous distinctive landforms are produced in lakes, including lacustrine plains, shoreline features and outlets. These have been used to identify similar features in Pleistocene lakes (Gilbert, 1891; Russell, 1885; Powers, 1939; Feth, 1955; Jones and Marsell, 1955). The use of these criteria is generally restricted, however, to the Pleistocene, because landforms are ephemeral.

Hydrology.—Although in many respects large lakes are similar to small seas, the differences in size result in differences in the hydrographic regimes. Lakes are virtually tideless, sensitive to air temperature, subject to freezing and can overturn periodically.

The presence or absence of tides is inferred only indirectly. As mentioned previously, the lack of tides is one factor responsible for restriction of the width of shore facies around lakes. Rapid fluvial-offshore facies changes are suggestive, therefore, of lacustrine environments.

Gebelein and Hoffman (1968) related the

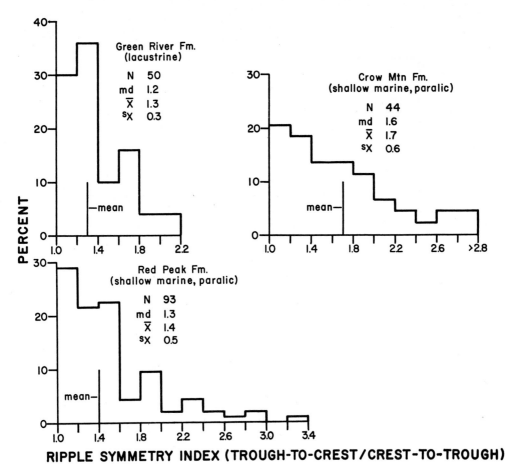

Fig. 21.—Comparison of ripple symmetry indices of linear asymmetric ripple marks in shallow marine and lacustrine deposits.

thickness of algal lamination to tidal flooding. At Cape Sable, Florida, the thickness of algal laminae is proportional to the duration of submergence by tidal water. The thickness varies, therefore, according to relative position on the tidal flat and a similar pattern in limestone may indicate tides. Another indicator of tides is the paleocurrent pattern of tidal exchange. This pattern is generally bimodal (upslope and downslope), but there are moderately complex modifications on extensive tidal flats (Klein, 1967, p. 370). If ancient tidal patterns are recognized, their existence precludes a lacustrine interpretation of the environment. However, Selley (1967, p. 220–222) doubted that all bimodal paleocurrent patterns are tidal, a point with which we concur (fig. 24).

Because lakes are limited bodies of water, they respond to seasonal changes in temperature. Annual fluctuations in water temperature in lakes should be less regular than in seas and the extremes should be greater. Such fluctuations possibly can be recognized through study of mineral equilibria (Smith and Haines, 1964, p. 45), fauna and oxygen isotopes. Some of the problems associated with oxygen isotope studies of lacustrine beds were discussed by Reeves (1968, p. 172–175).

Some lakes are subject to freezing because they are relatively small and shallow and are located in frigid climates. In contrast, large and deep lakes, because of their large heat budgets, may freeze only in the coldest years, if at all. Even in central New York, Lake Seneca does not freeze (Bradley, 1969, pers. commun.). At the present time, marine water freezes only in the harshest climates, which probably were not common during most of geologic history. The recognition of freezing episodes in sedimentary rocks is unlikely, although the occurrence of molds of ice crystals and ice-rafted materials, especially in thin cycles or varves, is indicative of frigid conditions.

A final hydrographic feature of lakes is the

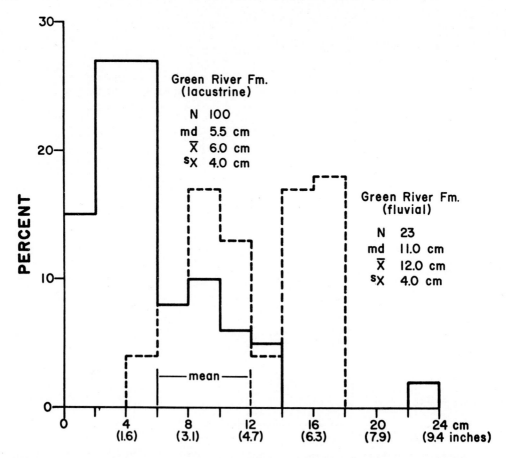

Fig. 22.—Comparison of thickness of cross-stratification from fluvial and lacustrine equivalents in Green River Formation. Fluvial cross-stratification is twice as thick as that of lacustrine cross-stratification.

annual or semiannual overturn that is characteristic of many lakes. Overturn also takes place in some marine basins but not so regularly. The role of overturn in relation to banded iron ores and glacial varves has been discussed. Nonglacial varves, in contrast, indicate a lack of overturn (i.e., long continued, oxygen-free hypolimnia). Detailed mineral, geochemical and faunal alternations might also suggest overturn.

Cyclic deposition.—Lacustrine deposits can display either base level or climatic cycles of deposition. Although such cycles also are found in marine units, the smaller size of lakes makes them more responsive to climatic or tectonic changes. Thus, local fluctuations that would cause only minor changes in marine regimes can lead to widespread and significant alternations in lacustrine deposition.

All closed lakes fluctuate greatly in water level regardless of depth. Because such lakes form regional base levels that rise or fall in response to climatic changes, their influence on surrounding areas is direct. The balance between erosion and deposition moves back and forth in response to changes in lake level. An example of such base level cycles is found in the Green River Formation. As the level of Lake Uinta rose and fell, regressive-transgressive cycles formed in shallow water areas at the same time that channel cutting and filling alternated on surrounding floodplains (Picard and High, 1968b).

Within lakes, extreme climatic fluctuations may cause the formation of cycles of authigenic minerals. Chemical cycles in the Lockatong Formation show decreasing amounts of pyrite and carbonate and increasing amounts of authigenic analcime during drying of the lake (Van Houten, 1964).

COMPARISON WITH OTHER ENVIRONMENTS

Throughout our discussion of lacustrine criteria, we have mentioned several of the ways in which lakes resemble or differ from fluvial and shallow marine settings. Direct comparisons are briefly discussed here in order to illustrate those properties that are expected to be similar (non-diagnostic of lakes) and those that are dissimilar (possibly diagnostic of lakes).

Physical features.—Lakes closely resemble shallow seas in physical processes of deposition. It is especially difficult to distinguish between ancient lakes and cut-off arms of the sea, and even in relatively young deposits it may not always be possible (Swain, 1966). General similarities in wave action and nearshore currents result in similarities in facies pattern, sequence, lithology, texture, bedding, sedimentary structures and paleocurrent patterns. The major difference between the two environments, which is the lack of tides in lakes, is poorly reflected in sedimentary rocks and is difficult to apply. Offshore, below wave-base, lake floors are quiet and undisturbed by physical processes, in contrast to seas where the bottom can be agitated below wave-base. However, where oxygenated, the bottom sediment of lakes can be reworked by organisms such as midge larvae (Bradley, 1969, pers. commun.).

In contrast to shallow seas, streams and lakes differ markedly from each other in physical processes. Currents in streams, though complex

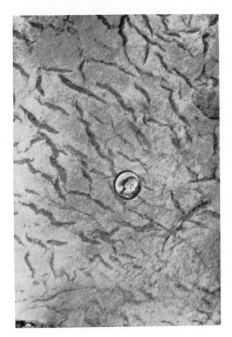

FIG. 23.—Oriented linear shrinkage cracks in Parachute Creek Member. Long axes of cracks approximate depositional strike. Quarter for scale.

and varied, are in general unidirectional and relatively strong, in comparison to nearshore lake activity. Consequently, differences are ex-

Green River Fm.
(lacustrine)
Picard (1967 b)
bimodal, unequal

Crow Mtn. Fm.
(shallow marine, paralic)
Tohill and Picard (1966)
bimodal, subequal

Red Peak Fm.
(shallow marine, paralic)
Picard and High (1968a)
trimodal, unequal

N

Salem Ls.
(shallow marine)
Pinsak (1957)
bimodal, unequal

Vermillion River
(fluvial)
Potter and Pettijohn (1963)
unimodal

PALEOCURRENT PATTERNS
OF LACUSTRINE, SHALLOW
MARINE, & FLUVIAL DEPOSITS

FIG. 24.—Comparison of paleocurrent patterns from fluvial, shallow marine and lacustrine rock units.

pected, and found, in texture, sedimentary structures and lithology. On the other hand, processes on floodplains are incompletely understood and resemble those in standing bodies of water.

Chemical features.—It is difficult to generalize about chemical features of lakes because they both resemble and differ from marine and stream water in their chemistry. As they have in the past, lakes range from fresh to hypersaline. Many large, long-lived lakes become saline at some time during their history. However, the duration of the lake and the amount of evaporation are not the critical factors, but rather the lack of overflow which causes salts to precipitate. Lake Victoria loses about 90 percent of its annual accession of water by evaporation, but the lake water contains only about 60 ppm dissolved solids (Bradley, 1969, pers. commun.).

The delicate balance in water chemistry achieved by buffers in marine water is absent in all but some saline lakes. As noted before, changes in lacustrine water chemistry and equilibria lead to distinctive mineral suites in lacustrine rocks. Many large lakes are similar to seas in pH and salinity, but in closed Na-rich lakes both pH and salinity can go beyond the range in sea water. Both large lakes and shallow seas differ from streams in pH and salinity. The comparatively small sizes and the local source areas of lakes cause the chemical evolution of each lake to be distinctive.

Biologic factors.—If fossils are abundant, marine and nonmarine deposits can generally be differentiated. However, specific identification of nonmarine subenvironments on the basis of organisms is uncertain.

Both lakes and streams are likely to be high stress environments. Populations of some streams are limited by the unreliability of water flow or by shifting channels. In some lakes rapid changes in water chemistry and size inhibit organisms; in other lakes the animal and plant populations rival or exceed those found in shallow seas. Shallow marine environments generally are characterized by large populations of organisms. Faunal restrictions, similar to those of some terrestrial environments, develop only locally in shallow seas.

Within a stream tract, conditions commonly are uniform and tolerant organisms flourish equally. However, large lakes are likely to have environmental gradients yielding abundant life in nearshore areas and only a sparse fauna offshore. As lakes become thermally and chemically stratified, the offshore bottom is less conducive to life, and foul stagnant bottoms are common in the centers of lakes.

Regional setting.—Both lakes and seas are associated with fluvial and deltaic environments. However, lacustrine units are enclosed by these deposits and seas are bordered by them. Both lakes and seas form base levels that control erosion and deposition in adjacent areas. In lakes such control is regional in extent, but in seas the control is generally subcontinental in scope.

ANCIENT LAKES

General

There are several outstanding examples of ancient lake deposits. A brief review of these rock units serves to illustrate the process of interpretation of a lacustrine environment. We have selected three examples for discussion because of our personal knowledge of the sequences and because they encompass a range of lacustrine characteristics. These lacustrine deposits are in the Late Triassic of New Jersey and Pennsylvania (Lockatong Formation) and Wyoming (Popo Agie Formation), and Eocene Lakes Uinta and Gosiute in Utah and Wyoming (Green River and lower Uinta Formations). Several features of these lacustrine deposits and of Great Salt Lake are compared in tables 10 and 11.

Triassic Lake, New Jersey and Pennsylvania

Geologic setting.—The Newark Group (Late Triassic) is a nonmarine sequence that fills basins in New Jersey and Pennsylvania. Three formations, the Stockton, Lockatong and Brunswick, are recognized together with several intrusive sills. The Stockton and Brunswick are dominantly fluvial redbeds; the Lockatong is composed mainly of dark gray, lacustrine argillite. The summary of the Lockatong that follows is based on recent studies by Van Houten (1964, 1965).

The Lockatong forms a lens 90 miles long, 30 miles wide and up to 3750 feet thick. Streams entered the basin from the east and southeast and only local fanglomerate was derived from the active border fault on the northwest. Lacustrine conditions persisted in the basin for about 5 m.y. The basin was sometimes open, and the lake drained into the sea. During these intervals, clastic deposits accumulated. At other times, the basin was restricted and interior drainage developed. Chemical deposits formed during periods of restriction. In addition to the alternations of through and interior drainage, climatic fluctuations led to the development of cyclic deposits in both clastic and chemical sequences.

General description.—Sedimentary cycles, about 15 ft. thick, are characteristic of the Lockatong. Detrital and chemical cycles are recognized. In general, each cycle consists of lower black shale, platy mudstone and marlstone, and upper massive mudstone. Throughout

TABLE 10.—CHARACTERISTIC MINERALS IN GREEN RIVER, LOCKATONG AND POPO AGIE FORMATIONS AND GREAT SALT LAKE

(t, trace; r, rare; c, common; a, abundant)

Mineral	Green River Fm.	Lockatong Fm.	Popo Agie Fm.	Great Salt Lake
organic material	a	r	r	r-c
chert (silicification)	c		r-c	r
clay minerals	r-c illite mixed-layered	c illite chlorite mixed-layered	a montmorillonite mixed-layered chlorite illite	c illite kaolinite montmorillonite
Na-feldspar	r-c	c	r	r
K-feldspar	c	t	t	c-a
analcime	c	a	c-a	
other Na-silicates	r-c (several)	t (amphibole)		
attapulgite, sepiolite	r	t		t(?)
dolomite, calcite	c-a	c	c-a	c
Na-salts	c (many kinds)			
halite, gypsum	r	?	t	r-c
fluorapatite	r	t		
pyrite	c	r-c	t	t-r
goethite			c	

Data from: Eardley, 1938; Eardley and Gvosdetsky, 1960; Milton and Eugster, 1959; Van Houten, 1965; High and Picard, 1965.

the Lockatong the following rock types, in approximate order of abundance, are present: 1) tough, massive, homogeneous, very fine-grained to aphanitic, medium- to dark-gray, detrital and colloidal mudstone (argillite); 2) platy, very dark gray to black, laminated, carbonate-rich mudstone and marlstone; 3) thin-bedded, calcareous, well-sorted, feldspathic siltstone, and minor, very fine-grained sandstone; 4) laminated, black silty, calcareous shale; and 5) dark grayish-red and greenish-gray, "tough," micaceous mudstone, siltstone, and minor fine-grained sandstone.

The chemical cycles are characterized by an upward increase in analcime and a corresponding decrease in carbonate. Detrital cycles commonly contain a lens of feldspathic siltstone in the upper massive sequence.

The sedimentary structures of the Lockatong are indicative of slow depositional rates. Cross-stratification and ripple marks are rare; fine lamination, burrows, shrinkage cracks and hydroplastically deformed beds are dominant.

Lacustrine criteria.—Numerous characteristics indicate that much of the Lockatong was deposited in a lacustrine environment. These include paleontology, bedding, sedimentary structures, mineralogy and regional setting. Other

TABLE 11.—ROCK AND SEDIMENT TYPES IN GREEN RIVER, LOCKATONG, AND POPO AGIE FORMATIONS AND GREAT SALT LAKE

Green River Fm.	Lockatong Fm.	Popo Agie Fm.	Great Salt Lake
brown shale (oil shale) black pyritic mudst. green mudst.	aphanitic mudstone a) calc. mudst. b) analcimic mudst. black pyritic mudst. red and gray micaceous mudst. black lam. silty shale	spherulitic clayey analcimolite analcimic silty claystone silty claystone	silty and sandy clay (black, tan, gray; carb.-rich)
carb.-rich sltst.	feld. siltstone	carb.-rich siltstone analcimic clayey sltst.	calcareous silt
carb.-rich v.f.-f. ss. v.f.-m. ss. cgl. sandstone	very fine sandstone	carb.-rich v.f. ss. analcimic v.f. sandstone	sand fanglomerate
oolite algal beds silty dolomite silty sdy. ls (sparite) pelleted carbonate micrite	dolomite	pseudo-oolite silty dolomite silty sdy. limestone limestone microcgl.	oolite algal bioherms faecal pellets

similar rock units, largely of lacustrine origin, compared with the Lockatong by Van Houten (1964, p. 508) are the Green River Formation; Dunkard Group (Pennsylvanian-Permian) in Pennsylvania, West Virginia and Ohio (Beerbower, 1961); Blomiden Formation (Triassic) of Nova Scotia and New Brunswick (Klein, 1962b); Albert Shale (Mississippian) of New Brunswick (Greiner, 1962); Keuper Marl (Bosworth, 1912, p. 51–116; Elliott, 1961; Klein 1962a); and the Caithness Flagstone Series (Devonian) of Scotland (Crampton, 1914; Rayner, 1963).

Marginal or shallow-water conditions during deposition of the Lockatong are indicated by rare footprints of phytosaurs and dinosaurs. Terrestrial conditions are indicated by the amphibians and fish. There are no marine fossils.

Deposition in quiet, standing water is suggested by the laterally persistent rock units, thin-stratification, varves, graded bedding, fine grain size and hydroplastic disruptions. Ripple-stratification and cross-stratification are both small-scale. Periodic exposure or partial evaporation is indicated by the presence of shrinkage cracks and salt casts. Thus, most of the Lockatong was deposited in quiet conditions in shallow stratified water (below wave base).

Thermal stratification of the water is suggested by the abundant pyrite and the lack of sessile bottom dwellers. Abundant, cyclically controlled authigenic carbonate and analcime suggest that deposition took place in a large body of water. Varve counts indicate that the cycles were climatically controlled and that a 21,000-year period is represented.

The Lockatong is a lens that is enclosed by fluvial units. Although the total size of the lake is unknown, it may have been restricted to the New Jersey-Pennsylvania graben or it may have been connected with other grabens along the East Coast. In either situation, the size was small in comparison with epicontinental seas.

Triassic Lake, Wyoming

Geologic setting.—The upper Chugwater Group (Middle? and Late Triassic) in Wyoming and adjacent Idaho, Utah and Colorado is a complex of fluvial-lacustrine rocks. The Jelm Formation (Middle ? Triassic) of Wyoming is disconformable on underlying marine units (Crow Mountain Formation) and is dominantly fluvial. Local ponds were present, however. In contrast, the overlying Popo Agie Formation is mainly lacustrine. The discussion of lacustrine rocks in the upper Chugwater that follows is drawn from our previous studies of these units (High and Picard, 1965, 1967, 1969a; High and others, 1969).

Lacustrine rocks in the Popo Agie extend over most of western Wyoming and parts of the adjacent states, an area of about 50,000 sq. mi. Shorelines have been identified only on the western and southern sides; the Popo Agie is truncated on the north and east by overlying units. The maximum size of the lake is unknown, but it is one of the largest lacustrine units that has been recognized.

General description.—The Popo Agie in Wyoming is subdivided into four informal rock units. In ascending order, these are the lower carbonate, purple, ocher, and upper carbonate. The lower carbonate unit is a carbonate pebble conglomerate that was deposited in a fluvial environment, but there were associated local ponds. The purple unit is a silty, montmorillonitic mudstone that was deposited on floodplains; local beds of analcimolite indicate that isolated lakes existed. The ocher unit is silty, analcimic, montmorillonitic mudstone and analcimolite, and represents a widespread and persistent lake. The general conditions continued during the deposition of the upper carbonate unit, which is dominantly argillaceous dolomite and dolomitic mudstone.

In Utah and Colorado, the Gartra Formation (Middle ? Triassic), which is fluvial, conformably underlies the Popo Agie. The lower carbonate unit of the Popo Agie is not present and and an additional fluvial unit, the sandstone and conglomerate unit, partially replaces the ocher unit in northwestern Colorado. Locally, the purple unit, which is also fluvial, thickens at the expense of the lacustrine ocher unit, indicating proximity to shorelines. Similar relationships are found in extreme eastern Idaho, and indicate the western extent of the lake.

Fossils, which are sparse in the Popo Agie, consist largely of lowland-dwelling reptiles and amphibians, coprolites, fresh-water pelecypods ("*Unio*") and "reed" stems. Burrows are common in the analcimolite, indicating an abundant fauna. Bedding and sedimentary structures are poorly preserved in the Popo Agie (table 6) and disturbed bedding is dominant.

Lacustrine criteria.—A lacustrine origin for the Popo Agie is interpreted primarily from paleontologic, mineralogic, geochemical and regional stratigraphic considerations.

The Popo Agie fauna is nonmarine. Although distinctions between fluvial and lacustrine environments are not definitive based on fossils, the presence in the Popo Agie of large predators such as phytosaurs and labyrinthodont amphibians is indicative of semipermanent water bodies. Therefore, a lacustrine environment is more probable for the Popo Agie than a fluvial environment.

The abundant analcime and carbonate in the

Popo Agie is indicative of a lacustrine origin, rather than a fluvial origin. Because streams are commonly acidic, and calcite is unstable at pH values less than about 7.8, fresh-water limestone is considered lacustrine. Similarly, authigenic analcime requires more basic conditions and higher concentrations of alkalis than are found in streams.

On the west and south, lacustrine beds of the Popo Agie grade into fluvial units. Marine rocks do not occur adjacent to the Popo Agie, which indicates that coastal lagoons are not a possible environment. On the north and east, the Popo Agie is truncated and shorelines have not been found. Former connections with marine seaways on the north and east are unlikely; the geosyncline was located on the west.

Eocene Lakes: Utah, Wyoming and Colorado

Geologic setting.—During the Eocene there were several large lakes in Wyoming and Utah. The largest and most persistent was Lake Uinta in Utah (table 11). In Wyoming, Lake Gosiute and a smaller lake were partially contemporaneous with Lake Uinta. The studies by Bradley (1925, 1926, 1929a,b, 1931, 1948, 1964) of Lakes Gosiute and Uinta have become classics. Lake Uinta is not as well known as Lake Gosiute; the recent papers on Lake Uinta are those of Cashion, 1967; Dane, 1954 and 1955; Picard 1955, 1957a,b,c, 1959, 1967b,c; and Picard and High 1968b. The discussion that follows is condensed from these sources.

General description.—The lacustrine rocks of Lake Uinta are found in the Green River and lower Uinta formations (Eocene). These deposits form a lens which is enclosed by the fluvial Wasatch, Uinta and Duchesne River formations.

Rock types in the Green River and lower Uinta are diverse. Lacustrine deposits that originated in the deeper central part of the lake are dominantly oil shale, dark brown and black calcareous and dolomitic shale, and sandy or argillaceous limestone and dolomite. The sandstone is calcareous and commonly tuffaceous. Shore phases of Lake Uinta are characterized by green and red silty claystone, calcareous, argillaceous siltstone and sandstone, oosparite, algal bio- and intramicrudite, algal biolithite, and intraformational and chert pebble conglomerate.

In the Parachute Creek Member of the Green River Formation, base-level cycles have been recognized in fluvial, nearshore and offshore facies. These cycles record the influence of fluctuations in the level of Lake Uinta on depositional processes within and adjacent to the lake. Cyclic deposits of shorter duration that are present in the saline facies are characterized by alternations of saline and nonsaline mudstone (fig. 11).

The bedding and sedimentary structures are divided into two suites. Varves, horizontal lamination, contorted and convolute bedding, syneresis cracks, salt casts and chert nodules and lenses are dominant in the central lake deposits. In contrast to the central lake deposits, current structures are characteristic of shore phases. Cross-stratification, ripple-stratification and ripple marks are common. Other common features are horizontal to wavy stratification, linear shrinkage cracks and algal bedding.

The mineralogy of the Green River Formation was discussed previously. Table 3 lists the authigenic minerals in the Green River, many of which are unique.

Fossils in the Green River and lower Uinta formations are sparse. Algae and ostracods are widespread; gastropods and spectacularly preserved fish are locally abundant. Bone fragments of turtles and birds are common in shore deposits. Fluvial units, which are correlative with parts of the Green River, contain a diverse mammalian fauna.

Lacustrine criteria.—Of the three examples briefly discussed, the Green River and lower Uinta formations are the easiest to identify as lacustrine. Regional relations, paleontology, sedimentary structures, mineralogy and geochemistry all furnish evidence of the existence of lacustrine conditions.

A large open body of water is indicated by facies patterns, paleocurrent patterns and sedimentary structures. "Quiet" water deposits are surrounded by a belt of rocks that are characterized by current structures. Within this belt, currents were bimodal, and moved up- and downslope. On the outside of this belt there are fluvial units that contain channels filled with coarse terrigenous clastic beds.

This large body of water is interpreted to be lacustrine, rather than marine, on the basis of paleontology, mineralogy and regional setting. The deposits of Lake Uinta are regional in extent, completely enclosed by fluvial units, contain fresh-water organisms, are far removed from marine deposits of the same age, and contain a mineral assemblage that is not characteristic of evaporation of marine water.

Comparisons

The three lacustrine units described here demonstrate the expected variability of lake deposits. They resemble each other only in approximate size and in their regional extent. In other parameters, such as stratigraphic sequence, sedimentary structures, lithology, mineralogy

TABLE 12.—SUMMARY AND EVALUATION OF LACUSTRINE CRITERIA

Characteristic	Remarks
Size	Local to regional (less than about 50,000 mi^2)
Shape	Overall "circular"; shape reflects position of bounding positive elements in structural basins
Facies	Narrow shore and nearshore deposits; fining of clastics toward center of basin
Lateral Continuity of Beds	Lacustrine beds generally more continuous than fluvial beds
Sequence	Regressive patterns dominated by offshore deposits
Regional Setting	Enclosed by fluvial units or disconformities
Biota	1. Fresh-water organisms; 2. "stress" communities: 3. diversity and quantity gradients
Authigenic Minerals	1. Saline minerals in terrestrial setting; 2. chert-banded, sedimentary, iron ores
Trace Elements	Further studies required; B, Li, F, and Sr may be higher in marine than in fresh-water samples; Ga may be higher in fresh-water samples
Isotopes	Further studies required; C^{13} may be enriched in marine carbonate samples and depleted in fresh-water samples; data on oxygen isotopes not consistent; lacustrine hydrocarbons marked by large variation in sulfur isotope ratios (S^{34}/S^{32}) in contrast to nearly constant ratios in marine hydrocarbons
Fresh-water Limestone	Lacustrine limestone probably cannot be distinguished from shallow water marine limestone on present petrographic criteria
Bedding Types	In general, lacustrine rocks not now distinguishable from shallow marine rocks on differences in bedding types; lacustrine deposits can be distinguished from fluvial deposits; varves suggestive of lakes
Sedimentary Structures	1. Ripple marks not more symmetrical in lacustrine than in shallow marine rocks; 2. scale of fluvial cross-stratification larger than nearshore lacustrine cross-stratification; 3. type and character of lacustrine shrinkage cracks not diagnostic
Paleocurrents	Bimodal opposed; similar to shallow marine paleocurrent patterns
Sedimentary Cycles	1. Base level; 2. climatic

and so forth, they show no consistent pattern. For each of them, however, a lacustrine interpretation can be made with considerable confidence. The sum of all properties, if considered together, strongly suggest a lacustrine origin.

SUMMARY

To summarize the large amount of diverse facts and interpretation presented here is difficult. It is helpful therefore to ask several questions. Is there a characteristic lacustrine sequence or facies pattern? Is there a lacustrine biota? Is there a lacustrine assemblage of minerals or sedimentary structures? Are texture, geochemistry, paleocurrent patterns or lithology ever definitive of lacustrine environments?

Our conclusion is that the answer to each of the foregoing questions is a qualified "no." There are no criteria which, by themselves, are sufficient to indicate ancient lacustrine deposits. Although some types of evidence, such as that furnished by mineralogic relations, may be better than others, such as lithology, all of the lines of evidence are imperfect. The interpretation of specific depositional environments is pleuristic and draws on both positive and negative evidence. Indeed, environmental reconstructions frequently are based on negative evidence. The information available for the interpretation of environment rather than affirming specific conclusions commonly only excludes some of the many possibilities. Several types of evidence are required before a reasonable interpretation is possible.

Despite these qualifications, environmental reconstructions can be made and with considerable confidence if sufficient facts are obtained. Lacustrine environments, perhaps, are more difficult to recognize than many other environments, but much of the difficulty is caused by the lack of specific information. When more information is available, environmental reconstruction will become a more precise, if not straightforward, task.

Table 12 summarizes significant criteria for the recognition of lacustrine deposits. In general, there are two steps in the identification of lacustrine deposits. In the first step, units deposited by flowing water (streams) are distinguished from those deposited in standing water (lakes and seas). The second step is to differentiate between lakes and epicontinental seas.

Deposition in flowing or standing water is best determined by the physical parameters of bedding, sedimentary structures, grain size, sorting and paleocurrent patterns. In addition, the presence or absence of carbonate is probably diagnostic of standing water. By establishing two separate classes of "flowing" and "standing" water, we do not mean to imply that currents do not exist in seas and lakes or that ponded water does not occur along stream courses. Rather, we are trying to establish generalizations that are useful for interpretation. Exceptions that present difficulties will be found. For example, floodplain deposition is possibly more similar to that of slack water bodies than it is to streams. Accordingly, floodplain deposits have been confused with lacustrine deposits in instances where only physical features were considered.

After the general mode of deposition has been established, lakes and seas can be differentiated. For this step, paleontologic, mineralogic, geochemical and regional stratigraphic information is the most useful. The presence of a fresh-water fauna that lacks marine forms, the close association with or even enclosure by fluvial units, and the occurrence of authigenic minerals that indicate local and restricted drainage areas are diagnostic of lakes.

ACKNOWLEDGMENTS

Acknowledgment is made to the donors of the Petroleum Research Fund, administered by the American Chemical Society, for major support of this research (Picard and High, grant 3217-A2). In addition, financial support was received from grants to Picard by the National Science Foundation (grant GA-12570) and the University of Utah Research Fund. W. H. Bradley, D. W. Boyd, J. H. Goodwin and R. A. Robison read preliminary drafts of the manuscript and offered suggestions for its improvement. Stephen Streeter, M. L. Jensen and Aurèle La Rocque read parts of the manuscript and suggested ways in which they could be improved. V. R. Picard, S. H. Curtis and V. P. Pabst typed several drafts of the manuscript. The figures were drafted by D. L. Olson.

REFERENCES

ANDERSON, R. Y., AND KIRKLAND, D. W., 1960, Origin, varves, and cycles of Jurassic Todilto Formation, New Mexico: Am. Assoc. Petroleum Geologists Bull., v. 44, p. 37–52.
BACKLUND, H. G., 1952, Some aspects of ore formation, Precambrian and later: Edinburgh Geol. Soc. Trans., v. 14, p. 302–308.
BAKER, F. C., 1928, The fresh water mollusca of Wisconsin: Wisc. Geol. and Nat. Hist. Survey Bull. 70, 976 p.
BASS, N. W., 1964, Relationship of crude oils to depositional environment of source rocks in the Uinta Basin, *in* Sabatka, E. F., ed., Guidebook to the Geology and Mineral Resources of the Uinta Basin: Intermountain Assoc. Geologists, p. 201–206.
BEERBOWER, J. R., 1961, Origin of cyclothems of the Dunkard Group (Upper Pennsylvanian-Lower Permian) in Pennsylvania, West Virginia, and Ohio: Geol. Soc. America Bull., v. 72, p. 1029–1050.
BELT, E. S., 1968, Carboniferous continental sedimentation, Atlantic Provinces, Canada, *in* Klein, G. deV., ed., Late Paleozoic and Mesozoic Continental Sedimentation, Northeastern North America: Geol. Soc. America Special Paper 106, p. 127–176.
BORCHERT, HERMANN, 1965, Formation of marine sedimentary iron ores, p. 159–204 *in* Riley, J. P., and Skirrow, G., eds., Chemical Oceanography, v. 2: Academic Press, London, 508 p.
——— AND MUIR, R. O., 1964, Salt Deposits, Van Nostrand Co., London, 338 p.
BOSWORTH, T. O., 1912, The Keuper Marl around Charnwood: Geol. Soc. London Quart., v. 68, p. 281–294.
BRADLEY, W. H., 1925, A contribution to the origin of the Green River Formation and its oil shale: Am. Assoc. Petroleum Geologists Bull., v. 9, p. 247–262.
———, 1926, Shore phases of the Green River Formation in northern Sweetwater County, Wyoming: U.S. Geol. Survey Prof. Paper 140-D, p. 121–131.
———, 1929a, The varves and climate of the Green River epoch: U.S. Geol. Survey Paper 158-E, p. 87–110.
———, 1929b, Algae reefs and oolites of the Green River Formation: U.S. Geol. Survey Prof. Paper 158-A, p. 1–7.
———, 1931, Origin and microfossils of the oil shale of the Green River Formation of Colorado and Utah: U.S. Geol. Survey Prof. Paper 168, 58 p.
———, 1933, Factors that determine the curvature of mud-cracked layers: Am. Jour. Sci., v. 26, p. 55–71.
———, 1948, Limnology and the Eocene lakes of the Rocky Mountain region; Geol. Soc. America Bull., v. 59, p. 635–648.
———, 1963, Paleolimnology, *in* Frey, D. G., ed., Limnology in North America: Univ. of Wisconsin Press, p. 621–652.
———, 1964, Geology of Green River Formation and associated Eocene rocks in southwestern Wyoming and adjacent parts of Colorado and Utah: U.S. Geol. Survey Prof. Paper 496-A, 86 p.
BROWN, R. W., 1957, Nonalgal megascopic marine plants, *in* Ladd, H. S., ed., Treatise on marine ecology and paleoecology: Geological Soc. America Memoir 67, v. 2, p. 729–730.
CAROZZI, A. V., 1962, Observations on algal biostromes in the Great Salt Lake, Utah: Jour. Geology, v. 70, p. 246–252.
CASHION, W. B., 1967, Geology and fuel resources of the Green River Formation, southeastern Uinta Basin, Utah and Colorado: U.S. Geol. Survey Prof. Paper 548, 48 p.
CLARKE, F. W., 1924, The data of geochemistry: U.S. Geol. Survey Bull. 770, 5th ed., 841 p.
CODY, R. D., 1970, Anomalous boron content of two continental shales in eastern Canada: Jour. Sedimentary Petrology, v. 40, p. 750–754.
CRAIG, H., 1953, The geochemistry of the stable carbon isotopes: Geochimica et Cosmochimica Acta, v. 3, p. 53–92.
CRAMPTON, C. B., 1914, Lithology and conditions of deposit of the Caithness Flagstone Series, *in* the Geology of Caithness: Geol. Survey Scotland Memoir, p. 80–103.

CROWLEY, A. J., 1939, Possible criterion for distinguishing marine and nonmarine sediments: Am. Assoc. Petroleum Geologists Bull., v. 23, p. 1716–1720.
CUTHBERT, L. F., 1944, Clay minerals in Lake Erie sediments: Am. Mineralogist, v. 29, p. 378–388.
CURRY, H. D., 1957, Fossil tracks of Eocene vertebrates, southwestern Uinta Basin, Utah, *in* Seal, O. G., ed., Guidebook to the Geology of the Uinta Basin: Intermountain Assoc. Petrol. Geol., p. 42–47.
DANE, C. H., 1954, Stratigraphic and facies relationships of upper part of Green River Formation and lower part of Uinta Formation in Duchesne, Uintah, and Wasatch Counties, Utah: Am. Assoc. Petroleum Geologists Bull., v. 38, p. 405–425.
———, 1955, Stratigraphic and facies relationships of the upper part of the Green River Formation and lower part of the Uinta Formation in Duchesne, Uintah, and Wasatch Counties, Utah: U.S. Geol. Survey, Oil and Gas Inv. Preliminary Chart OC 52.
DARTON, N. H., 1901, Preliminary description of the geology and water resources of the southern half of the Black Hills and adjoining regions in South Dakota and Wyoming: U.S. Geol. Survey, 21st Ann. Rept., pt. 4, p. 489–599.
DAVIS, R. A., JR., 1965, Underwater study of ripples, southeastern Lake Michigan: Jour. Sed. Petrology, v. 35, p. 857–866.
DAVIS, W. M., 1882, On the classification of lake basins: Proc. Boston Soc. Nat. Hist., v. 21, p. 315–381.
DEGENS, E. T., WILLIAMS, E. G., AND KEITH, M. L., 1957, Environmental studies of Carboniferous sediments, pt. 1: Geochemical criteria for differentiating marine from fresh-water shales: Am. Assoc. Petrol. Geologists Bull., v. 41, p. 2427–2455.
———, CHILINGAR, G. V., AND PIERCE, W. D., 1963, On the origin of petroleum inside freshwater carbonate concretions of Miocene age, p. 1–16 *in* Advances in Geochemistry: Pergamon Press, New York.
DINELEY, D. L., AND WILLIAMS, B. P. J., 1968, Sedimentation and paleoecology of the Devonian Escuminac Formation and related strata, Escuminac Bay, Quebec, p. 241–264 *in* Klein, G. deV., ed., Late Paleozoic and Mesozoic Continental Sedimentation, Northeastern North America: Geol. Soc. America Special Paper 106, 309 p.
DROSTE, J. B., 1961a, Clay mineral composition of sediments in some desert lakes of Nevada, California, and Oregon: Science, v. 133, p. 1928.
———, 1961b, Clay minerals in sediments of Owens, China, Searles, Panamint, Bristol, Cadiz, and Danby lake basins, California: Geol. Soc. America Bull., v. 72, p. 1713–1722.
EARDLEY, A. J., 1938, Sediments of Great Salt Lake, Utah: Am. Assoc. Petroleum Geologists Bull. v. 22, p. 1305–1411.
———, AND GVOSDETSKY, VASYL, 1960, Analysis of Pleistocene core from Great Salt Lake, Utah: Geol. Soc. America Bull., v. 71, p. 1323–1344.
ELLIOTT, R. E., 1961, The stratigraphy of the Keuper series in Nottinghamshire: York Geol. Soc. Proc., v. 33, p. 197–234.
EPSTEIN, SAMUEL, 1959, The variations of the O^{18}/O^{16} ratio in nature and some geologic implications, *in* Abelson, P. H., ed., Researches in Geochemistry: John Wiley and Sons, New York, p. 217–240.
EVANS, O. F., 1941, The classification of wave-formed ripple marks: Jour. Sed. Petrology, v. 11, p. 37–41.
FAHEY, J. J., 1962, Saline minerals of the Green River Formation: U.S. Geol. Survey Prof. Paper 405, 50 p.
FELTS, W. M., 1954, Occurrence of oil and gas and its relation to possible source beds in continental Tertiary of Intermountain region: Am. Assoc. Petroleum Geologists Bull., v. 38, p. 1661–1670.
FETH, J. H., 1955, Sedimentary features in the Lake Bonneville Group in the east shore area, near Ogden, Utah, *in* Eardley, A. J., ed., Tertiary and Quaternary Geology of the eastern Bonneville Basin: Utah Geol. Society, p. 45–69.
———, 1961, A new map of western conterminous United States showing the maximum known or inferred extent of Pleistocene lakes: U.S. Geol. Survey Prof. Paper 424-B, p. B110–B111.
———, 1963, Tertiary lake deposits in western conterminous United States: Science, v. 139, p. 107–110.
———, 1964, Review and annotated bibliography of ancient lake deposits (Precambrian to Pleistocene) in the Western States: U.S. Geol. Survey Bull. 1080, 119 p.
GARRELS, R. M., 1960, Mineral equilibria: Harper and Brothers, New York, 254 p.
GEBELEIN, CONRAD, AND HOFFMAN, PAUL, 1968, Intertidal stromatolites from Cape Sable, Florida (abs): Geol. Soc. America, Program with Abstracts, 1968 Annual Meeting, p. 109.
GILBERT, G. K., 1891, Lake Bonneville: U.S. Geol. Survey Monograph 1, 438 p.
GOULD, H. R., 1951, Some quantitative aspects of Lake Mead turbidity currents: Soc. Econ. Paleont. Mineralogists, Special Pub. No. 2, p. 34–52.
GOVETT, G. J. S., 1966, Origin of banded iron formations: Geol. Soc. America Bull., v. 77, p. 1191–1212.
GREINER, H. R., 1962, Facies and sedimentary environments of Albert Shale, New Brunswick: Am. Assoc. Petroleum Geologists Bull., v. 46, p. 219–234.
GRIFFIN, G. M., 1962, Regional clay-mineral facies-products of weathering intensity and current distribution in the northeastern Gulf of Mexico: Geol. Soc. America Bull., v. 73, p. 737–768.
GRIM, R. E., 1958, Concept of diagenesis in argillaceous sediments: Am. Assoc. Petroleum Geologists Bull., v. 42, p. 246–253.
———, KULBICKI, GEORGES, AND CAROZZI, A. V., 1960, Clay mineralogy of the sediments of the Great Salt Lake, Utah: Geol. Soc. America Bull., v. 71, p. 515–519.
HANSEN, S., 1940, Varuighed i damske og skaanske senglaciale afleijringer; Danmarks Geologioke Undersgelse, II Raekke, nr. 63, 478 p.
HARMS, J. C., 1969, Hydraulic significance of some sand ripples: Geol. Soc. America Bull., v. 80, p. 363–395.
HARRISON, A. G., AND THODE, H. G., 1958, Sulphur isotope abundances in hydrocarbons and source rocks of Uinta Basin, Utah: Am. Assoc. Petroleum Geologists Bull., v. 42, p. 2642–2649.
HAY, R. L., 1966, Zeolites and zeolitic reactions in sedimentary rocks: Geol. Soc. America Special Paper 85, 130 p.
HIGH, L. R., JR., AND PICARD, M. D., 1965, Sedimentary petrology and origin of analcime-rich Popo Agie Member, Chugwater (Triassic) Formation, west-central Wyoming: Jour. Sed. Petrology, v. 35, p. 49–70.

———, and ———, 1967, Stratigraphic relations of Upper Triassic units, northeastern Utah and Wyoming: Compass, v. 44, p. 88–98.
———, and ———, 1969a, Stratigraphic relations within Upper Chugwater Group (Triassic), Wyoming: Am. Assoc. Petroleum Geologists Bull., v. 53, p. 1091–1104.
———, and ———, 1969b, Sedimentary cycles in Green River Formation (Eocene): Modification of Walther's Law (abs.): Am. Assoc. Petroleum Geologists Bull., v. 53, p. 722–723.
———, Hepp, D. M., Clark, T., and Picard, M. D., 1969, Stratigraphy of Popo Agie Formation (Late Triassic), Uinta Mountain area, Utah and Colorado, in Lindsay, J. B., ed., Geologic Guidebook of the Uinta Mountains: Intermountain Assoc. Geologists, p. 181–192.
Hough, J. L., 1958, Fresh-water environment of deposition of Precambrian banded iron formations: Jour. Sed. Petrology, v. 28, p. 414–430.
Hunt, J. M., Stewart, Francis, and Dickey, P. A., 1954, Origin of hydrocarbons of Uinta Basin, Utah: Am. Assoc. Petroleum Geologists Bull., v. 38, p. 1671–1698.
Hutchinson, G. E., 1957, A treatise on limnology, v. 1: John Wiley and Sons, Inc., New York, 1015 p.
Jensen, M. L., 1963, The bearing of sulfur isotopes on Colorado Plateau uranium and petroleum deposits, in Crawford, A. L., ed., Oil and gas possibilities of Utah, Re-evaluated: Utah Geol. and Mineral. Survey Bull. 54, p. 275–284.
Johnson, J. H., 1961, Limestone-building algae and algal limestones: Johnson Publishing Company, Boulder, 297 p.
Jones, D. J., and Marsell, R. E., 1955, Pleistocene sediments of lower Jordan Valley, Utah, in Eardley, A. J., ed., Tertiary and Quaternary Geology of the eastern Bonneville Basin: Utah Geol. Society, p. 85–112.
Judson, Sheldon, and Murray, R. C., 1956, Modern hydrocarbons in two Wisconsin lakes: Am. Assoc. Petroleum Geologists Bull., v. 40, p. 747–750.
Keith, M. L., and Degens, E. T., 1959, Geochemical indicators of marine and freshwater sediments, in Abelson, P. H., ed., Researches in geochemistry: John Wiley and Sons, New York, p. 38–61.
Keller, W. D., 1956, Clay minerals as influenced by environments of their formation: Am. Assoc. Petroleum Geologists Bull., v. 40, p. 2689–2710.
Kindle, E. M., 1917, Some factors affecting the development of mud cracks: Jour. Geology, v. 25, p. 135–144.
Klein, G. deV., 1962a, Sedimentary structures in the Keuper Marl (Upper Triassic): Geol. Mag., v. 99, 137–144.
———, 1962b, Triassic sedimentation, Maritime Provinces, Canada: Geol. Soc. America Bull., v. 73, p. 1127–1146.
———, 1967, Paleocurrent analysis in relation to modern marine sediment dispersal patterns: Am. Assoc. Petroleum Geologists Bull., v. 51, p. 366–382.
Krauskopf, K. B., 1967, Introduction to geochemistry: McGraw-Hill, New York, 721 p.
Krumbein, W. C., and Sloss, L. L., 1951, Stratigraphy and sedimentation, 1st ed: W. H. Freeman and Co., San Francisco, 497 p.
Lahee, F. H., 1941, Field Geology, 4th ed: McGraw-Hill, New York, 853 p.
Langbein, W. B., 1961, Salinity and hydrology of closed lakes: U.S. Geol. Survey Prof Paper 412, 20 p.
Livingstone, D. A., 1963, Chemical composition of rivers and lakes: U.S. Geol. Survey Prof. Paper 440-G, 64 p.
Lohman, K. E., 1957, Diatoms, in Ladd, H. S., ed., Treatise on Marine ecology and paleoecology: Geol. Soc. American Memoir 67, v. 2, p. 731–736.
Longwell, C. R., 1928, Geology of the Muddy Mountains, Nevada, with a section through the Virgin Range to the Grand Wash Cliffs, Arizona: U.S. Geol. Survey Bull. 798, 152 p.
Lonka, Anssi, 1967, Trace elements in the Finnish Precambrian phyllites as indicators of salinity at the time of sedimentation: Bull. Commission Geologique Finlande, no. 228, 63 p.
McKee, E. D., 1954, Stratigraphy and history of the Moenkopi Formation of Triassic age: Geol. Soc. America Memoir 61, 133 p.
——— 1965, Experiments on ripple lamination, in Middleton, G. V., ed., Primary sedimentary structures and and their hydrodynamic interpretation: Soc. Econ. Paleont. Mineralogists Spec. Publ. 12, p. 66–83.
McLeroy, C. A., and Anderson, R. Y., 1966, Laminations of Oligocene Florissant lake deposits, Colorado: Geol. Soc. America Bull., v. 77, p. 605–618.
Milton, Charles, 1961, Mineralogy and petrology of the Green River Formation of Wyoming, Utah, and Colorado: Trans. New York Acad. Sciences, v. 23, p. 561–567.
———, and Eugster, H. P., 1959, Mineral assemblages of the Green River Formation, in Abelson, P. H., ed., Researches in geochemisty: John Wiley and Sons, New York, p. 118–150.
———, and Fahey, J. J., 1960, Classification and association of the carbonate minerals of the Green River Formation: Amer. Jour. Science, v. 258-A, p. 242–246.
———, Chao, E. C. T., Fahey, J. J., and Mrose, M. E., 1960, Silicate mineralogy of the Green River Formation of Wyoming, Utah, and Colorado: Internat. Geol. Congress, pt. 21, p. 171–184.
Moiola, R. J., and Weiser, D., 1968, Textural parameters: an evaluation: Jour. Sed. Petrology, v. 38, p. 45–53.
Moore, J. E., 1961, Petrography of northeastern Lake Michigan bottom sediments: Jour. Sed. Petrology, v. 31, p. 402–436.
Moussa, M. T., 1968, Fossil tracks from the Green River Formation (Eocene) in the Uinta Basin Utah: Jour. Paleont., v. 42, p. 1433–1438.
Müller, German, 1968, Exceptionally high strontium concentrations in freshwater onkolites and mollusk shells of Lake Constance, in Müller, German, and Friedman, G. M., eds., Recent Developments in Carbonate Sedimentology in Central Europe: Springer-Verlag, New York, p. 116–127.
Nelson, C. H., 1967, Sediments of Crater Lake, Oregon: Geol. Soc. America Bull., v. 78, p. 833–848.
Nightingale, W. T., 1930, Geology of Vermillion Creek gas area in southwest Wyoming and northwest Colorado: Am. Assoc. Petroleum Geologists Bull., v. 14, p. 1013–1040.
——— 1935, Geology of Hiawatha gas fields, southwest Wyoming and northwest Colorado, in Geology of Natural Gas: Am. Assoc. Petrol. Geologists, p. 341–361.

Pardee, J. T., and Bryan, Kirk, 1926, Geology of the Latah Formation in relation to the lavas of the Columbia Plateau near Spokane, Washington: U.S. Geol. Survey Prof. Paper 140-A, p. 1–16.
Peck, R. E., 1953, Fossil charophytes: Botanical Rev., v. 19, p. 209–227.
———, 1957, North American Mesozoic Charophyta: U.S. Geol. Survey Prof. Paper 294-A, p. 1–44.
Pettijohn, F. J., 1957, Sedimentary rocks, 2nd ed.: Harper and Bros., New York, 718 p.
Picard, M. D., 1955, Subsurface stratigraphy and lithology of Green River Formation in Uinta Basin, Utah: Am. Assoc. Petroleum Geologists Bull., v. 39, p. 75–102.
——— 1956, Summary of Tertiary oil and gas fields in Utah and Colorado: Am. Assoc. Petroleum Geologists Bull., v. 40, p. 2956–2960.
——— 1957a, Criteria used for distinguishing lacustrine and fluvial sediments in Tertiary beds of Uinta Basin, Utah: Jour. Sed. Petrology, v. 27, p. 373–377.
——— 1957b, Red Wash-Walker Hollow field, stratigraphic trap, eastern Uinta Basin, Utah: Am. Assoc. Petroleum Geologists Bull., v. 41, p. 923–926.
——— 1957c, Green River and lower Uinta Formations—subsurface stratigraphic changes in central and eastern Uinta Basin, Utah, in Seal, O. G. ed., Guidebook to the Geology of the Uinta Basin: Intermountain Assoc. Petrol. Geol., p. 116–130.
——— 1959, Green River and lower Uinta Formation subsurface stratigraphy in western Uinta Basin, Utah, in Williams, N. C. ed., Guidebook to the Geology of the Wasatch and Uinta Mountains transition area: Intermountain Assoc. Petrol. Geol., p. 139–149.
——— 1960, On the origin of oil, Eagle Springs field, Nye County, Nevada, in Boettcher, J. W. and Sloan, W. W., Jr., eds., Guidebook to the Geology of east central Nevada: Intermountain Assoc. Petrol. Geologists, p. 237–244.
——— 1962, Source beds in Red Wash-Walker Hollow field, eastern Uinta Basin, Utah: Am. Assoc. Petroleum Geologists Bull., v. 46, p. 690–694.
——— 1963, Duration of Eocene Lake, Uinta Basin, Utah: Geol. Soc. America Bull., v. 74, p. 89–90.
——— 1966, Oriented, linear-shrinkage cracks in Green River Formation (Eocene), Raven Ridge area, Uinta Basin, Utah: Jour. Sed. Petrology, v. 36, p. 1050–1057.
——— 1967a, Stratigraphy and depositional environments of the Red Peak Member of the Chugwater Formation (Triassic), west-central Wyoming: Contrib. Geology, v. 6, p. 39–67.
——— 1967b, Paleocurrents and shoreline orientations in Green River Formation (Eocene), Raven Ridge and Red Wash areas, northeastern Uinta Basin, Utah: Am. Assoc. Petroleum Geologists Bull., v. 51, p. 383–392.
——— 1967c, Paleocurrents and shoreline orientations in Green River Formation (Eocene), Raven Ridge and Red Wash areas, northeastern Uinta Basin, Utah: reply to discussion by R. A. Davis, Jr.: Am. Assoc. Petrol. Geologists Bull., v. 51, p. 2471–2475.
——— 1969, Oriented, linear-shrinkage cracks in Alcova Limestone Member (Triassic), southeastern Wyoming: Contrib. Geology, v. 8, p. 1–7.
——— and High, L. R., Jr., 1968a, Shallow marine currents on the Early (?) Triassic Wyoming shelf: Jour. Sed. Petrology, v. 38, p. 411–423.
———, and ———, 1968b, Sedimentary cycles in the Green River Formation (Eocene), Uinta Basin, Utah: Jour. Sed. Petrology, v, 38, p. 378–383.
———, and ———, 1969, Some sedimentary structures resulting from flash floods, p. 175–190 in Jensen, M. L., ed., Guidebook of northern Utah: Utah Geol. Mineralogical Survey Bull. 82, 266 p.
———, and ———, 1970, Sedimentology of oil-impregnated, lacustrine and fluvial sandstone, P. R. Spring area, southeast Uinta Basin, Utah: Utah Geol. Mineralogical Survey, Special Studies No. 33, 32 p.
Pinsak, A. P., 1957, Subsurface stratigraphy of the Salem limestone and associated formations in Indiana: Indiana Geol. Survey Bull. 11, 62 p.
Porter, Livingstone, Jr., 1963, Stratigraphy and oil possibilities of the Green River Formation in the Uinta Basin, Utah, in Crawford, A. L., ed., oil and gas possibilities of Utah, Re-evaluated: Utah Geol. Mineralogical Survey Bull. 54, p. 193–198.
Potter, P. E., and Pettijohn, F. J., 1963, Paleocurrents and basin analysis: Academic Press Inc., New York 296 p.
Potter, P. E., Shimp, N. F., and Witters, J., 1963, Trace elements in marine and fresh-water argillaceous sediments: Geochim. et Cosmochimica Acta, v. 27, p. 669–694.
Powers, W. E., 1939, Basin and shore features of the extinct lake San Augustin, New Mexico: Jour. Geomorphology, v. 2, p. 345–356.
Rayner, D. H., 1963, The Achanarras Limestone of the Middle Old Red Sandstone, Caithness, Scotland: Geol. Soc. Yorkshire Proc., v. 34, p. 117–138.
Reeves, C. C., Jr., 1968, Introduction to paleolimnology: Elsevier Publishing Co., New York, 228 p.
Rezak, Richard, 1957, Stromatolites of the Belt series in Glacier National Park and vicinity, Montana: U.S. Geol. Survey Prof. Paper 294-D, p. 127–154.
Rolfe, B. N., 1957, Surficial sediments of Lake Mead: Jour. Sed. Petrology, v. 27, p. 378–386.
Russell, I. C., 1885, Lake Lahontan: U.S. Geol. Survey Monograph 11, 288 p.
Sakamoto, Takao, 1950, The origin of the Pre-Cambrian banded iron ores: Amer. Jour. Science, v. 248, p. 449–474.
Sanders, J. E., 1968, Stratigraphy and primary sedimentary structures of fine-grained, well-bedded strata, inferred lake deposits, Upper Triassic, central and southern Connecticut, in Klein, G. deV., ed., Late Paleozoic and Mesozoic Continental Sedimentation, Northeastern North America: Geol. Soc. America Special Paper 106, p. 265–305.
———, and Friedman, G. M., 1967, Origin and occurrence of limestones, in Chilingar, G. V., Bissell, H. J., and Fairbridge, R. W., eds., Carbonate rocks: Origin, occurrence and classification: Elsevier Publishing Co., New York, p. 169–265.
Sarmiento, R., and Kirby, R. A., 1962, Recent sediments of Lake Maricaibo: Jour. Sed. Petrology, v. 32, p. 698–724.

SCHOPF, J. M., 1957a, Spores and related plant microfossils—Paleozoic, *in* Ladd, H. S., ed., Treatise on marine ecology and paleoecology: Geol. Soc. America Memoir 67, v. 2, p. 703–707.

———, 1957b, Spores and problematic plants commonly regarded as marine, *in* Ladd, H. S., ed., Treatise on marine ecology and paleoecology: Geol. Soc. America Memoir 67, v. 2, p. 709–717.

SCHÖTTLE, MANFRED, AND MULLER, GERMAN, 1968, Recent carbonate sedimentation in the Gnadensee (Lake Constance), Germany, *in* Muller, German, and Friedman, G. M., eds., Recent Developments in Carbonate Sedimentology in Central Europe: Springer-Verlag, New York, p. 148–156.

SELLEY, R. C., 1967, Paleocurrents and sediment transport in nearshore sediments of the Sirte Basin, Libya: Jour. Geology, v. 75, p. 215–223.

SILVERMAN, S. R., AND EPSTEIN, S., 1958, Carbon isotopic compositions of petroleums and other sedimentary organic materials: Am. Assoc. Petroleum Geologists Bull., v. 42, p. 998–1012.

SMITH, A. J., 1959, Structures in the stratified late-glacial clays of Windermere, England: Jour. Sed. Petrology, v. 29, p. 447–453.

SMITH, G. I., AND HAINES, D. V., 1964, Character and distribution of nonclastic minerals in the Searles Lake evaporite deposit, California: U.S. Geol. Survey Bull. 1181-P, 58 p.

SMITH, P. V., JR., 1954, Studies on origin of petroleum: Occurrence of hydrocarbons in recent sediments: Am. Assoc. Petroleum Geologists Bull., v. 38, p. 377–404.

SOLOHUB, J. T., AND KLOVAN, J. E., 1970, Evaluation of grain-size parameters in lacustrine environments: Jour. Sed. Petrology, v. 40, p. 81–101.

SWAIN, F. M., 1956a, Early Tertiary ostracode zones of Uinta Basin, *in* Peterson, J. A., ed., Geology and Economic Deposits of East Central Utah: Intermountain Assoc. Petrol. Geol. p. 125–139.

———, 1956b, Stratigraphy of lake deposits in central and northern Minnesota; Am. Assoc. Petroleum Geologists Bull., v. 40, p. 600–653.

———, 1964a, Tertiary fresh-water ostracoda of the Uinta Basin and related forms from southern Wyoming, western Utah, Idaho, and Nevada, *in* Sabtka, E. F., ed., Guidebook to the Geology and Mineral Resources of the Uinta Basin: Intermountain Assoc. Petrol. Geol., p. 173–180.

———, 1964b, Early Tertiary freshwater ostracoda from Colorado, Nevada and Utah and their stratigraphic distribution: Jour. Paleon., v. 38, p. 256–280.

———, 1966, Bottom sediments of Lake Nicaragua and Lake Managua, western Nicaragua: Jour. Sed. Petrology, v. 36, p. 522–540.

———, and PROKOPOVICH, N., 1954, Stratigraphic distribution of lipoid substances in Cedar Creek bog, Minnesota: Geol. Soc. America Bull., v. 65, p. 1183–1198.

———, BLUMENTALS, A., AND PROKOPOVICH, N., 1958, Bitumens and other organic substances in Precambrian of Minnesota: Am. Assoc. Petroleum Geologists Bull., v. 42, p. 173–189.

TANNER, W. F., 1964, Ripple mark indices and their uses: Sedimentology, v. 9, p. 89–104.

———, 1968, Shallow lake deposits, lower part of Morrison Formation (Late Jurassic), northern New Mexico: Mountain Geologist, v. 5, p. 187–195.

TANK, RONALD, 1969, Clay mineral composition of the Tipton Shale Member of the Green River Formation (Eocene) of Wyoming; Jour. Sed. Petrology, v. 39, p. 1593–1595.

TASCH, PAUL, 1964, Periodicity in the Wellington Formation of Kansas and Oklahoma, *in* Merriam, D. F., ed., Symposium on cyclic sedimentation: Kansas Geol. Survey Bull. 169, p. 481–496.

THODE, H. G., MONSTER, J., AND DUNFORD, H. B., 1958, Sulphur isotope abundances in petroleum and associated materials: Am. Assoc. Petroleum Geologists Bull., v. 42, p. 2619–2641.

TOHILL, BRUCE, AND PICARD, M. D., 1966, Stratigraphy and petrology of Crow Mountain Sandstone Member (Triassic), Chugwater Formation, northwestern Wyoming: Am. Assoc. Petroleum Geologists Bull., v. 50, p. 2547–2565.

TWENHOFEL, W. H., 1932, Treatise on sedimentation, 2d ed: The Williams and Wilkins Co., Baltimore, 926 p.

———, AND MCKELVEY, V. E., 1941, Sediments of fresh-water lakes: Am. Assoc. Petroleum Geologists Bull., v. 25, p. 826–849.

———, CARTER, S. L., AND MCKELVEY, V. E., 1942, The sediments of Grassy Lake, Vilas County, a large bog lake of northern Wisconsin: Amer. Jour. Science, v. 240, p. 529–546.

VAN HOUTEN, F. B., 1962, Cyclic sedimentation and the origin of analcime-rich Upper Triassic Lockatong Formation, west-central New Jersey and adjacent Pennsylvania: Amer. Jour. Science, v. 260, p. 561–576.

———, 1964, Cyclic lacustrine sedimentation, Upper Triassic Lockatong Formation, central New Jersey and adjacent Pennsylvania, p. 497–531 *in* Merriam, D. F., ed., Symposium on cyclic sedimentation: Kansas Geol. Survey Bull. 169.

———, 1965, Composition of Triassic Lockatong and associated formations of Newark Group, central New Jersey and adjacent Pennsylvania: Amer. Jour. Science, v. 263, p. 825–863.

VISHER, G. S., 1965, Use of vertical profile in environment reconstruction: Am. Assoc. Petroleum Geologists Bull., v. 49, p. 41–61.

WALKER, C. T., 1968, Evaluation of boron as a paleosalinity indicator and its application to offshore prospects: Am. Assoc. Petrol. Geologists Bull., v. 52, p. 751–766.

WALTHER, JOHANNES, 1893–94, Einleitung in die Geologie als historische Wissenschaft; Beobachtungen über die Bildung der Gesteine und ihrer organischen Einschlüsse, bd. 1: Jena, G. Fischer.

WEAVER, C. D., 1958, Geologic interpretation of argillaceous sediments, pt. 1: Origin and significance of clay minerals in sedimentary rocks: Am. Assoc. Petroleum Geologists Bull., v. 42, p. 254–271.

WICKMAN, F. E., 1952, Variations in the relative abundance of the carbon isotopes in plants: Geochimica et Cosmochimica Acta, v. 2, p. 243–254.

WILSON, L. R., 1957, Spores and pollen of the post-Paleozoic, *in* Ladd, H. S., ed., Treatise on marine ecology and paleoecology: Geol. Soc. America Memoir 67, v. 2, p. 719–728.

WILSON, R. F., 1958, Sedimentary facies of the Moenkopi Formation of Triassic age on the Colorado Plateau (abs.): Geol. Soc. America Bull., v. 69, p. 1749.

WULF, G. R., 1963, Bars, spits, and ripple marks in a Michigan lake: Am. Assoc. Petroleum Geologists Bull., v. 47, p. 691–695.

YEN, TENG-CHIEN, 1951, Fossil fresh-water mollusks and ecological interpretations: Geol. Soc. America Bull., v. 62, p. 1375–1380.

TIDAL FLATS

HANS-ERICH REINECK
Institut fur Meeres-Geologie und Meeres-Biologie, Senckenberg, Wilhelmshaven

ABSTRACT

Tidal flats are built in a marine shallow water environment in a specific vertical sequence intersected by runnels and channels. Bidirectional current marks, subaerial features, and horizons of bioturbation complete the environmental indicators.

PHYSIOGRAPHY AND MORPHOLOGY

Sediments in the intertidal zone lie between high and low water line over a vertical distance of 1–4 or more meters depending on the tidal range. Tidal range causes currents which form numerous gullies and channels (fig. 1). Currents of a high tidal range erode deeper channels and the channels in extensive tidal flats are deeper than in smaller ones.

Current velocity on the flats may reach one knot and form small scale current ripple marks on sandy bottoms. The current velocity in gullies and channels is up to 3 knots or more, so that underwater ripples and dunes are formed. (Lüders 1929, Häntzschel 1938, Hülsemann 1955, Reineck 1963).

The tidal flats are protected by barrier islands or sand bars or occur in sheltered bays. Wave action is not too strong but it is nevertheless an important factor.

Tidal flat deposits form a wedge-shaped body which is elongated parallel to the shore line, but may be intersected by channels or river estuaries.

FIG. 1.—Three types of tidal flats in the German Bay: A, to the west, sheltered by barrier islands with steep angle of foreshore; B, central, in a sheltered bay situation; C, to the north, unsheltered to the open sea, with gentle dipping foreshore. All are dissected by channels, runnels and rivers. Dark grey-intertidal zone, Light grey-subtidal zone (0–6 m depth), 10, 20, and 40 m line with one, two and four points.

FIG. 2.—Bedding types of mixed flats. Scale in centimeters. North sea, German Bay. Flaserbedding.

MATERIAL AND GRAIN SIZE DISTRIBUTION

Sediments accumulating on a tidal flat are mud (clay and silt) and sand (mostly fine-grained). Gravels are rare, but clay pebbles and shells often occur on channel bottoms.

Clay and silt are mainly: clay minerals, quartz, iron minerals, garnet, mica, feldspars, dolomite, and calcite and some minor ferromagnesium minerals.

Sand is mainly: quartz, feldspars, mica, heavy minerals, faecal pellets, broken shells, and foraminifera (van Straaten 1954).

Near the high water line (mud flats) the mud content is high especially on wind and wave-sheltered coast lines (van Straaten & Kuenen 1957, Emery & Stevenson 1957, Postma 1961, Thompson 1968). It decreases in the mixed flat and it is a minor constituent near the low water line in the sand flats (Reineck 1956). The mud content increases toward channels (Evans 1965) and below the low water line mud is especially abundant in the lateral deposits of big channels except for the channel bottom sediments.

The tidal flats in the Netherlands are less muddy than the tidal flats in the German Bay. Most of the tidal flats in Great Britain are very sandy. South of San Felipe, Gulf of California, the tidal flats are built of fine sand. North of San Felipe the mud content strongly increases towards the Colorado delta (Thompson 1968).

BEDDING TYPES

Cross-bedding of megaripples is rare in the intertidal zone but is common in the channels (Häntzschel 1938, van Straaten 1953, Reineck 1963, de Vries Klein 1967, Reineck & Singh 1967). On *sand flats* cross-bedding of small scale current ripples is very common. Sometimes the cross-bedding shows herring bone structures in section normal to the ripple crest axis and festoon bedding in section parallel to the ripples.

Laminated sand is not common. Climbing ripple structures are very rare.

Flaser bedding (fig. 2), wavy bedding, lenticular bedding (fig. 3), interbedding and interlamination of mud and sand or silt (fig. 4) are common bedding types in the *mixed flats* (Häntzschel 1936, Reineck 1960) and in the *lateral channel deposits* (Reineck & Wunderlich 1969). In *mud flats* there are thick mud layers with thin strips of sand (fig. 4). In arid climate mud near the extreme high tide level becomes chaotic by excessive evaporation whereby gypsum and halite become concentrated in the surface and nearsurface. Such crystal growth and the formation of mudcracks result in generation of a chaotic mud evaporite mixture (Thompson 1968).

Salt marsh deposits (intertidal and supratidal zone) are characteristically interfused by roots and by uneven nodule lamination (van Straaten

FIG. 4.—Bedding types of mixed flats. Scale in centimeters. North sea, German Bay. Interbedding and interlamination of mud and sand due to tidal currents and slack water. In the middle a trace of escape of *My-arenaria*.

FIG. 3.—Bedding types of mixed flats. Scale in centimeters. North sea, German Bay. Lenticular bedding.

1954, Reineck 1962, Redfield 1965). None of these bedding types is restricted to tidal flat deposits.

Microstructures include graded bedding in thin laminae generally less than 1 mm thick. Sometimes the laminae are graded upward from coarse to fine, and sometime the reverse. Small scale erosional features are very common (Reineck & Wunderlich 1969).

The origin of these bedding types is often related to the alternation of tidal currents and slack water. In a vertical column there are thicker sets which change in bedding type from set to set, resulting from the direction and force of wind and waves (fig. 4) (Reineck & Wunderlich, 1969).

Most of the layers are deposited in shallow depressions as flat erosional patches or shallow runnels, but most of them are laterally deposited in channels (point bars), and on other sheltered gently inclined places. On horizontal areas of flats there is hardly any net accumulation. Erosion and sedimentation always take place.

BIOTURBATION

Fauna of the tidal flat is plentiful, but only a limited number of species is represented. Most parts of the tidal flat surface layers are highly bioturbated by bottom-living animals (fig. 5) (van Straaten 1954, Reineck 1958, Schäfer 1962).

FIG. 5.—Bedding types of mixed flats. Scale in centimeters. North sea, German Bay. Bioturbated layers. Tidal surface with slow rate of sedimentation.

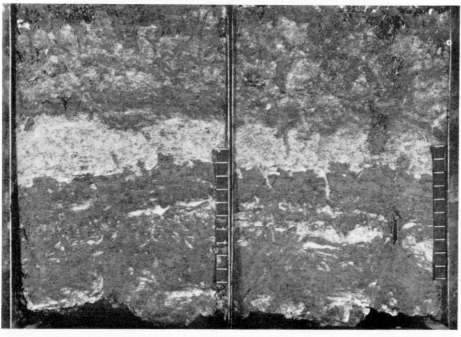

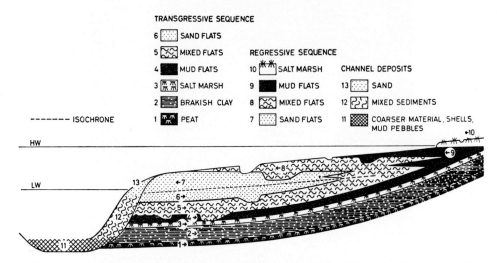

Fig. 6.—Full developed vertical sequences of transgression (transgressive deposits) and regression (progradational deposits). Instead of peat and brackish clay sometimes there may be found a transgression horizon.

All rapidly deposited layers are not bioturbated, especially lateral channel deposits and channel bottom sediments. Sometimes there are horizons with bottom living shells in living position and with traces of escape under them (fig. 4) (Reineck 1958). They were forced to follow the upward growing surface. In some layers there are concentrated faecal pellets and in others rolled algal mats are common in various climates.

SURFACE STRUCTURES

The most common feature on tidal flats is current ripple marks, but symmetrical and asymmetrical oscillation ripples are also present.

Subaerial marks are also important. Small to minor runnels and erosional depressions are often found (Häntzschel 1935). These flat depressions are often sculptured by oscillation ripples whereas the surrounding bottom is covered by algal mats. Tracks of birds and other land animals, raindrop and hail imprints, foam marks and desiccation cracks occur on the surface. Groove casts and flattened ripples are also common.

THE VERTICAL AND LATERAL SEQUENCE

A transgressive sequence (from top to bottom) (Grohne 1956) consists of:

sandflat deposits

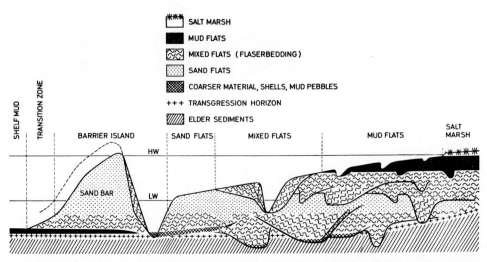

Fig. 7.—Vertical sequence of slowly deposited progradational deposits reworked by meandering channels. The lower part (subtidal zone) is built of channel deposits. Left side is open sea.

mixed flat deposits
mudflat deposits
brackish and freshwater clay
peat with Sphagnum
older sediments

A regressive sequence (from top to bottom) consists of:

peat
fresh water and brackish deposits
salt marsh deposits
mud flat deposits
mixed flat deposits
sand flat deposits

This sequence (fig. 6) is characteristic if there is a rich supply of sediments. Otherwise, the meandering channels rework the sediments, but at least the thick channel sediments rest at the base. Then follows the regressive sequence with sand, mixed sediments, mud and salt marsh deposits, but all are highly dissected by runnels and channels (fig. 7) (van Straaten 1964).

The regressive sequence of tidal flats on the Colorado River Delta, northwestern Gulf of California, as described by Thompson (1968) is from top to bottom:

chaotic muds
brown laminated silt
brown mottled mud
(crab zone)
gray burrowed clay
gray laminated silt and clay

In many cases of fossil and recent tidal flats the given sequence may not be fully developed.

In front of tidal flats lying parallel to a more or less straight coast there are sandbars or barrier islands. The landward side of the tidal flats is limited by land soils or by older sediments.

REFERENCES

EMERY, K. O., AND STEVENSON, R. E., 1957, Estuaries and lagoons: Geol. Soc. Amer. Mem., 67: p. 673–750, 3 pls., 30 figs., New York.
EVANS, G., 1965, Intertidal flat sediments and their environments of deposition in the Wash: Quart. Jour. Geol. Soc., v. 121, p. 209–245, pls. 16–20, 9 figs., London.
GROHNE, U., 1956, Die Geschichte des Jadebusens und seines Untergrundes: Natur u. Volk, v. 86. p. 225–233, 7 figs., Frankfurt am Main.
HÄNTZSCHEL, W., 1935, Fossile Schrägschichtungs-Bögen, "Fließ-wülste" und Rieselmarken aus dem Nama-Transvaal-System (Südafrika) und im rezenten Gegenstück: Senckenbergiana, v. 17, p. 167–177, 7 figs., Frankfurt am Main.
———, 1936, Die Schichtungs-Formen rezenter Flachmeer-Ablagerung im Jade-Gebiet: Senckenbergiana, v. 18, p. 316–356, 20 figs., Frankfurt am Main.
———, 1938, Bau und Bildung von Groß-Rippeln im Wattenmeer: Senckenbergiana, v. 20, p. 1–42, 21 figs., Frankfurt am Mainü.
HÜLSEMANN, J., 1955, Großrippeln und Schrägschichtungs-Gefüge im Nordsee-Watt und in der Molasse: Senckenbergiana lethea, v. 36, p. 359–388, 3 pls., 14 figs., Frankfurt am Main.
KLEIN DE VRIES, G., 1967, Paleocurrent analysis in relation to modern marine sediment dispersal patterns.—Amer. Assoc. Petrol. Geol. Bull., v. 51, p. 366–382, 7 figs., Tulsa.
LÜDERS, K., 1929, Entstehung und Aufbau von Großrücken mit Schillbedeckung in Flut-bzw. Ebbetrichtern der Außenjade: in Lüders and Trusheim: Beiträge zur Ablagerung mariner Mollusken in der Flachsee; Senckenbergiana v. 11, p. 123–142, Frankfurt am Main.
POSTMA, H., 1961, Transport and accumulation of suspended matter in the Dutch Wadden Sea.—Netherlands Jour. Sea Res., v. 1, p. 148–190, 10 figs., Groningen.
REDFIELD, A. C., 1965, Ontogeny of a salt marsh estuary: Sci., v. 147, p. 50–55, 4 figs.
REINECK, H.-E., 1956, Der Wattenboden und das Leben im Wattenboden: Natur u. Volk, v. 86, p. 268–284, 14 figs., Frankfurt am Main.
———, 1958, Wühlbau-Gefüge in Abhängigkeit von Sediment-Umlagerungen: Senckenbergiana lethea, v. 39, p. 1–24, 54–56, pls. 1–5, 3 figs., Frankfurt am Main.
———, 1960, Über Zeitlücken in rezenten Flachsee-Sedimenten: Geol. Rdsch., v. 49, p. 149–161, 3 figs., Frankfurt am Main.
———, 1962, Die Orkanflut vom 16. Februar 1962: Natur und Mus., v. 92, p. 151–172, 16 figs., Frankfurt am Main.
———, 1963, Sedimentgefüge im Bereich der südlichen Nordsee.—Abh. senckenberg. naturforsch. Ges., v. 505, 138, p. pls 1–12, 15 maps, 21 figs., Frankfurt am Main.
REINECK, H.-E. AND SINGH, I. B., 1967, Primary sedimentary structures in the recent sediments of the Jade, North Sea: Mar. Geol., v. 5, p. 227–235, pl. 1, 3 figs., Amsterdam.
———, AND WUNDERLICH, F., 1967, Zeitmessungen an Gezeitenschichten: Natur u. Mus., v. 97, p. 193–197, 3 figs., Frankfurt am Main.
REINECK, H.-E., AND WUNDERLICH, F., 1969, Die Entstehung von Schichten und Schichtbänken im Watt: Senckenbergiana marina, v. 1, p. 85–105, pls. 1–3, 14 figs., Frankfurt am Main.
SCHÄFER, W., 1962, Akuto-Paläontologie nach Studien in der Nordsee: 666 p. 36 pls., 277 figs., Frankfurt am Main. (Kramer).
STRAATEN, L. M. J. U. VAN, 1953, Megaripples in the Dutch Wadden Sea and in the Basin of Arcachon (France): Geol. en Mijnb., n. ser., v. 15, p. 1–11, 7 figs., Leiden.
———, 1954, Composition and structure of recent marine sediments in the Netherlands: Leidse geol. Meded., v. 19, p. 1–110, 10 pls., 26 figs., Leiden.

———, 1964, De bodem der Waddenzee: Het waddenboek, 76 p., figs., 32–104. Zutphen, (Thiane & Co).
———, AND KUENEN, Ph. H., 1957, Accumulation of fine grained sediments in the Dutch Wadden Sea: Geol. en Mijnb. n. ser., v. 19, p. 329–354, 19 figs. Leiden.
———, AND ———, 1958, Tidal action as a cause of clay accumulation:—Jour. Sed. Petrol., v. 28, p. 406–413 4 figs., Tulsa.
THOMAS, R. W., 1968, Tidal Flat Sedimentation on the Colorado River Delta, Northwestern Gulf of California: Geol. Soc. Amer. Special Paper 107, 133 p., 25 pls., 35 figs.

A SELECTIVE BIBLIOGRAPHY ON TIDAL FLATS

1. Summaries and reviews

GRIPP, K., 1956, Das Watt; Begriff, Begrenzung und fossile Vorkommen: Senckenbergiana leth., v. 37: p. 149–181, p.l 1–3, 5 figs., Frankfurt am Main.
HÄNTZSCHEL, W., 1939, Tidal flat deposits (Wattenschlick): *in* TRASK, P. D.: Recent Marine Sediments, reprinted 1955, p. 195–206, figs. 1–2, Amer. Assoc. Petrol. Geologists, Menasha.
REINECK, H.-E., AND SCHÄFER, W., 1956, Kleines Kusten-ABC: Natur u. Volk, v. 86: p. 261–284, 18 figs., Frankfurt am Main.
———, 1967, Layered sediments of tidal flats, beaches and shelf bottom: Amer. Assoc. Adv. Sci., v. 83: p. 191–206, 32 figs., Washington.
STRAATEN, L. M. J. U. VAN, 1950, Environment of formation and facies of the wadden sea sediments: Tijdschr. Kon. Nederl. Aardrijkskde. Genoot., v. 67: p. 94–108, 5 figs., Leiden.
———, 1959, Minor structures of some recent littoral and netritic Sediments: Geol. Mijnbouw N. S., v. 21: p. 197–216, 24 figs.
———, 1961, Sedimentation in tidal flats areas: Jour. Alberta Soc. Petrol. Geol., v. 9: p. 204–226.
———, 1964, De bodem der Waddenzee: Het Waddenboek, p. 75–151, Zutphen.
VERGER, F., 1968, Marais et wadden du littoral francais: 541 p., 230 figs. (Biscaye Frères) Bordeaux.

2. Hydrography

GOHREN, H., 1968, Triftströmungen im Wattenmeer: Mitt. Franzius Inst. für Grund- und Wasserbau Tech Univ. Hannover, v. 30, p. 142–270, 4 pls., 81 figs., Hannover.
KOCH, M., AND LUCK, G., 1966, Hydrometrische Untersuchungen im Bereich des Hohen Weges zur Frage der Wasservertriftung zwischen Jade und Weser. Jber. Forschungsstelle Norderney, v. 18, p. 57–70, 5 figs., Norderney.
KORITZ, D., 1957, Hydrometrische Untersuchungen auf dem Wurster Watt zwischen Weddewarten und Solthorner Buhne: Jber. Forschungsstelle Norderney, 1956, v. 8, p. 99–120, Norderney.
LÜNEBURG, H., 1952, Beiträge zur Hydrographie der Wesermündung: Inst. Meeresforsch. Bremerh., vol. p. 91–114, 8 pls., 3 figs., Bremen.
POSTMA, H., 1950, The distribution of temperatue and salinity in the Wadden Sea: Tijdschr. Kon. Nederl. Aardrijkskde. Genootsch., v. 67, p. 34–42, 9 figs., 2 photo., Leiden.
———, 1954, Hydrography of the Dutch Wadden Sea. A study of the relations between water movement, the transport of suspended materials and the production of organic matter: Proefschr. Rijksunivers., 106 p., 55 figs., Groningen.
———, 1961, Transport and accumulation of suspended matter in the Dutch Wadden Sea: Netherlands Jour. Sea Res., v. 1, p. 148–109, 10 figs., Groningen.
STRAATEN, L. M. J. U. VAN, 1961, Directional effects of winds, waves and currents along the Dutch North Sea Coast: Geol. Mijnbouw. v. 40, p. 333–346, 23 fig.
THIJSSE, J. TH., 1950, Veranderingen in waterbewegingen en bodemrelief in de Waddenzee: Tijdschr. Kon. Nederl. Aardrijkskde. Genootschl., v. 67: p. 326–333, Leiden.
VEEN, J. VAN, 1950, Eb- en Vleedschaar systemen in de nederlandse Getijwateren: Tijdschr. Kon. Nederl. Aardrijkskde. Genootsch. v. 66, p. 303–325, Leiden.

3. Petrography

BOURCARDT, J., 1954, Les vases de la Méditerranée et leur mécanisme de dépôt: Deep-sea Res., v. 1, p. 126–130, London.
CROMMELIN, R. D., 1940, De herkomst van het zand van de Waddenzee: Tijdschr. Kon. Nederl. Aardrijkskde. Gen., v. 62, p. 347–361.
———, 1943, De herkomst van het waddenslik met korrelgrootte boven 10 micron: Verhand. Geol. Mijnbouw. Gen. Nederl. Kolon., Geol. Ser., v. 13, p. 299–333.
DECHEND, W., 1950, Sedimentpetrologische Untersuchungen zur Frage der Sandumlagerungen im Watt Nordfrieslands: Dt. hydrogr. Z., v. 3, 294 p., Hamburg.
HANSEN, K., 1951, Preliminary report on the sediments of the danish wadden sea: Meddel. Dansk Geol. Foren., v. 12, p. 1–26, 21 figs., København.
KONING, A., 1950, Observations concerning sedimentation in the Wadden Sea area, in the light of some granular analyses: Kon. Nederl. Aardrijkskde. Genootsch., v. 67, p. 342–348, 4 figs., Leiden.
LÜDERS, K., 1939, Sediments of the North Sea: *in* TRASK, P. D.: Recent marine Sediments.—Amer. Assoc. Petrol. Geologists, p. 322–342, 4 figs. Menasha.
POSTMA, H., 1957, Size frequency distribution of sands in the Dutch wadden sea. A study in connection with ecological investigations in a tidal area: Arch. Néerl. Zool., v. 12, p. 319–349, 24 figs., Leiden.
SIMON, W. G., 1957, Sedimentpetrographische Kartierung des Neuwerker Watts im Sommer 1952: Die Küste, v. 6, p. 130–146, 5 figs., Heide (Holst.).
REINHOLD, TH., 1949, Over het mechanisme der sedimentatie op de Wadden: Med. Geol. Stichting, N. S., v. 3, p. 75–81.
SINDOWSKI, K.-H., 1957, Die synoptische Methode des Kornkurven-Vergleiches zur Ausdeutung fossiler Sedimentationsräume: Geol. Jb., v. 73, p. 235–275, 68 figs., Hannover.

STRAATEN, L. M. J. U. VAN, AND KUENEN, PH. H., 1957, Accumulation of fine grained sediments in the Dutch Wadden Sea: Geol. Mijnbouw. N. S., v. 19, p. 329–354, 19 figs.
———, AND ———, 1958, Tidal action as a cause of clay accumulation: Jour. Sed. Petrology, v. 28, p. 406–413, 4 figs., Tulsa.
VRIES KLEIN, G. DE, AND SANDERS, J. E., 1964, Comparison of sediments from Bay of Fundy and Dutch Wadden Sea tidal flats: Jour. Sed. Petrology, v. 34, p. 18–24, fig. 1, Tulsa.
WOLHENBERG, E., 1953, Sinkstoff, Sediment und Anwuchs am Hindenburgdamm: Die Küste, v. 2, p. 33–91, Heide.

4. Areal Publications

ANDREWS, R. S., 1965, Modern sediments of Willapa Bay, Washington: A coastal plain estuary: Techn. Report No. 118; 43 p., Seattle.
BOURCART, J., AND FRANCIS-BOEOF, CL., 1942, La Vase: Act. Scientif. Industr., v. 927, 67 p. (Hermann & Cie), Paris.
DEBYSER, J., 1957, La sédimentation dans le bassin d'Arcachon: Bull. Centre Etudes Rech. Scientif. Biarritz, v. 3, p. 405–418, Biarritz.
DIRCKSEN, R., 1951, Das Wattenmeer: 221 p., 97 figs., (Bruckmann) München.
DORJES, J., AND GADOW, S., REINECK, H.-E., AND SINGH, I. B., 1969, Die Rinnen der Jade (Südliche Nordsee). Sedimente und Makrobenthos: Senckenbergiana maritima, v. 1, p. 5–62, 15 figs., Frankfurt am Main.
ELLIS, CH. W., 1962, Marine sedimentary environments in the vicinity of the Norwalk Islands, Connecticut: State geol. Natural Hist. Surv. Connecticut Bull. 94, p. 1–89, 1 pl., 44 figs.
EMERY, K. O., AND STEVENSON, R. E., 1957, Estuaries and lagoons: Geol. Soc. Amer. Mem., 67: p. 673–750, 3 pls., 30 figs., New York.
EVANS, GR., 1959, The development of the coastline of the Wash, England: 2nd Coastal Geography Conference, p. 265–283, Washington.
———, 1965, Intertidal flat sediments and their environments of deposition in the Wash: Quart. Jour. Geol. Soc. London, v. 121, p. 209–245, pls. 16–20, 9 figs., London.
GELLERT, J. F., 1952, Das Aussenelbwatt zwischen Cuxhaven-Duhnen und Scharhörn. Ergebnisse einer geologisch-morphologischen Kartierung: Petermanns Geogr. Mitt., v. 2, p. 103–109, pl. 11, 2 figs., Gotha.
GIERLOFF-EMDEN, H. G., 1961, Luftbild und Küstengeographie am Beispiel der deutschen Nordseeküste: Bundesanst. Landeskde. Raumforsch., v. 4, p. 1–117, 22 figs., Godesberg.
GRIPP, K., 1956, Das Watt: Begriff, Begrenzung und fossile Vorkommen: Senckenbergiana lethaea, v. 37, p. 149–181, 3 pls., 5 figs., Frankfurt am Main.
GROHNE, U., 1956, Die Geschichte des Jadebusens und seines Untergrundes: Natur u. Volk, v. 86, p. 225–233, 7 figs., Frankfurt am Main.
GUILCHER, A., 1963, Estuaries, deltas, shelf, slope:—in The Sea, v. 3, p. 620–654, 22 figs. (Wiley & Sons), New York-London.
HÄNTZSCHEL, W., BRAND, E., BROCKMANN, CHR., OLDEWAGE, H., AND PFAFFENBERG, 1941, Zur jüngsten geologischen Entwicklung der Jade-Bucht: Senckenbergiana, v. 23, p. 33–122, pl. 1, 13 figs., Frankfurt am Main.
HAGEMAN, B. P., 1964, Blad Goeree en Overflakkee: Toelichtingen bij de Geologische Kaart van Nederland 1:50.000. Geologishce Stichting, Afd. Geologische Dienst, Haarlem.
HATAI, K., AND MII, H., 1955, Oberservations on the tidal flats in Uranouchi Bay, Kochi Prefecture, Shikoku, Japan: Rec. Oceanogr. Works, v. 2, p. 168–184, 1 pl.
JACOBSEN, N. K., 1964, Tondermarskens Naturgeografi med saeligt henblik pa Morfogenesen: Folia Geogr. Danica Tom., v. 7, 76 p., Kopenhagen.
KINDLE, E. M., 1930, The intertidal zone of the Wash, England: Comm. Sedimentation, v. 92, p. 5–21, Washington.
MCMULLEN, R. M., SWIFT, D. J. P., AND LYALL, A. K., 1967, A tidal delta with an ebb-flood channel system in the Minas Basin, Bay of Fundy: Preliminary Rept.:—Marit. Sediments, v. 3, p. 12–16, 4 figs., Dartsmouth, Nova Soctia.
MORGAN, J. P., LOPOK, J. R., VAN, AND NICHOLS, L. G., 1953, Occurrence and development of mud flats along the western Louisiana coast. Traff. & Nav., of Delta type coast: Traff. & Nav. of Louisiana coastal marshes: Tech. Rept. 29 p. 1–34.
PHILIPPONEAU, M., 1956, La Baie du Mont St-Michel. Etude de morphologie littorale: Mém. Soc. Geol. Mineral. Bretange, v. 11, p. 1–200, Rennes.
REDFIELD, A. C., 1965, Ontogeny of a salt marsh estuary: Sci., v. 147, p. 50–55, 4 figs.
REIMNITZ, E., AND MARSCHALL, N. F., 1965, Effects of the Alaska earthquake and tsunami on recent deltaic sediments: Jour. Geophys. Res., v. 70, p. 2363–2376, 12 figs.
REINECK, H.-E., MÜLLER, C. D., AND VOLLBRECHT, K., 1965, Der Knechtsand: Jber. Forschungsstelle Norderney 1964, v. 16, p. 143–201, 13 figs., Norderney.
REINHARDT, W., 1958, Zum Bodenaufbau des Quartärs, besonders des Holozäns der Ostfriesischen Küste von Juist bis Langeroog: Jber. Forschungsstelle Norderney 1957, v. 9, p. 11–30, Norderney.
RUMMELEN, F. F. F. E. VAN, 1965, Bladen Zeeuwsch-Vlaaderen West en Oost: Toelichtingen buj de Geologische Kaart van Nederland 1:50:000 Geologische Stichting, Afd. Geologische Dienst, Haarlem.
RUSSELL, R. J., AND MCINTIRE, W. G., 1965, Australian tidal flats: Australian Coastal Studies Techn. Rept. Nor 26 Part B.
SCHÄFER, W., 1954, Mellum: Inselentwicklung und Biotopwandel. Abh. naturwiss. Verein, v. 33, p. 391–406, 2 figs., Bremen.
SCHUSTER, O., 1952, Die Vareler Rinne im Jadebusen. Die Bestandteile und das Gefüge einer Rinne im Watt: Abh. senckenb. naturf. Ges., v. 486, p. 1–38, 14 figs., Frankfurt am Main.
SINDOWSKI, K. H., 1958, Das Eem im Wattgebiet zwischen Norderney und Spiekeroog, Ostfiesland: Geol. Jb., v. 76, p. 151–174, 10 figs., Hannover.

STEVENSON, R. E., AND EMERY, K. O., 1958, Marshlands at Newport Bay, California: Allan Hancock Foundation Pub. 20, p. 1–109, 50 figs., Los Angeles.
STRAATEN, L. M. J. U. VAN, 1950, Environment of formation and facies of the Wadden Sea sediments: Waddensymposium, p. 94–108.
——, 1954, Composition and structure of recent marine sediments in the Netherlands: Leids. Geol. Meded., v. 19, p. 1–110, 12 pls, 26 figs., Leiden.
——, 1957, The excavation at Velsen, general introduction. The Holocene deposits: Verhandel. Kon. Nederl. Geol. Mijnbouwkde. Genootsch., Geol. Ser., v. 17, p. 93–99, 158–183, pls. Taf. 1–2, 4 figs.
——, 1959, Origin of Dutch tidal flat formation: Salt Marsh Conference, v. 58, p. 9–21, Georgia.
——, 1963, Aspects of holocence sedimentation in the Netherlands: Geol. Inst., p. 149–169, Groningen.
——, 1965, Coastal barrier deposits in South- and North Holland in particular in the areas around Scheveningen and Ijmuiden: Meded. Geol. Stichting, N.S., v. 17, p. 41–75, pls. 1–14, 26 figs.
SWIFT, D. J. P., AND MCMULLEN, R. M., 1968, Preliminary studies of intertidal sand bodies in the Minas Basin, Bay of Fundy, Nova Scotia: Canadian Jour. Earth Sci., v. 5, p. 175–183, 1 pl., 17 figs.
VOORTHUYSEN, I. H. VAN, 1960, Das Ems-Estuarium (Nordsee): Verhand. Kon. Nederl. Geol. Mijnbouwk. Genootsch., v. 19, p. 1–300, Gravenhage.
VRIES KLEIN, G. DE 1964, Sedimentary facies in bay of fundy intertidal zone, Nova Scotia, Canada: Dev. Sedimentology, v. 1, p. 193–199, 1 fig., Amsterdam.
WRAGE, W., 1930, Das Wattenmeer zwischen Trischen und Friedrichtskoog: Archiv Deutsche Seekarte, v. 48, 128 p., Hamburg.
——, 1958, Luftbild und Wattforschung: Petermanns Geogr. Mitt., p. 6–12, Gotha.

5. Sedimentary Structures

GREENSMITH, J. T., AND TUCKER, E. V., 1966, Morphology and evolution of inshore shell ridges and mudmounds on modern intertidal flats, near Bradwell, Essex: Proc. Geol. Assoc., v. 77, p. 329–346, 6 figs.
HÄNTZSCHEL, W., 1935, Fossile Schrägschichtungs-Bögen, "Fließwülste" und Rieselmarken aus dem Nama-Transvaal-System (Südafrika) und ihire rezenten Gegenstücke: Senckenbergiana, v. 17, p. 167–177, 7 figs. Frankfurt am Main.
——, 1935, Rezente Eiskristalle in meerischen Sedimenten und fossile Eiskristall-Spuren: Senckenbergiana, v. 17, p. 151–167, 12 figs., Frankfurt am Main.
——, 1935, Erhaltungsfähige Schleifspuren von Gischt am Nordseestrand: Natur u. Volk, v. 65, p. 461–465, 6 figs., Frankfurt am Main.
——, 1936, Die Schichtungs-Formen rezenter Flachmeer-Ablagerungen im Jade-Gebiet: Senckenbergiana, v. 18, p. 316–356, 20 figs., Frankfurt am Main.
——, 1938, Bau und Bildung von Groß-Rippeln im Wattenmeer: Senckenbergiana v. 20, p. 1–42, 21 figs., Frankfurt am Main.
——, 1938, Senkrecht gestellte Schichtung in Watt-Ablagerungen: Senckenbergiana, v. 20, p. 43–48, 5 figs., Frankfurt am Main.
——, 1939, Schlick-Gerölle und Muschel-Klappen als Strömungs-Marken im Wattenmeer: Natur u. Volk, v. 69, p. 412–416, 6 figs., Frankfurt am Main.
——, 1941, Entgasungs-Krater im Watten-Schlick: Natur. u. Volk, v. 71, p. 312–314, Frankfurt am Main.
HÜLSEMANN, J., 1955, Großrippeln und Schrägschichtungs-Gefüge im Nordsee-Watt und in der Molasse: Senckenbergiana lethea, v. 36, p. 359–388, pls. 1–3, 14 figs., Frankfurt am Main.
LÜDERS, K., 1929, Entstehung und Aufbau von Großrücken mit Schillbedeckung in Flut- bzw. Ebbetrichtern der Außenjade: in LÜDERS, and TRUSHEIM: Beiträge zur Ablagerung mariner Mollusken in der Flachsee: Senckenbergiana, v. 11, p. 123–142, Frankfurt am Main.
——, 1930, Entstehung der Gezeitenschichtung auf den Watten im Jadebusen: Senckenbergiana, v. 12, p. 229–254, 10 figs., Frankfurt am Main.
——, 1932, Sediment und Strömung: Senckenbergiana, v. 14, p. 387–390, Frankfurt am Main.
——, 1934, Über das Wandern der Priele: Abh. naturwiss. Verein, v. 29, p. 19–32, pl. 1, 6 figs., Bremen.
MCKEE, E. D., 1957, Primary structures in some recent sediments: Bull. Amer. Assoc. Petroleum Geologists, v. 41, p. 1704–1747, 8 pls., 28 figs., Tusla.
PAPP, A., 1939, Beobachtungen über Sediment-Sonderung und Spülsäume an Binnenmeeren: Senckenbergiana, v. 21, p. 112–118, 3 figs., Frankfurt am Main.
REINECK, H.-E., 1955, Haftrippeln und Haftwarzen, Ablagerungs-formen von Flugsand: Senckenbergiana lethaea, v. 36, p. 347–357, 3 pls., 1 fig., Frankfurt am Main.
——, 1955, Eisblumen im Watt: Natur u. Volk, v. 85, p. 400–402, 1 fig., Frankfurt am Main.
——, 1955, Rippelartige Erosionsformen auf Schnittflächen: Senckenbergiana lethaea, v. 36, p. 191–194, pls. 1–2, 1 fig., Frankfurt am Main.
——, 1955, Schleif-Marken und Liege-Marken von strandendem Gischt: Senckenbertiana lethaea, v. 35, p. 357–359, 1 pl., Frankfurt am Main.
——, 1956, Abschmelzreste von Treibeis an den Ufersäumen des Gezeiten-Meeres: Senckenbergiana lethaea, v. 37, p. 99–304, pls. 5–9, 1 fig., Frankfurt am Main.
——, 1956, Wattenmeer im Winter: Senckenbergiana lethaea, v. 37, p. 129–146, pls. 1–6, Frankfurt am Main.
——, 1956, "Schlicksand", ein Aufarbeitungsprodukt von Grodenkanten: Senckenbergiana lethaea, v. 37, p. 125–129, Frankfurt am Main.
——, 1958, Longitudinale Schrägschicht im Watt: Geol. Rdsch., v. 42, p. 73–82, pl. 1, 14 figs., Stuttgart.
——, 1960, Über Zeitlücken in rezenten Flachsee-Sedimenten: Geol. Rdsch., v. 49, p. 149–161, 3 figs., Frankfurt am Main.
——, 1960, Über eingeregelte und verschachtelte Röhren des Goldköcher-Wurmes (*Pectinaria koreni*): Natur u. Volk, v. 90, p. 334–337, 5 figs., Frankfurt am Main.

———, 1961, Sedimentbewegungen an Kleinrippeln im Watt: Senckenbergiana lethaea, v. 42, p. 51–67, pls. 1–3, 5 figs., Frankfurt am Main.
———, 1962, Die Orkanflut vom 16. Februar 1962: Natur u. Mus., v. 92, p. 151–172, 16 figs., Frankfurt am Main.
———, 1962, Schichtungsarten in Wattenböden: Pflanzenernährung Düngung, Bodenkunde, v. 99, p. 154–159, 6 figs., Weinheim/Bergstr.
———, AND WUNDERLICH, F., 1967, Zeitmessungen an Gezeitenschichten: Natur u. Mus., v. 97, p. 193–197, 3 figs., Frankfurt am Main.
———, AND ———, 1969, Die Entstehung von Schichten und Schichtbänken im Watt: Senckenbergiana maritima, v. 1, p. 85–105, 3 pls., 4 figs., Frankfurt am Main.
RICHTER, R., 1922, Flachseebeochtungen zur Paläontologie und Geologie: Senckenbergiana, v. 4, p. 105–131, pl. 3, figs., 1–3, Frankfurt am Main.
———, 1924, Flachseebeobachtungen zur Paläontologie und Geologie VII–XI: Senckenbergiana, v. 6, p. 119–165, 8 figs., Frankfurt am Main.
———, 1926, Flachseebeobachtungen zur Paläontologie und Geologie, XV–XVI: Senckenbergiana, v. 8, p. 297–315, pls. 6–8, Frankfurt am Main.
———, 1935, Marken und Spuren im Hunsrück-Schiefer. I. Gefließ-Marken: Senckenbergiana, v. 17, p. 244–263, 12 figs., Frankfurt am Main.
———, 1936, Marken und Spuren im Hunsrück-Schiefer. II. Schichtung und Grundleben: Senckenbergiana, v. 18, p. 215–244, 4 figs., Frankfurt am Main.
———, 1942, Die Einkippungsregel: Senckenbergiana, v. 25, p. 181–206, 1 fig., Frankfurt am Main.
———, 1952, Fluidal-Textur in Sediment-Gesteinen und über Sedifluktion überhaupt: Notizbl. hess. L.-Amt Bodenforsch., v. 3, p. 67–81.
SCHÄFER, W., 1941, Zur Fazieskunde des deutschen Wattenmeeres. 1. Dangast und die Ufersäume des Jadebusens:—Taf. 1–8, Abb. 1–8; 2. Mellum, eine Düneninsel der deutschen Nordseeküste: Abh. senckenberg. naturf. Ges., v. 457, p. 1–54, pls. 9–12, 7 figs., Frankfurt am Main.
———, 1948, Zum Untergang der Oberahneschen Felder im Jadebusen: Senckenbergiana, v. 29, p. 1–16, 9 figs., Frankfurt am Main.
———, 1950, Klaffmuschel-Spülsäume am Wattenstrand: Natur u. Volk, v. 80, p. 173–176, 2 figs., Frankfurt am Main.
———, 1953, Schwimmende Verfrachtung von Muschelklappen: Natur u. Volk, v. 83, p. 355–357, 1 fig., Frankfurt am Main.
———, 1954, Dehnungsrisse unter Wasser im meerischen Sediment: Senckenbergiana lethaea, v. 35, p. 87–99, 12 figs., Frankfurt am Main.
———, 1954, "Geführte" Trockenrisse: Natur u. Volk, v. 84, p. 14–17, 3 figs., Frankfurt am Main.
———, 1956, Gesteinsbildung im Flachseebecken am Beispiel der Jade: Geol. Rdsch., v. 45, p. 71–84, 5 figs., Stuttgart.
SCHWARZ, A., 1929, Schlickfall und Gezeitenschichtung: Senckenbergiana, v. 11, p. 152–155, Frankfurt am Main.
———, 1930, Ein Seeigelstachel-Gestein: Natur. Mus., v. 60, p. 502–506, 3 figs., Frankfurt am Main.
———, 1932, Spannungsauswirkungen an raumschwündigen Stoffen: Senckenbergiana, v. 14, p. 300–331, 6 figs., Frankfurt am Main.
STRAATEN, L. M. J. U. VAN, 1950, Periodic patterns of rippled and smooth areas on water surfaces, induced by wind action: Kon. Nederl. Akad., v. 53, p. 1217–1227, 3 figs., 4 photos, Amsterdam.
———, 1950, Giant ripples in tidal channels: Kon. Nederl. Aardrijkskede. Genootsch., v. 67, p. 336–341, 3 figs., 2 photos., Leiden.
———, 1951, Texture and genesis of Dutch Wadden Sea sediments:—Proc. 3. internat. Congr. Sedimentology Groningen, p. 225–244, Wageningen.
———, 1951, Longitudinal ripple marks in mud and sand.—Jour. Sed. Petrology, v. 21, p. 47–54, 5 figs., Menasha.
———, 1953, Megaripples in the dutch wadden sea and in the basin of Arcachon: Geol. Mijnbouw. N.S., v. 15, p. 1–11, 7 figs., Leiden.
———, 1956, Composition of shell beds formed in tidal flat environment in the Netherlands and in the Bay of Archachon: Geol. Mijnbouw. N.S., v. 18, p. 209–226, 3 figs., Haag.
TRUSHEIM, F., 1929, Zur Bildungsgeschwindigkeit geschichteter Sedimente im Wattenmeer, besonders solcher mit schräger Parallel-schichtung. Senckenbergiana, v. 11, p. 47–55, 7 figs., Frankfurt am Main.
———, 1929, Trockenrisse mit Hydrobienfüllung im Schlickwatt: Natur u. Mus., v. 59, p. 52–54, 4 figs., Frankfurt am Main.
———, 1929, Rippeln im Schlick. Natur u. Mus., v. 59, p. 52–79, figs., Frankfurt am Main.
———, 1931, Spulsäume am Meeresstrand: Natur u. Museum, v. 61, p. 112–119, 8 figs., Frankfurt am Main.
———, 1936, Klein Beobachtungen an Rippeln. Natur u. Volk, v. 66, p. 288–293, 9 figs., Frankfurt am Main.
———, 1931, Versuche uber Transport und Ablagerung von Mollusken: in LÜDERS and TRUSHEIM: Beiträge zur Ablagerung mariner Mollusken in der Flachsee: Senckenbergiana, v. 13, p. 124–139, 3 figs. Frankfurt am Main.
———, 1936, "Wattenpapier", Natur u. Volk, v. 66, p. 103–106, 3 figs., Frankfurt am Main.
VRIES KLEIN, G. de, 1967:, Paleocurrent analysis in relation to modern marine sediment dispersal patterns: Amer. Assoc. Petroleum Geologists Bull., v. 51, p. 366–283, 7 figs. Tulsa.
WUNDERLICH, F., 1967. Feinblättrige Wechselschichtung und Gezeitenschichtung: Senckenbergiana lethasea, v. 48, p. 337–343, 2 pls., Frankfurt am Main.
———, 1967, Die Entstehung von "convolute bedding" an Platenrändern: Senckenbergiana lethaea, v. 48, p. 345–349, 1 Pl., Frankfurt am Main.
———, 1969, Studien zur Sedimentbewegung. 1. Transportformen und Schichtbildung im Gebiet der Jade: Senckenbergiana maritima, v. 1, p. 107–146, Pls. 1–6, 7 figs. Frankfurt am Main.

6. Biology and Aktuo-Paleontology

BARTENSTEIN, H., 1938, Die Foraminiferen-Fauna des Jade-Gebietes. 2. Foraminiferen der meerischen und brackischen Bezirke des Jade-Geibietes. Senckenbergiana, v. 20, p. 386–412, 15 figs, 2 maps, Frankfurt am Main.

——— AND BRAND, E., 1938, Die Foraminiferen-Fauna des Jade-Gebietes. 1. *Jadammina polystoma* n.g.n.sp. aus dem Jade-Gebiet: Senckenbergiana, v. 20, p. 318–385, 3 figs., Frankfurt am Main.

BENDEGOM, L. VAN, 1950, Enkele Beschousingen over de Vorming en vervorming von Wadden: Tijdschr. Kon. Nederl., p. 66–73.

BRADLEY, W. H. (1957): Physical and Ecologie Features of the Sagadahoc Bay Tidal Flat, Georgetown, Maine. —Geol. Soc. Amer. Mem. 67, p 641–682, 7 pfs., 10 figs.

BRAND, E. 1941, III. Die Foraminiferen-Fauna im Alluvlum des Jade-Gebietes: Senchkenbergiana lethaea, v. 23, p. 56–70, 1 fig., Frankfurt am Main.

BROCKMANN, CHR, 1935, Diatomeen und Schlick im Jade-Gebiet: Abh. senckenberg. naturf. Ges., v. 430, p. 1–64, 3 pls., 2 figs., Frankfurt am Main.

HÄNTZSCHEL, W. 1930, *Spongia ottoi* Geinitz, ein sternförmiges Problematikum aus dem sächsischen Cenoman: Senckenbergiana, v. 12. p. 261–274, figs. 3 Frankfurt am Main.

———, 1934, Sternspuren, erzeugt von einer Muschel: *Scrobicularia plana* (Da Costa): Senckenbergiana, v. 16, p. 325–330, 3 figs., Frankfurt am Main.

———, 1936, Ein Fisch (*Gobius microps*) als Erzeuger von Sternspuren: Natur u. Volk, v. 65, p. 562–569, 6 figs., Frankfurt am Main.

———, 1936, Seeigel-Spülsäume: Natur u. Volk, v. 66, p. 293–298, 4 figs. Frankfurt am Main.

———, 1937, Erhaltungsfähige Abdrücke von Hydro-Medusen: Natur u. Volk, v. 67, p. 141–144, 4 figs., Frankfurt am Main.

———, 1937, Seestern-Kriechspuren: Natur u. Volk, v. 67, p. 513–515, 3 figs., Frankfurt am Main.

———, 1938, Quergliederung bei rezenten und fossilen Wurmröhren: Senckenbergiana, v. 20, p. 145–154, 7 figs., Frankfurt am Main.

———, 1938, Quergliederung bei *Littorina*-Fährten, ein Beitrag zur Deutung von *Kechia anmulata* Glocker: Senckenbergiana, v. 20, p. 292–304, 6 figs., Frankfurt am Main.

——— 1939, Die Lebensspuren von *Corophium volutator* (Pallas) und ihre paläontologische Bedeutung. Senckenbergiana, v. 21, p. 215–227, 7 figs., Frankfurt am Main.

———,1940, Wattenmeer-Beobachtungen am Ringelwurm *Nereis*.—Natur u. Volk, v. 70, p. 144–148, 6 figs., Frankfurt am Main.

———, 1941, Gleichlauf und Verzweigung von Fährten der Strandschnecke *Littorina:*—Natur u. Volk, v. 71, p. 117–120, 2 figs., Frankfurt am Main.

———, 1951, Lebensspuren als Kennzeichen des Sedimentations-ramues: Geol. Rdsch., v. 43, p. 551–562, 2 figs., Stuttgart.

———, 1956, Rückschau auf die paläontologischen und neontologischen Ergebnisse der Forschungsanstalt "Senckenberg am Meer": Senckenbergiana lethaea, v. 37, p. 319–330, Frankfurt am Main.

HATAI, K. AND MII, H. H., 1955, Markings on a tidal flat in Ruanouchi Bay, Shikoku: Rec. Oceanographic Works, Japan, v. 2, p. 162–167, pl. 1.

HECHT, F. AND MATERN, H., 1930, Zur Okologie von *Cardium edule L:* Senckenbergiana, v. 12, p. 361–368, 3 figs., Frankfurt am Main.

——— AND ———, 1932, Der chemische Einfluß organischer Zersetzungsstoffe auf das Benthos, dargelegt an Untersuchungen mit marinen Polychaeten, insbesondere *Arenicola marina* L.: Senckenbergiana, v. 14, p. 199–220, 4 figs., Frankfurt am Main.

———, 1933, Der Verbleib der organischen Substanz der Tiere bei meerischer Einbettung: Senckenbergiana, v. 15, p. 165–249, 19 figs., 1 map, Frankfurt am Main.

KREJCI-GRAF, K., 1932, Definition der Begriffe Marken, Spuren, Fährten, Bauten, Hieroglyphen und Fucoiden: Senckenbergiana, v. 14, p. 19–39, Frankfurt am Main.

LINKE, O., 1936, Nachträglich veränderte Lebens-Spuren im Schlickwatt: Natur u. Volk, v. 66, p. 250–252, 4 figs., Frankfurt am Main.

———, 1936, Alpen—Strandläufer und Seepocken: Natur u. Volk, v. 66, p. 23–25, 2 figs., Frankfurt am Main.

———, 1939, Die Biota des Jadebusens: Helgol. Wiss. Meeresunters., v. 1, p. 201–348.

MEYER, K. O., 1953, Der Flohkrebs *Hyale nilssoni* im Jade-Gebiet: Natur u. Volk, v. 83, p. 319–322, 3 figs., Frankfurt am Main.

REINECK, H.-E., 1956, Der Wattenboden und das Leben im Wattenboden: Natur u. Volk, v. 86, p. 268–284, 14 figs., Frankfurt am Main.

———, 1957, Über Wühlgänge im Watt und deren Abänderung durch ihre Bewohner: Paläont. Z., v. 31, p. 32–34, Stuttgart.

———, 1957, Schnabel-Spuren im Watt: Natur u. Volk, v. 87, p. 46–50, 7 figs., Frankfurt am Main.

———, 1958, Wühlbau-Gefüge in Abhängigkeit von Sediment-Umlagerungen: Senckenbergiana, leth., v. 39, p. 1–24, 54–56, 5 pls., 3 figs., Frankfurt am Main.

———, 1959, Wenn eine Seehundspur versteinerte: Natur u. Volk, v. 89, p. 47–53, 6 figs., Frankfurt am Main.

RICHTER, G., 1967, Faziesbereiche rezenter und subrezenter Wattensedimente nach ihren Foraminiferen-Gemeinschaften: Senckenbergiana lethaea, v. 48, p. 291–335, 15 figs., Frankfurt am Main.

RICHTER, R., 1924, Flachseebeobachtungen zur Paläontologie und Geologie: Senckenbergiana, v. 6, p. 119–165, figs., Frankfurt am Main.

———, 1927, Die fossilen Fährten und Bauten der Würmer, ein überblick über ihre biologischen Grundformen und deren geologische Bedeutung: Paläont. Z., v. 9, p. 193–235, Pls. 1–4, 11 figs., Berlin.

———, 1928, Aktuopaläontologie und Paläobiologie, eine Abgrenzung: Senckenbergiana, v. 10, p. 285–292, Frankfurt am Main.

SCHÄFER, W., 1938 Paläontologische Beobachtungen an sessilen Tieren der Nordsee: Senckenbergiana, v. 20, p. 323–331, 11 figs., Frankfurt am Main.

———, 1938, Bewuchs-Verteilung von Seepocken (Balaniden) im Gezeiten-Gürtel: Natur u. Volk, v. 68, p. 564–569, 7 figs., Frankfurt am Main.
———, 1938, Die geologische Bedeutung von Bohr-Organismen in tierischen Hart-Teilen, aufgezeigt an Balaniden-Schill der Innenjade: Senckenbergiana, v. 20, p. 304–313, 8 figs., Frankfurt am Main.
———, 1939, Fossile und rezente Bohrmuschel-Besiedlungen des Jade-Gebiets: Senckenbergiana, v. 21, p. 227–254, 14 figs.,; Frankfurt am Main.
———, 1939, Polypen-Kolonien im Watt: Natur u. Volk, v. 69, p. 408–412, 5 figs., Frankfurt am Main.
———, 1939, Beobachtungen an sandwühlenden Flohkrebsen der Nordsee-Küste: Natur u. Volk, v. 69, p. 512–518, 7 figs., Frankfurt am Main.
———, 1941, Assiminea und Bembideon, Fazies-Leitformen für MHW-Ablagerungen der Nordseemarsch: Senckenbergiana, v. 23, p. 136–145, 9 figs., Frankfurt am Main.
———, 1941, Fossilisations-Bedingungen von Quallen und Laichen.—Senckenbergiana, v. 23, p. 189–216, 19 figs., Frankfurt am Main.
———, 1943, Weichkörper-Bewegungen von *Buccinum undatum*: Senckenbergiana, v. 26, p. 459–466, 3 figs., Frankfurt am Main.
———, 1948, Wuchsformen von Seepocken (*Balanus balanoides*): Natur u. Volk, v. 78, 1 fig., Frankfurt am Main.
———, 1949, Sandkorallen: Natur u. Volk, v. 79, p. 244–245, 1 fig., Frankfurt am Main.
———, 1950, über Nahrung und Wanderung im Biotop bei der Strandschnecke *Littorina littorea*: Archiv. Molluskenkde., v. 79, p. 1–8, 5 figs., Frankfurt am Main.
———, 1950, Nahrungsaufnahme und ernährungsphysiologische Umstimmung bei *Aeolis papillosa*: Arch. Moll., v. 79, p. 9–14, 5 figs., Frankfurt am Main.
———, 1950, Der "Sipho" der Klaffmuschel (*Mya arenaria*): Natur u. Volk, v. 80, p. 142–145, 3 figs., Frankfurt am Main.
———, 1951, Fossilisations-Bedingungen brachyurer Krebse: Abh. senckenb. naturf. Ges., v. 485, p. 221–238, pls. 53–54, 12 figs., Frankfurt am Main.
———, 1952, Biogene Sedimentation im Gefolge von Bioturbation: Senckenbergiana, v. 33, p. 1–12, 10 figs., Frankfurt am Main.
———, 1952, Biologische Bedeutung der Ortswahl bei Balaniden-Larven: Senckenbergiana, v. 33, p. 235–246, 4 figs., Frankfurt am Main.
———, 1953, Zur Unterscheidung gleichförmiger Kot-Pillen meerischer Evertebraten: Senckenbergiana, v. 34, p. 81–93, 6 figs., Frankfurt am Main.
———, 1954, Form und Funktion der Brachyuren-Schere: Abh. senckenb. naturf. Ges., v. 489, p. 1–65, 128 figs., Frankfurt am Main.
———, 1955, über die Bildung der Laichballen der Wellhorn-Schnecke: Natur u. Volk, v. 85, p. 92–97, 6 figs., Frankfurt am Main.
——— 1955, Fossilisations-Bildungen der Meeressäuger und Vögel: Senckenbergiana lethaea, v. 36, p. 1–25, pls. 1–2, 31 figs., Frankfurt am Main.
———, 1956, Wirkungen der Benthos-Organismen auf dem jungen Schichtverband: Senckenbergiana lethaea, v. 37, p. 183–263, pls. 1–2, 35 figs., Frankfurt am Main.
———, 1962, Aktuo-Paläontologie nach Studien in der Nordsee: 666 p., pls. 1–36, 277 figs., (Kramer); Frankfurt am Main.
———, 1963, Biozönose und Biofazies im marinen Bereich: Aufsätze u. Red. senchenb. naturf. Ges., v. 11, p. 1–36; Frankfurt am Main.
SCHLOSSER, A., 1949, *Sagartia* und *Metridium*, "zwei Seerosen" der Gezeitenzone: Natur u. Volk, v. 79, p. 237–234, 4 figs., Frankfurt am Main.
SCHWARZ, A., 1929, Die Ausbreitungsmöglichkeiten der Hydrobien (Die Bildung eines einheitlichen Hydrobien-Sedimentes): Natur u. Volks, v. 59, p. 50–51, Frankfurt am Main.
———, 1931, Insektenbegräbnis im Meer: Natur u. Museum, v. 61, p. 453–465, 19 figs., Frankfurt am Main.
——— 1932, Der tierische Einfluß auf die Meeressedimente (besonders auf die Beziehungen zwischen Frachtung, Ablagerung und Zusammensetzung von Wattensedimenten): Senckenbergiana, v. 14, p. 118–172, 28 figs., Frankfurt am Main.
———, 1932, Mövengengewölle: Natur u. Museum, v. 62, p. 305–310, 7 figs., Frankfurt am Main.
———, 1932, Spülsaum aus Exuvien von Garnelen (Crangon): Senckenbergiana, v. 14, p. 428–429, 1 fig., Frankfurt am Main.
SCHUSTER, O., 1951, Die Lebensgemeinschaften aus dem Südwatt der Nordsee-Insel Mellum: Senckenbergiana, v. 32, p. 49–65, 6 figs., Frankfurt am Main.
STRAATEN, L. M. J. U. VAN, 1952, Biogene textures and the formation of shell beds in the Dutch Wadden sea: Kon. Nederl., Akad. Wet. Proc. B, v. 55, p. 500–516, 6 figs., Amsterdam.
TRUSHEIM, F. 1929, Massentod von Insekten: Natur u. Mus., v. 59, p. 54–61, 3 figs., Frankfurt am Main.
———, 1930, Sternförmige Fährten von *Corophium*: Senckenbergiana, v. 12, p. 254–260, 3 figs., Frankfurt am Main.
———, 1931, Beiträge zur Ablagerung mariner Mollusken in der Flachsee: Senckenbergiana lethaea, v. 13, p. 124–139, 3 figs., Frankfurt am Main.
———, 1932, Paläontologisch Bemerkenswertes aus der Ökologie rezenter Nordsee-Balaniden: Senckenbergiana, v. 14, p. 70–87, 11 figs., Frankfurt am Main.
WACHS, H., 1930, Schwimmfährten von Plattfischen: Natur u. Museum, v. 60, p. 459–461, 2 figs., Frankfurt am Main.
WOHLENBERG, E., 1937, Die Wattenmeer-Lebensgemeinschaften im Königshafen von Sylt. Helgol. Wiss. Meeresunters., v. 1, p. 1–92, 67 figs., Kiel.

7. Ancient Tidal Flats

GOLDRING, R. B., 1962, The trace fossils of the Baggy Beds (Upper Devonian) of North Devon, England. Paläont. Z., v. 36, p. 232–251, pls. 22–23, 5 figs., Stuttgart.

GRIPP, K., 1956, Das Watt; Begriff, Begrenzung und fossile Vorkommen: Senckenbergiana lethaea, v. 37, p. 149–181, pls. 1–3, 5 figs.. Frankfurt am Main.
GUTSTADT, A. M., 1968, Petrology and depositional environments of the Beck Spring Dolomite (Precambrian) Kingston Range, Calif.: Jour. Sed. Petrology, v. 38, p. 1280–1289, 7 figs., Tulsa.
HÄNTZSCHEL, W., 1953, Zur Frage der Kennzeichen fossiler Watten-Ablagerungen: Natur u. Volk, v. 83, p. 255–262, 4 figs., Frankfurt am Main.
———, 1955, Rezente und fossile Lebensspuren, ihre Deutung und geologische Auswertung: Experentia, v. 11, p. 373–382, 12 figs., Basel.
HESSLAND, I., 1955, Studies in the lithogenesis of the Cambrian and basal Ordovician of the Böda Hamm sequence of strata. Bull. Geol. Inst. Upsala, v. 35, p. 35–109, Upsala.
HUNGER, R., 1954, Zur Petrogenese des unteren Muschelkalks: Naturwiss. Rdsch., v. 2, p. 61–65, 11 figs., Stuttgart.
JENSSEN, W., 1954, Frühdiagenetische und spätere Veränderungen der Sedimente des Ruhrkarbons (Feinstrandigraphische Beobachtungen an Pyrit und Toneisenstein): Geol. Jb., v. 69, p. 195–206, pl. 21, Hannover.
KOPSTEIN, F. P. H. W., 1954, Graded bedding of the Harlech Dome: Proefschrift Rijksuniversiteit, 97 p., 47 figs, 3 maps, Groningen.
KUTSCHER, F., 1962, Beiträge zur Sedimentation und Fossilführung des Hunsrückschiefers. 2. Die Chondriten als Lebensanzeiger: Notizbl. hess. L.-Amt Bodenforsch., v. 90, p. 494–498, Wiesbaden.
MACKENZIE, D. B., 1968, Studies for students: Sedimentary features of alameda avenue cut, Denver, Colorado. Mountain Geol., v. 5, p. 3–13, 8 figs.
MASTERS, CH. D. (1967): Use of sedimentary structures in determination of depositional environments, Mesaverde formation, Williams Fork Mountains, Colorado: Amer. Assoc. Petroleum Geologists Bull., v. 51, p. 2033–2043, fig. 22.
MATTER, A., 1967, Tidal flat deposits in the Ordovician of Maryland: Jour. Sed. Petrology, v. 37, p. 601–609, 5 figs.
PICARD, M. D. AND TOHILL, B., 1966, Stratigraphy and petrology of Crow Mountain Sandstone Member (Triassic), Chugwater Formation, northwestern Wyoming. Bull. Amer. Assoc. Petroleum Geologists, v. 50, p. 2547–2565, 18 figs.
POTTER, P. E. AND GLASS, H. D., 1958, Petrology and sedimentation of the Pennsylvanian sediments in Southern Illinois: a vertical profile: Illinois State Geol. Surv., Rep. Invest. 204 60 p., pl. 1–6, 19 figs., Urbana.
ROEHL, P. O., 1967, Stony mountain (Ordovician) and Interlake (Silurian) facies analogs of recent low-energy marine and subaerial carbonates, Bahamas: Amer. Assoc. Petroleum Geologists Bull., v. 51, p. 1979–2032, 48 figs.
RÜCKLIN, H., 1953, Die Grenzschichten Buntsandstein/Muschelkalk im Saarland- ein fossiles Watt: Jber. Mitt. Oberrhein, Geol. Ver., v. 35, p. 26–42, 1 pl., 5 figs., Stuttgart.
———, 1955, Das Holzer Konglomerat im Saarkarbon. Eine geröllanalytische Studie: Geol. Jb., v. 70, p. 435–510, pl. 31, 30 figs., Hannover.
RUSNAK, G. A., 1957, A fabric and petrologic study of the Pleasantview Sandstone: Jour. Sed. Petrology, v. 27, p. 41–55, 8 figs., Tulsa.
SMITH, N. D., 1968, Cyclic sedimentation in a Silurian intertidal sequence in eastern Pennsylvania: Jour. Sed. Petrology, v. 38, p. 1301–1304, 4 figs., Tulsa.
STRAATEN, L. M. J. U. VAN 1954, Sedimentology of recent tidal flat deposits and the psammites du Condroz (Devonian): Geol. Mijnbouw, N. S., v. 16, p. 25–47, pls. 1–2, 15 figs.
TANNER, W. F., 1963, Permian shoreline of central New Mexico:—Bull. Amer. Assoc. Petroleum Geologists, v. 47, p. 1604–1610, 9 figs., Tulsa.
TRUSHEIM, F., 1934, Eine bedeutsame Schichtfläche aus dem Muschelkalk und ihre Auswertung durch die Meeresgeologie: Natur u. Volk, v. 64, p. 333–340, 6 figs., Frankfurt am Main.
VRIES KLEIN, G. DE, 1967, Paleocurrent analysis in relation to modern marine sediments dispersal patterns: Amer. Assoc. Petroleum Geologists Bull., v. 51, p. 366–382, 7 figs.
———, 1967, Comparison of recent and ancient tidal flat and estuarine sediments: Amer. Assoc. Adv. Sci., v. 83, p. 207–218, 7 figs.
WEST, I. M., BRANDON, A. AND SMITH, M., 1968, A tidal flat evaporitic facies in the Viseán of Ireland: Jour. Sed. Petrology, v. 38, p. 1079–1093, 14 figs., Tulsa.
WUNDERLICH, F., 1966, Genese und Umwelt der Nellenköpfchenschichten (oberes Unterems, rhein. Devon) am locus typicus im Vergleich mit der Küstenfazies der Deutschen Bucht: Dissertation zur Erlangung der Doktorwürde. Frankfurt.

8. Miscellaneous

BUCHER, W. H., 1938, Key to papers published by an institute for the study of modern sediments in shallow sea:—Jour. Geol., v. 46, p. 726–755, 6 figs.
HÄNTZSCHEL, W. 1956, Rückschau auf die paläontologischen und neontologischen Ergebnisse der Forschungsanstalt "Senckenberg am Mee": Senckenbergiana lethaea, v. 37, p. 319–330.
RICHTER, R., 1929, Gründung und Aufgaben der Forschungsstelle für Meeresgeologie "Senckenberg" in Wilhelmshaven: Natur u. Mus., v. 59, p. 1–30, 22 figs., Frankfurt am Main.
———, 1929, Grundsätzliches zur Erweiterung der Forschungsanstalt für Meeresgeologie und Meerespaläontologie "Senckenberg" in Wilhelmshaven: Natur u. Mus., v. 59, p. 250–253, Frankfurt am Main.
SCHÄFER, W., 1957, Aufgaben und Ziele der Meerespaläontologie: Die Naturwiss., v. 44, p. 294–299, Berlin—Göttingen—Heidelberg.
———, 1961, Zur Erweiterung der Forschungsanstalt fur Meeresgeologie und Meeresbiologie "Senckenberg" in Wilhelmshaven: Natur u. Volk, v. 91, p. 145–140, 6 figs., Frankfurt am Main.

9. Methods

CHMELIK, F. B., 1967, Electro-osmotic core cutting: Marine Geol., v. 5, pl. 321–325, 2 figs., Amsterdam.
EBERLE, G., 1929, Ein einfaches Mittel um Tierspuren zu erhalten: Natur u. Mus., v. 59, p. 200–204, 5 figs., Frankfurt am Main.
GOEMANN, H. B., 1937, Ein Sediment-Härtungsverfahren mit Dioxan und Paraffin: Senckenbergiana, v. 18, p. 77–80, 1 fig., Frankfurt am Main.
HÄHNEL, W., 1961, Die Lackfilmmethode zur Konservierung geologischer Objekte: S.–A. Präparator, Z. für Museumstechnik, v. 7, p. 243–264, 14 figs.
HÄNTZSCHEL, W., 1938, Die Anwendung der Lackfilms in der Meeres-Geologie: Natur u. Volk, v. 68, p. 117–119, 1 fig., Frankfurt am Main.
HERTWECK, G. AND REINECK, H.-E. 1966, Untersuchungsmethoden von Gangbauten und anderen Wühlgefügen mariner Bodentiere: Natur und Mus., v. 96, p. 429–438, 9 figs., Frankfurt am Main.
MCMULLEN, R. M. AND ALLEN, J. R. L., 1964, Preservation of sedimentary structures in wet unconsolidated sand using polyester resins: Marine Geol., v. 1, p. 88–97, pls. 1–2, 2 figs., Amsterdam.
REINECK, H.-E., 1957, Stechkasten und Deckweiß, Hilfsmittel des Meeresgeologen: Natur u. Volk, v. 87, p. 132–134, 2 figs., Frankfurt am Main.
——, 1958, Über das Härten und Schleifen von Lockersedimenten: Senckenbergiana lethaea, v. 39, p. 49–56, Frankfurt am Main.
——, 1960, Über Hilfsmittel und Methoden der marinen Aktuo-Geologie: 2. Deutsche Geol. Ges., v. 112, p. 531, Stuttgart.
——, 1961, Die Herstellung von Meeresboden-Präparaten im Senckenberg-Institut Wilhelmshaven: Museumskde. 1961, p. 87–89, 5 figs., Berlin.
——, 1961, Versteinerte Nordsee: Natur u. Volk, v. 91, p. 151–162, 10 figs., Frankfurt am Main.
——, 1962, Reliefgüsse ungestörter Sandproben. Pflanzenernährung, Düngung, Bodenkunde, v. 99, p. 151–153, 3 Figs., Weinheim Bergstr.
——, AND ROSENBOOM, W., 1969, Stechkästen zur Entnahme von Watten- und Unterwasserproben: Natur u. Mus., v. 99, p. 209–214, 4 figs., Frankfurt am Main.
SCHÄFER, W., 1956, Arbeitsmethoden und technische Mittel des Meerespaläontologen: Natur u. Volk, v. 86, p. 284–290, 5 figs., Frankfurt am Main.
SCHWARZ, A., 1929, Ein Verfahren zum Härten nichtverfestigter Sedimente: Natur u. Mus., v. 59, p. 204–207, 2, figs., Frankfurt am Main.
VOIGT, E., 1949, Die Anwendung der Lackfilmmethode bei der Bergung geologischer und bodenkundlicher Profile: Mitt. Geol. Staatsinst., v. 19, p. 111–129, Hamburg.

RECOGNITION OF EVAPORITE—CARBONATE SHORELINE SEDIMENTATION

F. J. LUCIA

ABSTRACT

Evaporitic carbonate shoreline sediments are deposited in an arid or semiarid climate by tidal currents which transport the sediment from the marine environment to the shore. The sediments accumulate primarily as tidal-flat deposits which normally prograde seaward, producing a vertical sedimentary sequence from marine to supratidal. The supratidal sediments are the most easily recognized. They are characterized by irregular laminations, desiccation features, and a general lack of fossil material. The intertidal sediments are best recognized by their position immediately below the supratidal sediments. Intertidal sediments are commonly pelleted carbonate muds with numerous burrows, a restricted fossil assemblage, and may contain algal stromatolites.

The arid or semiarid climate allows sea water to evaporate and become saturated with respect to gypsum or anhydrite when the water circulation is sufficiently restricted from the open ocean. The low permeability of the tidal-flat sediment can produce this restriction to circulation. A channel connecting a lagoon to the open ocean can also produce this restriction to circulation, but the channel must be so small compared to the size of the lagoon as to be inconsequential. Gypsum (anhydrite?) crystallizes out of the interstitial hypersaline water within the sediment and out of standing bodies of hypersaline water. Nodules, replacement crystals, and pore-filling crystals are formed within the sediment, and laminated evaporite sediments are formed from standing bodies of water. The precipitation of gypsum (anhydrite?) produces a dolomitizing fluid which moves down through the underlying sediment. Extensive reflux dolomitization is characteristic of evaporitic shoreline carbonate deposits.

Evaporites are very susceptible to removal by shallow ground waters. Therefore, they are rarely well preserved in outcrops. Comparisons of subsurface and outcrop evaporitic carbonate sequences commonly show that the anhydrite or gypsum has been leached from the outcrop samples leaving molds and occasionally producing solution collapse breccias. Calcification of anhydrite or gypsum and dolomite is common. The recognition of evaporitic shoreline sedimentation from outcrop samples, then, often necessitates establishing whether or not dissolution of evaporites and calcification of dolomite and anhydrite or gypsum have occurred.

INTRODUCTION

Knowledge of carbonate sedimentation has increased considerably in the past 15 years, and, as knowledge has increased, water depth has decreased. We have moved from deep water to shallow water and from shallow water to the carbonate coastal plain. The change from the shallow water marine environment to the coastal plain environment is across an important diagenetic and sedimentologic boundary. The geographical name for this boundary is the shoreline. It is the intention of this paper to present and discuss the criteria and approaches considered most useful at present to recognize the shoreline in ancient carbonate rocks, paying particular attention to evaporitic shoreline sediments. The criteria necessary to recognize evaporite-shoreline carbonate sedimentation depends upon the diagenetic state of the rocks. Therefore, we will begin with Recent carbonate shorelines which contain the least amount of diagenesis but an incomplete sedimentary record, proceed to the subsurface where the best preserved sedimentary records are found, and lastly move to the outcrop with all its diagenetic problems.

RECENT SHORELINE CARBONATE SEDIMENTS

Tidal-flat Environment

The most common carbonate shoreline environment is the tidal-flat environment. The primary source of the sediment deposited on the tidal flat is the marine environment. The sediment is carried by tidal and storm currents from the ocean onto the land, and, in this respect, the direction of transport is just opposite of that during alluvial deposition. The sediment trapped on the tidal flat causes a net addition of sediment to the tidal-flat environment. If the rate of accumulation of the tidal flat is greater than the rate of relative sea-level rise, the tidal-flat environment will tend to prograde seaward. If the relative sea-level rise is faster than the rate of sedimentation, or if the relative sea level falls, relatively little tidal-flat sediment is likely to accumulate.

Since the tidal flat reflects the action of tidal currents on a shoreline, it is proper to divide the tidal-flat environment into general environments of deposition based on the action of the tide (fig. 1). One environment is the *intertidal environment*, the limits of which are defined by mean low-tide and mean high-tide levels. The second environment is the *supratidal environment*, which is defined as that area of tidal-flat sedimentation out of reach of the daily tides. Its elevation could be higher or lower than the mean high-tide mark. This environment is covered only by spring or storm tides and lies exposed subaerially for long periods of time. Some of the material may be deposited as a precipitate from the sea water. Thus, gypsum precipitated out of a lake of marine water in the supratidal environment is considered part of the supratidal sedi-

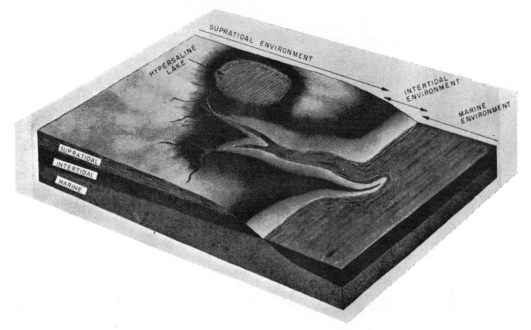

Fig. 1.—Diagrammatic view of a prograding evaporitic tidal flat showing general depositional environments and vertical sequence.

ment. All sediment deposited below mean low tide is referred to as *marine* sediment in this paper. Subtidal, however, is a perfectly acceptable synonym.

A tidal flat also reflects the manner in which the tidal currents flow on and off the tidal flat. Basically, there are two types of flow on a tidal flat: the regular on and off lap of the tides, and the funneling of the tides into channels. The presence of these two types of flow often results in a topography of flats and channels. The channels may range in size from a few inches to several hundreds of feet wide and may be found in the supratidal and intertidal environments, or, for that matter, in the marine realm of deposition.

If the tidal-flat environment progrades seaward, marine sediment should be overlain by intertidal sediments, which in turn should be overlain by supratidal sediments. This sequence should be found laterally as well as vertically. Repetition of prograding tidal-flat sedimentation cycles should result in an environmental distribution similar to that illustrated in Figure 2.

Vertical sequences described from the Recent are summarized in Table 1.* Van Straaten's

* For an excellent description of intertidal-marine sediments and particularly tidal channels, the reader is referred to Davies (1970) description of Shark Bay which was published after this paper was submitted.

(1954) sequence from the Wadden Zee is included because, although it is composed of clastic sediments, it has been used as a basis to interpret carbonate sequences.

It is apparent from Table 1 that the supratidal sediments are more distinctive than the intertidal sediments. Therefore, the recognition of the tidal-flat environment can best be accomplished through the recognition of supratidal rocks. The supratidal environment is exposed subaerially for weeks if not years at a time, and the resulting sedimentary structures reflect the periodicity of sedimentation as well as the effects of exposure. Irregular laminations caused by a combination of periodic sedimentation, dessication, filamentous algal layers, and lack of extensive life are most characteristic. Mud cracks are diagnostic. Dessication also produces "birdseye" texture. Alternating wetting and drying lithifies and tends to break the supratidal sediment into lithoclasts. Although they are indicative of the supratidal environment, such lithoclasts may be transported and thus deposited in other environments.

Once the supratidal environment has been recognized, the marine and intertidal environments of deposition are inferred to be present. The depositional sequence indicates that the sediments immediately below the supratidal should be intertidal sediments. In some cases, however, there may not have been a daily tide

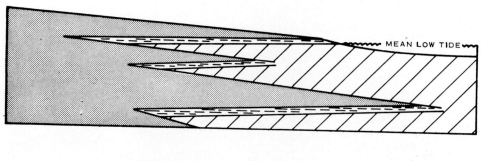

Fig. 2.—Diagrammatic cross section showing the postulated distribution of sediments from three cycles of tidal-flat sedimentation.

present during deposition, so no intertidal sediments would be present. In this case, marine sediments would directly underlie the supratidal. The differentiation between marine and intertidal sediments can be very difficult because of the similarity of the sedimentary structures. Both depositional environments can have enough burrowing organisms to rework the sediment completely and produce burrowed and churned sedimentary structures. The vertical and lateral sequence of organic communities, however, may be helpful in differentiating between marine and intertidal sediments.

The biota present in the intertidal environment should be different from the biota in the marine environment. Therefore, the transition from intertidal rocks down into marine rocks might be represented by a recognizable change in the fossil assemblage. The algal stromatolites are the most diagnostic organic feature of the tidal-flat sediments (Logan, Rezak, and Ginsburg, 1964). In Recent tidal flats algal stromatolites are found primarily in the high intertidal or the low supratidal environments (Illing, et al., 1965; Logan, 1961). Recent carbonate intertidal sediments commonly appear to be pellet packstones with gastropod (Cerithidea) shells common. The marine environment can sometimes be directly recognized by the fossil assemblages. The presence of a normal marine fossil assemblage, such as a coral, bryozoan, echinoderm, and stromatoporoid assemblage, is excellent evidence for a marine deposit. The lack of a normal marine assemblage, however, does not necessarily mean that the rock was not deposited in a marine environment.

TABLE 1.—CHARACTERISTIC FEATURES OF RECENT TIDAL FLAT SEQUENCES

	Clastics	Carbonates			
	Wadden Sea (Van Straaten, 1954)	Andros Island (Shinn, et al, 1965)	Qatar (Illing, et al, 1965)	Trucial Coast (Curtis, et al, 1963)	Bonaire (Deffeyes, et al, 1965) (Lucia, 1968)
Supratidal	Disrupted laminations	Laminated packstone Algal mats Mud cracks Lithoclasts Birdseye Dolomite	Mudstone to packstone Birdseye Some disrupted laminations Few marine organisms Gypsum crystals Dolomite	Mudstone Anhydrite nodules and lenses Gypsum crystals Lack of marine organisms Dolomite	Bedded gypsum Thin algal mats Lithoclasts Dolomite Mud cracks
Intertidal	Upper-burrowed & churned Lower-cross and horizontal laminations	Burrowed pellet packstone Marine gastropods and forams	Burrowed pellet packstone Cerithid gastropods Algal mats Gypsum crystals Dolomite	Mudstone and packstone Algal mats Dolomite	Burrowed pellet packstone Cerithid gastropods and forams Dolomite
Marine Channels	Large-scale cross laminations Numerous current ripple structures	Present but no data	Not present	?	Not present
Marine	?	Burrowed pellet packstone Marine gastropods and forams	Burrowed packstone and wackestone Small lamellibranches, gastropods, and peneroplid forams	"Lagoonal sediments"	Halimeda grainstone

Shoreline Dolomitization

Modern carbonate tidal flats commonly contain modern dolomite. The amount varies from about 1 foot of 100 percent dolomite in the Persian Gulf tidal flats to scattered dolomite crystals in the Florida Bay area. This dolomite is believed to be formed out of sea water which has been concentrated through evaporation to gypsum saturation (Deffeyes, Lucia, and Weyl, 1965). This construction of a supratidal environment by tidal-flat sedimentation produces the necessary conditions for sea-water evaporation to the point where gypsum will precipitate and thus produce a dolomitizing water. The heavy hypersaline water moves down from the supratidal surface and dolomitizes the underlying sediments. The shoreline marks the boundary between a dolomitizing environment and a nondolomitizing environment. Reflux dolomitization is part of the geologic cycle inherent with tidal-flat sedimentation, particularly in arid and semiarid climates. However, Florida Bay has a more tropical climate and small amounts of modern dolomite along with interstitial waters having salinities 5 to 6 times that of normal sea water have been reported by Shinn (1964) from the supratidal of that area.

Evaporite Deposition

Modern shoreline deposits often contain evaporite minerals, principally gypsum, anhydrite, and salt. These evaporites are generally found in arid and semiarid environments. Gypsum occurs in four main textural modes: 1) bedded or laminated, 2) displacement "axe head" crystals within the sediment, 3) filling pore space, and 4) replacing preexisting carbonate (Illing et al., 1965; Lucia 1968). Anhydrite is found only in the Arabian Trucial Coast, where it is present as nodules, coalesced nodules, and discontinuous beds within the supratidal sediment (Curtis, et al., 1963). Shoreline salt deposits are usually in the form of thin ephemeral crusts.

Nodular anhydrite and displacement, pore filling, and replacement gypsum are clearly forms produced by precipitation within the sediment out of the interstitial water and represent a diagenetic, not a sedimentary, environment (Murray, 1964; Kerr and Thomsom 1963). Bedded or laminated gypsum, however, does form out of a standing body of water and, therefore, does represent a sedimentary environment. With burial, the gypsum will convert to anhydrite. The texture of the anhydrite or gypsum must be considered first before the paleogeographic significance of the evaporites can be determined. The diagenetic forms of anhydrite or gypsum only infer the presence of hypersaline water "within" the sediment. Hypersaline water "within" the sediment occurs associated with hypersaline bodies of water or under supratidal flats. Figure 3 illustrates the distribution of hypersaline water under the Persian Gulf tidal flat in Qatar, which has been described by Illing, et al., (1965). Hypersaline water is present in supratidal, intertidal, and marine sediment under the supratidal surface. The chlorinity gradually changes across the intertidal zone until, under the marine environment, the interstitial water is similar to the marine water. Displacement, replacement, and pore-filling forms of gypsum are present under the supratidal surface.

Bedded or laminated anhydrite and gypsum imply deposition out of a standing body of water

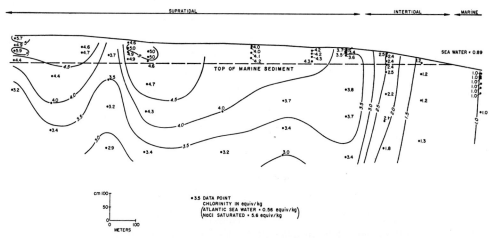

Fig. 3.—Distribution of chlorinity in the sediments under the tidal flat described by Illing, et al (1965), located in Qatar, Persian Gulf.

that is "restricted" from the ocean by some means. The question of what this "restriction" means must be answered before the geographic significance of the bedded anhydrite can be determined. On Bonaire, Netherlands Antilles, as well as along other coast lines, bedded gypsum is accumulating in coastal hypersaline lakes "restricted" from the ocean by a barrier composed of permeable sediment. The source of the sea water is flow through the sediment and not through a channel.

A hypersaline lagoon which was restricted from the ocean by a discontinuous barrier is a more commonly used paleogeographic model than the hypersaline lake. Some geologists, however, refer to evaporites being formed at the distal end of a long arm of the sea without any physical barrier except gradual shallowing. To determine the size and nature of the surface connection between a hypersaline body of water and the ocean, a study was made of the restricted marine bodies of water described in the literature. The results of this study are presented in the following discussion.

Sea water must be evaporated to about one-third its original volume in order for it to become oversaturated with respect to gypsum. This corresponds to a salinity of about 100 °/₀₀ as opposed to 35 °/₀₀ for normal sea water. Whether or not a body of marine water will reach this salinity depends upon the amount of interchange between the restricted body of water and the open ocean. Under steady-state conditions, the amount of water flowing into an evaporite basin per unit time must be equal to the amount of water flowing out plus the amount lost to evaporation minus the amount of rain water and runoff, all for the same unit of time, or

$$V_i = V_r + D, \quad (1)$$

where

V_i = volume of inflow per unit time,
V_r = volume of outflow per unit time,
D = net loss to evaporation per unit time (evaporation—rain and runoff).

If there is no precipitation of salt (NaCl)—and this is a reasonable assumption for this discussion, since sea water must be evaporated to about one-tenth its original volume for salt precipitation—the amount of salt transported in must equal the amount of salt transported out per unit time, or

$$V_i S_i = V_r S_r, \quad (2)$$

where S_i is the salinity of the inflow (in this case always taken as 35 °/₀₀, and S_r is the salinity of the outflow. The salinity of the outflow should approximate the maximum salinity reached. Substituting equation (2) into (1) yields

$$\text{Volume inflow } V_i = D \frac{S_r}{S_r - S_i}, \quad (3)$$

$$\text{Volume outflow } V_r = D \frac{S_i}{S_r - S_i} \quad (4)$$

$$S_r = \frac{V_i S_i}{V_i - D}$$

However,

$$V_i = A_i \overline{V}_i$$
$$D = A_0 E,$$

Therefore,

$$S_r = \frac{A_i \overline{V}_i S_i}{A_i \overline{V}_i - A_0 E}, \quad (5)$$

where (see fig. 4)

A_i = cross-sectional area of inflow,
$\overline{V}_i$ = velocity of inflow,
A_0 = surface area of water body,
E = net rate of evaporation.

For evaporitic environments $A_i \overline{V}_i > A_0 E$. By factoring equation (5), we have

$$S_r = S_i + \frac{S_i A_0 E}{A_i \overline{V}_i - A_0 E}$$

Rearranging gives

$$\frac{S_r - S_i}{S_i} = \frac{\dfrac{A_0 E}{A_i \overline{V}_i}}{1 - \dfrac{A_0 E}{A_i \overline{V}_i}}$$

Thus, for arid climates the salinity reached in the restricted basin (S_i) is related to the A_0/A_i ratio. This ratio can be evaluated by two approaches. With the first approach we shall use data available to us from modern restricted basins.

Data that are readily available from modern restricted basins include the surface area of the water body, the cross-sectional area of the inlet, the salinity ranges, some tide data, some depth figures, the rainfall, and some evaporation data. Thus, we can measure A_0 rather accurately and get a good idea of the salinity ranges. A_i is somewhat smaller than the cross-sectional area of the inlet (A_t) because a small part of the inlet should be occupied by the outflowing water. Some of the outflow will probably be through the sediment (subsurface reflux), which will tend to increase the amount of the inlet used by the inflow. Because we can obtain data only on the cross-sectional area of the inlet, and this is what we really want to know in a paleogeographic sense, we shall assume that A_t is a good approximation of

BARRED – BASIN MODEL

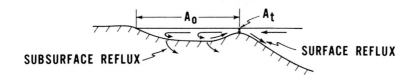

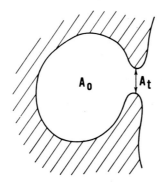

Fig. 4.—Diagram of Barred-Basin Model showing the relationship between the ocean, restricted basin, inlet, inflow, surface reflux, and subsurface reflux.

A_i. Variation in E can be held to a minimum by restricting our observation to dry climates, i.e., rainfall of less than about 60 cm per year as an arbitrary limit. We can do little about the velocity variation except to assume that a maximum value will be reached which is the same in all cases.

Table 2 lists some data from bodies of marine water in dry climates. The cross-sectional areas of the inflow for the Bocana de Virrial in Peru and for Shark Bay had to be estimated from very meager data but are believed to be within a factor of 10 of the true area. If the ratio A_0/A_t is plotted against the salinity ranges for bodies of water with an average depth of 10 meters or less in a dry climate, a rapid increase in salinity occurs when this ratio is about 10^6, as shown in Figure 5. Data from the Red Sea and the Persian Gulf fall to the left of the curve. The field of gypsum precipitation is shown on the plot and intersects the line at about $A_0/A_t = 10^6$. Thus, shallow basins in dry climates with ratios higher than this value will precipitate gypsum.

The order of magnitude for A_0/A_t needed to precipitate gypsum is supported by a second approach. If we substitute into equation (4) some reasonable values for flow velocities and for the net rate of evaporation, we can approximate a value for A_0/A_i. Because we are interested in the value of this ratio at the point where gypsum will precipitate from sea water, S_i will equal 35 ‰, and S_r will equal 100 ‰. From the available data, a reasonable value for the net rate of evaporation (E) is considered to be 200 cm/yr. cm². Actually, the value which we use for E can vary within a factor of 10 (from 30–300) and be within the limits of error of our answer. Let us consider three velocities—10, 100, and 1000 cm/sec—and assume that the maximum velocity falls within this range.* Substituting these values into equation (5) gives A_0/A_i ratios of 10^6, 10^7, and 10^8 respectively, which agree with the A_0/A_t values from Figure 5 within a factor of 10.

Thus, it is concluded that gypsum deposition means "restriction" to an A_0/A_t value in the order of 10^6 or higher in an arid or semiarid climate. While this indicates that bedded gypsum can be produced in standing bodies of water which have a connecting channel to the ocean, this channel would generally be so small as to be geographically insignificant.

There is little evidence to support the theory of evaporite deposition at the distal end of a narrow shallow sea. Present understanding of shoreline sedimentation processes indicates that the shoreline is a sediment trap. The shoreline

---

* Some maximum current velocities that have been reported are: Shark Bay—95 cm/sec (Davies, 1970), Florida Straits—150 cm/sec (Neuman and Ball, 1970), Ojo de Liebre Lagoon—200 cm/sec (Phleger, 1969).

TABLE 2.—DATA FROM RESTRICTED MARINE AREAS

Body of Water	Salinity Range (°/oo)	Daily Tides Recorded	Area of Surface A_0 (km²)	Area of Inflow A_t (km²)	A_0/A_t	Average Depth (m)	Evaporation Rate (cm/yr)	Rainfall (cm/yr)
Red Sea (Sverdrup et al., 1946)	37–41	Yes	0.4×10^6	1.5	3×10^5	491	?	0–20
Persian Gulf (Emery, 1956)	39–41	Yes	0.2×10^7	2.7	10^5	25	?	2.5–20
Ojo de Liebre (Phleger and Ewing, 1962)	36–48	Yes	2.56×10^2	2.3×10^{-2}	10^4	6	180	3
Sharks Bay	36–65	Yes	1.28×10^4	5×10^{-1}	2×10^4	10	220	20
Gulf of Salwa (Sugden, 1963)	45–59	Yes	6.7×10^3	3×10^{-1}	2×10^4	6	?	2.5–20
Bocana de Virrila (Morris and Dickey, 1957)	36–150	?	2×10	5×10^{-5}	4×10^5	2	?	10–20
Laguna Madre, South (Fisk, 1969; Hedgepeth, 1947)	12–108	No	4.60×10^2	5×10^{-4}	9×10^5	5	?	50–75
Karabugas Gulf (Grabau, 1924)	160–289	No	1.44×10^4	7.5×10^{-4}	2×10^7	10	?	2.5–10
Pekelmeer	70–360	No	3	0		1/2	150	51
Larnaca (Bellamy, 1900)	102–360	No	5.4	0		1	?	62
Persian Gulf Salina (Bramkamp and Powers, 1955)	90–200	?	?	0		?	?	2.5–20

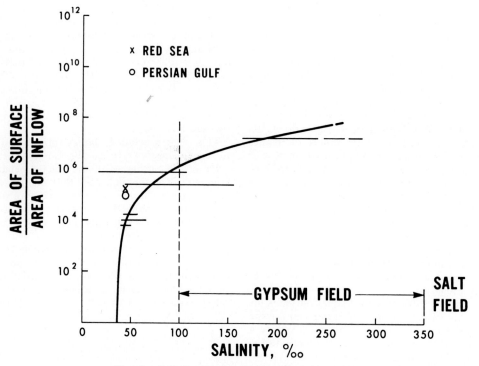

FIG. 5.—Relationship between A_0/A_i and salinity.

progrades out to fill in the shallow sea, rather than the sea bottom filling evenly as proposed by Sugden (1963). As the shoreline progrades seaward, the evaporite deposits follow.

If bedded gypsum and anhydrite infer a hypersaline lake or a large body of water with a minor inlet to the ocean, salt deposition must certainly indicate complete surface disconnection from the ocean. This certainly is true of the modern salt deposits associated with the Ujo de Liebre Lagoon in Baja California (Phleger, 1969). Here 2 to 3 meters of salt have been deposited in a salt basin which is flooded by the ocean once every 20 years. The source of the NaCl for salt deposits may be from ground water or sea water springs as well as from periodic flooding. Therefore, a bedded salt deposit infers either a salt lake on a coastal plain (supratidal) or an inland salt sea.

SUBSURFACE SHORELINE EVAPORITE CARBONATE ROCKS

The areas of Recent carbonate sedimentation provide important but incomplete data on the depositional sequence and rock textures and mineralogy to be found in evaporite-carbonate shoreline sediments. The depositional sequence is incomplete because it is in the process of accumulating, the very fact that makes studies of Recent sediments so important. The rock textures and mineralogy change with time and as new realms of temperature, pressure, and water chemistry are encountered.

The study of ancient shoreline carbonates, guided by knowledge of the Recent, provides data on the nature of the completed sedimentary sequence and the resulting rock textures and mineralogy. The best preserved sequences and textures seem to be found in the subsurface. Perhaps this is because they are "pickled" in salt water. Although there are numerous differences between the Recent and the subsurface, such as the loss of pore space, aragonite and magnesium calcite converted to low magnesium calcite or dolomite, and gypsum converted to anhydrite, the rock textures remain amazingly similar. In this section, four subsurfaces studies will be discussed to determine the criteria useful for identification of evaporite-carbonate shoreline environments in ancient rocks.

Upper Clearfork (Leonard), Flanagan and Robertson Fields, West Texas

The Flanagan field is located in Gaines County, Texas (fig. 6). Paleogeographically, it is located on the eastern side of the Permian Central Basin Platform, some ten miles from the west edge of the Midland Basin. The study of seven cores, numerous logs, and many ditch cuttings of the upper Clearfork of Leonardian (Permian) age in this area has shown the presence of thick sections of supratidal rock with interbedded marine rock, all of which are completely dolomitized. Anhydrite is present as nodules (fig. 7) as well as pore-filling and replacement crystals. No bedded anhydrite or salt was found.

The criteria for recognizing tidal-flat sedimentation as outlined from the Recent have been applied to the upper Clearfork dolomites in the Flanagan field. The common sequence of sedimentary features found is shown in Table 3.

The supratidal rocks show characteristic irregular laminations. The type of lamination varies from wispy and poorly defined to very distinct (figs. 8 and 9). In some instances, vertical breaks are found crossing laminations and appear to be desiccation cracks (fig. 9). Quartz silt is often found concentrated on top of the laminations as well as in these vertical breaks. In a few intervals, beds of current-bedded quartz-silt-rich dolomite several inches thick are found in the intervals of supratidal rocks. Only one algal stromatolite was found in the seven cores examined (fig. 10). Lithoclasts are very abundant and are found in beds as well as scattered throughout the irregularly laminated rocks. They range in size from several centimeters to several millimeters long and are texturally similar to the irregularly laminated supratidal rocks. In thin section, the texture of the supratidal rocks varies considerably. Floored cavities, small "birdseye" structures, pisoliths (?), brecciation, and other features probably resulting from early diagenesis make thin sections of supratidal rocks distinctive. The dolomite crystal size in the matrix is less than 5 microns, with larger dolomite crystals in the larger pores. Fossils are scarce, and when present are primarily small forams, ostracods, and mollusks. The top of the supratidal section is characterized by a sharp break with lithoclasts of supratidal rocks in the overlying sediment (fig. 11).

The sediment immediately underlying the supratidal sediment is characteristically a lime mud with a few fine sand-size pellets and widely scattered small fossils (figs. 12 and 13). The fact that the contact with the supratidal is transitional suggests a depositional contact. Occasionally thin beds of quartz silt are found in these rocks. Abundant discontinuous fractures filled with anhydrite and pieces of the surrounding rock are commonly but not exclusively found in this rock type. The pattern of these fractures suggests formation while the sediment was cohesive but soft and pliable. Burrows, floored cavities, and a wispy mottled structure are also

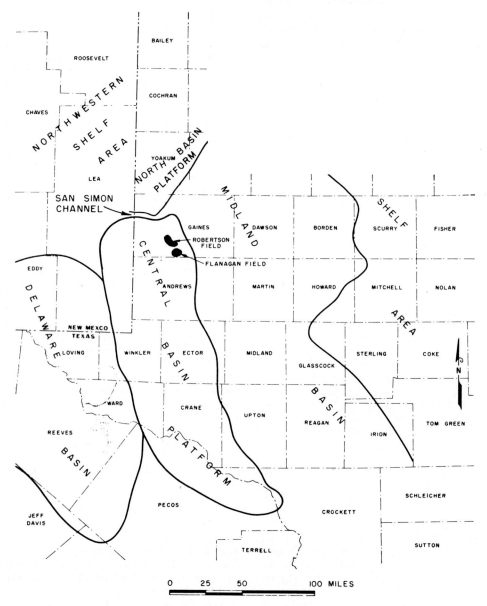

Fig. 6.—Location of Flanagan and Robertson Fields with respect to Permian paleogeography.

common. The dolomite crystals average about 15 microns in size.

Rocks with current features are found primarily within and immediately below these rocks. These rocks contain very few fossils. Dolomite crystal size is about 20 microns. The most common type of current feature present is the current lamination (figs. 14 and 15). The current-laminated rocks have alternating laminae of sorted fine-pellet sand and muddy fine-pellet sand. Figure 15 shows a channel lag composed of supratidal lithoclasts. Only one good example of cross-bedding was found. These rocks probably represent sediments deposited in channels or gullies located near the intertidal-marine boundary.

Three types of rocks are found beneath the current-laminated rocks. The most prevalent is a churned lime mud with scattered medium sand-size pellets and a few crinoid and molluscan fragments (wackestone) (fig. 16). The next most abundant type is a medium- to coarse-grained,

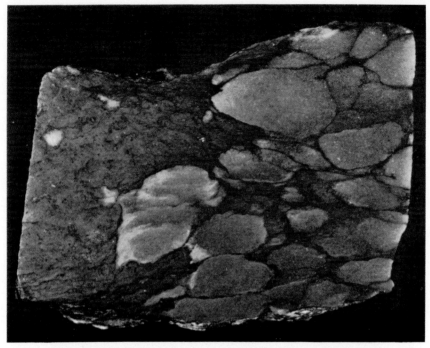

Fig. 7.—Anhydrite nodules, both isolated and coalesced, in marine dolomite, Flanagan Field, Upper Clear Fork Formation. Core is 3 inches wide.

churned lime sand with interparticle lime mud (packstone) which contains crinoids, bryozoans, and large fusulinids (fig. 17). The third type is an algal-foram (?) lime sand with some interparticle lime mud (packstone). All of these types have an average dolomite crystal size of 40 microns.

The depositional environment of the rocks described above is interpreted from the sequence and from a knowledge of tidal-flat sedimentation. There does not appear to be any evidence of lengthy subaerial exposure except within the supratidal rocks, and it is therefore reasonable to assume that this is a depositional sequence.

TABLE 3.—SEQUENCE OF SEDIMENTARY FEATURES IN THE UPPER CLEARFORK, FLANAGAN FIELD, TEXAS

Interpreted Sedimentary Environment	Sedimentary Structure	Fossils	Particle Size
Supratidal	Irregular laminations Lithoclasts Desiccation features Quartz silt beds	*Rare* Thin-shelled small forams, ostracods, molluscans.	Lithoclasts to lime mud.
Intertidal	Distinct burrows Churned-to wispy-mottled structures Quartz silt beds Algal stromatolites Discontinuous fractures	*Very few* Thin-shelled small forams, ostracods, molluscans. Filamentous algae.	Fine sand-size pellets to lime mud.
(channel)	Current-laminated rocks Cross-bedding	*Very few* Echinoids. Small molluscans.	Fine sand-size pellets to mud with some lithoclasts.
Marine	Churned rocks Burrowed rocks	*Locally abundant* Echinoids, bryozoans, large fusulinids, molluscans, algal-forams (?)	Coarse sand-size pellets to lime mud.

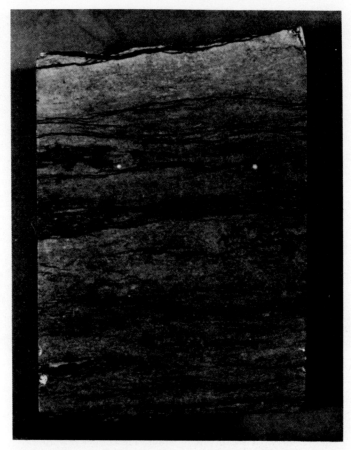

Fig. 8.—Supratidal dolomite, Flanagan Field, Upper Clear Fork Formation showing wispy laminations. Core is 3 inches wide.

Thus, the sediments immediately underlying the supratidal probably represent the intertidal zone. Since rocks of this type are only a few feet thick, they probably represent a daily tidal fluctuation of a few feet. The transition to marine rocks must be beneath this rock type and should be represented by a change in the nature of the rocks. Immediately under the intertidal sediments are current-bedded rocks which are in turn underlain by rocks with crinoids, bryozoans, and large fusulinids, i.e., a more marine fossil assemblage than is found in the intertidal rocks. The grain size is coarser than that found in the intertidal or current-feature rocks. Here, then, is a change in the nature of the rock and a suggestion of more marine conditions based on the interpretation of the biota. It is most likely, therefore, that the rocks with current features represent channel or gully deposits formed at the transition from marine to intertidal deposition and that the three rock types underlying the current-bedded rocks represent marine deposition.

Within the section of the upper Clearfork studied in the Flanagan field, there are two main marine intervals herein called the upper and lower marine wedges (fig. 18). While the upper marine interval shows little thickness variation within this area, the lower marine wedge almost completely pinches out into tidal-flat sediments (fig. 19). The marine pinchout represents the precise location of the shoreline at the maximum transgression of the sea before the progradation of the overlying tidal-flat cycle.

The paleogeographic picture reconstructed from this study is an exposed supratidal coastal plain on the Central Basin Platform, an intertidal environment or shoreline paralleling the eastern edge of the Central Basin Platform, and the shallow marine environment east of the shoreline. The reflux dolomitization and anhyrite (gypsum?) was produced in response to the prograding supratidal deposits. This reconstruction differs from that of Adams and Rhodes (1960) in emphasising the importance of the tidal flat environment.

Stony Mountain (Ordovician) and Interlake (Silurian), Williston Basin, Montana

P. O. Roehl (1967) described the sedimentary and diagenetic facies from the Stony Mountain (Ordovician) and Interlake (Silurian) formations from the Willison Basin. The rocks are essentially all dolomite and represent shallow marine to supratidal sedimentary environments. Roehl uses one unique subenvironment name, *infratidal*, which is essentially synonymous with shallow marine. Roehl's environmental criteria are listed below.

Environment	Criteria
Supratidal	1. Laminated dolomite.
	2. Algal mats and stromatolites.
	3. Flat-pebble breccias or conglomerates
Intertidal	1. Argillaceous burrows.
Marine (Infratidal)	1. Churned.
	2. Bioclasts.

The anhydrite beds present in these sequences are generally confined to the supratidal facies and are texturally coalesced nodular rather than bedded anhydrite. My experience, gained from examining anhydrite units from Ordovician cores in this area, is similar. The thickness of these anhydrite beds ranges from 1 to 20 feet. Some bedded anhydrite is present but is usually in beds not over 6 inches to a foot thick (fig. 20). Therefore, the "restriction" which produced the hypersaline water necessary for the formation of the anhydrite (gypsum) beds was most likely the permeability of the tidal-flat sediment, and the paleogeography was more coastal plain with saline or hypersaline lakes rather than an evaporite lagoon or a barred-basin.

Castile Formation (Ochoan), West Texas

The Castile evaporite fills the Delaware Basin and consists of about 1,800 feet of laminated anhydrite and calcite with some beds of salt. The laminated texture of the anhydrite indicates

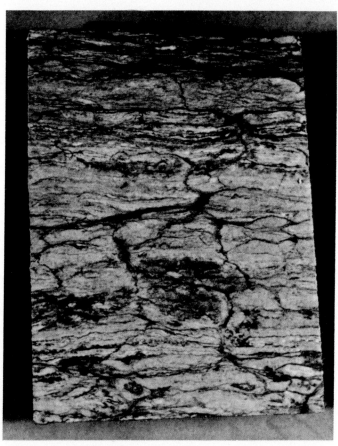

Fig. 9.—Supratidal dolomite, Flanagan Field, Upper Clear Fork Formation showing irregular laminations and mud cracks. Core is 3 inches wide.

Fig. 10.—Dolomitized algal stromatolite, Flanagan Field, Upper Clear Fork Formation.

deposition out of a standing body of water, and the thickness suggests that this body of water persisted for a long period of time. The lateral distribution and thickness indicates that the body of water was about 165 miles long, 96 miles wide, and hundreds if not thousands of feet deep (King, 1947). The climate must have been semi-arid or arid.

The largest cross-sectional area of a channel to the sea that would allow such a body of water to produce the bedded anhydrite deposits can be calculated from our previous analysis:

$$\frac{A_0}{A_t} = 10^6$$

$A_0 = 12,000$ mi² (assumes an ellipse)

therefore:

$$A_t = \frac{12,000}{10^6} = 0.12 \text{ mi}^2$$

If we assume that the inlet channel was 20 feet deep, then the channel width would be 3.2 miles (fig. 21). This, of course, is a *maximum* size. If there were more than one inlet, the size of each inlet would be proportionally smaller. It is possible that there was no surface connection at all, but that the body of water was fed by seepage through the sediment barrier and the outflow was through the sediment by seepage refluxion. It would appear that the Castile was deposited in a body of water more like an inland sea than a hypersaline lagoon or a silled basin.

Buckner Formation (Upper Jurassic), East Texas and Arkansas

The thick Upper Jurassic evaporite sequence called the Buckner is the subject of a recent paper by Dickinson (1968). The evaporite sequence is some 1100 feet thick. As opposed to the laminated texture of Castile anhydrite, the texture of the Buckner evaporite is predominantly nodular anhydrite. Some laminated anhydrite is present in the lower member along with some salt. While no mention is made of the thickness of the beds of bedded anhydrite, Dickinson does say that it is associated with nodular anhydrite and microcrystalline dolomite, indicating that these beds are not too thick. The salt bed is reported to be 34 feet thick.

The Buckner, therefore, must represent evaporite deposition in a tidal-flat, supratidal environment. The paleogeography was a coastal plain with shallow saline lakes in an arid or semiarid climate rather than the inland sea or salt lake of the Castile. To call the Buckner evaporite a hypersaline lagoon or barred-basin deposit misrepresents the paleogeography.

OUTCROP SHORELINE EVAPORITE-CARBONATE ROCKS

Evaporite-carbonate shoreline sediments which crop out have been subjected to more diagenesis than most surface examples. Besides the usual weathering problems encountered in outcrops, exposures of evaporite-carbonate shoreline sequences commonly have the added difficulty that the evaporite has been dissolved. This dissolution probably results from the influx of fresh water near the surface of the earth. In the following discussion of several outcrop studies, the carbonate textural criteria will be considered first and the problem of evaporite removal second.

Manlius Formation (Lower Devonian) of New York State

The study of the Manlius formation in New York State by LaPorte (1967) indicates environments from marine to supratidal. LaPorte uses the term "subtidal" for marine. In this paper, however, we will use the term "marine." LaPorte's environmental criteria are summarized below.

Supratidal:
1. Dolomitic laminated mudstone
2. Mud cracks
3. Birdseye texture
4. Fossils scarce (ostracods)
5. Algal laminae
6. Burrows

FIG. 11.—Marine dolomite overlying laminated supratidal dolomite. Notice lithoclasts of supratidal dolomite in base of marine dolomite. Flanagan Field, Upper Clear Fork Formation. Core is 2 inches wide.

Fig. 12.—Intertidal dolomite, Flanagan Field, Upper Clear Fork Formation. Notice discontinuous fractures.

Intertidal:
1. Interbedded pellet mudstone and skeletal calcarenite
2. Some pebble conglomerates and mudcracks
3. Fossil types few but individuals abundant
4. Ostracods, tentaculitids, brachiopods
5. Algal stromatolites and oncolites

Marine:
(Subtidal)
1. Pellet mudstone and reefy biostromes
2. Biota relatively abundant and diverse stromatoporoids, rugose corals, brachiopods, ostracods, snails, and codiacean algae.

In this example, the only dolomite is found scattered in the supratidal rocks, and there is no evidence for evaporites having been present. The marine sediment represents a protected lagoon environment based on regional facies distribution.

The amount of intertidal sediment appears to be over-estimated in this study. The intertidal facies (LaPorte, 1967, fig. 33) is overlain by marine sediment rather than supratidal sediment, as would be expected from the normal sedimentary sequence. Of course, intertidal sediment does not have to be overlain by supratidal sediment, but unless it is, one of the most convincing criteria is missing. The primary criteria used to identify this sediment as intertidal are algal stromatolites and mud cracks. The algal stromatolites are oncolites and branching, digitate types, not the mat or domed-mat types. They are found at the base of the intertidal unit overlying the Rondont Dolomite. Mud cracks are apparently found throughout the intertidal sedimentary unit.

These rocks are too fossiliferous to be supratidal, so the interpretive problem is between intertidal and marine. LaPorte in part circumvents this problem by leaving the period of his tidal fluctuations open: daily? monthly? yearly? Thus, the mud cracks could have been produced during a seasonal low tide. There does not seem to be any convincing evidence that these rocks, with the exception of the algal stromatolites at the base of the unit, were deposited between the daily tidal range (the intertidal of this paper). A more realistic environmental description might be shallow-restricted marine, which occasionally was exposed by abnormally low tides, rather than simply a tidal-flat environment. The algal stromatolites might well be associated with an overall transgressive, rather than a prograding, regressive, tidal flat environment.

El Paso Group (Lower Ordovician), West Texas

The El Paso Group, Lower Ordovician in age, has recently been subdivided based on units composed of shoreline sediments (Lucia, 1968). One of these units is called the Cindy Formation, and the sedimentary sequence in this formation is summarized below and in Figure 22. The formation is a sandy dolomite.

	Environment	Characteristics
Upper Member (70 feet) (See Figures 23 and 24)	Supratidal	Irregular laminations Lithoclasts LLH algal stromatolites Some mud cracks Thin sandy beds Lack of fossils
Middle Member (30 feet) (Figure 25)	Near-shore Bars and channels	Medium to small scale Sandy lithoclastic to peletoidal grainstone Fossils scarce
Lower Member (15 feet)	Supratidal	Discontinuous thin beds Current laminations Lithoclasts Burrows Lack of fossils

This package of shoreline sediments is very widespread and easily correlated. It represents a major shoreline progradation and indicates the presence of a widespread carbonate coastal plain (fig. 26). The fact that this package of shoreline sediments is dolomite suggests evaporite conditions. However, no evaporites are presently found associated with these shoreline sediments. Whether or not they were once present and have been removed is not known.

FIG. 13.—Intertidal dolomite, Flanagan Field, Upper Clear Fork Formation. Notice burrow structures. Core is 3 inches wide.

Fig. 14.—Current lamination in dolomite, Flanagan Field, Upper Clear Fork Formation. Core is 3 inches wide.

The algal stromatolites found in the supratidal are of the LLH type (fig. 23): Club-shaped algal stromatolites (SH-V type) are also found in the El Paso Group and are commonly associated with fossiliferous limestone. One occurrence is shown in Figure 27. Here the algal stromatolite appears to be growing on an exposure surface and is filled in with pellet packstone and fossil wackestone. This suggests a general transgression with intertidal algal stromatolites being covered with marine sediments. The sequence seems to be similar to the intertidal unit described by LaPorte (1967).

New Market Limestone (Middle Ordovician), Western Maryland

Matter (1967) describes some tidal flat features from the Middle Ordovician of Western Maryland. His results are summarized below.

Environment	Characteristics	Lithology
Marsh	Lumpy limestone Ripples Mud cracks Some fossil fragments	Dolomites and Limestone
Supratidal	Discontinuous laminations Mud cracks Very scarce fossils	Dolomite
Intertidal	Stromatolite (LLH) Ribboned Mud cracks Ripples Birdseye texture Edgewise conglomerates Very few fossils	Dolomite and Limestone
Marine	Intraclastic beds Bioclastic beds Oncolites, burrows, fossiliferous	Limestone and Dolomite

Matter obtained the term "marsh" from Van Straaten (1954) who used the term in the same sense as supratidal is now used. However, the lumpy limestones described by Matter appear to be too fossiliferous to be placed in the supratidal environment. The difficulty in placing these rocks in the marine environment is that the origin of the disrupted, lumpy, texture could involve desiccation. The origin of this texture is not clear at present, and until such time as the origin is better understood, the depositional environment will be in doubt.

Jeffersonville Limestone (Middle Devonian), Indiana

The Jeffersonville limestone (Middle Devonian), which outcrops in Indiana, has been described recently by R. D. Perkins (1963) and B. J. Bluck (1965). The laminated zone of Perkins and microfacies 1, 2, and 3 of Bluck appear to be the same. Perkins interprets the environment as "shelf lagoon" while Bluck interprets the environment as lacustrine and mud flat, which, within the definition used in this paper, would be supratidal. There was no separation of intertidal sediments from supratidal sediments. The textures in these supratidal rocks are listed below.

1. Irregular laminations
2. Birdseye structure
3. Mud cracks
4. Pebble conglomerates or breccias
5. Small collapse breccias
6. Barren of fossils
7. Algal mats and stromatolites

The tidal flat and lacustrine facies contain more dolomite than the marine facies and is reported to be all dolomite in the outcrop belt (Perkins, 1963). Both Perkins and Bluck described calcite pseudomorphs after gypsum in these tidal flat rocks, and Bluck indicates one occurrence of calcite replacement of dolomite. Collapse breccias, which are interpreted to be due to the removal of evaporites, are present.

Outcrop Evaporites

Dolomite and anhydrite are commonly associated with shoreline sediments in the subsurface. In the outcrop this association is not often observed because of the removal of evaporites by fresh water flows. The effects of fresh water on evaporite-carbonate shoreline rocks is illustrated in Figure 28. Dissolution of anhydrite and gypsum produces cavities, some of which collapse to produce collapse breccias. These

FIG. 15.—Current laminated dolomite with lithoclastic lag deposit at base. Notice anhydrite nodules at top of core. Flanagan Field, Upper Clear Fork Formation. Core is 3 inches wide.

Fig. 16.—Dolomitized marine wackestone, Flanagan Field, Upper Clear Fork Formation. Notice churned texture.

cavities are sometimes filled with calcite and internal sediment. Gypsum (anhydrite?) and dolomite are calcitized to produce calcitic dolomite or limestone in place of an anhydritic or gypsiferous dolomite.

The fresh-water flows encountered near the surface of the earth commonly dissolve beds of anhydrite, gypsum, and salt, producing cavities which collapse to produce collapse breccias. As Stanton (1966) has pointed out, this collapse may well occur during the evaporite dissolution stage rather than after a large cavity has been produced by the complete dissolution of an evaporite bed. Evaporite dissolution and collapse presents two main problems: 1) recognizing that the breccia is a product of evaporite removal, and 2) determining the nature of the removed evaporite. Correlation with subsurface evaporites is the best evidence for an evaporite-dissolution breccia. Other useful criteria are stratigraphic conformity associated with restricted marine or tidal flat carbonates, and other evidence of evaporites (as outlined below). The nature of the removed evaporite is lost except as can be deduced from indirect evidence.

Isolated nodules and replacement or pore-filling crystals are removed to produce vugs. Figure 29 illustrates leached anhydrite nodules in the outcrop of the San Andres (Permian) formation in New Mexico. These vugs have forms similar to the anhydrite nodules found in the subsurface (see fig. 7). Particularly notice the lobate outline and internal septa. Murray (1964) discussed anhydrite molds formed by the leaching of replacement anhydrite crystals, and an example is illustrated in Figure 30. Recognition is based on the straight sides and rectangular outlines typical of replacement anhydrite crystals. These vugs, of course, can be filled with any type of pore-filling material.

The calcitization process associated with evaporite leaching is not as well documented as the leaching effects. Lucia (1961) illustrated calcitization of both dolomite and anhydrite/gypsum in the outcrop of the Tansill formation (Permian). Braddock and Bowles (1963) illustrate calcitization of dolomite in the Minnelusa formation of the Black Hills associated with low Mg/Ca waters. Here, the high Ca/Mg waters are formed by the dissolution of gypsum near the earth surface. They discuss dedolomitization

Fig. 17.—Dolomitized marine packstone, Flanagan Field, Upper Clear Fork Formation. Core is 3 inches wide.

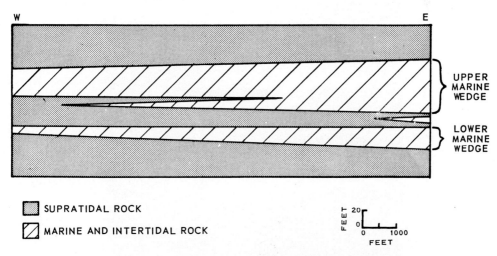

Fig. 18.—Generalized cross section of the Flanagan Field showing environment distribution in the upper part of the Upper Clear Fork Formation.

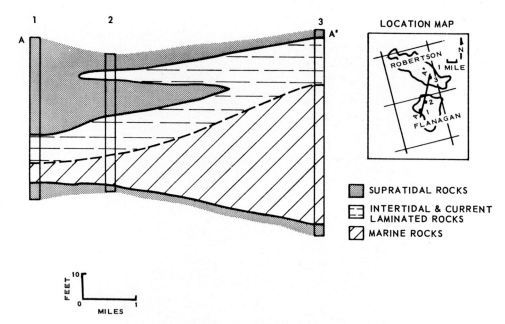

FIG. 19.—Cross section from Flanagan to Robertson Field showing lateral facies changes in Lower Marine Wedge.

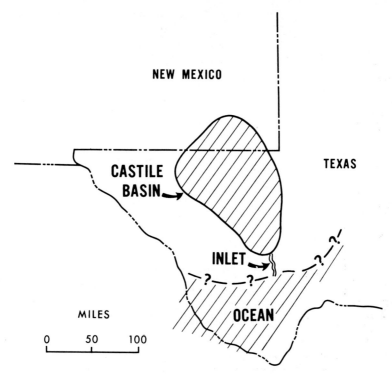

Fig. 21.—Generalized map of the Permian Castile Sea showing the calculated maximum size of the inlet which would allow gypsum deposition in the Castile Sea.

but not calcitization of anhydrite or gypsum, even though their photomicrographs suggest it to be present. De Grott (1967) shows that high Ca/Mg ratios resulting from the dissolution of gypsum by fresh water can produce calcite from dolomite. His experiments indicate three conditions necessary for calcitization of dolomite.

1. A high rate of water flow to keep the Mg/Ca ratio low.
2. A partial pressure of carbon dioxide considerably lower than 0.5 atm.
3. Temperature not higher than 50°C.

These conditions indicate that near surface conditions are necessary for the process to proceed (De Groot, 1967).

Textural evidence for dolomite calcitization involves recognizing dolomite pseudomorphs. Evamy (1967) illustrates some dolomite rhombs replaced by calcite retaining the clear rim-cloudy center relationship, as well as relict sedimentary textures of the predolomitization rock. As Evamy points out, a distinction must be made between replacement calcite and pore-filling calcite. The replacement calcite is a mosaic of equigranular calcite while the pore-filling calcite shows an increase in calcite crystal size from the pseudomorph wall to the center (Bathurst, 1958). An example of calcitized dolomite from the Gypsum Springs Formation of Wyoming is illustrated in Figure 31. Here relict rims of dolomite are preserved in a mosaic of calcite. Is the calcite replacing the dolomite or vice-versa? Dolomitization of large calcite crystals begins at the outside and works in as either enlarging dolomite crystals or as pseudomorphic replacement of the calcite crystals (Lucia, 1962). Because the dolomite shows neither of these relationships to the calcite, the calcite must be replacing the dolomite in this case.

The textural evidence for calcitized anhydrite or gypsum is based on the texture of replace-

---

Fig. 20.—Bedded anhydrite and dolomite, Red River Formation, Williston Basin, Montana. Core is 3 inches wide.

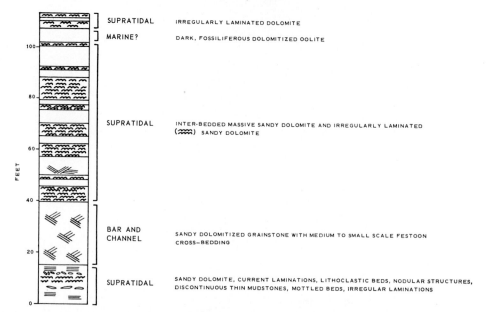

FIG. 22.—Generalized section of the Cindy Formation, El Paso Group, El Paso, Texas. An example of an outcrop of shoreline carbonate sedimentation.

ment anhydrite or gypsum. Replacement anhydrite/gypsum commonly has inclusions of small dolomite crystals and an outline which contains rectangular outlines and straight sides. Much replacement and pore-filling anhydrite/ gypsum, however, does not have this outline. Calcite crystals or mosaics of calcite crystals with rectangular outlines and dolomite inclusions are interpreted to have been produced by the calcitization of replacement anhydrite/

FIG. 23.—Supratidal dolomite, Cindy Formation, El Paso Group, El Paso, Texas. Notice laminations and LLH stromatolite.

Fig. 24.—Supratidal dolomite, Cindy Formation, El Paso Group, El Paso, Texas. Notice laminations.

gypsum. An example of calcitized replacement anhydrite is illustrated in Figure 32. Anhydrite/gypsum nodules may also be calcitized. In Figure 33 the lobate outline, wisps of dolomite, and included dolomite crystals along the edge are used as criteria to conclude that this area of mosaic calcite is a calcitized anhydrite/gypsum nodule. Calcitized anhydrite/gypsum is characteristically a mosaic of equigranular calcite or a single crystal of calcite. The texture of pore-filling calcite in anhydrite/gypsum molds is similar to that described by Bathurst (1958). Figure 34 illustrates pore-filling calcite (as evidenced by the increase in crystal size from the wall toward the center) in an anhydrite/gypsum mold (as evidenced by the square angles and straight sides).

Two recent papers which deal with the problem of the outcrop expression of evaporites are Armstrong's (1967) interpretation of the Arroya

Fig. 25.—Cross-bedded sandy dolomite, Cindy Formation, El Paso Group, El Paso, Texas.

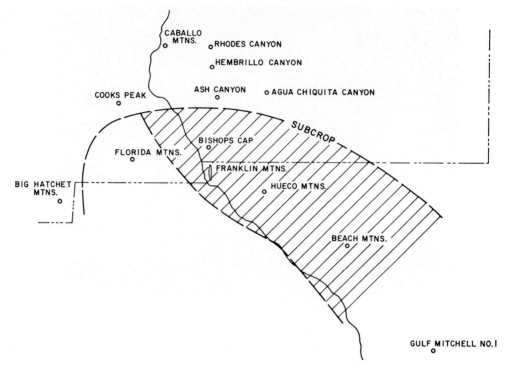

Fig. 26.—Map showing the distribution of tidal-flat sediments of the Cindy Formation and lateral equivalents, El Paso Group.

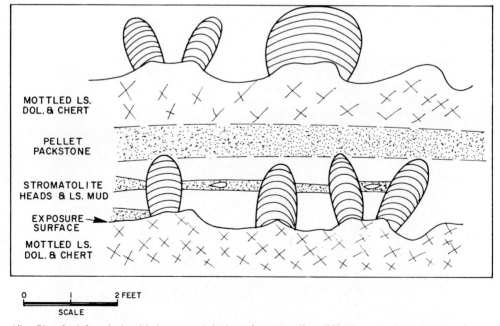

Fig. 27.—Sketch of the relationship between club-shaped stromatolites (SH-V), exposure surfaces, and enclosing sediments in lower part of the El Paso Group, El Paso, Texas.

CARBONATE SHORELINE SEDIMENTATION

EFFECTS OF FRESH WATER ON EVAPORITIC–CARBONATE SHORELINE DEPOSITS

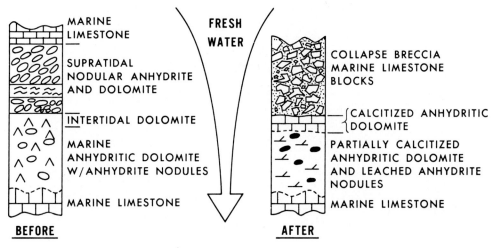

Fig. 28.—Diagrammatic illustration of the effects fresh water would have on an evaporitic-carbonate shoreline sequence.

Fig. 29.—Leach anhydrite nodules in San Andres dolomite outcrop, New Mexico. Notice lobate outline and internal septa.

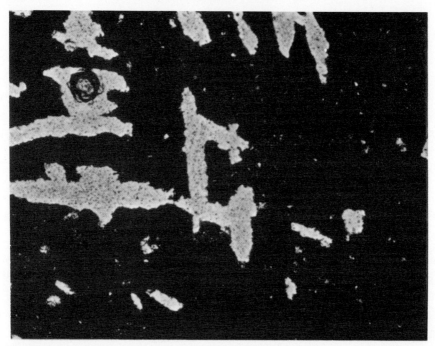

Fig. 30.—Leached replacement anhydrite (gypsum?) crystals in a limestone, Mississippian, Montana. Notice straight sides and rectangular re-entrants.

Penasco Formation (Mississippian) in north-central New Mexico and Fisher and Roddas' (1967) interpretation of the Edwards Formation (Lower Cretaceous) of Texas. Armstrong concludes that the collapse breccias in the Arroyo Penasco Formation were formed by the dissolution of gypsum beds deposited in the supratidal environment. The source of the dissolving fresh water was the overlying Pennsylvanian unconformity. After becoming saturated with calcium sulfate, these waters flowed through the underlying anhydritic or gypsiferous dolomite and calcitized or dissolved the gypsum or anhydrite and partially calcitized the dolomite. This produced the interesting relationship of a limestone under an evaporite collapse breccia which is actually a calcified anhydritic or gypsiferous dolomite.

The calcification process is the reverse of the reflux dolomitization process. Instead of marine water evaporating and precipitating gypsum to produce a high Mg/Ca water capable of dolomitization, fresh water dissolves gypsum to produce a low Mg/Ca water capable of calcitizing dolomite. The dolomite model produces evaporite beds. The dedolomite model produces breccia beds. The refluxing hypersaline marine water produces nodular, replacement, and pore-filling gypsum (anhydrite?). The fresh water leaches and calcitizes the gypsum/anhydrite, leaving replacement and pore-filling calcite. Reflux tends to convert a lime-sediment into a gypsiferous dolomite; the fresh waters tend to return the gypsiferous dolomite back to a limestone.

Fisher and Rodda's (1967) interpretation of the "Kirchberg Lagoon" in Central Texas is based principally on the nature and distribution of several evaporite-solution collapse units. There are four evaporite-solution units in the center of the lagoon and only one or two at the margin. These authors conclude that this Cretaceous evaporite was deposited in a hypersaline lagoon, while Armstrong envisions a supratidal environment for his Mississippian example. Without any data on the texture of the evaporite either conclusion is possible. Armstrong (1967) does report algal stromatolites associated with his evaporite collapse breccia indicating a tidal-flat environment. Fisher and Rodda (1967) suggest that the distribution of dolomite supports the lagoonal theory because it is similar to that described by Adams and Rhodes (1960) for the Permian of West Texas. My experience, as illustrated in the Flanagan field study, indicates that the Permian represents supratidal dolomitization rather than the lagoonal configuration envisioned by Adams and Rhodes. Whether or

Fig. 31.—Skeletal rims of dolomite rhombohedrons (light color) in a mosaic of calcite (darker color), Gypsum Springs Formation. Wyoming. Calcite is interpreted to be replacing the dolomite.

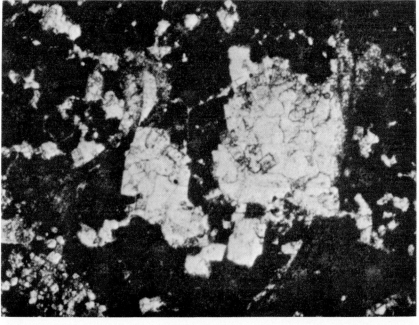

Fig. 32.—Calcite mosaic (clear areas) produced by calcitization of replacement anhydrite (gypsum?) crystals; Mississippian, Montana. Notice straight sides and rectangular re-entrants of the calcite mosaic.

FIG. 33.—Calcitized anhydrite nodule in a dolomite; Cretaceous, Texas. Notice mosaic texture of calcite, the lobate outline of the calcite, and the dark wisps of dolomite in the calcite.

not the "Kirchberg Lagoon" ever really existed remains to be determined. It is in danger of becoming the Kirchberg Island or Coastal Plain.

CONCLUSIONS

The shoreline can be identified in many carbonate sequences. It is located immediately below the supratidal environment, which is easily recognized by the following criteria, listed in order of importance:

1. Irregular laminations
2. Scarcity of or lack of fossils
3. Birdseye textures formed by dessication
4. Mud cracks
5. LLH algal stromatolites
6. Lithoclastic conglomerates
7. Association with shallow marine sediments

The intertidal environment, which is the shoreline, is more difficult to recognize directly due to its marine affinities. In a sea with no tides, there would be no intertidal sediments at all. If present, however, the following criteria, listed in order of importance, are useful for identification.

1. Position immediately below the supratidal environment
2. Algal stromatolites
3. Restricted marine organisms
4. Numerous burrows

Evaporites, most commonly gypsum and anhydrite, are associated with carbonate shoreline deposits. Gypsum and anhydrite are formed 1) by precipitation out of a standing body of water, and 2) by precipitation out of interstitial water. Criteria are available to distinguish between these two types of evaporite deposits. Gypsum and anhydrite precipitated out of a standing body of water has a bedded or laminated texture. Gypsum and anhydrite precipitated out of interstitial water has several different textural forms. These textural forms are individual or coalesced nodules, replacement crystals, and pore-filling crystals.

The restriction of sea-water circulation needed to allow evaporation of sea water to gypsum or anhydrite saturation can occur in two ways. First, the low permeability of the tidal-flat sediment provides a restriction to circulation. The sediment in and under the tidal-flat environment is filled with marine water, whose circulation with the ocean is restricted due to the low permeability of the sediment. Evaporation of this marine water at the supratidal surface produces a concentrated brine capable of precipitating all the interstitial textural forms of gypsum (and anhydrite?). Hypersaline lakes are also produced in low areas on the tidal flat in response to the sediment-permeability barrier. Bedded and laminated gypsum (anhydrite) deposits are produced in response to evaporation of sea water and ground water at the surface of the hypersaline lake.

The second form of restriction is the constric-

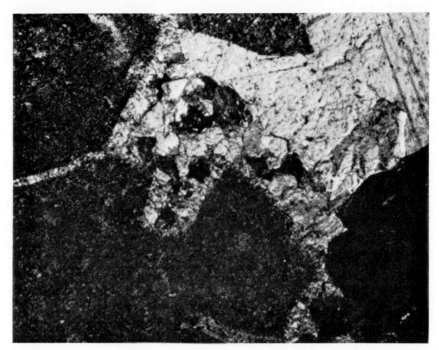

Fig. 34.—Pore-filling calcite in anhydrite mold; Mississippian, Montana. Notice increase in crystal size away from the wall and the straight sides and rectangular re-entrants in the outline.

tion of the channel connecting the ocean with a standing body of water. The ratio of the surface area of the restricted water to the cross-sectional area of the oceanic opening necessary to concentrate sea water to gypsum saturation is $10^6/1$ or greater. As an example, the maximum size of the channel which connected the Delaware Basin to the open ocean during the deposition of the Castile anhydrite was in the order of 3.2 miles wide and 20 feet deep. The areal size of the Castile sea was about 12,000 mi².

Observations on subsurface anhydrites indicate that most occurrences of anhydrite and gypsum were formed either within and under the supratidal carbonate sediment or in lakes on the supratidal surface in response to the sediment-permeability barrier. These evaporites are part of the supratidal, coastal-plain environment. However, thick sequences of bedded anhydrite/gypsum, like the Castile anhydrite, as well as thick sequences of salt were apparently deposited in large bodies of fairly deep water. It would be a mistake to include these in the supratidal, coastal-plain environment. Thick-bedded gypsum deposits probably imply the presence of a relatively small open channel to the ocean. Thick salt deposits probably imply a large body of relatively deep water that was isolated from the ocean. The source of the salt would have to be from either ground water or sea water springs.

The evaporites, so common in the subsurface, are commonly expressed in the outcrop as 1) collapse breccias, 2) molds of anhydrite/gypsum nodules, pore-filling crystals, and replacement crystals, and 3) calcitized anhydrite/gypsum. The molds are sometimes infilled with pore-filling materials. In addition, the dolomite associated with evaporite-shoreline carbonates is commonly calcitized in the outcrop. All of these features can be formed in response to the removal of gypsum and anhydrite by fresh water. The leaching of gypsum or anhydrite by fresh water produces molds of gypsum or anhydrite as well as cavities large enough to allow the formation of collapse breccias. This dissolution also produces a water with a high Ca/Mg ratio which is capable of calcitizing the preexisting dolomite. The association of calcitized dolomite and calcitized anhydrite or gypsum indicates that this water is also capable of calcitizing anhydrite or gypsum. The flow of fresh water through evaporite-shoreline carbonate rocks, therefore, tends to convert the anhydrite, gypsum, and dolomite sequence to a limestone sequence.

ACKNOWLEDGMENTS

I gratefully acknowledge the contribution of the following people to the ideas presented in this paper. Dr. R. N. Ginsburg and Mr. E. A.

Shinn for liberal discussions about their work in the Recent. Dr. R. C. Murray, Dr. A. Thomson, and Mr. S. D. Kerr whose work with the anhydrite-gypsum problem is basic to much of this paper. Drs. K. S. Deffeyes, P. K. Weyl, and R. M. Lloyd for a working knowledge of evaporite geochemistry. Dr. R. J. Stanton for discussions about collapse brecciation and Dr. R. Rezak for discussions on algal stromatolites. The work was done at Shell Development Exploration and Production Research Laboratory in Houston, Texas.

BIBLIOGRAPHY

ADAMS, J. E., AND RHODES, M. L., 1960, Dolomitization by seepage refluxion: Am. Assoc. Petroleum Geologists Bull., v. 44, p. 1912–1920.
ARMSTRONG, A. K., 1967, Biostratigraphy and carbonate facies of the Mississippian Arroyo Penasco Formation, north-central New Mexico: Memoir 20, New Mexico State Bureau of Mines and Mineral Resources, Socorro, New Mexico.
BALL, M. M., SHINN, E. A., STOCKMAN, K. W., 1963, Geologic effects of hurricane "Donna" (abs): Am Assoc. Petroleum Geologists Bull., v. 47, p. 349.
BATHURST, R. G. C., 1958, Diagenetic fabrics in some British Dinantian limestones. Liverpool and Manchester Geol. Jour. v. 2, pt. 1.
BELLAMY, C. F., 1900, A description of the salt-lake of Larnaca in the island of Cyprus: Quart. Jour. Geol. Soc. London, v. 56, p. 747–558.
BLUCK, B. J., 1965, Sedimentation of middle Devonian carbonates, southeastern Indiana: Jour. Sedimentary Petrology, v. 35, p. 656–682.
BOWLES, C. G., AND BRADDOCK, W. A., 1963, Solution breccias of the Minnelusa Formation in the Black Hills, South Dakota and Wyoming: U. S. Geol. Survey Prof. Paper 475-C, p. C91–C95.
BRADDOCK, W. A. AND BOWLES, C. G., 1963, Calcitization of dolomite by calcium sulfate solutions in the Minnelusa Formation, Black Hills, South Dakota and Wyoming: U. S. Geol. Survey Prof. Paper 475-C, p. C96–C99.
BRAMKAMP, R. A., AND POWERS, R. W., 1955, Two Persian Gulf lagoons (abs): Jour. Sedimentary Petrology, v. 25, p. 139.
BRIGGS, L. I., 1958, Evaporate facies: Jour. Sedimentary Petrology, v. 28, p. 46–56.
CAMPBELL, C. F., 1962, Depositional environments of Phosphoria Formation (Permian) in southeastern Bighorn Basin, Wyoming, Am. Assoc. Petroleum Geologists Bull., v. 46, p. 478–503.
CURTIS, R., EVANS, G., KINSMAN, D. J. J., AND SHEARMAN, D. J., 1963, Association of dolomite and anhydrite in the Recent sediments of the Persian Gulf: Nature, v. 197, p. 679–680.
DAVIES, G. R., 1970, Carbonate bank sedimentation, eastern Shark Bay, Western Australia *in* Carbonate sedimentation and environments, Shark Bay, western Australia, by B. W. Logan et al: Am. Assoc. Petroleum Geologists Memoir 13, p.
DEFFEYES, K. S., LUCIA, F. J., AND WEYL, P. K., 1965, Dolomitization of Recent and Plio-Pleistocene sediments by marine evaporite waters on Bonaire, Netherlands Antilles, *in* Dolomitization and limestone diagenesis—a symposium: Soc. Econ. Paleontologists and Mineralogists, Spec. Pub. 13, p. 71–88.
DE GROOT, K., 1967, Experimental dedolomitization: Jour. Sed. Petrology, v. 37, p. 1216–1220.
DELLWIG, L. F., 1955, Origin of the saline salt of Michigan: Jour. Sedimentary Petrology, v. 25, p. 83–110.
DICKINSON, K. A., 1968, Petrology of the Buckner Formation in adjacent parts of Texas, Louisiana, and Arkansas: Jour. Sedimentary Petrology, v. 38, p. 555–567.
EMERY, K. O., 1956, Sediments and waters of Persian Gulf: Am. Assoc. Petroleum Geologists Bull., v. 40, p. 2354–2383.
EVAMY, B. D., 1967, Dedolomitization and the development of rhombohedral pores in limestones, Jour. Sed. Petrology, v. 37, p. 1204–1215.
FISCHER, A. G., 1964, The lofer cyclothems of the Alpine Triassic, pp. 107–105 *in* Symposium on cyclic sedimentation: Geological Survey Kansas Pub. Bulletin 169.
FISHER, W. L., AND RODDA, P. U., 1967, Stratigraphy and genesis of dolomite, Edwards Formation (Lower Cretaceous) of Texas: Proceedings of the Third Forum on Geology of Industrial Minerals, 1967, Geological Survey Kansas Special Distribution Publication 34, p. 52–75.
FISK, H. N., 1959, Padre Island and the Laguna Madre flats, coastal south Texas: National Assoc. Sci.—National Res. Coun., 2nd Coastal Geography Conference.
GINSBURG, R. N., 1960, Ancient analogues of recent stromatolites: VXI International Geol. Cong., pt. 22, p. 26–35.
GOLDBERG, M., 1967, Supratidal dolomitization and dedolomitization in Jurassic rocks of Hamakhtesh Hagatan, Israel: Jour. Sed. Petrology, v. 37, p. 760–773.
GRABAU, A. W., 1924, Principles of Stratigraphy: Dover Pub., Inc., New York.
HEDGPETH, J. W., 1947, The Laguna Madre of Texas (Ethel M. Qull. Ed.): Trans. Twelfth North American Wildlife Conf.
HOFMANN, H. J., 1964, Mud cracks in the Ordovician Maysville Group: Jour. Geology, v. 72, p. 638–641.
ILLING, L. V., A. J. WELLS, AND J. C. M. TAYLOR, 1965, Penecontemporary dolomite in the Persian Gulf, *in* Dolomitization and limestone diagenesis—a Symposium: Soc. Econ. Paleont. Mineralogists Spec. Pub. 13, p. 89–111.
KERR, S. D., JR., AND A. THOMSON, 1963, Origin of nodular and bedded anhydrite in Permian shelf sediments Texas and New Mexico: Am. Assoc. Petroleum Geologists Bull., v. 47, p. 1726–1732.
KING, R. H., 1947, Sedimentation in Permian Castile Sea, Am. Assoc. Petroleum Geologists Bull., v. 31, p. 470–477.
LAPORTE, L. F., 1967, Carbonate deposition near mean sea-level and resultant facies mosaic; Manlius Formation (Lower Devonian) of New York State: Am. Assoc. Petrol. Geologists Bull., v. 51, p. 73–101.

LOGAN, B. W., 1961, Cryptozoan and associate stromatolites from the Recent, Shark Bay, western Australia: Jour. Geology, v. 69, p. 517–533.
———, REZAK, RICHARD, AND GINSBURG, R. N., 1964, Classification and environmental significance of algal stromatolites: Jour. Geology, v. 72, p. 68–83.
LUCIA, F. J., 1961, Dedolomitization in the Tansill (Permian) Formation: American Bull., v. 72, p. 1107–1110.
———, 1962, Diagenesis of a crinoidal sediment: Jour. Sed. Petrology, v. 32, p. 848–865.
———, 1968, Recent Sediments and Diagenesis of South Bonaire, Netherlands Antilles: Jour. Sed. Petrology, v. 38, p. 848–858.
———, 1968, Sedimentation and paleogeography of the El Paso Group: West Texas Geological Society Guidebook, Delaware Basin Exploration, West Texas Geological Society Publication No. 68-55, p. 61–75.
MATTER, ALBERT, 1967, Tidal flat deposits in the Ordovician of western Maryland: Jour. Sed. Petrology, v. 37, p. 601–609.
MORRIS, R. C., AND DICKEY, P. A., 1957, Modern evaporite deposition in Peru: Am. Assoc. Petroleum Geologists Bull., v. 41, p. 2467–2474.
MURRAY, R. C., 1964, Origin and diagenesis of gypsum and anhydrite: Jour. Sed. Petrology, v. 34, p. 512–523.
NEUMANN, A. C., AND BALL, M. M., 1970, Submersible observations in the Straits of Florida: geology and bottom currents: Geol. Soc. America Bull., v. 81, pp 2861–2874.
PERKINS, R. D., 1963, Petrology of the Jeffersonville Limestone (Middle Devonian) of southeastern Indiana: Geol. Soc. America Bull., v. 74, p. 1335–1354.
PHLEGER, F. B., AND EWING, G. C., 1962, Sedimentology and oceanography of coastal lagoons in Baja California, Mexico: Geol. Soc. America Bull., v. 73, p. 145–182.
PHLEGER, FRED B., 1969, A modern evaporite deposit in Mexico: Am. Assoc. Petroleum Geologists Bull., v. 53, pp 824–829.
ROEHL, P. O., 1967, Stony Mountain (Ordovician) and Interlake (Silurian) facies analogs of recent low-energy marine and subaerial carbonates, Bahamas: Am. Assoc. Petroleum Geologists Bull., v. 51, p. 1979–2032.
SCHENK, P. E., 1967, The Macumber Formation of the Maritime Provinces, Canada—A Mississippian analogue to recent strand-line carbonates of the Persian Gulf: Jour. Sed. Petrology, v. 37, p. 365–376.
SHINN, E. A., 1964, Recent dolomite, Sugarloaf Key, South Florida carbonate sediments: Guide Book Trip No. 1, Geol. Soc. America Annual Convention, p. 62–67.
———, GINSBURG, R. N., AND LLOYD, R. M., 1965, Recent supratidal dolomite from Andros Island, Bahamas, in Dolomitization and limestone diagenesis, a symposium: Soc. Econ. Paleont. Mineralogists Spec. Pub. 13, p. 112–123.
———, 1968, Practical significant of birdseye structures in carbonate rocks: Jour. Sedimentary Petrology, v. 38, p. 215–223.
STRAATEN, L. M. J. VAN, 1954, Composition and structure of recent marine sediments in the Netherlands: Leidse Geol. Mededel, v. 19, p. 1–110.
STANTON, R. J., JR., 1966, The solution brecciation process: Geol. Soc. America Bull., v. 77, p. 843–848.
SUGDEN, W., 1963, The hydrology of the Persian Gulf and its significance in respect to evaporite deposition: Am. Jour. Sci., v. 261, p. 741–755.
SVERDRUP, H. U., JOHNSON, M. W., AND FLEMING, R. H., 1946, The oceans—their physics, chemistry, and general biology: Prentice-Hall, Inc., New York.

CRITERIA FOR RECOGNIZING ANCIENT BARRIER COASTLINES*

KENDELL A. DICKINSON, HENRY L. BERRYHILL, Jr., AND CHARLES W. HOLMES

ABSTRACT

Recognition of ancient barrier coastlines is largely dependent on criteria derived from the study of modern barrier coastlines, especially where interpretations are based on limited data.

Ancient barrier coastlines are of special interest to petroleum geologists, because potential reservoir rocks, source beds, and cap rock are found in close proximity.

Barrier islands are formed on broad, gently sloping coastlines which have an abundant supply of sand. The formation of the islands and their diverse character are controlled by the action of the wind and the sea, together with tectonism and sediment supply. Barrier islands are long and narrow and nearly straight or gently curved in outline. They generally are lenticular in cross section unless altered after deposition.

Shoreface sediments which form the seaward margin of barrier islands are composed mainly of shelly and clayey sand that lacks lamination because of the activity of burrowing organisms. Beach sediments are generally composed of clean, well-sorted, gently dipping bedded sand. Sand dunes, if preserved, are important criteria, but their recognition is difficult because most of their sedimentary structures are not unique and are easily destroyed by erosion and vegetation. Tidal flats, which border the lagoon or bay margins, contain partly laminated sediments consisting mostly of clayey, silty sand. Algal mats, shells, and a variety of sedimentary structures, including ripple marks, slump structures, and mud cracks, are present in tidal flat sediments. Sediments deposited in tidal inlets and tidal deltas are diverse because these areas receive sediments from the lagoon, bay or salt marsh, and the open sea. Various current structures are common in the inlet sediments. Washover fans and flats may contain uniquely associated beds consisting of shelly sand on the bottom, pond deposits in the middle, and eolian sand on top. Sediments of eolian flats, which are generally covered with vegetation, consist mostly of structureless, fine-grained sand with some mottling caused by burrows and decaying plant roots. Soil horizons are occasionally found in these deposits. Lagoon, bay, and sound sediments are diverse in character, but generally are a mixture of sand, silt, and clay. They are finer near the center of the depositional basin, and they may contain peat, oyster reefs, abundant shells, and evaporite deposits.

Sedimentary environments of the modern barrier coastline may be distinguished by their biota. Similar interpretations are possible in ancient deposits if the fossils are preserved.

The internal structure of a barrier-island sand body depends on climate, sediment supply, tidal range, and tectonism; and the interplay of these factors produces islands of three basically different types: (1) prograding, (2) stationary, and (3) migrating landward.

Ancient barrier islands or related features have been recognized largely by their characteristic geometry and relation to strand lines. More recent attempts have utilized to a greater degree paleontologic and sedimentary analyses.

INTRODUCTION

As the petroleum geologist views a map of the world today and studies characteristics of modern coastlines, his attention is directed to those segments commonly referred to as barrier coastlines, where a combination of features seems especially favorable for the formation and trapping of oil. These features are the offshore island and the adjacent inshore lagoon or bay that separates it from the mainland. Found in proximity are a potential reservoir body, the porous and permeable sand of the barrier islands, and a potential cap rock and hydrocarbon source rock, the organic-rich, fine-grained sediments of both the lagoon and the open sea. The proximal relationship of such diagnostic features would seem to be definable both in the subsurface and on the outcrop, yet recognition of the barrier coastline, and in particular sandstone bodies that represent ancient barrier islands, still remains a difficult task. Three reasons can be given for the illusiveness or obscurity of the ancient barrier coastline in the subsurface: (1) the coastline at a given time is an extremely narrow zone; (2) the islands of sand that most typically form coastline barriers are relatively fragile features in the high energy of the littoral zone and are soon modified in a geologic sense, often beyond easy recognition; and (3) barrier islands themselves are diverse in character and occur with various thick clastic sequences of coastal plains and continental shelves where they can be largely obscured by much thicker and more grossly defined sedimentary bodies.

Our discussion will be focused primarily on these linear sandy islands of the barrier coastline known variously as barrier island, offshore bar, barrier bar, and outer bank; excluded are the organic reef barriers such as the Florida keys. We shall not discuss in detail the genesis of barrier islands. Rather we shall describe those features believed to be most typical of barrier islands and attempt to develop usable criteria.

To establish our criteria of recognition, we have drawn upon field studies of sedimentary environments on Padre Island, South Texas, and criteria published by earlier workers. The

---

* Publication authorized by the Director, U.S. Geological Survey.

field studies were strongly supplemented by aerial photography, including black and white, color, and color-infrared. And for comparison of coastlines features, the study was supplemented by an aerial photographic study of the Outer Banks of North Carolina. Sampling of surface deposits, trenching, and drill-hole pattern within the Padre Island barrier complex was laid out so as to cover all depositional environments. Sand grain size was analyzed for most samples, and 14 wells were drilled through Padre Island in the South Bird Island quadrangle. Shell counts and size analyses were made on the well samples. Internal sedimentary structures were described and photographed in the trenches. The field data and the aerial photos were used to construct environmental sediment maps at a scale of 1:24,000 for selected segments of Padre Island.

The framework of our discussion is built around the characteristics of the various sedimentary deposits that make up the modern barrier island and the nature of sediments that accumulate in environments closely associated with the barrier island. We shall begin by relating the barrier island to general physiography and then describe the geometry, the individual types of deposits, and the internal structure of barrier islands. The individual types of deposits in most barrier islands are shoreface, beach, dunes, overwash features, eolian plain, tidal inlets and flats, and lagoon sediments. For comparison, we shall follow the discussion of modern barrier islands with descriptions of several ancient deposits that have been interpreted as barrier islands.

CHARACTERISTICS OF MODERN BARRIER ISLANDS

Physiographic Setting

Modern barrier islands lie along broad, gently sloping coastlines that have an abundance of sand. They are also present around some large deltas, such as the Mississippi and the Niger. Approximately 3,550 miles of the world's coastlines today are of the barrier type, distributed approximately as follows: North America, 2,000 miles; Europe, 500 miles; South America, 400 miles; Africa, 300 miles; Australia, 200 miles; and Asia, 300 miles. The barrier coastline is characteristic of the extensive and broad coastal plain region of the eastern and southern United States.

The building of barrier islands is controlled by the action of wind and sea. The configuration of the barrier island and both its surface and internal structure are molded by the interactions of these two elements plus rainfall, sediment supply, and tectonism. In humid or subhumid areas, the surface of barrier islands is stabilized by vegetation, and such features as beach ridges are preserved. For example, on the surface of Galveston Island, Texas, and locally on the Outer Banks of North Carolina, beach ridges record the progradation of the seaward shore. The lagoons or bays of the humid areas are characterized by salt marshes, humate deposits, or peat bogs bordering or replacing parts of the lagoon or bay deposits. In such areas the lagoon or bay deposits contain much more fragmental and fibrous plant material and lenticular deposits of peat. In more arid regions, sand is drifted more extensively by wind, and dunes and dune migration, though common to most barrier islands, is a more prominent surface phenomenon. Evaporite deposits may form locally in small salt pans along the inshore side of the lagoon, as along Laguna Madre, Texas.

Geometry of Barrier Islands

Barrier islands generally are long and narrow and most have a length-width ratio exceeding 10, although a high ratio is not a necessary part of the definition of a barrier island. For example, the types listed by Shepard (1960) are: long, straight or smoothly curved, segmented, cuspate, and lobate. Padre Island, Texas, the longest barrier island in the world, is 110 miles long and ranges from 1/2 to 4 miles wide. The general shape of Padre Island as shown in Figure 1 is typical of most barrier islands. Along other coasts, such as those of southern New Jersey and Virginia, the islands are much more discontinuous. The barrier islands of the North Carolina coast are marked by prominent capes.

In plan view, obvious characteristics of barrier islands are the generally smooth, gently curved, ocean-facing shore and the irregular lagoon-facing shore.

In cross section barrier islands are lenticular, and they tend to be convex upward. As stated previously, barrier islands and related features result from a number of conditions, and the cross-sectional shape of an individual barrier island lens depends greatly on the combination of factors building and molding it. In the subsurface, the simple cross section of a barrier may not be easily recognized where a series of barrier islands is stacked vertically, or where the sand was spread out in sheetlike forms by a transgressing sea.

Depositional Environments and Sedimentary Characteristics

Depositional environments or facies of barrier islands and associated sediments, though characteristic of a particular segment of a bar-

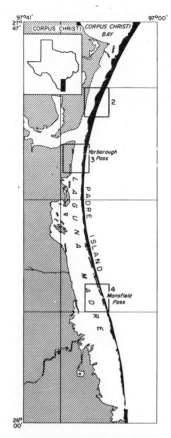

Fig. 1.—Map of modern barrier coastline, South Texas. 2, South Bird Island quadrangle; 3, Yarborough Pass quadrangle; 4, South of Potrero Lopeno, SE'quadrangle (see figs. 2, 3, and 4).

rier coastline, vary in both magnitude of development and proximal relationship from one part of a coastline to another. Maps of three segments of Padre Island, Texas, have been prepared to illustrate the features that are believed to be common to barrier islands in general and the variations in the association of the different types of deposits within a single barrier island (figs. 2, 3, and 4). The locations of the segments are shown on Figure 1. Study of the several depositional environments on Padre Island is the principal source of our field data and forms the outline for the descriptions that follow.

Shoreface Sediments

Shoreface sediments are those deposited along the front of the barrier island seaward from the low tidal level to a water depth of about 30 to 40 feet; for most places, quantitative data do not exist for accurately fixing the seaward limit of the toe of the island. The structures preserved in shoreface sediments of barrier islands may vary greatly, depending on the level of organic activity, wave energy, character of the sediments, and slope of the sea floor. Along Padre Island these sediments seem to consist largely of poorly laminated sand containing much shelly debris. Along Galveston Island, Texas, internal layering in the sand of the shoreface has been largely destroyed by the churning of burrowing organisms (Bernard, LeBlanc, and Major, 1963, p. 200). Glauconite and echinoid fragments locally are prominent in these sediments (Shepard and Moore, 1955). Grain size becomes finer seaward, and the very fine sand along the toe of the shoreface sediments may be intercalated with layers of silty clay. Graded bedding has been reported in some shoreface sediments (Barnard, LeBlanc, and Major, 1962; Hayes, 1967). According to Hayes (1967), the graded bedding results from hurricane-generated turbidity currents that originate in the lagoons and flow seaward through island inlets cut by storms. Studies by Bradley (1957) indicate that the Gulf of Mexico sediments have lower heavy-mineral content than barrier-island sediments.

Beach Sediments

Beach sediments, which generally are composed of clean, well-sorted, bedded sand, are one of the best-known constituents of a barrier island. However, these sediments may make up a relatively small part of the barrier island. For example, in the South Bird Island quadrangle, South Texas, where Padre Island is about 2 miles wide, the beach occupies only 2.4 percent of the total area of a segment of the island (fig. 2). General features of a modern beach and its sediments are shown in Figures 5 and 6.

The barrier islands of the Gulf of Mexico and Atlantic coasts of North America consist mostly of quartz sand. According to Shepard (1960), beaches of the western Gulf of Mexico contain 95 percent terrigenous materials, 2 percent shell material, and 1 percent Foraminifera. These percentages do not apply to southern Padre Island, where shell content exceeds 50 percent in places (Watson, 1958, p. 34). According to Rogers and Strong (1959) and Friedman (1967), beach sands generally are negatively skewed, but Hayes (1967) reported that positively skewed beach sands occur in the central part of Padre Island, Texas, where sands from two different provinces are mixed by converging long-shore currents.

Perhaps the most universal characteristic feature of the beach sands is bedding. Types of bedding in beach sands are shown in Figure 5. Bedding in beach sands has been summarized by McKee (1957). Bedding generally is even but variable in thickness. Banding is common and

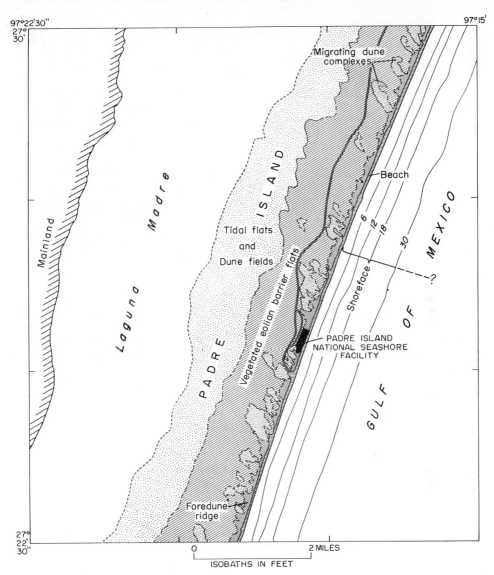

Fig. 2.—Depositional environments, South Bird Island quadrangle.

may be imparted by laminae of heavy minerals (fig. 5a) or layers of coarse shell fragments (figs. 5b and 5d). Scattered unbroken shells are locally prominent. The initial dip of the bedding, which may be as much as 10°, generally is seaward. Interbedding between eolian and beach sand is common in back-beach areas (figs. 5a and 5c). On the shoreward side of the longshore bars (Shepard, 1950; fig. 6c), a gentle shoreward or reverse dip may be found.

Sand Dunes

Sand dunes are common to most barrier islands and their recognition can be a key to recognizing the sediments of an ancient coastline. The moundlike character of the dunes is shown in Figure 7. The extent of the dunes, their position on the surface of the barrier island, and their rate of migration varies from place to place and from time to time, primarily as a function of climate. Dunes typically form a belt or foredune ridge along the front of the barrier island immediately inland from the back side of the beach (fig. 6). Other dunes may be scattered in patches farther inland and over larger areas as dune fields. Mound shape, eolian ripple marks, and internal planar crossbedding are universal characteristics (fig. 7).

Theoretically, ancient deposits of eolian sand should be recognizable by sedimentary char-

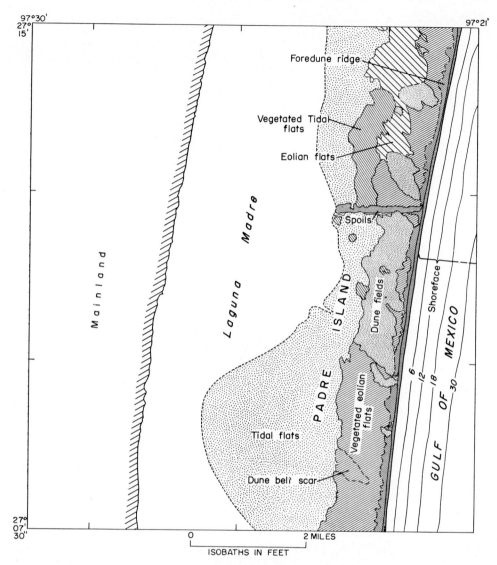

Fig. 3.—Depositional environments, Yarborough Pass quadrangle.

acteristics, but some criteria such as crossbedding, frosted grains, and ripple marks (McKee, 1957) are common to other environments. Statistical parameters of grain size, composition, particle shape, and fossil content may be more reliable indicators.

Geologists have long claimed some degree of proficiency in recognizing eolian crossbedding, and much literature is available on the subject. Unfortunately, some types of water-formed crossbedding cannot be readily distinguished from the eolian variety. Eolian crossbedding in foredune ridges generally is characterized by wedge-shaped crossbeds and by variability in direction of dip (figs. 7b and c). Regarding modern south Texas barriers, Shepard (1960, p. 208) states,

> "wherever it is possible to find bedding in the sand of a barrier, the structure is the best means of distinguishing dune and beaches."

McBride and Hayes (1963) have described the eolian crossbedding on Mustang Island, where the mean angle of inclination is 24°. Dips of this magnitude have been reported on sediment slopes from tidal inlets by Hoyt and Henry (1965).

Frosting of grains has usually been attributed to eolian action. However, Heald (1956) claims that frosting is more likely a result of diagenetic alteration than eolian abrasion, and Kuenen (1959) attributes frosting to corrosion. No differences in frosting between beach and dune sands were recognized in Gulf Coast barrier islands by Shepard (1960).

Statistical grain-size parameters have been useful in differentiation of environments of modern barrier islands, and such studies may also apply to rocks. A study by Mason and Folk (1958) indicated that plots of skewness versus kurtosis may effectively differentiate beach, dune, and eolian environments. According to them, beach sands form normal curves, dunes are positively skewed and mesokurtic, and eolian flat sands are positively skewed and leptokurtic.

No differences in skewness and kurtosis were noted in the Gulf Coast barrier island or environments by Shepard (1960), but according to him dune sand contains some silt. According to Friedman (1961; 1967), beach sands tend to lack a fine-grained tail. For ancient deposits evidence of this type is lacking.

Composition can be a valuable criterion for the differentiation of dune and beach sands if the sediments have not been drastically altered. Dune sand generally contains fewer shells, and on islands where beach sediments are shelly, eolian sand may be differentiated on this basis. According to Shepard (1960), dune sand contains more plant fibers and heavy minerals than

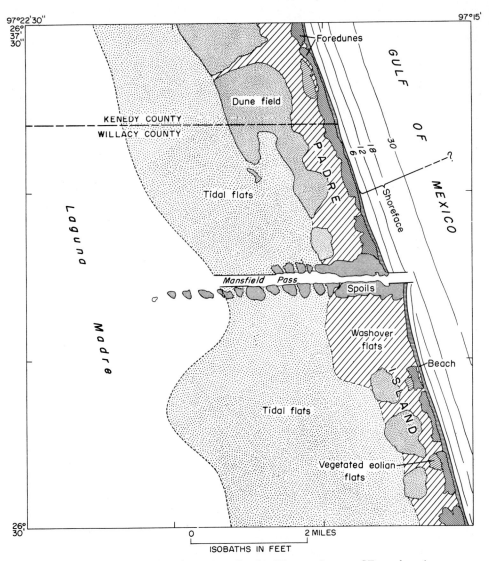

Fig. 4.—Depositional environments, South of Potrero Lopeno, SE quadrangle.

Fig. 5.—Bedding of beach deposits, Padre Island, Texas. a) Lamination in nonshelly beach deposit caused by variation in heavy-mineral content. b) Trench in shelly beach near beach-foredune boundary; layers A and C are laminated beach sediment and layer B is predominantly eolian sand (grooves are caused by removal of shells during excavation). c) Trench in nonshelly beach near beach-fore-dune boundary; layers A and C are of eolian origin and layers B and D of beach origin. d) Lamination in depositional cusp in shelly beach.

beaches. Rounded pumice fragments up to several inches in diameter are common in the dunes of Padre Island. The pumice apparently floated in from southern Mexico or Central America. Pumice of this origin may have been reaching Gulf of Mexico shores as early as the Mesozoic and may yield evidence for Gulf-facing beaches and aid in the recognition of ancient barrier islands in this area.

Shape of grains may offer a valuable clue to recognition of dune sediments. Ventifacts or other wind-polished pebbles are direct evidence of the eolian environment, and Shepard (1960) found that dune sands have greater roundness where onshore winds are prevalent.

The presence of land fossils such as terrestrial snails or vertebrate tracks is further criteria for distinguishing eolian sand from subaqueous environments.

Actually, dunes tend to be ephemeral, and in the high-energy zone of the coastline, the sand in dunes is highly susceptible to reworking by water, especially during storms. Thus eolian features may be destroyed by transgressing seas.

Tidal Inlets and Deltas

Tidal inlets are bodies of water connecting the lagoon or bay behind the barrier with the open sea. Tidal deltas are formed both in the lagoon or bay side of the inlet and on the ocean side (figs. 8a, 8b, and 8c). The deltas on the ocean side are generally smaller because of the erosive power of longshore currents. Tidal inlets may be intermittent or relatively permanent, although even the permanent ones generally migrate in response to longshore drift (Hoyt and Henry, 1965). Inlets make up a small proportion of the barrier complex on coasts such as the south

Texas coast, where tidal range is small and the supply of sediment is large. Padre Island, for instance, is unbroken by permanent natural inlets for 110 miles. However, Shepard (1960, p. 208) pointed out that the number of inlets will be reduced by filling as the islands grow. Also, where inlets migrate, a large part of the island may be composed of earlier inlet sediment. Along the coast of Georgia, where inlets are numerous, the islands are from 7 to 18 miles long; the inlets separating them are as much as 2 miles wide and 50 feet deep. The inlets are asymmetrical in cross section because of the longshore drift, and they may migrate across a strip 6 to 8 miles long (Hoyt and Henry, 1965).

Sediments that fill inlets are characterized by diverse faunal and mineral content because they receive sediments from the lagoon, bay, or salt marsh, and from the open sea. Crossbedding, and ripple marks, including megaripples, and other current structures, are common in the inlet sediments. Beds generally dip toward the center of the inlet perpendicular to the inlet margins, some as much as 30°. Sediment that may include lag gravel near the center of the channel generally decreases in grain size toward the lagoon or marsh (Hoyt and Henry, 1965). Sediments are built above sea level in parts of the tidal delta at Corpus Christi Pass. On the surface, and to a depth of about 3 feet, these deposits are primarily composed of sand and clayey sand that overlie lagoonal sediments. On subaerial vegetated parts of the delta, the sand is generally structureless and scour channels and algal mats are common on the surface. Ripple marks and bedding are very common on parts of the delta covered by shallow water. Shells from the bay and the beach are intermixed. Sponge-cake structure, apparently formed by emission of gas from underlying algal mats, has been noted on the tidal delta at Corpus Christi Pass.

Washover Fans and Flats

Washover fans and washover flats that consist of coalescing deposits provide a characteristic sedimentary association that is of great value in recognizing certain barrier island environments in rocks. Washover features are common on hurricane-dominated barrier coasts. Typical

FIG. 6.—Geomorphic forms of beach surface. a) Flat winter beach showing foredune ridge (A), heavy minerals (B), and seaweed flotsam (C). b) Summer beach with many shells and small wave-cut cliff on berm; dark area is heavy-mineral concentration. c) Longshore bar and trough on shelly beach at low tide. Current direction in trough parallels beach. d) Back-beach shell pavement with small lee dunes indicating effect of wind.

Fig. 7.—Sand dunes on Padre Island, Texas. a) Foredune ridge showing vegetation, ripple marks on surface, and shell pavement at base. b) Cut in foredune ridge showing nearly parallel bedding near base and large-scale crossbedding in upper part. c) Crossbedding on wind-scoured dune. d) Flooded dune field on wind-tidal flat, aerial view.

examples are on the barrier islands of the south Texas coast and on the Outer Banks of North Carolina.

The washover fan is a thin deposit spread inland in a semicircular area surrounding a breach or pass cut through the foredune or beach ridge by the flood tide of severe storms (fig. 8c). The initial deposit formed during the storm is a graded basal shell layer (fig. 9) that is thin or discontinuous around the periphery of the fan. The shell layer subsequently is covered by eolian sand around the margins and by pond sediments in the central area of the fan. The pond is usually formed within a week or two after the hurricane, when the breach to the sea is closed by longshore drift of sediment, and the pond may persist for a year or more (fig. 8d). The pond persists longer during wet periods, but it is generally encroached

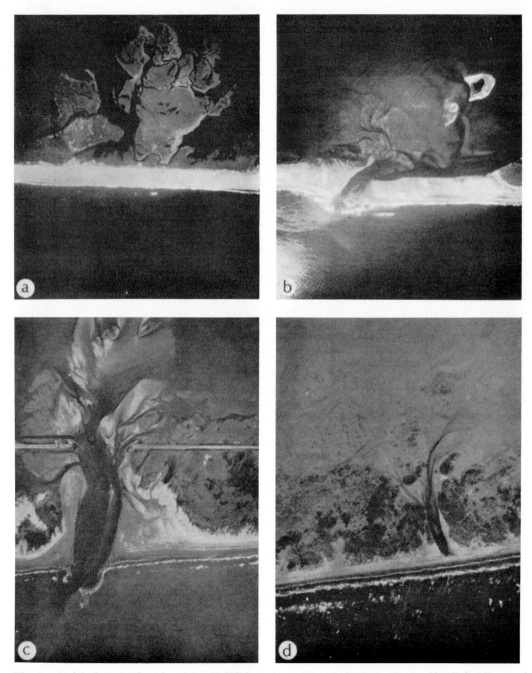

Fig. 8.—Aerial photographs taken from 12,000 feet of segments of the Outer Banks, North Carolina, and Padre Island, Texas. a) Print from infrared color photograph of relict tidal delta, Outer Banks. b) Print from infrared color photograph of Drum tidal inlet and delta, Outer Banks. c) Corpus Christi Pass, Texas, three days after hurricane Beulah (September 1967), delta building and inlet closing in progress. d) Ponded washover pass, central Padre Island, one year after hurricane Beulah.

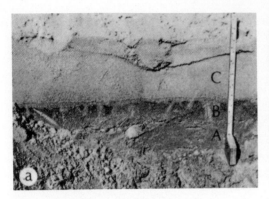 

FIG. 9.—Washover fan sediments. a) Types of deposits in washover fans: (A) shelly layer, (B) organic pond sediments with polychaete worm tubes, and (C) eolian sand. b) Graded sand deposited during hurricanes beginning with shells at (A) and ending with clay at (B).

upon by eolian sand and becomes gradually smaller until completely filled. The pond sediments are dark-colored, organic-rich layers in which reducing conditions resulting from decay of organic material in the underlying shell layer probably prevail. The pond, which may become hyper- or hypo-saline at a late stage, supports fauna, including crabs, small fish, and an abundance of polychaete worms. The final layer of the pond is in some cases a thin layer of clay. The pond deposits usually become covered by eolian sand (fig. 9a). The washover sediments interfinger with beach, dune, barrier flat, tidal flat, and in some cases lagoonal beds.

A fan measuring 4 by 4-1/2 miles was described from northern Mustang Island by Andrews (1967). Here the sediments interfinger with bay sediments on the back side of the island. Andrews pointed out that the washover fan sediments rest on a scoured surface.

Eolian Flats

Eolian flats are characterized by the hummocky surface left behind by a migrating dune belt or dune field. The older the flat, the more vegetated it becomes. Eolian features may be destroyed on a dune flat before a new dune belt migrates across it. The eolian flat may contain ephemeral lakes that form in deflation basins during periods of high rainfall.

The sediments of the eolian flat are non-laminated (figs. 10a, b, c), but generally show gray mottling that results from the decay of root plants. Soil horizons representing soil formation between dune migrations are common (fig. 10a). Eolian flats that form directly behind shell beaches contain shell detritus from the open-sea beach (fig. 10c). The composition of sediment on the eolian flats is nearly the same as for the sand dunes. The ephemeral lakes, which generally contain fresh water, support luxuriant green algae that later form matlike layers in the sediment (fig. 10b). Black-colored beds in the lake-bottom sand are the apparent result of reducing conditions and organic decay.

Tidal Flats

Tidal flats border the lagoon side of the island in areas periodically flooded by waters from the lagoon (figs. 2, 3, and 4). On the tidal flat, sediment transport and deposition are controlled by lagoonal waters during hurricanes and at other times of exceptionally high tides, and by the wind during dry periods. Eolian sand occasionally migrates across the tidal flats.

The sediments in the tidal flats are predominantly sand and clayey sand that may or may not be laminated (fig. 11). The sediments commonly contain algal mats, slump structures, ripple marks, megaripple marks, and mud cracks (fig. 11). The slumping occurs during periods of high wind activity, when loosely packed eolian sand slumps into the lagoon (fig. 11b). Megaripple marks are produced by rapid water flow across the tidal flats during hurricanes. The location of former strand lines along the lagoon are indicated by partially buried flotsam. Where sediments of the tidal flat are predominantly of barrier-island origin, heavy minerals are conspicuous (fig. 12).

Lagoon, Bay, and Sound Sediments

Great diversity in bay and lagoonal sediments results from many factors such as climate, source areas, and the location of tidal inlets, washovers, and river mouths. Generally the sediment consists mostly of silt and sand with lesser amounts of chemical precipitates and organic materials that modify its character. Sorting results in a trend toward concentration of the finer grains in the center of the basin. Sediments close to the barrier island are strongly influenced by the composition of the island. Sediments on the landward side are diverse and more closely related to the mainland sediment. River mouths may contribute wood fragments and an abundance of ostracods. Tidal inlets allow some admission of sediments, including echinoid plates, glauconite and planktonic forams; and where the tidal range is great, tidal channels may develop (Allen, 1965). Washovers also add normal marine sediments to the lagoon, especially on the islandward side. Some shells, such as *Barnea* sp., owing to shape and light weight, are transported across the barrier and into the lagoon by the wind.

The sediments of Laguna Madre, South Texas, which is a hypersaline lagoon in a semiarid environment, are discussed in detail by Fisk (1959) and Rusnak (1960). In Laguna Madre, the sediments are shell bearing clayey sand or sandy clay in the central, deeper parts and sand along the barrier island. Interlayering of sand and lagoonal mud is shown in Figure 10d. Along the mainland shore where sedimentation is slow, sand, gravel, and beach rock are found. Some of the sand consists of microcoquina and oolites; the gravel is mostly shell gravel composed of pelecypods (*Mulinia* sp. and *Anomalocardia* sp.), and the beach rock is white or cream-colored coquina cemented with microcrystalline aragonite (Rusnak, 1960). Twenty three samples of northern Laguna Madre sediments averaged 10% calcareous material, mostly shells and shell detritus.

FIG. 10.—Barrier island and lagoon sediments. a) Trench in vegetated eolian flats, Padre Island, Texas: older surface is indicated by soil horizon at arrow. b) Trench in washover flat sediment: ripple-marked thin clay laminae at arrow. c) Trench in vegetated eolian flats: small white grains are shell detritus. d) Cores of lagoonal sediments: irregularly interbedded layers are ripple-marked sand; dark layers are organic-rich clay and silt.

Fig. 11.—Tidal flat, Padre Island, Texas. a) High-water shoreline on wind-tidal flat. b) Eolian sand from barrier island migrating into lagoon; note slumping along shore. c) Trench in tidal sediment; lower layer (A) is lagoonal clayey sand containing abundant crab fragments; middle layer (B) is mixed lagoon and eolian sand; top layer (C) is eolian sand. d) Tidal flat surface showing weathered mud-cracks, ripple marks, and coyote tracks.

Some irregular bedding is found in the mudflats and in the deeper parts of the lagoon (fig. 10d). Diagenetic gypsum nodules and beds are found beneath the surface (Masson, 1955; Fisk, 1959). The size of nodules increases with depth.

The bays behind the barrier islands of the subhumid central Texas coast are receiving sediments from rivers, from barrier islands, and from the continental shelf (Shepard and Moore, 1955). The typical bay sediment is silty clay without lamination except near river mouths. Oyster reefs are common in the bays. Near river mouths the bay sediment is characterized by wood fragments, small ferruginous aggregates, and a dearth of mollusk shells.

Farther north, in more humid areas, salt marshes (Hoyt, Weimer, and Vernon, 1964) and other highly organic deposits such as peat and humate are associated with lagoon and bay deposits. The lagoon and bay sediments are interfingered with, and in some cases—perhaps at a later stage—replaced by salt marsh sediments. Salt marshes contain abundant sand, clay, and marine mollusks. Lagoon deposits with salt marshes, peat, or humate deposits along the shore also contain a large fragmental organic component.

Charleston Pond along the Rhode Island coast is an example of a lagoon in a temperate climate where the sediment source on the landward side is glacial till. As reported by William R. Dillon (unpublished data, 1969), the lagoon sediments consist of a sand-silt mixture. Plant fragments may make up as much as 16 percent of the deposit. Polychaeta worm pellets are also reported. During hurricanes, sand and a few marine mollusk shells from the beach are dumped over the barrier and into the lagoon. The sediments along the landward side of the lagoon consist mostly of lag gravel from glacial outwash sediments. The lagoon sediments are generally nonlaminated, probably because of reworking by organisms. Earlier peat and silt deposits underlie the present lagoonal sediment.

Fauna and Flora

The shallow water of the coastal margin is the habitat for an overwhelming majority of the marine invertebrate forms whose shell remains can be preserved for the fossil record. The benthonic life of the coastal margin is extremely abundant.

Fossils when properly collected and studied probably are the most direct means of interpreting ancient environments, and recognition of the ancient barrier-island coastline presents an ideal problem to be solved in this manner. Although much difference exists between ancient and recent fossil forms, the application of general paleoecological principles can be a most valuable tool in interpreting ancient fossiliferous environments. Studies of modern fauna are an invaluable aid in such studies. Important works on paleoecology and ecology are those of Ager (1963) and Phleger (1960).

In general, the barrier coastline complex can be divided into three subaqueous biologic environments: (1) freshwater lakes and river mouths; (2) brackish to hyper-saline water of the marshes, lagoons, or bays; and (3) normal marine.

In the modern environment of the northern Gulf of Mexico, Parker (1959, 1960) differentiated the following macro-invertebrate assemblages:

1. Freshwater and low salinity marsh
2. River influenced, low salinity
3. Delta-front distributory and interdistributory
4. Low salinity oyster reef
5. Enclosed lagoon or inter-reef
6. Open sound or open lagoon
7. Open sound or open lagoon center
8. High salinity oyster or mollusk reef
9. Open shallow hypersaline lagoon near inlet
10. Enclosed hypersaline lagoon
11. Inlet and deep channel

Similar environmental divisions for Foraminifera were made by Phleger (1960), Shenton (1957), Lehman (1967), and Grossman (1967). Siler and Scott (1964) also discussed the biologic assemblages from the northern Gulf of Mexico.

Freshwater biota in the fossil record generally can be recognized because the forms are markedly different from those of the marine environment. In the barrier coastline complex, freshwater forms may be found mixed with lagoon and bay forms near river mouths, in freshwater lakes on the island, or on the mainland, thus providing a clue to lagoonal proximity.

Fossils of the marine environment are well known for the geologic periods, and a detailed listing of invertebrate forms is unnecessary here, as the reader has an abundant literature to

FIG. 12.—Tidal flat sediments, Padre Island, Texas. a) Current marks and heavy mineral concentrations on surface. b) Trench showing laminated storm tidal flat sediments; dark layers are heavy mineral concentrations; sediments above disconformity at arrow were deposited after hurricane Beulah (September 1967).

draw upon. The abundance, variety, and location of marine invertebrate forms within the sand body of the barrier island are the key to the environment. Thus, the beach is perhaps the most easily recognized zone within the barrier island sand. Certain forms that seem out of place within a marine suite offer strong clues to environmental recognition. For example, echinoid fragments are abundant in modern Gulf of Mexico sediments and are present in the islands only in sands deposited in inlets of hurricane washover. Also, organisms such as *Callianassa major* (Weimer and Hoyt, 1964) that form burrows in the beach may provide valuable clues for the recognition of that environment. Coral fragments may be mixed with other forms in beach sand, but coral reefs do not form along sand-barrier coastlines. The ecology of Foraminifera in recent sediments along barrier coastlines has been covered in detail by Phleger (1960).

The fauna of lagoons and bays presents the most complex and interesting problem in barrier coastline recognition because of the large variation in physical and hydrographic conditions in these environments. Salinity, for example, may vary from nearly fresh to several times that of normal ocean water, not only from one part of a coastline to another but from one year to another. In bays and lagoons with connection to the open sea, normal marine forms invade from the seaward side. Wherever the lagoon receives very little inflow from the sea but receives a normal inflow of fresh water from rivers, salinity is lower and fresh-water and brackish-water forms predominate. Simmons (1957, p. 190), in relating fauna to salinity of lagoonal waters, reports that with increases in salinity above normal marine salinity in Laguna Madre, Texas,

> 1. the fewer the species. 2. the greater the number of individuals of each species available. 3. the larger the average individual of each vertebrate species. 4. the smaller the average individual of many invertebrate species, i.e., blue crabs, barnacles.

In the North Sea–Baltic Sea transition area, Sorgenfrie (1958) has shown that the number of molluscan species decreases in the direction of lower salinity as it falls below normal salinity. Changes in biologic diversity in the ancient environment have been discussed in detail by Tappan (1960) and Ager (1956).

Some general principles may be applied to fauna of ancient barrier coastlines: arenaceous Foraminifera generally are typical of marshes; mollusk shells with abraded surfaces are characteristic of the high-energy beach environment; delicate forms are more common in the lagoon environment; and marine invertebrates are smaller in hypersaline lagoons; in the lagoonal environment, many bivalve forms are preserved in the articulated condition.

The flora of the barrier coastline includes the grasses and mangroves of the marshes, sea grasses, and algae. Extensive marshes, such as the Great Dismal Swamp of North Carolina, lie along the inner margin of some barrier coastlines and contribute plant debris that eventually accumulates as beds of peat. Beds of peat not uncommonly become intercalated with fossiliferous sandy sediments along a barrier coastline. Algae grow in profusion in parts of lagoons and in ponds both on barrier islands and on the inshore side of lagoons. The algae contribute organic laminae to the sediment and are an effective trapping mechanism for clayey sediments. Carbonaceous algal laminae are common both in lagoonal and tidal and wind-tidal flat sediments.

Internal Structure of Barrier-Island Sand Bodies

The internal structure of a barrier-island sand body depends on several factors, including climate, sediment supply, tidal range, and tectonism. The interplay of these factors produces islands of different basic types in regard to both the shape in cross section and the vertical relations of the various types of deposits. We recognize three basic types: a) prograding, b) stationary, and c) migrating landward (see fig. 13). These types may be further modified by the effects of tidal inlets and dynamics of sediment supply.

Galveston Island, Texas, is an example of the prograding type (Bernard, LeBlanc, and Major, 1962). In this type of island, offlap occurs as the seaward edge of the island progrades; beach sand is deposited progressively seaward over silty or clayey gulf sediments. The history of the progradation is recorded by the series of parallel beach ridges that are apparent in aerial views of the barrier island. In the prograding process, the age of the deposits is younger seaward (fig. 13a). Few details about the lateral contacts of the prograding type of barrier sand body with other sediments are known, but one would expect the offshore contact with shoreface sediments to be gradational and the inshore contact with lagoon sediments more abrupt. However, lagoonal sediments locally, may encroach upon sand of the barrier island.

In Galveston Island, Texas, grain size decreases downward, according to Bernard, LeBlanc, and Major (1962), owing to the depositional relationship to shoreface sediments. However, input of sediment from diverse sources

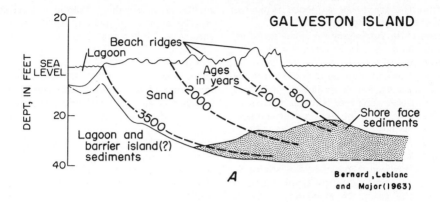
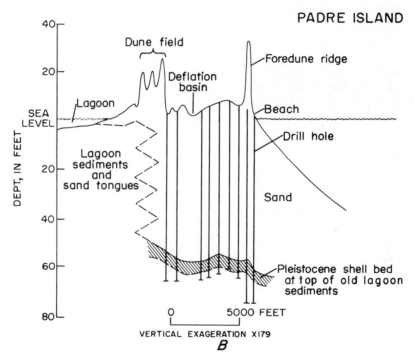
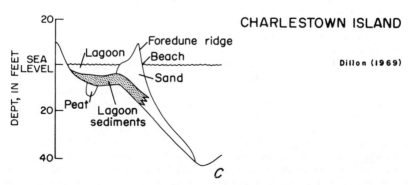

Fig. 13.—Cross sections of barrier islands showing basic types. a) Galveston Island, Texas; prograding island. b) Northern Padre Island, Texas (South Bird Island quadrangle); generally stationary. c) Charlestown, Rhode Island; island migrating landward.

may disrupt the grain-size sequence typical of islands of the prograding type.

In islands of the stationary type, the facies are uniform, or nearly so, in a vertical sense, and the lateral positions remain more constant. Changes in sediment size and content are dependent entirely on source dynamics, and little vertical grain-size change, except for interspersed shell layers, can be expected if the sediment source has been stable. On northern Padre Island, Texas, which at least for most of its history has been a stationary island, sediment size increases downward in some areas where the sediment source has switched from south to north along the coast. In a stationary island gradational lateral change to shoreface sediment occurs in a seaward direction; the relationship to the lagoonal sediments is complexly intertonguing (fig. 13b). Since the 1880's, Northern Padre Island has migrated into Laguna Madre and has increased its width by about one-third to one-half a mile. The rate of migration during the past 20 years has been about 50 feet per year. Northern Padre Island rests with sharp contact on a Pleistocene shell bed (fig. 13b).

Charlestown Island, Rhode Island, is a typical example of a barrier island that is migrating into the adjacent lagoon. The diagnostic feature of this type of island is the sharp contact of the sand and the underlying lagoonal sediments (fig. 13c). Within the sand body of the migrating barrier island, particularly in the lower parts, beach sediments are likely to lie on eolian sand.

It should be pointed out that the occurrence of numerous tidal and storm-cut inlets may also greatly influence the internal structure of a barrier island of any type. As discussed under Tidal Inlets, the migration and filling of inlets may superimpose channel-fill deposits across large segments of the island.

ANCIENT BARRIER ISLANDS

The first recorded interpretation of a barrier island in the geologic literature of the United States is that of Rich (1923). However, he later changed his interpretation (Rich, 1926). The classic work of Bass (1934) on the Bartlesville shoestring sands of Oklahoma was the first unequivocal interpretation of an ancient barrier island, although this work was based mainly on the geometry of the sandstone. Since the work of Bass, sandstone bodies at other places, both in outcrop and in the subsurface, have been reported in the literature as offshore bars, barrier islands, or related features. Brief descriptions of some of these are given to point out criteria that have been used for identifying ancient barrier islands and to illustrate the problems encountered in recognizing ancient barrier islands. Hopefully, they will aid the student of recent sediments in establishing diagnostic criteria that can be applied to ancient rocks. Maps of four sandstone bodies that have been referred to as barrier islands or related features are shown as Figure 14. The distinction between related features such as barrier islands, offshore bars, longshore bars, spits, cheniers, and other beaches is very difficult when the dimension of geologic time is applied because a given feature may pass from one stage to another during formation and subsequent modification. On economic grounds, distinction between these related features is academic because any one of them may possess the necessary characteristics for oil and gas accumulation. All of these narrow linear features, including fluvial deposits, have been included over the years under the general term "shoestring sands."

Gallup Sandstone, Bisti Oil Field, New Mexico

The Bisti bar complex has been described as an example of a longshore bar rather than a barrier island (Sabins, 1963). Although this feature is much larger than longshore bars, a brief description is included here because the form is closely related, and because it is well documented mineralogically, texturally, paleontologically, and geometrically. The following sedimentary facies were described by Sabins (1963, p. 193):

1. Main Gallup Sandstone—non-productive widespread regressive marine sheet sandstone consisting of:

    a. Offshore sand facies—very fine-grained sandstone with abundant primary dolomite grains and little or no glauconite. Southwest of Bisti field this facies is overlain by the Beach sand facies.

    b. Beach sand facies—medium-grained sandstone with low to moderate content of primary dolomite grains and little or no glauconite.

2. Low SP facies—poorly sorted sandy shale and shaly sand with low to moderate content of primary dolomite grains and little or no glauconite. This facies underlies each of the productive bar sands and separates them from the underlying offshore sand facies.

3. Bar sand facies—medium-grained sandstone with little or no primary dolomite grains and very abundant glauconite. The three individual bars are here designated as the Marye, (fig. 14a), Huerfano, and Carson Members [sic] of the Gallup Formation. On their seaward (northeast) margin, the bar sands grade into the Fore-Bar facies; on their landward (southwest) margin they grade into the Back-Bar facies.

4. Fore-Bar facies—marine shale characterized by an open marine fauna of planktonic Foraminifera, collophane, and *Inoceramus*.

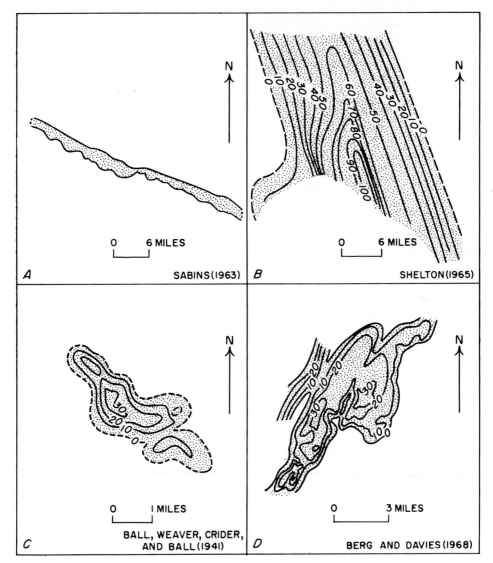

Fig. 14.—Isopach and outline maps of ancient oil producing barrier island or barrier bar structures. a) Marye bar of Late Cretaceous age in San Juan Basin, New Mexico. b) Barrier bar in lowermost unit of the Upper Cretaceous Eagle Sandstone, Billings, Montana. c) Barrier island or bar in the Mississippian Michigan Formation, the Austin Gas Field, Mecosts County, Michigan. d) Barrier bar in the Lower Cretaceous Muddy Sandstone, member of the Thermopolis Shale, Bell Creek oil field, Powder River Basin, Montana.

5. Back-Bar facies—marine shale, but of a restricted marine environment lacking the open marine fauna; pyrite-filled benthonic Foraminifera are common.

6. "Upper" Mancos Shale—shale of open marine environment overlying the entire Gallup Formation.

The glauconite content of the bar sand, which may reach 10 percent, distinguishes the bar from the mainland beach. The sand ranges in grain size from $1.0\ \phi$ to $3.0\ \phi$, in sorting index from 0.5 to 1.0 (moderate), and in grain shape from subangular to subrounded. Grain size decreases slightly downward, a characteristic that may have resulted from an upward increase in the winnowing effect of the depositional currents rather than from a vertical facies change as on Galveston Island, Texas.

Eagle Sandstone, Billings, Montana

The lowermost unit of the Upper Cretaceous Eagle Sandstone in the vicinity of Billings, Montana, has been interpreted by Shelton (1965; 1967) to be an ancient barrier island on the basis of comparison with modern Galveston Island,

Texas (Figure 14b). Although the lowermost unit of the Eagle Sandstone is approximately three times wider and two times thicker than Galveston Island, it is composed of an offlapping sequence of well-sorted, fine-grained sandstone whose grain size increases upward similar to the sand in Galveston Island.

The dominant structure in the upper part of the sand body is large-scale low-angle inclined bedding, whereas in the lower part the sand is mottled or poorly stratified because of burrowing organisms. Shelton (1965) states that

> Chert and rock fragments are common constituents. Glauconite is present throughout the sandstone body, and muscovite and carbonaceous material are accessories in the lower part.

No fossils were reported and neither lagoonal nor seaward facies were described. Gradational lower and lateral contacts and a sharp upper contact are reported.

Michigan Strays, Michigan Basin

Twenty-three shoestring sands, the so-called "Michigan strays," have been reported near the base of the Mississippian System in the Michigan Basin by Ball, Weaver, Crider, and Ball (1941). According to them,

> It is hard to conceive of a shoestring field in which the evidence of offshore sandbar origin is more conclusive, or in which the mechanics of the formation of the bar can be more clearly traced.

Their conclusion was based on the geometry, porosity distribution, and the relation of the sand bodies to sea-floor topography. The largest of the sandstone bodies described by Ball and others (1941) is 9 miles long and 3 miles wide, and in cross section the bars are convex upward (fig. 14c). Although no information on petrology is available, the Michigan strays are an important example because they seem to be related to sea-floor topography, which in turn is related to structural highs. An outline of one of the Michigan strays is shown in Figure 14c. The question left unanswered is whether the Michigan strays were formed as a result of normal sedimentological processes along a coastline or whether they developed as a result of mild tectonism.

Muddy Sandstone Member of the Thermopolis Shale, Bell Creek Field, Wyoming

The Lower Cretaceous Muddy Sandstone Member of the Thermopolis Shale of the Bell Creek oil field in the Powder River Basin in Montana has been interpreted to be a barrier bar by Berg and Davies (1968). This example is particularly important because it is one of the few examples for which a complete petrologic study was made and then related to facies in a modern barrier island. The sandstone has an average thickness of 20 feet and a maximum thickness of about 36 feet. The roughly oval sand body is 12 miles long and 4 miles wide (fig. 14d). The mineralogical composition, as determined by Berg and Davies, is 86 percent quartz, 1.7 percent mica, and 1.8 percent calcite cement. The average grain size is 0.16 mm. The grain size generally increases upward to a point about 5 feet from the top and then decreases. Four units were described by Berg and Davies as follows:

	Feet
1. Eolian and beach sand, fine- to very fine-grained, weathered, massive, dark, pseudolaminae near base	6
2. Beach and shoreface sand, fine- to very fine-grained, well laminated with shells	6
3. Middle shoreface sand, very fine-grained, burrowed, laminated, and interbedded silt and clay	18
4. Lower shoreface shelly sand, burrowed, laminated, and interbedded silt and clay	6
Total	36

Berg and Davies identified barrier-island facies as well as associated and lagoonal facies. The arrangement of the four units in the barrier-bar facies suggest an offlapping sequence such as in modern Galveston Island. The sequential arrangement may simply represent shoaler waters, in which case the barrier bar would have been a barrier island only in its late stage. Nevertheless, the characteristics as described by Berg and Davies strongly support their interpretation of the muddy sandstone in the Bell Creek oil field as a barrier coastline feature.

Bartlesville Shoestring Sands, Kansas

The Bartlesville shoestring sands in eastern Kansas, of Middle Pennsylvanian age, are an interesting example of ancient barrier islands. The linear sand lenses are enclosed in the Cherokee Shale, which contains both marine and nonmarine intervals. These ancient barrier islands or offshore bars, described by Bass (1934) range from 50 to 125 feet in thickness, from 0.5 to 1.5 miles in width, and from 2 to 6 miles in length. Two systems of islands 25 to 45 miles long lie in gently curved trends (Bass, 1934, fig. 2). The lenses tend to be convex upward in cross section. The sand is predominantly fine to very fine and angular to subangular, but in part of the area, coarse to very coarse, rounded to well-rounded grains are reported. The sand is finely laminated and exhibits sharply defined layers. The interpretation of the Bartlesville shoestring sands as ancient barrier islands is based heavily on the geometry and on parallelism with the ancient shoreline.

Verden Sandstone Member of the Marlow Formation, Oklahoma

The Verden Sandstone Member of the Marlow Formation, of Permian age, crops out in a belt that varies from 200 to 1,500 feet in width and extends for 75 miles through central southern and southwestern Oklahoma (Bass, 1939). Its maximum thickness is 15 feet, and it is composed of dolomitic, rounded, medium- to coarse-grained sandstone. Interbedded with the sandstone are thin layers of fine-grained gypsiferous sandy shale up to 3 inches thick that apparently are the source of intraformational cobbles that occur near the top of the unit. The sandstone is medium bedded with very prominent crossbeds that commonly dip at angles of 10° to 20°. At many localities, normal and giant ripple marks are found in the sandy shale. The axes of both kinds of ripple marks generally parallel the long axis of the sandstone belt. Grain size within the sandstone units ranges from clay and silt to small pebbles (5 mm), but within each bed the sand is moderately well sorted. Marine gastropods and pelecypods were observed at many localities.

The Verden Sandstone Member has been interpreted by Bass (1939) as some form of barrier beach. The exact nature of the origin of the Verden is not completely clear, but it seems certain to have been a barrier island or a closely related feature. Bass' conclusion is based primarily on the geometry, the marine fossils, and the relation of the trend to the axes of the ripple marks.

The length and alignment of the chain of sandstone bodies in the Verden Sandstone Member and their narrow width and thinness strongly support Bass' interpretation.

Ancient Gulf Coast Barrier Islands

Ancient Gulf Coast barrier islands have been reported from rocks of various epochs, beginning with the Late Jurassic and extending to the Pleistocene. This is to be expected in view of the gently sloping plain, the abundance of terrestrial sediments, and the probable preponderance of onshore winds in much of the Gulf of Mexico during this period.

Sand bodies in the Upper Jurassic Cotton Valley Group of the northern Gulf coast have been interpretated as barrier islands by Sloane (1958) and Thomas and Mann (1966). In this area shallow seas and abundant supply of sand during the Late Jurassic created favorable conditions for barrier island formations. Nevertheless, we wish to express some doubt concerning this interpretation, because the so-called lagoon averages about 70 miles in width, the lagoonal sediments appear to more closely resemble normal marine sediments, and few beaches or island characteristics have been reported as substantiation.

A complex of ancient barrier islands from the Tertiary Frio Clay of south Texas (Boyd and Dyer 1964; 1966) demonstrates the difficulty of recognizing the characteristic geometry and other relations of a single barrier where a number of barrierlike sandstone units lie in close proximity. According to Boyd and Dyer (1964)

> The Frio barrier bar consists of coarse to fine grained, well sorted, porous quartzose sands which grade updip into the lagoonal shales and downdip into inner continental shales.

Their interpretation of the shales as lagoonal is based primarily on faunal content. The lagoonal shale is red and green and contains brackish-water fossils. Eolian sand is reported from the bar sands (Boyd and Dyer, 1966, p. 173).

We feel it inconceivable that a single barrier island will remain stationary and accumulate sand to a thickness of nearly 3,000 feet. Almost certainty the Frio Clay is a composite sandstone made up of numerous barrier coastline sand bodies and very probably deltaic sediments as well.

Other ancient barrier bars reported in the Gulf Coast are from the Eocene Carrizo Sand (D. K. Davies, personal communication), the lower Miocene Catahoula Sandstone (Fisher and McGowen, 1969), the Pleistocene Ingleside barrier, southern Texas (Price, 1951), and the Holocene Pine Island barrier in southern Louisiana (Weide, 1968). Excellent dune and lagoonal facies are exhibited in the outcropping Carrizo Sand just southwest of Hearne, Texas.

CONCLUSIONS

We preface our concluding statements by emphasizing that the search for ancient barrier-island deposits in the subsurface requires a knowledge of environmental relationships in Holocene coastline deposits. Only after a large number of holes have been drilled is it possible to distinguish the linear shape and straightness, the most obvious features of a barrier island. Consequently, the subsurface geologist must employ a thorough understanding of the facies relations that might be encountered along a barrier coastline. He must also remember that barrier islands are modified, at times greatly, as sea level changes. The sand may become largely incorporated with marine sheet sands as a rising sea encroaches shoreward, or it may become dissected and intermixed with fluvial deposits during regression. The most obvious sedimentary feature, eolian crossbedding, is largely destroyed in the reworking process. An understanding of modern marine processes and environmental relations along the coastline are paramount to interpreting ancient environments. The key to the search for any ancient coastline is a regional paleofacies or paleotectonic map that suggests positions of ancient coastlines. Only from some sort of regional synthesis that outlines broad facies interrelations can target areas be reasonably selected for detailed exploration.

In the detailed study of a target area, the geologist also must bear in mind that in applying his knowledge of modern coastline environments, a single characteristic, even though seemingly insignificant in the overall environmental regime, might be a clue in a specific situation. An example might be a fossil type that seems out of place in an otherwise diagnostic suite. We have pointed out the role played by storms in sweeping marine forms typical of the continental shelf onto the barrier island and sometimes into the lagoon behind. Noteworthy is the fact that the ancient barriers we have discussed have been defined on the basis of considerable drilling and on a substantial amount of subsurface data. What is desired is the ability to identify the environmental position on the basis of a minimum of drill holes.

The recognition of a specific environment of the barrier coastline involves tabulating sedimentary characteristics of the ancient deposit and comparing them with known characteristics of modern coastal sediments. Following are the significant criteria, classified as to environment.

Beach

1. *Sediments.*—Beach sediments consist of well-sorted sand that locally contains shell detritus and heavy minerals. These sediments are commonly banded in layers of variable thickness with initial dips that are generally less than 10°.
2. *Fossils.*—Although a large variety of marine forms are present, some forms, such as *Donax* sp., which inhabit the intertidal zone, may be predominant. Shell detritus and highly abraded shells may also be common.

Dunes and Eolian Flats

1. *Sediments.*—The sediments are generally fine-grained to very fine-grained, well-sorted sand containing more plant fibers and heavy minerals than adjacent sediments in the beach or lagoon. Eolian crossbedding may be preserved in the dunes, but may be difficult to recognize. The eolian flats are generally structureless though a few burrows, root forms, or old soil horizons may be preserved. Algal layers from fresh-water lakes that form in deflation basins may be present.
2. *Fossils.*—A few marine shells and shell detritus may be found in the eolian deposits. Land snail burrows and plant fibers may also be present.

Washover Fans and Flats

1. *Sediments.*—Graded shell layers are characteristic. These, together with associated washover pond sediments and eolian deposits, are diagnostic.
2. *Fossils.*—The basal shell layers contain a normal marine fauna. The pond deposits overlying the normal marine shell layers contain worm tubes and carbonaceous laminae.

Tidal Flats

1. *Sediments.*—Tidal-flat sediments consist of interbedded lagoonal and island sediments.
2. *Fossils.*—Beds containing lagoonal fauna are intercalated with the relatively unfossiliferous barrier island sediments. Algal laminae may be common.

Lagoon, Bay, or Sound

1. *Sediments.*—Typically composed of clayey silt and sand, becoming finer-grained toward the center of the basin. Locally oolites, shell gravels, beach rock, algal layers, and peat deposits may be common.
2. *Fossils.*—Fossil forms characteristic of brackish or hypersaline conditions are of prime importance. Also significant in the mixture of lagoon fossils with those from open sea or river fauna. Among indicators that may aid in recognition of the lagoon, bay, or sound environment are the presence of smaller individuals that may indicate hypersalinity conditions or preservation of articulated pelecypod shells or fragile shells without abrasion that may indicate a low-energy environment. The kinds and distribution of Foraminifera are especially good indicators because many strongly represent their environments of deposition.

BIBLIOGRAPHY

AGER, D. V., 1963, Principles of paleoecology: McGraw-Hill Book Co., New York, 371 p.
———, 1956, The geographic distribution of brachiopoda in the British Middle Lias: Quart. Jour. Geol. Soc., London, v. 112, p. 157–187.
ALLEN, T. R. L., 1965, Coastal geomorphology of eastern Nigeria: Beach Ridge Barrier Islands and vegetated tidal flats: Geologie en Mijnbouw, v. 44, p. 1–21.

ANDREWS, P. B., 1967, Facies and genesis of a washover fan, St. Joseph Island, central Texas coast (Ph.D. Thesis): Texas Univ., Austin, Texas, 238 p.

BALL, M. W., WEAVER, T. J., CRIDER, H. H., AND BALL, D. S., 1941, Shoestring gas fields of Michigan, in Levorsen, A. I., (Ed.), Stratigraphic type oil fields: Am. Assoc. Petroleum Geologists Bull., p. 237–266.

BASS, N. W., 1934, Origin of Bartlesville Shoestring sands, Greenwood and Butler Counties, Kansas: Am. Assoc. Petroleum Geologists Bull., v. 18, p. 1313–1345.

———, 1939, Verden Sandstone of Oklahoma, an exposed shoestring sand of Permian age: Am. Assoc. Petroleum Geologists Bull., v. 23, p. 559–581.

BERG, R. G., AND DAVIES, D. K., 1968, Origin of Lower Cretaceous Muddy Sandstone at Bell Creek Field, Montana: Am. Assoc. Petroleum Geologists Bull., v. 52, p. 1888–1898.

BERNARD, H. A., LEBLANC, R. J., AND MAJOR, C. F., 1963, Recent and Pleistocene geology of southeast Texas, in Geology of the Gulf Coast and central Texas, and guidebook excursions: Rainwater, E. H., and Zingula, R. P., (Ed.), Houston Geol. Soc., Houston, Texas, p. 175–224.

BOKER, T. A., 1956, Sand dunes on northern Padre Island (M.S. thesis): Kansas Univ., Lawrence, Kansas, 100 p.

BOYD, D. R., AND DYER, D. F., 1964, Frio bar system of south Texas: Gulf Coast Assoc. Geol. Soc. Trans., v. 14, p. 309–322.

———, 1966, Frio barrier bar system of south Texas: Am. Assoc. Petroleum Geologists Bull., v. 50, p. 170–178.

BRADLEY, J. S., 1957, Differentiation of marine and sub-aerial sedimentary environments by volume percentage of heavy minerals, Mustang Island, Texas: Jour. Sed. Petrology, v. 27, p. 116–125.

DILLON, W. R., 1969, Submergence of a glaciated coast recorded in the stratigraphy of a Rhode Island Lagoon: Unpubl. manuscript, 22 p.

FISHER, W. L., AND MCGOWEN, J. H., 1969, Depositional systems in Wilcox Group (Eocene) of Texas: Am. Assoc. Petroleum Geologists Bull., v. 53, p. 30–54.

FISK, G. N., 1959, Padre Island and Laguna Madre Flats, coastal south Texas: Second Coastal Studies Inst., Louisiana State Univ., Baton Rouge, La., 1959, p. 103–152.

FRIEDMAN, G. M., 1961, Distinction between dune, beach, and river sands from their textural characteristics: Jour. Sed. Petrology, v. 31, p. 514–529.

———, 1967, Dynamic processes and statistical parameters compared for size frequency distribution of beach and river sands. Jour. Sed. Petrology, v. 37, p. 327–354.

GROSSMAN, STUART, 1967, Ecology of Rhizopodea and Ostracoda of Southern Pamlico Sound region, North Carolina, Pt. 1, Living and sub-fossil rhizopod and ostracoda populations: Kansas Univ. Paleont. Contr. 44, Ecology, art. 1, p. 1–82.

HAYES, M. O., 1967, Hurricanes as geological agents: Case studies of hurricanes Carla, 1961, and Cindy, 1963: Bur. Econ. Geol., Texas Univ., Inv. Rept. 61, 54 p.

HEALD, M. T., 1956, Cementation of Simpson and St. Peter Sandstones in parts of Oklahoma, Arkansas, and Missouri: Jour. Geol., v. 64, p. 16–30.

HOYT, J. H., AND HENRY, V. J. J., 1965, Significance of inlet sedimentation in the recognition of ancient barrier islands: Wyoming Geol. Soc., 19th Field Conf., Field Trip Guidebook, p. 190–194.

HOYT, J. H., WEIMER, R. J., AND VERNON, J. H., 1964, Late Pleistocene and Recent sedimentation, Central Georgia Coast, U.S.A., in Deltaic and shallow marine deposits, Van Straaten, L. M. J. U., (Ed.): Sixth Internat. Sedimentological Cong., The Netherlands and Belgium, 1963, Proc., p. 170–176.

JOHNSON, D. W., 1959, Shore processes and shoreline development: New York, John Wiley and Sons, Inc., 584 p.

KUENEN, P. H., 1959, Experimental abrasion and frosting of sand (abs.): Am. Assoc. Petroleum Geologists and Soc. Econ. Petrologists and Mineralogists, Ann. Mtg., Dallas, Texas, 1959, Program.

LEBLANC, R. T., AND HODSON, W. D., 1959, Origin and development of the shoreline: Second Coastal Geog. Conf., Coastal Studies Inst., Louisiana State Univ., Baton Rouge, La., 1959, p. 57–101.

LEHMAN, E. P., 1967, Statistical study of Texas Gulf Coast recent foraminiferal facies: Micropaleontology, v 3, p. 325–356.

MASON, C. C., AND FOLK, R. L., 1958, Differentiation of beach dune and eolian flat environments by size analysis, Mustang Island, Texas: Jour. Sed. Petrology, v. 28, p. 211–226.

MASSON, P. H., 1955, An occurrence of gypsum in southwest Texas: Jour. Sed. Petrology, v. 25, p. 72–77.

MCBRIDE, E. F., AND HAYES, M. O., 1963, Dune crossbedding on Mustang Island, Texas: Am. Assoc. Petroleum Geologists Bull., v. 46, p. 546–551.

MCKEE, E. D., 1957, Primary structures in some recent sediments: Am. Assoc. Petroleum Geologists Bull., v. 41, p. 1704–1747.

PARKER, R. H., 1959, Macro-invertebrate assemblages of central Texas coastal bays and Laguna Madre: Am. Assoc. Petroleum Geologists Bull., v. 43, p. 2100–2166.

———, 1960, Ecology and distributional patterns of marine macro-invertebrates, northern Gulf of Mexico, in Shepard, F. P., Phleger, F. P., and van Andel, T. H., (Eds.), Recent sediments, northwest Gulf of Mexico: Tulsa, Am. Assoc. Petroleum Geologists, p. 302–337.

PHLEGER, F. B., 1960, Ecology and distribution of Recent Foraminifera: Baltimore, Md., Johns Hopkins Press, 297 p.

PRICE, W. A., 1951, Barrier island not offshore bar: Science, v. 113, p. 487–488.

RICH, J. L., 1923, Shoestring sands of eastern Kansas: Am. Assoc. Petroleum Geologists Bull., v. 7, p. 103–113.

———, 1926, Further observations on shoestring oil pools of eastern Kansas. Am. Assoc. Petroleum Geologists Bull., v. 10, p. 568–580.

ROGERS, J. W., AND STRONG, CYRUS, 1959, Textural differences between two types of shoestring sands: Gulf Coast Assoc. Geol. Soc. Trans., v. 9, p. 167–170.

RUSNAK, G. E., 1960, Sediments of Laguna Madre, Texas, in Shepard, F. P., Phleger, F. B., and van Andel, T. H., (Eds.), Recent sediments, northwest Gulf of Mexico: Tulsa, Am. Assoc. Petroleum Geologists, p. 153–196.

SABINS, F. F., JR., 1963, Anatomy of stratigraphic traps, Bisti oil field, New Mexico: Am. Assoc. Petroleum Geologists Bull., v. 47, p. 193–228.

SHELTON, J. W., 1965, Trend and genesis of lowermost sandstone unit of Eagle sandstone at Billings, Montana: Am. Assoc. Petroleum Geologists Bull., v. 49, p. 1385–1397.
———, 1967, Stratigraphic models and general criterion for recognition of alluvial, barrier bar, and turbidity-current sand deposits: Am. Assoc. Petroleum Geologists Bull., v. 51, p. 2441–2461.
SHENTON, E. A., 1957, A study of the Foraminifera and sediments of Matagorda Bay, Texas: Gulf Coast Assoc. Geol. Soc. Trans., v. 7, p. 135–150.
SHEPARD, F. P., 1950, Longshore bars and longshore troughs: U.S. Army Corps of Engineers, Beach Erosion Board, Tech. Mem. 15.
———, 1960, Gulf Coast barriers in Recent sediments, northwest Gulf of Mexico: Tulsa, Am. Assoc. Petroleum Geologists, 394 p.
SHEPARD, F. P., AND MOORE, D. G., 1955, Sediment zones bordering the barrier islands of central Texas coast: Soc. Econ. Paleontologists and Mineralogists Spec. Pub. 3, p. 78–98.
SILER, W. L., AND SCOTT, A. J., 1964, Biotic assemblages, south Texas coast, *in* Depositional environments, south-central Texas coast: Gulf Coast Assoc. Geol. Soc., Ann. Mtg. 1964, Field Trip Guidebook, p. 137–157.
SIMMONS, E. G., 1957, An ecological survey of the upper Laguna Madre of Texas: Inst. Marine Sci. Pub., v. 4, no. 2, p. 156–200.
SLOANE, B. J., JR., 1958, The subsurface Jurassic Bodcow sand in Louisiana. Louisiana Geol. Survey Bull. 33, 33 p.
SORGENFRIE, T., 1958, Molluscan assemblages from the marine middle Miocene of South Jutland and their environments. Reitzel, Copenhagen, 2 v., 503 p.
TAPPAN, H., 1960, Cretaceous biostratigraphy of northern Alaska. Am. Assoc. Petroleum Geologists Bull., v. 44, p. 273–297.
THOMAS, W. A., AND MANN, C. J., 1966, Late Jurassic depositional environment, Louisiana and Arkansas. Am. Assoc. Petroleum Geologists Bull., v. 50, p. 170–182.
WATSON, R. L., 1968, Origin of shell beaches, Padre Island, Texas (M.S. Thesis): Texas Univ., 121 p.
WEIDE, A. E., 1968, Bar and Barrier-island sands (abs.): Gulf Coast Assoc. Geol. Soc. Trans., v. 18, p. 405–415.
WEIMER, R. J., AND HOYT, J. H., 1964, Burrows of *Callianassa major* Say, geologic indicators of littoral and shallow neritic environments: Jour. Paleontology, v. 38, p. 761–767.

TRACE FOSSILS AS CRITERIA FOR RECOGNIZING SHORELINES IN STRATIGRAPHIC RECORD

JAMES D. HOWARD
University of Georgia Marine Institute, Sapelo Island, Georgia*

ABSTRACT

Biogenic sedimentary structures offer a new and exciting approach to the interpretation of ancient sedimentary environments. Although trace fossils have been studied extensively by European geologists since early in this century, it is only in recent years that they have received much more than passing mention in North America. The increased interest in tracks, trails, burrows, and borings is due primarily to the environmental or facies approach to the study of sedimentary rocks.

Appreciation of biogenic sedimentary structures as facies indicators has been influenced significantly by the emphasis on studies of physical sedimentary structures which in the past two decades have introduced many new keys to paleoenvironment interpretation. Additional impetus to the utilization of trace fossils has come from detailed studies of modern sediments which illustrate clearly the important relations that exist between the animals and sediments in a particular environment.

In the study of ancient and present-day nearshore sedimentary environments, the facies significance of biogenic sedimentary structures can be demonstrated readily. Field studies which utilize trace fossils in conjunction with physical sedimentary structures, lateral and vertical changes in the sedimentary sequence, offer new opportunities in the search for stratigraphic accumulations of oil and gas.

INTRODUCTION

In order to demonstrate the usefulness of trace fossils in the interpretation of ancient sedimentary environments, this paper describes a clastic shoreline sequence from the Upper Cretaceous of the Western Interior. The units are exposed in the Book Cliffs and the Wasatch Plateau of east-central Utah. The interpretation of these rocks as representing a regional, cyclic, regressive sequence was made by Spieker (1949), and subsequent, more detailed, studies were made by Young (1955, 1957).

The well-exposed and accessible Book Cliffs outcrops serve as an excellent example for the utilization of trace fossils because of their abundance and because of the relative scarcity of body fossils in the siltstones and sandstones. Furthermore, these Upper Cretaceous units exhibit rapid facies changes which contain a significant variety of physical as well as biogenic sedimentary structures.

The prospect of applying the results of this study to other areas of clastic shoreline sedimentation is considered excellent, and the same types of trace fossil ichnogenera should be expected in similar sedimentary facies.

ACKNOWLEDGMENTS

Considerable assistance in the identification of facies and their contained physical and biogenic sedimentary structures has been gained through discussions with numerous colleagues.

* Present address: Skidaway Institute of Oceanography, Savannah, Georgia 31406.

Drs. W. K. Hamblin, J. Keith Rigby, Hugh W. Dresser, Charles V. Campbell and Robert W. Frey have been especially helpful, and their generous assistance is gratefully acknowledged. Drawings of trace fossils were made by Roger B. Williams. Research support for this study came from National Science Foundation Grant GA-719. This paper is University of Georgia Marine Institute Contribution No. 226.

DISCUSSION

The principal rock unit cited in this paper is the Upper Cretaceous Blackhawk Formation which is exposed in the Book Cliffs and the Wasatch Plateau of east-central Utah. In this unit the writer has studied 12 sedimentary cycles in 4 regressive sequences. This cyclic sequence has provided an opportunity to investigate the repetitive nature of the trace fossil facies suites. In each of the Upper Cretaceous sequences examined, the vertical sequence of trace fossils has been duplicated with no significant changes between them. Pleistocene and Recent sediments from the Georgia coast exhibit similar types of trace fossil assemblages and have been studied to aid in interpretation of the Upper Cretaceous sedimentary sequence.

The nature of the biogenic sedimentary structures in all three study areas shows a similar trend from shoreline into deeper water. One trace fossil, *Ophiomorpha* (Weimer and Hoyt, 1964), is present in the Upper Cretaceous and in the Pleistocene, and its modern counterpart, the burrow of *Callianassa*, is present in the Recent. As expected, however, the specific trace fossils are not all the same in units so widely separated in time and geographic location; furthermore, facies are not necessarily the same. Comparisons of the trace fossil assemblages between the Recent and Pleistocene, however, are direct, and with practically no exceptions the same biogenic structures have been found in similar facies of both units.

Offshore Facies

Outcrops of the Upper Cretaceous offshore facies are composed of very thin-bedded gray siltstones (fig. 1). Other than the somewhat shaley appearance, there are few physical sedimentary structures. The most obvious feature of this facies is that the sediment was almost completely reworked and churned by burrowing organisms. Some fecal castings can be identified, but much of the sediment texture is a heterogeneous array of biogenic structures. A few specific organic structures can be described because of their characteristic morphology. Some of these have been assigned the precise names of ichnogenera, such as *Asterosoma*, *Arthrophycus*, *Teichichnus* and *Scolicia*, and others have been given working field names, such as "smooth tubes," "chevron burrows," and "small white tubes," based on their unique morphology (Howard, 1966). From the standpoint of the stratigrapher attempting to establish the lateral relationships of nearshore facies, they could be referred to as trace fossil 1, 2, 3, or A, B, C; what is more important is the association of certain traces with different textures. In this case they are either the random markings or organic reworking, continuous castings of ingested sediments, or recognizable burrows. Nearly all of these possess an orientation chiefly in the horizontal plane. Only *Asterosoma* is commonly more vertical than horizontal.

The foregoing are characteristics of the offshore facies which can be observed in outcrop or from drill cores. In summary, they are:

1. Grain size—silt
2. Physical sedimentary structures—very thin bedded with few current structures or other obvious physical sedimentary structures.
3. Biogenic sedimentary structures—thoroughly bioturbated sediments with a few definite trace fossils, most of which are the castings of animals moving through the substrate.

From these data it is determined that this was a relatively low-energy environment with no indication of marked current or wave effect. The sediments were rich in organic matter which supported an active and varied group of deposit-feeding organisms. Traces of suspension-feeding organisms are absent, and, therefore, the overlying waters were probably too turbid or too slow moving to support other than substrate feeders. It is assumed that sedimentation rates were relatively slow and continuous, because the sediment is extensively churned and alternations of highly reworked and relatively undisturbed intervals are lacking. If there had been pulses of sedimentation, or if storms had occasionally resuspended the sediments there would be gaps in the biogenic record and in their place would be sharp bedding planes separating beds showing relatively little reworking at their base and dense reworking at their tops. It is further assumed that the animals were shallow burrowers working in the upper oxygenated portion of the substrate; in fine-grained, organic-rich environments in the Recent, sediments are oxygen deficient at a few centimeters in depth.

Near the middle of the offshore gray siltstone facies, thin sandstone beds begin to appear which become gradually thicker and more abundant upward (fig. 2). These thin beds are important because they indicate the transition to the next facies and contain a new trace fossil assemblage. These beds are composed of very clean, fine-grained sand containing asymmetrical ripple lamina. When the thickness of these ripple-laminated sand beds increases to an inch or more, they contain *Ophiomorpha* and a second burrow-type of *Asterosoma*. *Asterosoma*, as mentioned previously, is abundant in the interbedded gray siltstones; however, its burrowing habit in the clean sand is much different than in the siltstone. In the sand beds *Asterosoma* has lost most of the characteristic concentric wall structure it possesses in the gray siltstones. It now contains almost no wall structures and is always oriented in the bed as a nearly vertical, slightly curving tube. This change in habit is considered due to the lack of organic detritus in the sand, or in the consistency of the substrate which necessitated a change in feeding habit from a deposit feeder to a suspension feeder. From a purely descriptive field approach it is not especially important that it is a variety of *Asterosoma*; it could as easily be considered a trace-fossil characteristic of cleaner sands. On the other hand, if such relationships can be recognized, they are beneficial in establishing further details of the environment. *Ophiomorpha*, the other burrow found in the interbedded sandstones, is also considered to be the burrow of a suspension-feeding organism, based on its similarity to the modern-day burrow of callianassid shrimp and because of its common presence in clean sands.

The situation just described—gray siltstone with a characteristic trace-fossil assemblage and thin sand beds containing different burrows or a change in the burrowing habit—represents the way in which trace fossils may be useful to paleoecologic studies. A rather minor difference in the lithology may immediately change the nature of the biologic community. With the appearance of a thin clean sandstone bed in the siltstone sequence, a new burrow type (*Ophiomorpha*) appears, several ichnogenera are eliminated, and one form (*Asterosoma*), which suc-

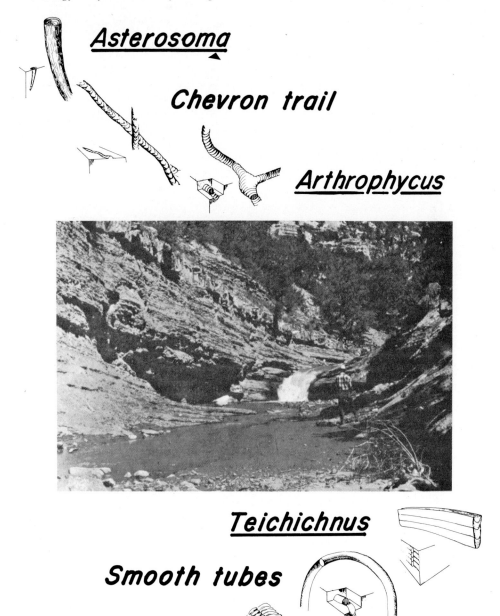

FIG. 1. Offshore facies. This facies is composed of very thin-bedded, highly burrowed siltstone. The most abundant trace fossils are illustrated and they show no obvious preferred vertical segregation.

Fig. 2. Upper part of offshore facies. Toward the top of the offshore facies, sandstone beds begin to appear. These sand beds contain large-scale ripple laminae and are burrowed only by *Ophiomorpha* and *Asterosoma*. This photograph shows about 18 feet of section.

cessfully makes the lithologic transition, changes its burrowing habit and the bioturbate textures so abundant in the gray siltstone are replaced by asymmetrical ripple-laminae. All this occurs in the thin beds in the midst of the gray siltstone sequence. It is unlikely that the salinity and temperature changed appreciably, but the composition and texture of the substrate did change for a brief time and thereby eliminated certain organisms and permitted others to survive.

Offshore-Shoreface Transition

Continuing upward in each regressive cycle the sand beds increase in thickness and become the dominant lithology, although thin interbeds of gray siltstones still persist (fig. 3). The physical sedimentary structures change as the small ripple lamina are replaced by crossbedded sands. In the Blackhawk Formation this facies may continue as alternations of crossbedded sandstones up to two or three feet in thickness with interbedded siltstones, or it may become a thick-bedded sand sequence of truncated megaripples in which preferred paleocurrent directions are difficult to establish. In either case *Ophiomorpha* and *Asterosoma* are present, and in addition one finds simple snail trails on bedding-plane surfaces. A summary of the outcrop data is:

1. Grain size—fine sand.
2. Physical sedimentary structures—crossbedded sands with interbedded gray siltstones.
3. Biogenic sedimentary structures—burrows of *Ophiomorpha*, *Asterosoma*, and simple snail trails; no bioturbate textures.

The truncated sets of sands suggest an environment of constantly shifting sediments effected by relatively high-energy conditions such as storms or strong tidal forces. These structures lack any preferred current direction. The shifting substrate and clean sand precludes the presence of a deposit-feeding infauna. Instead only filter feeders such as *Ophiomorpha* and *Asterosoma* can successfully inhabit this environment, and even they are not abundant; the only surface feeders to leave trails are some snails.

The shallow shelf of the Georgia coast probably represents similar conditions, although here a special situation exists due to the presence of relict Pleistocene sediments. Studies now in progress indicate that shelf sediments at least 30 or 40 miles from shore are clean sands with sand waves or megaripples as the principal sedimentary structures.

Lower Shoreface

The lower shoreface is one of the most distinct and interesting facies in the Upper Cretaceous units (fig. 4). It contains characteristic sedimentary structures and bedding types and a definitely zoned trace-fossil assemblage. The unit is composed of fine- to medium-grained sand, which is somewhat dirty due to the visible presence of fine, dispersed organic detritus and some clay. Bedding units characteristically average 12 to 18 inches in thickness and bedding plane boundaries are sharp. There are commonly a few lenses of very clean, megarippled sand within this unit.

In this facies large, well-defined trace fossils are abundant and exhibit definite vertical zonation. Every bed, except those which contain crossbedded sand is partially bioturbated but always retains some remnant of primary lamination. *Ophiomorpha* and *Asterosoma* continue landward from the offshore-shoreface crossbedded sands below and increases in density in the shoreface. Commonly a bed near the base of the shoreface will contain an extremely dense accumulation of vertical to nearly vertical *Asterosoma* burrows which flare slightly at the bedding surface. Above this, *Asterosoma* changes its burrowing form to that of a helicoid funnel, which is very abundant up to about midway through the shoreface, where it disappears.

Ophiomorpha continues to be present in great abundance in the shoreface, where it is represented by two forms which occur contemporaneously. One is the same burrow type which appears first in the offshore facies, and continues through the offshore-shoreface transition, and upward into the foreshore. A second variety of *Ophiomorpha*, restricted to the shoreface, is considerably different. It is larger, more robust, and primarily in the horizontal plane; it was probably the burrow of a different species of a callianassid-type shrimp. This burrow is restricted to the lower shoreface, where it is the most obvious and abundant trace fossil present.

Another burrow type, referred to as "plural curving tubes," is also very characteristic of the shoreface and is likewise vertically zoned. It begins just below the middle shoreface and continues up into the upper shoreface.

As pointed out previously, the bedding units in the lower shoreface are distinct and consistent in thickness. In nearly every bed there is a distinct pattern to the biogenic fabric. At the base of the bed, lamination is obvious with only a few burrows present. Upward, within the bed, the abundance of burrow structures gradually increases, and at the top of the bed the burrows are extremely dense (fig. 5). The upper bounding surface is erosional and the next bed repeats this pattern.

In the lower shoreface, the following information is found at the outcrop:

1. Grain size—fine- to medium-grained dirty sand.
2. Physical sedimentary structures—bedding units 12 to 18 inches thick containing even parallel laminae. Occasional crossbedded sand beds 1 to 3 feet thick which are lens-shaped.
3. Biogenic sedimentary structures—assemblage of abundant and distinct trace-fossil types which show definite species zonation upward within the facies. Biogenic fabric in each bed shows a vertical change of clean to burrowed.

From these data some conclusions can be reached. Trace fossils suggest a predominance of suspension-feeding organisms. Organic detritus trapped in the sediments is certainly greater than in the underlying crossbedded sands but is apparently not sufficient to support an abundant detritus-eating community, as is true of the offshore gray siltstone. The abundance of large constructed burrows suggests that this is a suspension-feeding assemblage, and the morphology of most of the burrow systems substantiates this. The vertical zonation of trace fossils is distinct. This is probably a depth zonation, reflecting the ability of animals to withstand increasingly greater wave energy as seen in Recent situations. At the top of the Cretaceous shoreface only *Ophiomorpha* and "plural curving tubes" suggest a form capable of withstanding these conditions, owing to their thick wall structures.

The nature of the beds and their biogenic content permit some conclusions with respect to the nature of sedimentation in the shoreface. Bedding thickness at a particular outcrop is remarkably constant, and the laminations within the beds are mostly parallel and subhorizontal. The "parallel to burrowed" vertical sequence of each bed shows progressive increase in the density of

 Snail trails

 **Asterosoma** (straight tubes)

 **Ophiomorpha**

Fig. 3. Offshore-shoreface transition. This facies is characterized by thick beds of crossbedded fine sand. Burrowing is much less than in facies above and below. The trace fossils shown are generally the only forms and these are not abundant.

burrows upward, and the top of nearly every bed has been eroded (fig. 5). Without the presence of the burrows, the erosional nature of these beds could be easily overlooked. Yet in this case the truncations indicate that the top of each bedding unit is an erosional surface. From this it is concluded that the *preserved record* of sedimentation in the lower shoreface has been controlled by periodic storms or strong tidal action which penetrated to a depth of probably 24 inches or less, during which time the sediments were resuspended and deposited. Following this burrowing animals reoccupied the substrate and thereby left an uneven record of bioturbation.

Ophiomorpha

Plural curving tubes

Asterosoma (helicoid funnel)

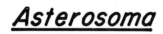
Ophiomorpha (horizontal)

Asterosoma

Teichichnus

Fig. 4. Lower Shoreface. This facies is characterized by fine- to medium-grained, dirty sand and an abundance of trace fossils. The vertical arrangement of trace fossils illustrated here represents the general vertical sequence of traces in the lower shoreface.

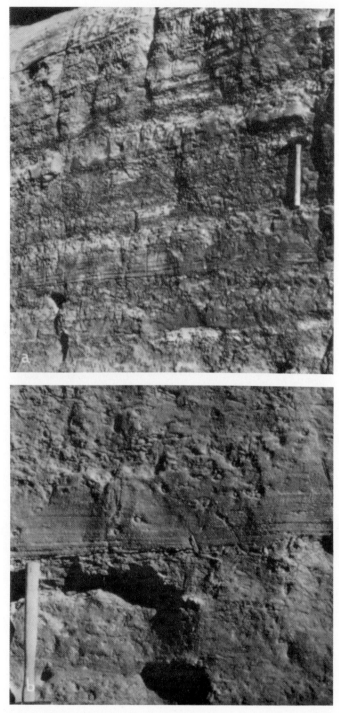

FIG. 5a and b. Bedding and bioturbate textures in the lower shoreface. Beds are marked by sharp erosional contacts. The basal part of each bed contains well-defined lamina which is only slightly burrowed. Upward the bed contains progressively more burrows. Figure 5a shows several of the "parallel to burrowed" beds. Figure 5b is a detailed view of the contact of two beds. *Ophiomorpha* is the most abundant burrow shown in these photographs.

Lower shoreface comparisons can be made between the Recent and Pleistocene of Georgia and the Upper Cretaceous of Utah. The characteristic "parallel to burrowed" bedding unit of the Cretaceous has not been found in the Recent and Pleistocene and the highly churned and well-burrowed textures which are preserved are more complex in the younger units. Inshore core stations in Recent sediments are so highly bioturbated that there is still considerable work to be done to determine whether burrowing species zonation will be as obvious here as in the Cretaceous. On the other hand, the Recent cores do show the presence of mottled textures, a high organic content, and the presence of clay, so that in the Pleistocene, Recent, and Upper Cretaceous the same general lateral changes in facies features are seen. Moving from the offshore-shoreface transition into the shoreface, the sediments change from clean, megarippled sands with relatively little burrowing to more organic-rich sands with thin laminae sets which are well burrowed and contain a predominance of suspension feeders.

The shoreface in both modern and ancient sediments is a very characteristic facies. It appears to be one of the most biologically rich marine environments, yet it is bracketed both to landward and seaward by cleaner sediments containing a less-varied fauna. The reason for this in Recent sediments of the Georgia coast is the presence of an abundance of nutrient detritus transported from the nearby tidal marsh. Another factor which makes this environment biologically attractive is that the waters are kept well oxygenated by wave action.

Upper Shoreface

The boundary between the upper and lower shoreface in the Upper Cretaceous (fig. 6) is sharp, where the highly burrowed beds of the lower shoreface are replaced in the upper shoreface by clean, medium- to fine-grained sand in distinct laminae which contain a very restricted variety of trace fossils. In the upper shoreface as in the lower shoreface, bedding planes which separate laminae sets are erosional surface, and the burrows serve to show the erosional character. The most characteristic physical sedimentary structures are low-dipping laminae which represent truncated, wedge-shaped sets (C. V. Campbell, personal communication). The very low-angle, wedge-shaped features in this facies are seen throughout the hierarchy of bedding units. Individual laminae, laminae sets, and beds all possess this feature. Ripple laminae and trough cross-laminae are occasionally present.

Biogenic sedimentary structures are greatly reduced in variety. Only *Ophiomorpha* and "plural curving tubes" continue from the lower to the upper shoreface. The latter are present only in the basal upper shoreface of the Upper Cretaceous, and even *Ophiomorpha* diminishes upward and is absent near the top.

In summary, the characteristics of this facies are:

1. Grain size—fine- to medium-grained, clean, well-sorted sands.
2. Physical sedimentary structures—low-angle distinct laminae making up wedge-shaped laminae sets and bed sets. Also present in minor amounts are ripple laminae and trough cross-laminae.
3. Biogenic sedimentary structures—very restricted variety of ichnogenera; may be abundant in terms of individuals.

The lower shoreface exhibits a great variety of burrow types in detritus-rich and somewhat clayey sediments, whereas the upper shoreface contains a very restricted biogenic assemblage in very clean, well-sorted sands. The truncated nature of bedding planes, indicated by eroded burrow tops and dipping laminae sets, is one of the most important physical distinctions of this environment. If these are overlooked and the sets are considered to be even, parallel laminae, the facies might be interpreted as one in which there was a continuous addition of sand. The truncation of physical and biogenic structures, however, points to the influence of storms during which time the bottom is eroded to some depth and then rebuilt. The absence of a community of deposit-feeding infauna in this facies, just as in the offshore-shoreface transition, is an indication of the absence of organic detritus present in any significant abundance in the substrate. The only burrows which are present, therefore, are suspension feeders, and even these, as is indicated by numerous truncated burrow tops, indicate an environment of significantly higher energy than seen previously.

Shoreface-Foreshore Transition

Trough crossbedded sands, interpreted to represent the plunge zone at the foot of the beach, are present near the top of each complete vertical sequence in the Upper Cretaceous units studied (fig. 6).

The trough crossbedded sands are clean and well-sorted and possess the same grain-size parameters as the upper shoreface. The biogenic record in this facies is likewise distinct and different from the shoreface. Burrowing is not abundant in the Cretaceous exposures, but at most outcrops there is some evidence of animal activity. Several types of burrows are found in this facies in the Recent and Pleistocene but no

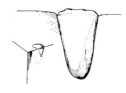 **Plug-shaped burrow**

 Ophiomorpha

 **Plural curving tubes**

FIG. 6. Upper shoreface, shoreface-foreshore transition and foreshore. The lower two-thirds of the outcrop shown in the photograph is interpreted to be upper shoreface and the upper one-third the shoreface-foreshore transition and foreshore. *Ophiomorpha* is found throughout the shoreface and plural curving tubes in the upper shoreface. Plug-shaped burrows are present, though not abundant, in the shoreface-foreshore transition. Clean, fine-grained sand in well-defined laminae is present throughout the upper one-third of this outcrop.

burrow that is characteristic of the Cretaceous outcrops is found in Recent and Pleistocene backshore deposits.

In summary, the shoreface-foreshore transition shows the following:

1. Grain size—clean, well-sorted, fine- to medium-grained sand.
2. Physical sedimentary structures—with trough cross laminae common.
3. Biogenic sedimentary structures—burrows are not abundant but may indicate that several different organisms have burrowed in this facies.

Upper Cretaceous shoreface-foreshore transition deposits examined in the course of this study have produced some unexpected results. These structures were first thought to represent channel systems cutting through the beach, but paleocurrent measurements show that they consistently reflect a flow direction parallel to, rather than across, the depositional strike. The characteristic is repeated at the top of the beach sequence in more than 12 beaches. The Upper Cretaceous backshore sediments consist of multiple sets of trough cross-laminae, 6 to 18 inches thick and usually about 12 inches thick. It is believed that this represents a strong longshore drift component along the top of the beach.

Foreshore

Low dipping, parallel to subparallel sand beds occupy the top few feet of nearly all of the complete sandstone cycles of the Book Cliff Cretaceous units. Only where channel-fill deposits have replaced the top of the sequence are they missing.

Sands in this facies are clean and well-sorted and contain the same grain size as the underlying trough crossbedded sands. Bedding plane surfaces are not erosional and when traced laterally the foreshore beds appear to grade laterally into the trough units.

Biogenic sedimentary structures are extremely rare in this facies and can be considered absent. Physical sedimentary structures are limited to occasional beds containing convolute lamination and more rarely, flame structures.

Thus, this facies is characterized by:

1. Grain size—Very clean, well-sorted, fine- to medium-grained sand.
2. Physical sedimentary structures—low dipping parallel to subparallel beds with some convolute lamination.
3. Biogenic sedimentary structures—absent.

This facies is interpreted to represent a foreshore beach deposit. The absence of deposit feeders is not surprising in such a clean sand. However, most beaches do contain some burrowing fauna which filter feed. The absence of evidence for any burrowing fauna in this facies cannot be explained unless the fauna consisted exclusively of very small organisms, such as amphipods which left no significant signature in the sedimentary record.

CONCLUSION

The foregoing represents one example of the use of trace fossils in combination with other primary structural and textural features as keys to facies interpretation. Their significance in this instance is greatly enhanced because of the paucity of body fossils. However, even when other fossil evidence does exist, the importance of trace fossils as primary environmental structures should be considered.

REFERENCES

HOWARD, J. D., 1966, Characteristic trace fossils in Upper Cretaceous sandstones of the Book Cliffs and Wasatch Plateau: Central Utah: Utah Geol. Min. Survey, Bull. 8, p. 35–53.
SPIEKER, E. M., 1949, Sedimentary facies and associated diastrophism in the Upper Cretaceous of central and eastern Utah: Geol. Soc. America, Memoir 39, p. 55–81.
WEIMER, R. J. AND HOYT, J. H., 1964, Burrows of *Callianassa major* Say, geologic indicators of littoral and shallow neritic environments: Jour. Paleont., v. 38, p. 761–767.
YOUNG, R. G., 1955, Sedimentary facies and intertonguing in the Upper Cretaceous of the Book Cliffs, Utah and Colorado: Geol. Soc. America, Bull., v. 66, p. 177–202.
———, 1957, Late Cretaceous cyclic deposits, Book Cliffs, Utah: Amer. Assoc. Petrol. Geol., Bull., v. 41, p 1770–1774.

RECOGNITION OF ANCIENT SHALLOW MARINE ENVIRONMENTS

PHILIP H. HECKEL
Kansas Geological Survey, Lawrence, Kansas 66044*

ABSTRACT

Modern shallow marine environments encompass a great variety of conditions from shoreline to a depth of about 600 feet. In sedimentary rocks, these environments are most readily inferred from diverse assemblages of fossils whose modern relatives are marine. Some sparse and restricted biotas may represent fully marine environments in which certain factors were unfavorable to many types of organisms. Many unfossiliferous black shales represent a foul environment that supported no benthonic life and are inferred to be marine mainly by stratigraphic relations. Marine environments that lack significant sedimentation would be represented in the record only by a submarine paraconformity.

Recognition of marine subenvironments is possible through direct lithic analogy to distinctive modern sediments of known depositional environments, such as oolite, sea-margin laminated calcilutites, and organism-controlled features such as reefs. In less distinctive marine facies, subenvironments are difficult to discriminate because visible differences may have resulted from a complex interplay of many variable factors that did not coincide to produce unique subdivisions. Ecologic consideration of fossil assemblages may distinguish clear-water from turbid-water, or soft-substrate from hard-substrate environments. Petrographic considerations also allow environmental inference. For example, calcilutite indicates a quiet-water environment that might be either shallow and protected from water agitation by a physical barrier, or deep and protected by water depth itself. The presence of calcarenite composed of whole shells exhibiting little fragmentation or abrasion might indicate only local organic proliferation or lack of dilution by fine sediment. In contrast, calcarenite composed of fragmented, abraded, well-sorted skeletal grains indicates water turbulence and winnowing of fines, processes that are more probable in shallow water.

Environmental syntheses based on stratigraphic, petrographic and paleontologic criteria may bring into focus those aspects of ancient marine environments (such as water depth) that are difficult to determine from the record. On a local scale, detailed facies mapping in undeformed rocks may allow detection of original topography that controlled facies changes. On a larger scale, systematic lithic variation along outcrop of an entire stage of rocks may provide a regional picture of the lateral succession of ancient marine environments across an epicontinental basin. Perhaps one of the best modern laboratories to study analogs of ancient marine epicontinental deposition is the Sahul-Arafura Shelf and Gulf of Carpentaria between orogenic New Guinea and cratonic Australia.

INTRODUCTION

Definitions

Bathymetric position.—Shallow marine sedimentary environments correspond to the sublittoral benthonic zone that floors the coastal shelf off major land masses. This zone of deposition often has been referred to as "neritic" by geologists (e.g., Dunbar and Rodgers, 1957), but others of a more ecologic inclination prefer to restrict that term to the water mass above the bottom (Ekman, 1953; Hedgpeth, 1957). As generally accepted, the sublittoral or shelf zone (lying beneath neritic waters) extends from the line of lowest tides to the edge of the continental shelf (Ekman, 1953).

The edge of the shelf is commonly considered to occur at the 100 fathom (600 foot) contour, but the break in slope that defines the physiographic shelf edge varies somewhat on either side of this depth from one area to another. Because the actual break in slope is likely to differentiate sedimentary environments of divergent natures, the suggestion of Dunbar and Rodgers (1957) to

---

* Present address: Department of Geology, Iowa University, Iowa City, Iowa 52240.

recognize the break in slope as the lower limit is more useful for our purposes (fig. 1).

Thus, the sublittoral zone with shallow water environments is bounded by the littoral zone with marginal environments on one side and by the bathyal zone with slope environments on the other.

Salinity.—The foregoing definition must be modified further so that the term "marine" can be meaningfully applied. Many marginal environments such as the intertidal are so closely related to the sea that they are treated further in this article. Restricted bodies of coastal water, however, such as estuaries which are conduits for fresh water or lagoons which are subject both to freshwater runoff and to excessive evaporation, must be excluded from the shallow marine environment when their salinities deviate greatly from normal marine salinity for long periods of time, because their biota then will deviate from what is typical for marine environments. Exclusion of associated but restricted bodies of predominantly fresh or strongly hypersaline water permits a more coherent definition of major criteria for marine environments.

Normal salinity for marine water in the open

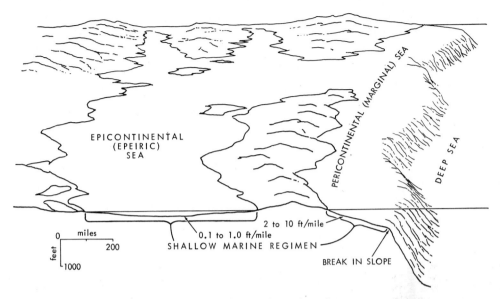

Fig. 1.—General characteristics of shallow marine regimen in two types of shallow seas.

ocean averages between 35 and 36 parts dissolved salts per thousand parts of water. The number of species of organisms is greatest at normal marine salinity and declines into either higher or lower salinities (Pearse and Gunter, 1957).

The points at which the greatest reduction in number of species takes place away from normal marine salinity ideally should be used to define the limits of the marine environment because composition and diversity of fossil assemblages are major tools for recognizing ancient marine environments. These points, however, are difficult to determine. Salinity tolerance of marine organisms differs from taxon to taxon. It is also dependent on the period or duration of less favorable conditions in transitional environments which typically are extremely variable because of interaction of factors such as wind, tides, evaporation, and fresh-water runoff. As a result, no single salinity values can be chosen as limits of the marine environment. Therefore for our purposes, transition zones within which the number of species declines significantly define the upper and lower limits of the marine environment.

Toward lower salinities, a transition zone can be recognized between 20 and 30 ppt. Percival (1929, *in* Pearse, 1950) noted that the number of marine species declines greatly when salinity decreases below 30 ppt but that some species persist in salinities down to 21 ppt. Hedgpeth (1957) shows that the region between 17 (or 20) and 30 ppt is classified as "marine" or "brackish" by different workers. Ekman (1953) points out that this region is not "brackish" with respect to the composition of the biota it supports, but that only more tolerant marine forms predominate. Experiments with corals, a strictly marine group, show that all readily survive one-day exposure to salinities of 28 ppt but none could live long in a salinity of 17 ppt (Vaughan and Wells, 1943).

Toward higher salinities, a transition zone seems reasonably recognized between 40 and 50 ppt. Only unusually tolerant organisms normally live in salinities above 40 ppt (Carpelan, 1967), and these are limited in number. Experiments on corals show that a few do not endure even 40 ppt for one-half day and that most are killed above 48 ppt for similar periods of exposure; few corals from normal marine water survive even 40 ppt when exposed for periods of a month or more (Vaughan and Wells, 1943). The Red Sea with a salinity of 41 ppt, however, does support a small coral fauna along with other typically marine forms such as echinoderms and sponges. Because open marine waters rarely exceed 41 ppt even in partially restricted arms like the Red Sea and because more completely restricted hypersaline lagoons experience such strong variations in salinity, it is difficult to narrow the range within which the greatest reduction of faunal diversity takes place above normal marine salinity. Moreover, the recent report by Kinsman (1964a) of corals living in a salinity of 48 ppt in parts of the Persian Gulf, shows that data on organic tolerances of higher salinities appear less consistent than those concerning tolerance of lower salinities.

To generalize, the marine environment in its broad sense extends from 20 to 50 ppt in salinity. It encompasses the normal marine environment, ranging from 30 to 40 ppt, and includes at either end two transitional zones of restricted marine environments ranging from 20 to 30 ppt and from 40 to 50 ppt.

Types of shallow seas.—Shallow marine environments, so defined, occur both in what I shall term *pericontinental* (or marginal) seas that occupy shelf areas around the continents, and in broad, shallow *epicontinental* (or epeiric) seas that lie upon the continents (fig. 1). The latter are rare today, but they provided much, perhaps a majority on some continents, of the preserved and accessible stratigraphic record. Important distinctions between pericontinental and epicontinental seas lie in dimensions, depth, and bottom slope, which have profound effects on water circulation, environment, and ultimately nature of sedimentation (Shaw, 1964; Irwin, 1965). Modern pericontinental seas have maximum depths of 600 to perhaps 1000 feet at the edge of the shelf and maximum widths of 100 to 300 miles, giving an average bottom slope of 2 to 10 feet per mile. Epicontinental seas on the other hand, spread for perhaps 1000 to even 2000 miles in breadth over the craton. As they probably only rarely reached depths of 600 feet and generally may have attained only 100 feet (Shaw, 1964, p. 5), bottom slopes would have been less than 1 foot per mile, probably averaging 0.1 to 0.6 foot per mile. Although recent work by Lineback (1969) indicates depths of up to 1000 feet and thus also steeper slopes for the Illinois Basin, this is only a local area of a much larger epicontinental sea in the greatest portion of which depths and slopes suggested by Shaw probably prevailed.

Scope

The entire spectrum of shallow marine environments encompasses a great variety of conditions in both types of seas. The great heterogeneity of these environments allows establishment of few clearcut criteria covering the entire range. Furthermore, the criteria that can be established are primarily unidirectional. That is, a criterion may differentiate the marine environment from nonmarine or transitional environments, or it may differentiate a shallow marine from a deep marine environment, but few single criteria can unequivocally define a shallow marine environment. Differentiating shallow from deep marine environments is not critical when dealing with most epicontinental sediments, but it must be considered when dealing with pericontinental seas that probably were developed in complex geosynclines such as recorded in the Tethyan deposits of Eurasia or Cordilleran deposits of western North America.

In subsequent sections this article evaluates criteria based on lithology, mineralogy, fossil content, and sedimentary structures by showing the modern distributions of these features, because these provide the basis for paleoenvironmental interpretation. Furthermore, determining why certain features are restricted to particular environmental ranges helps to evaluate how well present distribution can be applied to the stratigraphic record. Various criteria useful for recognizing subenvironments of the shallow marine regimen are discussed, and some are synthesized into a general model of shallow marine sedimentation. Examples of environmental interpretations from the Pennsylvanian of Kansas and Devonian of New York show what can be derived from integrated stratigraphic and paleontologic syntheses of certain stratigraphic units whose origins are not obvious. Modern analogies are utilized where possible, and observations on modern epicontinental seas are stressed.

Limitations

Another effect of the broad extent and heterogeneity of shallow marine environments is that most detailed investigations have been focused only on limited, more readily analyzed aspects of the deposits in this regimen. Therefore, in this article, certain aspects of the record may be neglected, others may be overgeneralized, and much of what is covered may be treated superficially. Nevertheless, in a symposium of this sort, a general review should be valuable in bringing together, summarizing (however incompletely), and evaluating many of the criteria and some of the basic ideas involved in interpreting ancient shallow marine environments in the stratigraphic record. Finally, in consideration of the voluminous amount of literature on this subject, I ask the forbearance of those authors whose pertinent work I have inadvertently overlooked.

Acknowledgments

I am indebted to W. K. Hamblin of Brigham Young University and D. F. Merriam of the Kansas Geological Survey for initial persuasion to undertake this project and for continuing encouragement during its course. J. K. Evans of Atkinson Petroleums Ltd., Calgary, kindly permitted use of material from his unpublished doctoral dissertation on a Midcontinent Pennsylvanian black shale. E. W. Behrens of the University of Texas Marine Institute, R. Evans of Kansas University and P. Tasch of Wichita

State University, generously provided information and references on Texas Coast faunal diversity, evaporite deposition, and branchiopods respectively. J. J. Veevers of Macquarie University, C. V. G. Phipps of Sydney University and J. N. Casey of the Australian Bureau of Mineral Resources kindly provided information and references concerning sedimentation in the shallow sea between Australia and New Guinea. I greatly appreciate critical review of the manuscript by L. F. Laporte of Brown University and J. M. Cocke of Iowa University (now at East Tennessee State University) whose suggestions resulted in material improvement of the presentation. Finally, I wish to gratefully acknowledge the hard work of R. B. Williams who drafted the illustrations and of (Mrs.) L. E. Page and (Mrs.) J. M. Combes who organized and checked the list of references.

GENERAL LITHOLOGIC EVIDENCE

One need only peruse a modern textbook on stratigraphy and sedimentation (Dunbar and Rodgers, 1957; Krumbein and Sloss, 1956; Weller, 1960) to determine that gross lithology of a deposit usually does not provide sufficient information to delineate nature of the depositional environment to the extent that is useful for our purposes (fig. 2).

Terrigenous Clastics

Terrigenous mud and sand are presently being deposited extensively in fluvial and alluvial situations, such as the Mississippi delta, as well as in sublittoral environments along the Louisiana and Texas coasts. Thus sandstones and shales can just as readily be nonmarine as marine. Furthermore, recent corings from oceanic depths show that sandstones and shales can be deep-sea as well as shallow marine deposits. Conglomerates also are being deposited in a wide range of environments from nonmarine to both shallow and deep marine, although they are probably less common outside nonmarine and transitional environments. Clearly, other evidence such as fossil content and sedimentary structures must be examined before depositional environments of terrigenous detrital rocks can be determined.

Carbonates

Carbonate sediments are primarily marine in origin, but a large amount also is deposited in marginal intertidal and supratidal environments which deviate to various degrees from marine conditions. As much or more carbonate is presently being deposited in the deep sea as in shallow marine environments (Rodgers, 1957). Some carbonate is deposited inland in nonmarine environments such as lakes, caves, and hot springs. Again, fossil evidence and sedimentary structures must be examined in order to determine environment of deposition.

Cherts

Many bedded and nodular cherts seem to be marine deposits, but this is determined from faunal and stratigraphic evidence. Moreover, conditions under which chert is formed in marine deposits are still subject to controversy. Chert is known to be forming today by precipitation of hydrous silica as beds in a nonmarine lake in east Africa (Eugster, 1967, 1969) and as nodules in ephemeral lakes associated with a restricted coastal lagoon in South Australia (Peterson and von der Borch, 1965). Some bedded chert apparently also has formed from chemical weathering as a soil on the Miocene peneplain of Australia (Dunbar and Rodgers, 1957).

Redbeds

Terrigenous redbeds of all grain sizes are mainly nonmarine deposits. Dehydrated ferric oxide, hematite, which is responsible for the red pigment is formed from weathering of iron-rich minerals in soils and alluvial deposits in warm, humid to arid climates (Walker 1967; Van Houten, 1968). Red deposits also must be well drained in order to maintain not only oxidizing conditions but also dehydrating conditions (see Schmalz, 1968). Red coloration is known to form in place after deposition in nonmarine environments (Walker 1967; Van Houten, 1968). Likewise, if transported red sediments are deposited in a well-drained (and necessarily nonmarine) basin, the red color will be preserved (Dunbar and Rodgers, 1957). On the other hand, if red sediments are deposited in marine waters, the hematite probably becomes either hydrated to "limonite" or reduced to the ferrous state and loses the red color. Loss of red color upon introduction into the marine environment is shown to occur with modern red sediments by Cartwright (1928) on the California coast and by Dawson (in Raymond, 1927) on the coast of Nova Scotia, and is inferred to have occurred in the Mississippian Red Bedford Delta of Ohio (Dunbar and Rodgers, 1957). In a few places today, however, red sediments are being deposited intact in the marine environment off the mouth of the Amazon River (Dorsey, 1926) and along parts of the tropical coasts of Africa and China (Raymond, 1927). Evidently, either oxidizing conditions are being maintained for a distance into the sea, or the sediment is being deposited rapidly before it can be reduced (Dunbar and Rodgers, 1957). Marine redbeds are known in the

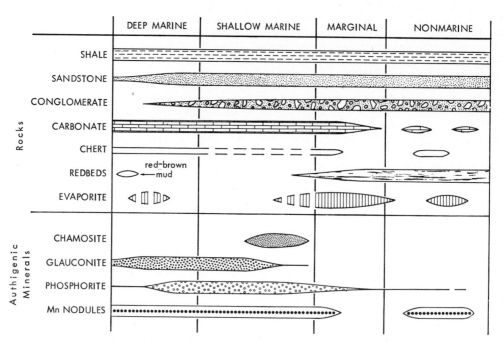

FIG. 2.—Modern distribution of rock types and certain authigenic minerals among major depositional environments. Dashed lines show regions of potential or inferred formation in the past.

geologic record, for instance in the Pennsylvanian of Colorado (Walker, 1959), but the criterion for establishing them as marine is fossil content, which is a more powerful tool for distinguishing between marine and nonmarine environments.

Marine red-brown mud also occurs widespread in deep, open ocean basins. If a reddish deposit from this environment were incorporated into the readily accessible stratigraphic record, its marine fossil content like that reported in Emery, Tracey and Ladd (1954, p. 20) would reveal its marine origin. Indeed, red limestone recognized by Garrison and Fischer (1969) to be of deep-water marine origin in the Alpine Jurassic contains a substantial fauna of marine organisms.

Evaporites

Evaporite deposits such as gypsum, anhydrite, and halite are forming today in restricted coastal waters and adjacent tidal flats in arid climates where evaporation greatly exceeds fresh-water replenishment, and where some type of barrier impedes free circulation of water (Krumbein and Sloss, 1956; Weller, 1960). In restricted estuaries and lagoons, water lost by evaporation is replenished by marine inflow through gaps in the barrier, but the combination of the incoming hydrostatic head and restrictive nature of the barrier prevent outflow of brine concentrated by evaporation, and salinity increases to the point at which salts precipitate. Evaporite formation of this kind is known from a sea-coast estuary in Peru (Morris and Dickey, 1957) and from the Gulf of Kara-Bogaz, where salts are precipitated out of brine concentrated from brackish water (13 ppt) of the Caspian Sea (Garbell, 1963). Salt formation by means of a modification of this mechanism is known in Baja California (Phleger and Ewing, 1962) and along the Trucial Coast of the Persian Gulf (Kinsman, 1966) where the barrier involves position above normal high-tide line, and salts precipitate within supratidal sediment after periodic flooding by abnormally high tides.

Evaporite formation also occurs in arid continental regions where salts accumulate in undrained basins, as in the western United States. Typically, such continental deposits are rich in certain salts, such as borates, which are not present in great amounts in sea water.

Evaporites also could form within a marine basin away from the coastal margin if there is an arid climate and a deeper local area in the basin into which brine concentrated by excess evaporation can flow because of its greater density, according to the model described by Richter-Bernburg (1955) and Schmalz (1969). The critical factor in this case is the presence around the deeper area of a sill high enough to prevent the dense brine from flowing out and comingling again with normal marine water. Although the sill performs the same function as the subaerial barrier along coastal lagoons, it may be entirely submarine, and normal marine neritic waters can be maintained continuously over the top as the water concentrated by evaporation continually sinks and is replaced by more water from the surrounding open sea. Occurrence of ancient thick salt deposits in the middle of epicontinental basins with marine deposits around the edges suggests that large-scale density currents may have occurred in places like the Michigan Basin during the Silurian; greater subsidence of the deeper part of such a basin would increase effectiveness of the sill and ultimately increase intensity of the evaporite environment (Sloss, 1953).

A model for marine evaporite deposition in shallow epicontinental seas has been outlined by Shaw (1964, p. 4–71). In contrast to the open free circulation with great currents and measurable tidal range observed in modern oceans and pericontinental seas, the shallowness and breadth of epicontinental seas would hamper development of great currents, and diurnal tidal exchange would be effectively damped around the periphery of the sea by bottom friction. Such reduction of natural water circulation would give rise to hypersaline conditions around the periphery of an epicontinental sea in an arid climate where water lost through evaporation would not be replaced rapidly enough to prevent development of a permanent salinity gradient. Density currents set up by the salinity gradient and currents resulting from inflow of water from the open sea to replace that lost by evaporation would be ineffective in reducing the peripheral hypersalinity because of bottom friction and also because hypersalinity must exist in the first place before such currents can be set up. Wind-driven currents would tend primarily to pile up water periodically on the downwind side of the sea without reducing evaporation or altering salinity. Only fresh water influx would be effective in reducing salinity in peripheral portions of such seas. In arid climates, therefore, evaporites could form around the periphery of a marine environment with no more noticeable a barrier than shallowness of water that increases the effects of bottom friction and ultimately reduces water circulation. This model should be kept in mind also during the following discussion of paleontological evidence, for even if the climate is not dry enough for salinity to reach a sufficiently high value to precipitate evaporite minerals, salinity still could be high enough to restrict the biota from that characteristic of a normal marine environment.

Although evaporites form only in water that by definition is too saline (>72 ppt) to be "marine," they can form in environments within a general marine regimen or at least marginal environments derived from and closely associated with the marine regimen. Thus, evaporites do not necessarily imply a nonmarine basin or even barred coastal lagoons marginal to a marine basin. They may record a local, more rapidly subsiding area within a marine basin, or the entire periphery of a shallow marine epicontinental sea that is unrestricted by subaerial physical barriers.

Salinities recorded by various marine-derived evaporite associations have been summarized by Scruton (1953), using data originally obtained from the experiments of Usiglio on evaporation of sea water. Calcium carbonate begins to precipitate from sea water evaporated to about half its original volume, with a salinity of 72 ppt, and continues up to a salinity of about 200 ppt. At that value calcium sulphate starts to deposit and continues beyond the point where sodium chloride begins to precipitate at 353 ppt, which corresponds to evaporation to one tenth of the original volume of the sea water. Environmental terms have been given by Sloss (1953) to evaporite associations. The "penesaline" environment, defined by evaporitic carbonates interbedded with anhydrite and gypsum, records salinities from 72 to 353 ppt. (Dolomite apparently is a common evaporitic carbonate in the stratigraphic record, even though it was not noted in Usiglio's experiments.) Penesaline salinities are in excess of those tolerated by most aquatic life. The "saline" environment, defined by major accumulation of halite along with anhydrite, records salinities in excess of 353 ppt. Similar lithologic associations observed in the modern Peruvian evaporites coexist with water of salinities within or close to the ranges determined experimentally (Morris and Dickey, 1957).

MINERALOGIC EVIDENCE

Detrital minerals depend mainly on provenance and thus can be deposited in almost any environment. Certain authigenic minerals, however, are known to be forming solely or in large amounts only in the marine environment, and thus are marine indicators in sedimentary rocks (fig. 2).

Chamosite and Glauconite

Authigenic iron minerals can be divided into four major facies groups based on dominance of oxide, silicate, carbonate, or sulphide. Each facies group reflects different degrees of availability of oxygen in the bottom environment (Porrenga, 1967). Many studies show that modern occurrences of the iron silicates, chamosite and glauconite, are exclusively marine. Both minerals occur as green pellets and fillings of small shells, and this similarity of appearance and occurrence has caused much chamosite to be mistaken for glauconite. Recent studies summarized by Porrenga (1967), particularly of the areas off the Niger and Orinoco deltas and northern Borneo, show that each mineral seems to be characteristic of a certain temperature and depth zone. Chamosite occurs only in tropical areas in shallow waters about 30 to 500 feet deep; in shallower water it is commonly replaced by goethite ("limonite") which may be an oxidation product of the chamosite. Glauconite forms mainly in deeper water ranging from 30 to 6000 feet and occurs in greatest abundance from 100 to 2400 feet. Both minerals occur mainly in regions of little or no detrital sedimentation. Depth restrictions of the two minerals apparently are temperature-dependent, with chamosite the warmer-water mineral and glauconite the cooler-water form. Thus the depth ranges indicated by these minerals are only as good as the extent to which temperature varies regularly with depth.

Phosphorite

The most favorable conditions for the formation of phosphorite in large amounts apparently occur in shallow marine water between 100 and 1000 feet in depth (Bromley, 1967). Upwelling of phosphate-rich water results in either direct precipitation of calcium phosphate as nodules or laminae, or replacement of various forms of calcium carbonate. Where such upwelling is particularly common, as along the west coasts of continents, phosphate enrichment of the surface waters typically results in an excessive phytoplankton bloom that poisons the water and causes mass mortality of fish which contribute more phosphorite in the form of organic remains to the bottom sediment (Brongersma-Sanders, cited in Bromley, 1967). Furthermore, upwelling also is responsible for excessive aridity of nearby land, which results in reduction of terrigenous detrital influx that might otherwise dilute and mask the phosphorite.

The environment of phosphorite formation extends into estuaries where high concentrations of dissolved phosphate provide favorable conditions for replacement of calcium carbonate (Pevear, 1966). Degens (1965) points out that because continents do not have the immense phosphorus reservoir of the oceans, nonmarine phosphorites are minor in abundance. Because they consist mainly of vertebrate bones, they are not difficult to distinguish from marine phosphorite. The deep sea, once thought to be the site of much phosphorite formation, is now considered to be insignificant compared to shallower water, because most phosphate is released to the water from descending organic matter above depths of 3000 feet (Bromley, 1967); however, nodules formed in shallow environments apparently have moved down slopes to final resting places in deeper water from which they occasionally have been dredged. In any case, most workers agree that occurrence of readily noticeable amounts of phosphorite indicates slowdown or absence of terrigenous detrital sedimentation for a substantial period of time.

Manganese Nodules

Although manganese nodules traditionally have been associated with the deep-sea environment (Ladd, 1957), they also are known to occur in shallow marine environments and, along with iron nodules, in lacustrine and paludal environments (Price, 1967). In fact, lacustrine nodules locally are so common in Green Bay (off Lake Michigan) that they may be an economic source of manganese (Rossman and Callender, 1968). Thus manganese nodules by themselves are not good environmental indicators. As with other authigenic minerals, however, they indicate slowdown of deposition and were used by Garrison and Fischer (1969) in conjunction with other criteria to strengthen an interpretation of slow deposition in deep water for certain European Mesozoic limestones.

Geochemical Indicators

Price (1967) shows that concentrations of iron and certain minor elements in manganese nodules vary systematically from one environment to another. Thus ratios of various elements in manganese nodules should yield estimates of paleosalinity. Other suggested indicators for paleosalinity include gallium and boron in illite (Degens et al., 1957; Walker and Price, 1963) and calcium-iron ratio in sedimentary phosphorite (Nelson, 1967). Such geochemical indicators are, however, not apparent on outcrop and thus are outside the scope of this report. Moreover, there have been problems in using several of these methods outside of the original studies. (See Eagar and Spears, 1966; Muller, 1969; Guber, 1969.) Thus the extent of their applicability is questionable and may be quite limited.

Recent discovery by Lowenstam and McConnell (1968) of fluorite in hard parts of certain crustaceans and other invertebrates that are common in shallow marine waters but rare at greater depths suggests another potential mineralogical or geochemical indicator of the shallow marine environment. It is not known, however, how widely fluorite is preserved in the stratigraphic record or how easily it can be detected.

BIOTIC EVIDENCE

Observations in Modern Environments

In order to discuss adequately the evidence provided by fossils in determining depositional environment, general ecologic tolerances of living organisms in modern seas must be considered. Although certain aspects of ecologic tolerance may have changed through geologic time, many fundamental aspects probably have not; thus there can be no better starting place for paleoenvironmental interpretation than evaluation of modern environmental observations. Environmental tolerances of those organic groups having preservable hard parts that are significant in the sedimentary record are summarized in Appendix 1. So that this presentation will be of more general use, only higher taxa are dealt with. Nevertheless, some references to more detailed environmental differentiation within certain groups are included.

Toleration of organisms to salinity is emphasized in the following discussion because this factor is the critical difference between marine and nonmarine environments. In order to live in an aqueous environment, most organisms must maintain a certain concentration of nutrient salts in their body fluids. Because salt concentration within the body fluids of most marine invertebrates is close to that of normal sea water, they do not need to osmoregulate, that is, control the inflow and outflow of fluid through the semipermeable membranes of their body walls when in sea water (Pearse and Gunter, 1957). But they are restricted to marine environments of a certain salinity range because they will absorb more fluid than their cells can hold in brackish or fresh water, and will dehydrate by losing body fluids into the surrounding medium in hypersaline water. In either case death will result. Such organisms with a limited salinity tolerance are termed stenohaline. Different marine organisms have tolerance ranges of different breadths but all are centered around normal marine salinity (fig. 3). Organisms which have a much wider range of salinity tolerance are termed euryhaline. These either can tolerate great changes in salt concentration in their body fluids, as with certain algae, or have evolved various methods of osmoregulation to maintain constant concentrations of body fluids in highly variable or nonmarine environments, as with certain groups of arthropods, molluscs, and worms (Pearse and Gunter, 1957).

Light requirements of organisms also are emphasized because light penetration is directly related to depth and is therefore a critical difference between shallow- and deep-water environments. Temperature and water turbulence are less directly related to depth but seem to be important controlling factors for distribution of minor taxa within major organic groups. Turbidity and nature of substrate vary greatly within a general environment, and these factors have value in differentiating subenvironments within the shallow marine regimen. Oxygenation of water and availability of nutrients are necessary for all organisms considered here, and these factors are of interest mainly in the adverse effects on the environment caused by their absence.

Application to the Stratigraphic Record

Most geologists are familiar with the problems of preservation even of hard parts of organisms in the stratigraphic record, and this aspect will not be elaborated here. Suffice it to say that those fossils that do become preserved may yield valuable information for paleoenvironmental interpretation subject to limitations imposed by problems of transport into alien environments which will be elaborated later.

Fossil indicators for the marine environment.— Because distribution of aquatic organisms is closely limited by salinity and variations in salinity, fossils are a powerful tool for distinguishing marine from nonmarine environments in the stratigraphic record. Furthermore, because salinity tolerances—particularly of stenohaline organisms—do not seem to change readily through time, an association of fossils whose modern descendants and/or relatives are marine indicates that the ancient environment in which they lived was marine.

A fossil assemblage that indicates exclusively a marine environment within the broad limits shown on Figure 3 would contain members of groups that have restricted salinity tolerance. These include calcareous red and green algae, radiolarians, silicoflagellates, coccoliths, most calcareous foraminifers, calcisponges and hyalosponges, corals, bryozoans, brachiopods, echinoderms, barnacles and several molluscan groups: monoplacophorans, chitons, scaphopods, and cephalopods (fig. 3). Only rarely do all occur together on account of different preference of substrate or other environmental variables within the marine regimen, but, typically, rep-

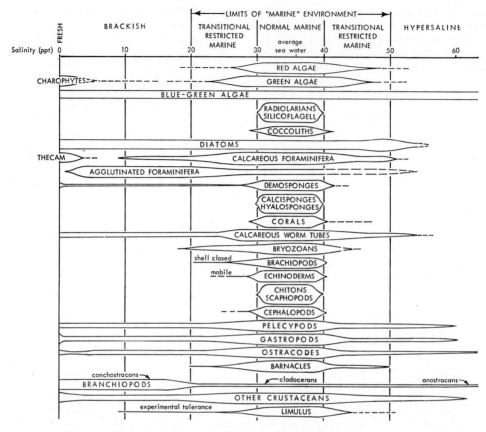

FIG. 3.—Modern distribution of major fossilizable nonvertebrate groups relative to salinity. Thickness of bar roughly reflects diversity of taxa.

resentatives of several of these groups will occur in a normal marine assemblage. Nevertheless, presence of only a few untransported members of a strictly normal marine group, such as corals, should indicate a normal marine environment. On the other hand, a few bryozoans, calcareous foraminifers, and, to a lesser extent, either of the algal groups, lingulid brachiopods, or mobile echinoderms and cephalopods can live in or move intermittently into environments where the salinity deviates from the normal range into the transition zones of the marine environment. Thus any of these fossils alone, or with a few other members of these or of the more tolerant groups to be mentioned shortly, might signify a marine environment that is marginal to the open sea and slightly restricted for some reason. Large parts of epicontinental seas might be populated with this type of restricted but marine assemblage. Also occurring in a marine assemblage may be many types of blue-green algae, diatoms, agglutinated foraminifers, demosponges, calcareous worm tubes, pelecypods, gastropods, ostracodes and the rarer arthropod groups, but because these more tolerant groups include members that can live also in nonmarine environments, they do not by themselves necessarily indicate a marine environment.

For comparison to the marine assemblage, a typical brackish-water assemblage might consist of a limited number of taxa of several of the previously mentioned more tolerant groups: pelecypods, gastropods, ostracodes, branchiopods, malacostracans, agglutinated foraminifers, diatoms, worm tubes and blue-green algae. A typical hypersaline assemblage would not differ much in gross taxonomic composition up to the point that salinities become so high that only anostracan branchiopods, and perhaps blue-green algae and ostracodes as well as certain types of bacteria and other microscopic organisms, can survive. (See Tasch, 1963.) The apparent similarity between brackish and hypersaline biotas can best be explained by recognition that either environment is also one of great salinity variation. Within the major tolerant groups, usually only a few families or genera are euryhaline, and these are capable of adjusting actually to changes in salinity and therefore can tolerate the wide range of salinities encountered in either extreme (Pearse and Gunter, 1957). In fact, both brackish and hypersaline environments can become established in the same body of water in different seasons.

Most organisms that have become adapted to the more stable environment of fresh water, however, are stenohaline and thus are not found in other environments (Pearse and Gunter, 1957). Fresh-water assemblages are characterized by charophytes, thecamoebinids, and a few specific pelecypods, gastropods, ostracodes, conchostracan branchiopods, demosponges, diatoms, blue-green algae and worm tubes, in various combinations depending on other aspects of the environment. In some cases only one or two groups are present.

Many euryhaline organisms that dominate brackish water assemblages can also inhabit the normal marine environment but are eliminated or kept to low numbers because of predation or increased competition for space. As an illustration, oysters can live in marine waters but are greatly subdued through voracious predation by a marine snail (*Thais*, the oyster drill) which, however, cannot tolerate lower salinities. Oysters therefore are abundant only in brackish estuaries and lagoons. The much greater number of oysters in brackish water also reflects a general rule observed among all inhabitants of restricted environments, that of increase in numbers of individuals among a reduced number of taxa. This effect in the general case results from reduction of competition from other taxa for a similar amount of nutrients and space per unit area. It must be kept in mind, however, that such demonstrably euryhaline ("brackish-water") forms that can numerically dominate a restricted environment do not indicate by their mere presence such an environment. In fact, if they are present in reduced numbers with even a few members of stenohaline marine groups, probably only a normal marine environment is indicated.

Such a "mixed" assemblage also could result from a rapid succession of environments of different salinities in a restricted marginal body of water. As an example, Baffin Bay and Laguna Madre on the south Texas Gulf Coast are normally hypersaline (50 to 60 ppt), but torrential rains accompanying a recent hurricane reduced the salinity to 2 to 10 ppt after which the bays transformed back to hypersaline through normal marine conditions with a rapid succession of organism assemblages typical for the different salinity ranges (E. W. Behrens, personal communication, 1968). If sedimentation accompanying such a change were insufficient to separate the assemblages, then a "mixed" assemblage would result.

Another problem of environmental interpretation from fossil assemblages is that of fossil transport, which has been discussed much in the literature. (See Ladd, 1957, p. 38–40, 49–50; Johnson, 1960; Shepard, 1964; Fagerstrom, 1964.) It is possible that fossil transport could be needlessly invoked to explain faunal mixing due to a natural cause, such as rapid succession of different environments. Nevertheless, fossil transport is common enough to be seriously considered when making paleoenvironmental interpretation. Mechanical transport generally is not particularly difficult to determine with larger fossils because they will be broken up, abraded, and the pieces sorted in movement sufficiently strong to transport them into an environment of enough difference to invalidate a general interpretation. Disarticulation of such jointed forms as certain algae, echinoderms, and arthropods, however, cannot be used to suggest transport unless the pieces also are abraded, because simple disarticulation can take place in a quiet environment merely upon decay of connecting organic matter and movement by burrowing or scavenging organisms. Transport is a more serious problem with microfossils both because they are smaller and easier to move large distances, and because they are lighter and more difficult to fragment or abrade. However, as Merrett (1924) points out, whereas nonmarine fossils may readily be carried into a marine environment, marine organisms are rarely transported into nonmarine environments. For example, small fresh-water charophyte oogonia and large but floatable land plant leaves may be carried down rivers into the sea, and, certain fresh-water diatoms can be wind-blown into the sea (Kolbe, 1954). Thus a minor amount of exclusively nonmarine fossils in a marine assemblage does not invalidate a fully marine interpretation of the environment. On the other hand, truly fresh water deposits should contain no marine fossils. Marine fossils might be moved, nevertheless, into restricted environments marginal to the sea. This could result from hurricane-induced floods stranding organisms on mud flats above shoreline. Fossils introduced in such a way to form mixed assemblages should be smaller than a certain critical size for mechanical transport. In contrast, a natural mixing of assemblages in a similar marginal area, such as the Texas bay example resulting from rapid succession of environments, would be characterized by absence of signs of mechanical transport and by presence of the entire range of shell sizes in a particular frequency distribution. (See Fagerstrom, 1964.)

Underlying the previous discussion has been an implicit assumption that salinity tolerance of the various groups has been consistent throughout the time that they have occurred in the fossil record. How valid is this assumption? One need only peruse the annotated bibliographies in the Treatise on Marine Ecology and Paleoecology (G.S.A. Memoir 67) to see that composition of fossil associations by major groups not only is similar to living organism associations but also has remained much the same through the fossiliferous record. In particular, the consistency of the most diverse assemblage which we recognize in modern seas as "normal marine", must reflect an environment that has not varied much throughout the ages. Because the normal marine environment is the least variable of modern regimens, it is reasonable to conclude that the modern marine assemblage has been "marine" ever since the constituent organisms have appeared. This does not rule out, of course, the possibility of slow progressive change in the exact nature of the marine environment, such as the amount or ratio of dissolved salts; but if this has occurred, it has been sufficiently slow that the organism assemblage has not been affected, and the problem does not concern us here.

Because all invertebrates seem to have evolved ultimately from marine ancestors, the lineages of those inhabiting only restricted environments today must have left the normal marine environment sometime in the past. Nevertheless, on account of the apparent general stability of environmental preference of many organisms, paleontologists for years have been utilizing knowledge of salinity tolerances of certain genera of pelecypods, gastropods, foraminifers, and ostracodes for paleoenvironmental interpretation in the more recent portions of the stratigraphic record. Such applications can be carried back into the record only as far as the lineage, usually a genus, can be traced, and this is rarely into the Mesozoic. Moreover, one cannot be certain that the genus did not leave the marine environment for its present habitat at some point during its existence. Other aspects of ancient assemblages such as diversity, numbers of individuals, dominance of certain taxa, and types of associates must be considered to determine whether or not a particular environment-marking genus may have inhabited a different environment in the past.

Fossil indicators of water depth.—Only red, green, and blue-green algae, hermatypic corals, calcisponges and a few members of other major groups are restricted to modern shallow marine environments (fig. 4). Of these, only the algae and indirectly the corals (through their symbiotic association with unicellular algae) are restricted to living in shallow water on account of known intrinsic ecologic control. The algae are restricted to the zone of sunlight penetration because solar energy is necessary for maintenance of their metabolic food-producing process of photosynthesis. Other factors such as increased turbidity and latitude (ray angle) only serve to reduce light penetration from its apparent maximum of about 520 feet for algal growth in clear tropical water.

Certain planktonic protistid groups, for example, coccoliths and diatoms, also live only in near-surface waters because they too require sunlight for photosynthesis. Because they live in greatest numbers over the open oceans, however, most settle out after death to great depths where, relatively undiluted by shallow-water organisms, they may be the most conspicuous members of the biota incorporated into the sediment. Thus, along with other planktonic organisms such as radiolarians, pteropods, and certain foraminifers, they strongly suggest deep water when found in great numbers as fossils.

In addition to calcisponges, other readily identifiable minor groups of organisms known only from shallow marine water today include the brachiopod *Lingula*, flat echinoids, and the merostome arthropod *Limulus*. Because of difficulty investigating deeper water, it is entirely possible that these forms do in fact live at greater depths and merely have not been collected yet. Certainly, as far as known, there is nothing intrinsically "shallow-water" about the ecologic requirements of these organisms. *Lingula*, *Limulus*, and flat echinoids may be mostly controlled by preference for soft sandy substrate, and the particular texture or consistency that is optimum for them may be best developed today in nearshore shallow waters. It is obvious from Figures 3 and 4 that invertebrate groups have achieved a far greater tolerance of different depths than of different salinities. Moreover, as *Limulus* has been shown experimentally to be not utilizing its entire range of salinity toleration (Pearse, 1928), it may also not be utilizing parts of other toleration ranges. This could apply to the other organisms as well with respect to depth-toleration ranges. We know why most organisms die outside a restricted salinity range, but we do not know much about the control of depth over their life processes. Thus occurrences of the nonalgal groups now seemingly restricted to shallow water are only weak suggestions of shallow-water environments in the fossil record.

So far as known, only hyalosponges as a major group are now restricted to living in deep

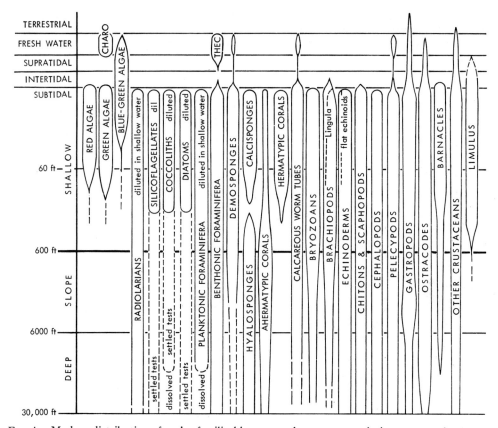

Fig. 4.—Modern distribution of major fossilizable nonvertebrate groups relative to water depth.

marine environments. Precautions similar to those outlined in the previous paragraphs must be applied also to the reliability of hyalosponges in the record, and indeed, they have been considered denizens of the shallow marine environment in the Paleozoic. For example, Devono-Mississippian hyalosponges of the northern Appalachians are associated with lingulid, orbiculoid and spiriferid brachiopods, crinoids, ophiuroids, echinoids, cephalopods, wood and leaf fragments, and "flabellate algae" (Caster, 1941). This association has been considered by many as part of the "typical shallow-water assemblage." The land plant material is undoubtedly washed in, and if the "flabellate algae" are not benthonic (or not even algal as many old reports of this abused group have turned out), then, as there is nothing intrinsically "shallow-water" about any invertebrates, these Paleozoic hyalosponges may have lived in the deeper part of this epicontinental sea not far ecologically from their present habitat. I bring this up not to show that the hyalosponges have always lived in deep water, but to emphasize the caution that must accompany any attempt to interpret water depth from benthonic invertebrate assemblages. Certainly the occurrence of "shallow-water" lingulids and "deep-water" hyalosponges in the same deposit underlines the problems of applying modern descriptive analogies to interpretation of ancient water depth. The distinction between a "typical shallow-water biota" and a "deep-water biota" on the taxonomic level at which we are working, and without considering algae, probably is attainable only from ratios of planktonic versus benthonic forms. To reiterate, therefore, only benthonic algae (including boring forms), which are dependent on sunlight, can give positive evidence (when found in place) for shallow water depth of an ancient environment.

The observation that great numbers of planktonic fossils in a deposit strongly suggest a deep-water origin is supported by several considerations. Planktonic organisms theoretically could live anywhere in the sea but are most numerous away from shore over deep water due to less competition from benthonic organisms for nutrients in the thicker (deeper) mass of water. Furthermore, the plankton multiplies in great numbers when more nutrients become available due to periodic upwelling near the break in slope at the edges of continental shelves. One must keep in mind, however, that at times, large numbers of planktonic foraminifers are carried locally into shallow water (Crickmay et al., 1941). Nevertheless, it has generally been recognized that abundance and ratio of planktonic foraminifers increase markedly in an offshore direction in modern pericontinental sediments. (See Walton, 1964.) This observation has been used in conjunction with other lines of evidence to indicate offshore direction and deep-water environments in ancient pericontinental regimens in the Mesozoic of Texas (Griffith, et al., 1969) and the Alps (Garrison and Fischer, 1969). It is not known how abundant the shelled plankton were in shallow epicontinental seas where there would be less thickness (depth) of water for providing nutrients to both benthonic and planktonic organisms. Abundance of planktonic foraminifers and coccoliths in the Cretaceous Greenhorn Limestone of the Western Interior (Eicher, 1969; N. Minoura, personal communication, 1970) and of styliolines in the Devonian Geneseo Shale and Genundewa Limestone of New York indicates that at least at times certain pelagic, perhaps entirely planktonic, organisms did populate such seas extensively. Although the greater living space available for plankton in thicker columns of water might suggest that such deposits originated in deeper parts of these seas, by no means are depths on an oceanic scale, with their unwarranted tectonic implications, indicated.

Some ancient water-depth determinations have been made from analogy to modern genera of known restricted depth distribution, particularly among foraminifers (Walton, 1964) and molluscs (Woodring, 1957). Again, such analogies are meaningful only as far back as the genera are extant. In addition, many workers have realized that this distribution is apparently related causally to temperature which, although generally decreasing with depth, can vary independently of depth. Extrapolating depth interpretations of such distributions very far geographically is unwarranted because of the phenomenon of "tropical submergence" noted in many invertebrates, including molluscs and foraminifers (Ekman, 1953; Schenck, 1928). Acting in accordance with this phenomenon, many cold-water forms occur in shallow water at higher latitudes but inhabit only deep water of similar temperature in the tropics. Thus their value as depth indicators diminishes in directions perpendicular to the equator.

Other problems arise in paleodepth determinations based on fossil assemblages: 1) Turbidity currents can transport shallow-water microfossil assemblages down continental slopes into abyssal ocean depths with no signs of abrasion (Shepard, 1964). Although primarily a phenomenon of pericontinental seas today, such currents undoubtedly operated in parts of epicontinental seas where sedimentary slopes were steep enough. 2) The ability of many organisms to encrust or hold in some way to a firm substrate also allows them to attach to floating vegetation and become epiplankton. When parts of the floating mat die and decay, the epiplanktonic organisms, which can include preservable red algae, sink to the bottom which can lie at any depth. 3) The Recent rapid sea level rise has created a problem of relict biotas in places where sedimentation is insufficient to completely bury fossil assemblages of lower sea-level stands. For example, off the Texas coast, a sample from nearly 280 feet of water contained foraminifers characteristic of the present depth and also macroinvertebrates characteristic of much shallower water (Shepard, 1964); such a mixed, partly relict assemblage would be difficult to interpret correctly very far back in the record.

Environmental evaluation of extinct groups.— At this juncture it is desirable to evaluate the probable relations of extinct groups of organisms to the shallow marine environment. Because no extinct organic group can be considered intrinsically marine or nonmarine, the basic criteria used in such an evaluation must involve the nature of the assemblage of associated fossils for which reasonable environmental interpretation has been established from observation of modern descendents. Brief environmental resumes of several extinct groups are given in Appendix 2.

Archaeocyathids, stromatoporoids, hyolithids, trilobites, conularids, tentaculites, styliolines, conodonts and graptolites occur throughout their entire portions of the stratigraphic record in association with fossils from living groups whose modern descendants are both strictly or largely marine. Furthermore, none of these groups occur in assemblages or rocks reasonably considered to be nonmarine. Certain groups such as tentaculites and conularids appear to be more numerous with only a limited number of associates, which suggest a somewhat restricted environment. Individual abundance in these groups, however, may have been controlled by turbidity or substrate grain size rather than salinity, because they are typically most abundant in shales or fine-grained limestones. As pelagic organisms, styliolines, conodonts, and graptolites could settle onto any kind of substrate and may occur alone in black shales that represent substrates unfavorable to benthonic life. Apparent near-restriction of chitinous graptolites and chitinophosphatic conularids to certain shales probably reflects mainly difficulty of preservation of noncalcareous material in more agitated or aerated environments.

Eurypterids, on the other hand, were apparently wholly marine in the early Paleozoic, but became euryhaline by the Silurian when they inhabited the entire range of environments from brackish through marine to hypersaline. By the Devonian, they were greatly reduced in the marine environment, and by the later Paleozoic they were restricted to brackish and perhaps even fresh-water environments within terrestrial sequences. Because eurypterids had poorly calcified chitinous skeletons, their occurrence only in fine-grained carbonates and shales may result from preservation only in lutitic sediments. Nevertheless, the fact remains that they are found in Ordovician, Silurian, and rarely Devonian marine lutites, but do not occur in Upper Paleozoic marine lutites. As they are found only in nonmarine Upper Paleozoic lutites, eurypterids seem to have emigrated from the marine environment during the course of the Paleozoic.

Possible depth restrictions are suggested only for the archaeocyathids and stromatoporoids, both of which commonly are closely associated with algae. While they may have been predominantly shallow-water inhabitants, lack of definite observation that they are always associated with algae leaves open the possibility that they lived in deep water as well. Based on available evidence, the other groups may have lived and been deposited at any water depth.

A comment seems appropriate here on the fusulines, an extinct branch of foraminifers, because they have been ascribed as a group to signify the deepest-water (150–180 feet) phase of deposition in late Paleozoic epicontinental seas by analogy to nondescendent modern foraminifers of similar size and shape (Elias, 1937). This general interpretation for deepest water in such a sea has persisted even though Elias himself (1964) points out that the various types of fusulinids undoubtedly inhabited different environments from moderately shallow to deep water. I have observed that fusulines occur locally with abundant phylloid red or green algae in certain Midcontinent Upper Pennsylvanian algal-mound complexes and are common in small channels in the tops of several of these mound complexes. Both occurrences seem to indicate very shallow water for these particular fusulines as determined by the shallow optimum depth for profuse algal growth and the nature of the mound complexes as topographic highs.

Significance of Biotic Diversity

Biotic diversity is most simply defined as the absolute number of different species in a particular sample (Hessler and Sanders, 1967). Diversity of different organic assemblages can be compared readily in samples of roughly the same size from different environments. Although alluded to several times in the previous discussion, this attribute of biotic assemblages needs further elaboration with respect to its significance in the interpretation of fossil assemblages.

Variation in organic diversity within certain regimens has long been noted on a regional scale. For instance, diversity of several marine groups such as molluscs and corals increases substantially from cold climates into the tropics, a pattern that Fischer (1960) has related to greater constancy of the warm-water environment. Such observations have been developed into a general theory which relates biotic diversity directly to environmental stability. (See Hessler and Sanders, 1967; Buzas and Gibson, 1969.) Apparently the greatest number of different kinds of organisms live in an environment that is stable, that is, one in which all factors such as temperature, salinity, oxygen, sedimentation, *et cetera*, vary only within a narrow range over a long period of time. What most organisms cannot tolerate is rapid and extreme change of physical environmental factors. Therefore, in a stable environment, greater numbers of species are able to evolve into a delicately interacting community without having to maintain a tolerance to changes in the physical environment.

Relation of diversity to salinity.—In relating biotic diversity to recognition of shallow marine environments, it is readily apparent that it can help distinguish fully marine environments from restricted marginal or nonmarine environments. As already shown, most organisms cannot tolerate the salinity variation characteristic of lagoons and estuaries. These environments also undergo wide variation in other physical factors such as temperature, and far fewer organisms can withstand the rigors of combined variation of all factors. It is not well known how much control is exerted by each factor, but salinity variation is certainly one of the more important because it forces most organisms to osmoregulate in order to survive. In any case, it is well documented that number of species decreases from the more stable normal marine environment into restricted marginal bodies of water with highly variable salinities. As an example, on the Texas Gulf Coast Odum and others (1960) show that after collecting about 1000 individuals

TABLE 1.—MODERN DIVERSITY CLINES RELATED TO SALINITY IN BALTIC AND BLACK SEA REGIONS; FROM EKMAN (1953, P. 93, 119).

Baltic Sea:	Entrance 30–35 ppt	Danish Isles 10–33 ppt	Main part 7–15 ppt	Gulfs at head <7 ppt
Pelecypod species	87	34	24	5
Echinoderm species	35	8	2	—

Locality:	Aegean Sea	Marmara Sea	Bosporus	Black Sea
Salinity:	normal marine	intermediate		~18 ppt
molluscan genera	157	103	86	56
molluscan species	410	240	151	91

in each environment, 36 molluscan species were recovered from the normal marine open Gulf, 20 were obtained from more brackish Aransas Bay and 14 were found in generally hypersaline Laguna Madre. Gunter (1947) has suggested that such "diversity clines" related to salinity ought to be recognized in the stratigraphic record.

Diversity clines related to salinity on a large scale are observed among certain modern shelled invertebrate populations upon entering partly restricted arms of the sea such as the Baltic or Black Seas (Table 1), where salinities are lower than in the open ocean but probably more constant than in smaller marginal bodies of water. One can note also from Table 1 that the more generally euryhaline molluscs reduce in diversity only gradually into brackish water whereas the more stenohaline echinoderms reduce abruptly in diversity away from the normal marine environment. The two echinoderm species reported for the main part of the Baltic Sea are present at salinities below that generally considered to be tolerable to the group as a whole; these are probably mobile forms straying into an environment in which they will not remain viable for long.

Clines in biotic diversity are also known to be related to temperature (Fischer, 1960), and it may be difficult to distinguish between salinity and temperature effects in the fossil record. Because average temperature varies only on a large scale, diversity clines noticeable on a small scale are more likely to be due to factors such as salinity and so forth which change rapidly across small areas at the margins of the sea. In application to the stratigraphic record, Stevens (1968) has related decreasing diversity of certain elements of the invertebrate fauna to entering a stress environment along an ancient shoreline in the Pennsylvanian of Colorado.

It should be emphasized at this point that small-scale diversity clines can be caused by changes in other environmental factors that are not necessarily related to distance from shoreline. As elaborated in a later section, certain biotic elements within a marine environment are controlled by nature of substrate, turbidity, sedimentation rate, and oxygenation. One may note what type of organisms are disappearing in a cline to give clues as to which factors a particular diversity cline may be related. For example, disappearance of attached organisms or of suspension feeders would suggest decrease in firmness of substrate or increase in turbidity or rate of deposition, respectively.

Another point that must be emphasized with respect to applying changes in faunal diversity to the record is adequacy of collection. Different samples must be roughly of equal sizes if they are to be compared, and they must be large enough to include most of the faunal elements in a particular population. In a Texas Gulf Coast study, Odum and others (1960) show that samples of even 60 to 200 individuals were not sufficient to separate the normal marine environment unequivocally from the brackish-bay environment (fig. 5). This may be because salinities periodically become nearly normal marine in the bay, and a transitory population of many exclusively marine forms is established for a while. This increases the diversity of preserved remains in an otherwise generally restricted regimen. Because some elements are present in only small numbers in the normal marine environment but add greatly to its biotic diversity, and not all of these will become established in a bay only temporarily marine, only large samples adequately reflect the differences in diversity between the two regimes.

Relation of diversity to water depth.—Faunal diversity cannot be used as readily, however, to differentiate shallow from deep marine environments. Contrary to the longstanding general impression that the deep sea supports a fauna of much lower diversity than shallow water (Ekman, 1953; Bruun, 1957), the stable deep-

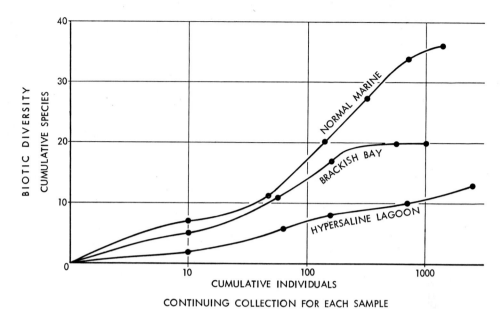

Fig. 5.—Relationship between increasing amount of collection and apparent biotic diversity in samples from 3 different environments in Texas Gulf Coast (after Odum et al., 1960).

water fauna is actually more generally diverse. Recently improved collection methods have enabled Hessler and Sanders (1967) to show that faunal diversity in samples from depths of 4500 to 14000 feet in the North Atlantic is as great or greater than in shallow tropical seas where diversity previously had been considered to be maximum (Table 2). Moreover, the poorest deep-water sample exhibits more than twice the diversity in a temperate shallow-water sample of equivalent size. The average number of species in Hessler and Sanders' five single deep-water hauls using the improved collection device is 267 from over 10,000 individuals, which is more than 3 times the average of 73 species in nine single previous deep-water hauls cited in their article. Previous hauls made with inadequate collecting devices provided less than 1000 individuals per haul, and such incomplete samples have been responsible for the misconception of the depauperate nature of the deep-water fauna. Fossilizable organic groups so far shown to be more diverse than previously realized at great depths are molluscs, echinoderms, arthropods, and benthonic foraminifers (Buzas and Gibson, 1969).

It is hardly surprising that faunal diversity is as great or greater in deep water than in shallow water if one considers that diversity seems directly related to environmental stability. The deep-water environment probably has been more stable with respect to all physical factors for much longer periods of time than most any shallow-water environments which are closer to, and thus more strongly influenced by, atmospheric changes in temperature, wind-induced water turbulence, and river discharge, which exerts control over turbidity, sedimentation rate, and, in nearshore areas, salinity. Shallow-water environments are even less stable over long periods of time in view of recent glacially induced changes of sea level of a type that seems relatively common throughout the record of ancient epicontinental seas as well. Thus deep water is a much more favorable place for establishment and maintenance of delicately interacting com-

TABLE 2.—COMPARISON OF BIOTIC DIVERSITY IN DEEP AND SHALLOW MARINE ENVIRONMENTS USING EQUIVALENT LARGE SAMPLES OBTAINED BY AN IMPROVED COLLECTING DEVICE; DATA FROM HESSLER AND SANDERS (1967)

	Shallow water <100 feet		Deep water 4500–14,000 feet				
	Tropical	Temperate	North Atlantic				
No. of species	201	90	365	257	310	208	196
No. of individuals	5897	3845	25242	13425	12083	5897	3737

munities of large numbers of highly specialized forms that do not need to maintain the tolerances that would prevent them from being exterminated by extreme weather-induced variations in physical environment (Buzas and Gibson, 1969).

Biotic diversity nevertheless must be applied to the fossil record with care. High diversity does not necessarily indicate the most stable environments; for example, restricted circulation can lead to bottom stagnation which produces a stable but foul euxinic environment that supports no benthonic fauna. High biotic diversity thus should indicate the most stable life-supporting environment, which within an epicontinental sea would be most likely the deeper offshore parts away from weather influence and overwhelming sediment influx. Furthermore, it should be emphasized that interpretation of biotic diversity as reflecting stability refers to single environments, because considering large areas, more species inhabit the changeable shallow-water regimen on account of the much greater variety of microenvironments (such as in reefs) typically available there than in the deeper sea (Hessler and Sanders, 1967).

EVIDENCE IN SEDIMENTARY STRUCTURES

Various types of structures, from large-scale stratification features to organic burrowing, are apparent in most exposures of sedimentary rocks. Some kinds of structures are usually available for possible paleoenvironmental interpretation where fossil remains are not. Moreover, even when fossils are present, sedimentary structures reflect the last processes operating on the sediments in their place of deposition and may provide evidence as to whether or not a fossil assemblage might be transported. For example, oolitic calcarenites with high-angle cross bedding in ridges that form the axes of some Bahamian islands are apparently of aeolian origin (Ball, 1967). Also, turbidity currents can transport microfossil assemblages from shallow water into the deep sea (Shepard, 1964). Thus fossil shell assemblages in beds interpreted by sedimentary features to be turbidites are automatically suspect for environmental inference about the final resting place.

Stratification and Other Inorganic Features

Evidence available from stratification and associated features gives information on the nature of the medium of transport. Scale, shape and orientation of many of these features reflect direction, turbulence and other conditions of sediment movement. (See Harms, 1969.)

Unfortunately for our purposes, however, sedimentary processes are about the same for aqueous media of different salinities, and so structures resulting therefrom cannot be used per se to distinguish marine from nonmarine environments (fig. 6). In short, many similar structures can be produced in quite different environments (Stanley, 1968). The same is true in determination of water depth. Although theoretically various characteristics of waves are related to water depth, Allen (1967) points out that in practice the effective flow depth of a bottom current between the sediment surface and an interface within the water column may be a very small fraction of the absolute depth. Thus current features typical of shallow water can form in very deep water. Hulsemann (1968) emphasizes the potential importance of such internal currents and associated waves and notes that ripple marks and cross stratification which are characteristic of shallow-water deposits have been found also in deep-water slope deposits. In fact, some ripple marks photographed in water 10,000 ft deep look identical to those formed at shoreline (Dangeard and Giresse, 1968, p. 85). Furthermore, other factors that control formation of certain structures, such as velocity or turbulence of the depositing medium or consistency of the sediment, vary independently of depth.

In the past, modern sedimentary structures have been studied mainly in easily accessible areas that are periodically subaerially exposed, such as tidal flats, beaches, and river channels. Here certain suites of structures have been found to have environmental significance for interpretation of the stratigraphic record as is pointed out in other papers in this symposium. Only recently have advances in techniques of observation and sampling allowed more thorough study of the structures in continually subtidal environments (Shepard, 1964). Perhaps in time certain structures or associations of structures will be found that are exclusively characteristic of the shallow marine environment, but at present no unequivocal indicators are known. Nevertheless Figure 6 shows that certain combinations of structures seem to be more common in the marine environment or certain portions of it, whereas others, so far as known, are more common in nonmarine environments. Potter and Pettijohn (1963) emphasize that differences in abundance rather than kinds of sedimentary structures reflect differences in environment. Once a marine environment has been indicated, however, certain sedimentary structures and particularly associations of structures (Sutton, 1969) can aid in identifying subenvironments within the marine regimen. A brief general resume of environmental significance of most sedimentary structures summarized on Figure 6 is given in Appendix 3.

Certain sedimentary features require conditions of at least intermittent subaerial exposure which occurs only in marginal or nonmarine environments and thus indicate absence of a continually subtidal marine environment. *Raindrop prints* are the most unequivocal, because drops would retain their shape and mark sediment through only the thinnest film of water. *Mudcracks* result from desiccation of muddy sediment, and this occurs by far most commonly under subaerial conditions in intertidal, supratidal and nonmarine environments of transitory water coverage. Although desiccation of a muddy sediment might occur subaqueously when overlying water becomes more saline and draws interstitial water out of the sediment, such changes in salinity occur mainly in marginal lagoons or nonmarine lakes. *Birdseye* structures resulting from entrapment of gas bubbles and desiccation in carbonate mud are preserved in supratidal and less commonly intertidal sediments and are not known from continuously subtidal sediments (Shinn, 1968a). *Salt crystal casts*, indicate greatly increased salinity and should occur in the same range of environments in which evaporites can form. Scattered salt casts not directly associated with large-scale evaporite deposits, however, would suggest only short duration of evaporating conditions in a changeable environmental regimen and thus probably would be most common in transitorily hypersaline lagoons or lakes in the marginal or nonmarine regimen. *High-angle* and *trough and festoon* types of *cross bedding* also are features that apparently occur only in marginal or nonmarine environments and thus also constitute negative evidence as far as the shallow marine environment is concerned.

Organic Features

Bioturbation.—Both soft-bodied and shelled organisms disturb the sediment in their search for food and shelter wherever the substrate is favorable to life. Such conditions are present from oxygenated marine and brackish-water environments to alluvial deposits which provide habitats for terrestrial air-breathing burrowers. Thus, as with most other sedimentary structures, bioturbation by itself does not distinguish the shallow marine from other regimens (figs. 6, 7). Even though sediment-disturbing organisms are restricted in diversity outside of the marine environment, the number of euryhaline individuals may be greater in brackish estuarine

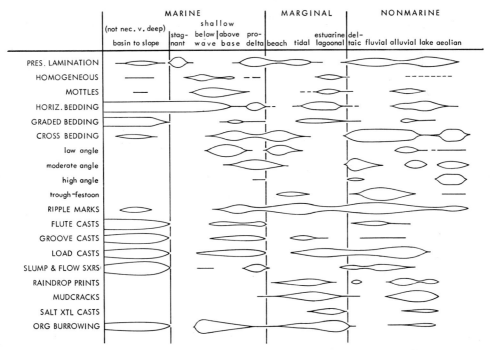

Fig. 6.—Distribution of various types of sedimentary structures among major depositional environments. Summarized largely from Potter and Pettijohn (1963), McKee (1964), Moore and Scruton (1957), and Hulsemann (1968).

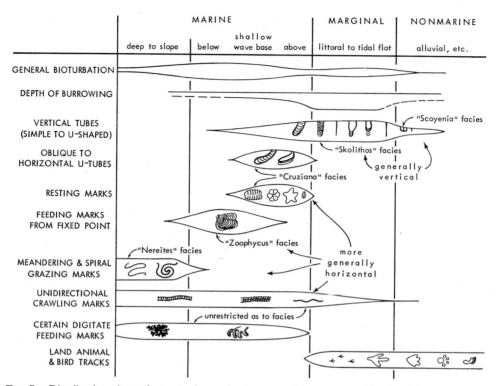

Fig. 7.—Distribution of certain types of organic structures (lebensspuren, ichnofossils) and "ichnofacies" among major depositional environments. Summarized from Seilacher (1964, 1967); McAlester and Rhoads (1967).

environments because of the higher organic content of the sediment (Emery, Stevenson, and Hedgpeth, 1957); thus intensity of burrowing gives little clue to salinity conditions.

Differences in amount of bioturbation apparent in sedimentary rock can, however, give clues to other aspects of environment and aid in subdividing the marine regimen once its general identity is established. Also, activity of burrowers and sediment-ingesters has great control over preservation of primary inorganic sedimentary structures.

Bioturbation is indicated in completely burrowed deposits by swirled fabric apparent in the orientation of grains and by faint color mottling, and is suggested by total lack of sedimentary fabric in homogeneous deposits if it can be determined that the sediment probably did not come from a completely homogeneous source. Less complete burrowing of sediment may leave more distinct mottling and definite burrow structures that stand out in contrast to the unburrowed portions. (See Moore and Scruton, 1957.)

Lack of bioturbation is best indicated by presence of delicate primary stratification features such as lamination. Thus preservation of laminated sediments suggests one of the following sets of conditions that would prevent establishment of benthonic organisms in a marine or marginal environment: 1) Bottom stagnation may result under reducing conditions inimical to life in and above the substrate and ultimately give rise to deposition of laminated, typically black shales. 2) Supratidal muds, particularly in carbonate areas, are inundated so rarely that the environment may be too dry, or the salinity too high in isolated evaporating pools of sea water, or too low in pools of rain water for a burrowing fauna to exist. 3) In prodelta environments deposition may be so rapid that burrowers cannot completely churn all the sediment and destroy primary laminations (Moore and Scruton, 1957). 4) In areas of rapidly shifting sands, such as offshore bars, the substrate may be too mobile and perhaps the sediment size too coarse for a great number of burrowers, and thus not only cross laminations but also ripple structures are often preserved (See Imbrie and Buchanan, 1965.)

Trace fossils.—In many cases, environments may be interpreted from specific burrow, track, and trail forms that are left in the sediment by both hard and soft-bodied organisms. Investigation of organic structures, variously called lebensspuren or ichnofossils, has been centered primarily in Europe around the extensive work of R. Richter and his colleagues. In recent years, American workers have begun systematic study of trace fossils, particularly in modern sediments (eg., Rhoads, 1966, 1967; Shinn, 1968b; Howard, this symposium). Because modern tracks, trails and burrows have been investigated thoroughly only in the littoral and very shallow marine environments, one must heed the warning of Häntzschel (1962) that it is dangerous to infer that similar trace fossils in the record indicate only shallow water when it is not known whether or not they also occur at greater depths in modern environments. More recent work with the underwater camera is alleviating the lack of knowledge about deeper-water organic markings. Although some shallow-water markings occur also at depth, others do not, and several distinctive trails are being discovered only at greater depth (Seilacher, 1967).

The most useful published summaries in English for our purposes on general environmental significance of trace fossils are those of Seilacher (1964, 1967) who has recognized several distinct environmental groupings of trace fossils termed "ichnofacies" (fig. 7). Environmental interpretation of these facies has been made chiefly from study of ichnofossils in the stratigraphic record because trails and burrows are difficult to study in modern unconsolidated sediments of the subtidal environment. Nevertheless, Seilacher has used ecological principles derived from study of modern environments to formulate his interpretation of ichnofacies.

Trace fossils provide certain advantages over other fossils particularly in that they cannot be reworked or transported like actual remains of hard parts. Also trace fossils are most commonly found in deposits where other fossils are only rarely preserved, such as thick alternations of sandstone and shale. Usefulness of trace fossils among rocks of different ages is based on the tenet that while organisms themselves have evolved through time, their behavioral reaction to the same environments has remained essentially the same. Thus ichnofossils should have long time ranges. Also, it has been found that many different animals living in the same environment have generally similar behavioral reactions to the environment and thus leave roughly similar markings. As a result, many ichnofossils have a narrow facies range regardless of specific organic origin.

For example, in very shallow marine and marginal environments such as beaches and tidal flats, protection is a chief concern for organisms, and a great variety of different mobile animals and certain sessile suspension feeders produce deep vertical burrows, ranging from simple to U-shaped, which constitute what Seilacher terms the "Skolithos" facies (fig. 7). Organisms responsible include crustaceans, worms, clams, and lingulid brachiopods. Studies by Rhoads (1967) and McAlester and Rhoads (1967) show that burrows tend to be deeper, commonly up to three times, in marginal environments than in subtidal environments. This apparently results from the greater ecologic stress in unstable marginal environments where factors such as temperature and salinity vary widely, and the organism must go deeper into the substrate for protection from ecologic extremes than it would in the more stable subtidal environment. Raising and lowering of certain distinctive U-shaped burrows termed *Diplocraterion yoyo* has been shown to reflect reaction of an organism to alternation of episodes of relatively rapid deposition and erosion in an unstable marginal environment by Goldring (1964). Seilacher (1967, citing 1963) has defined a group of certain types of burrows characteristic of nonmarine deposits, particularly redbeds, as the "Scoyenia" facies. Land animal and bird tracks also indicate nonmarine as well as certain marginal environments.

In the shallow subtidal marine environment, burrows are not only less deep but also tend to be more oblique or horizontal than in marginal environments. Oblique burrows function not only as protection, but also as "feeding mines" for deposit feeders which become more common in quieter water where finer organic material settles out. Such feeding burrows typically exhibit faint traces of previous positions of the structures. Also characteristic of the shallow marine environment are various distinctive resting marks of a variety of organisms that rest on or bury themselves slightly in the sediment either temporarily or permanently. Such resting marks illustrated on Figure 7 include from left to right those of a trilobite, anemone, asterozoan, and pelecypod. Together, these two major groups of organic structures are termed the "Cruziana" facies. Also supposedly characteristic of the shallow marine environment, and occurring in areas where sediments are poorly sorted is a distinctive set of feeding burrows constituting the "Zoophycos" facies. These markings have been determined by Plička (1968) to be formed by sabellid (annelid) worms in shallow-water environments of rapid deposition.

In the extremely stable environment of deep water, organisms no longer require protective burrows. Rather, as many deep-water organisms are mobile deposit feeders, they are more concerned with "mining" as efficiently as possible the food-rich sediment layers in their immediate vicinity. Thus they have evolved systematic feeding programs which result in trails with complex reticulate patterns and meandering and spiral shapes (fig. 7). This group of ichnofossils is termed the "Nereites" facies.

There are also many organic traces, particularly unidirectional crawling marks of mobile benthonic organisms such as snails and worms and certain types of digitate feeding marks, that apparently occur throughout the range of ichnofacies and thus have little value in suggesting a particular environment. Although Seilacher does not mention what types of organic traces might be expected in nonmarine aqueous environments such as lakes, markings made by strictly marine groups such as asterozoans certainly would not be expected there. Traces made by clams and snails, however, should be common as well outside of the marine environment.

Even though Seilacher's original interpretations of trace fossils were based mainly on environmental inference from the stratigraphic record, recent work in modern sediments is corroborating many of his observations. Certain U-shaped burrows are being excavated in tidal flats and other shallow environments; and photographs made with the underwater camera have revealed spiral and meandering fecal pellet trails typical of the inferred deep-water "Nereites" facies at depths of 10,000 to 15,000 feet (Seilacher, 1967).

SUBDIVISION OF THE SHALLOW MARINE ENVIRONMENT

After an interpretation of shallow marine origin has been established for a particular suite of deposits, a great deal more can be determined in many cases about what part, or subenvironment, of the shallow marine regimen a particular unit represents. Lithologic and paleontologic features, especially in combination with sedimentary structures and grain attributes, can be used in refining environmental interpretation.

Direct Lithic Analogy

Certain rock types have sufficiently narrow and well-delineated conditions of formation and are distinctive enough to be unequivocally recognized in the record, that they can be interpreted by direct analogy to comparable modern sediments of known depositional environment. (See Laporte, 1968.) This is particularly true among carbonate rocks which have a variety of distinctive grain types that are easy to recognize and which, in many cases, record only a limited range of environmental conditions.

Oolite.—Formation of typical carbonate ooids requires water supersaturated with calcium carbonate along with strong and consistent agitation to provide proper physiochemical conditions for precipitation around a nucleus as well as to promote formation of even laminae around the nucleus (Newell, Purdy and Imbrie, 1960). Modern ooid formation on a large scale is known only from very shallow, warm-water environments, such as on the Bahama banks, the Mediterranean coast of Africa (Lucas, 1955, cited in Rusnak, 1960) and the south shore of the Persian Gulf (Kinsman, 1964b). In these areas, tidal currents keep the grains in nearly constant agitation, and deposits of oolite sand occur in the form of tidal deltas, elongate bars ("bores"), barchan dunes, and lobes. (See Ball, 1967.) Accumulations of ooids are known to be at least 10 feet thick in places in the Bahamas and may perhaps attain several tens of feet in thickness judging from Figure 6 in Purdy (1961).

Considering that thick extensive oolite sands require conditions that are likely to occur only in very shallow water subject to constant agitation, it is reasonable to interpret extensive Pennsylvanian oolites up to 17 feet thick and typically resting on broad carbonate buildups in southeastern Kansas as very shallow-water deposits. Moreover, the two major current directions 180° apart determined in several of these oolites strongly suggest that tidal currents were responsible for the agitation involved in their formation (Hamblin, 1969). Likewise, similarity of the overall elongate form cut by transverse depressions of the Pleistocene Miami Ooolite of south Florida to that of modern oolite shoals cut by transverse tidal channels on nearby Great Bahama Bank has allowed Hoffmeister, Stockman and Multer (1967) to reasonably interpret a similar depositional environment for the older deposit.

Care must be exercised, nevertheless, in the environmental interpretation of ooids and oolite. For example, ooids forming on a shoal with steep sides, such as the Great Bahama Bank, can be transported easily on a large scale into the adjacent deep-water environment. Although an admixture of fine material with planktonic fossils which settle out only in quieter deep water generally distinguishes such a deposit from the typically mud-free and eventually spar-cemented autochthonous oolite remaining on the shoal, up to 7-inch layers of well-sorted oolite containing very little pelagic mud were collected below 6000 feet off Great Bahama Bank by Andrews and others (1970).

It should be pointed out that ooids can form in fair numbers also in environments other than the classical Bahamian type of submarine shoal. Rusnak (1960) shows that ooids presently are forming along an agitated shoreline in the generally terrigenous regimen of hypersaline Laguna Madre along the south Texas coast. These ooids rarely consitute more than 30 percent of the beach and adjacent subtidal sediment and are associated with a substantial amount of superficial, single and multiple nucleate ooids as well as uncoated grains. This association distinguishes the deposit from that of the classical shoal, in which ooids constitute 80 to 90 percent of the grains in the sediment (Purdy, 1963), although the environmental conditions of supersaturation and agitation in shallow water are similar.

On the other hand, Freeman (1962) has shown that ooids are forming also in quiet water in subtidal portions of Laguna Madre. These, however, are asymmetrical, often with oolitic coating on only one side of the nucleus. Lack of symmetry reflects lack of agitation and distinguishes quiet-water ooids from those originating in turbulent environments at shoreline or on the classical shoal.

Another point to be kept in mind is that oolite also can form in agitated environments in large nonmarine lakes that are supersaturated with calcium carbonate, such as modern Great Salt Lake (Eardley, 1938). Indeed, oolite is known also from lake deposits in the Eocene Green River Formation of Utah (Picard and High, 1968). Perhaps only the associated fauna might distinguish a marine from a lacustrine oolite.

Laminated calcilutites.—Laminated carbonate muds that carry little or no fauna are characteristic of low-lying shorelines from extremely shallow subtidal to supratidal environments. These are sea-margin transitional deposits with significant paleogeographic implications and are distinctive enough to be useful in fixing the limits of ancient marine environments. The lamination in these muds results from intermittent deposition of sediment of slightly differing texture or composition. Generally, the type of deposition responsible is controlled by sticky mats of blue-green algae which flourish only in extremely shallow water and on sediment that may be emergent for long periods of time. Blue-green algae can withstand the extremes of temperature, salinity and desiccation inherent in such an exposed situation and are not commonly preserved in more open marine environments. Perhaps most significantly, these same ecological extremes prohibit much development of sediment ingesters and burrowers that typically destroy sedimentary lamination in a more open marine environment.

Recent work is bringing to light the environmental significance of associated structures and features in laminated shoreline carbonate mud. Birdseye (Shinn, 1968a), fine-grained dolomite (Illing *et al.*, 1965; Shinn *et al.*, 1965), and stromatolites (Logan, *et al.*, 1964; Gebelein, 1969) aid in differentiating intertidal, shallow subtidal, and supratidal environments, and in some cases may distinguish microenvironments within each. (See also F. J. Lucia, this symposium, for modifications brought about by evaporite formation in these environments.) Dolomite, birdseye, and flat or gently undulating, often mudcracked lamination generally are more characteristic of the supratidal environment, whereas stromatolitic heads and mud-pebble conglomerates typically form in the intertidal to very shallow subtidal environments. Many authors have applied these types of criteria in paleoenvironmental evaluation of shoreline deposits of all ages in the stratigraphic record (e.g., Laporte, 1967; Roehl, 1967; Friedman, 1969).

A note of caution in dealing with laminated calcilutites attributed to activity of blue-green algae in the shoreline environment is that algal mats can flourish also along the shores of large fresh-water lakes. Thus, just as with oolite, the general sedimentary regimen must be established as marine or nonmarine by the associated fossils, as has been emphasized both in this paper and by Picard and High in this symposium. The more specific lithologic criteria merely distinguish which parts of the sea or lake are under consideration.

Reefs and other organic buildups.—In regions of the sea where conditions are optimum, certain types of shelled organisms flourish and secrete and bind sufficient calcium carbonate to modify the environment such that it maintains the optimum conditions for these organisms. In part, this entails providing firm substrate for attachment of massive and encrusting organisms that cannot live on soft mud bottoms or shifting sands. Most importantly, such organisms produce enough sediment, firm or otherwise, that the sea bottom is built up and kept at the optimum level for proliferation of these particular organisms in an area undergoing slow subsidence. This results in the formation of thick, skeletal, bound and unbound carbonate buildups variously termed reefs, bioherms, banks, and mounds. Once initiated, these buildups tend to perpetuate themselves through organic proliferation until a drastic change in environment eliminates a majority of the contributors. These buildups also typically provide a great variety of microenvironments for many organisms which by themselves may not contribute much material to the buildup, but which do augment the biota and provide many reefy environments with the greatest biotic diversity known in the shallow marine regimen.

Most organisms that contribute substantial material to modern buildups are marine, and include corals, red and green algae, and certain foraminifers, molluscs, and bryozoans. Most buildups contain algae and thus are restricted to shallow water. Modern nonalgal coral buildups, however, are known from deep water (Teichert, 1958). Some buildups result from proliferation mainly of one type of organism such as the Upper Paleozoic phylloid algae which formed mound complexes that perpetuated an environment less favorable to most members of the nonmound biota (Heckel and Cocke, 1969). A few organisms that have reef-building capacity thrive outside of the marine environment. It is well known that modern oysters cement to one another and form extensive reefs by themselves in brackish bodies of water. Certain blue-green algae apparently promote formation of hard massive ledges of calcium carbonate in nonmarine lakes, such as fresh-water Green Lake in central New York and hypersaline Great Salt Lake in Utah (Eardley, 1938). Again, biotic composition must be considered in order to determine whether or not carbonate buildups are marine.

Biotic Considerations

In less distinctive lithofacies, subenvironments may be difficult to discriminate because preserved visible differences have resulted from a complex interplay of many independently varying factors that did not coincide to produce unique, easily delineated subdivisions. Ecologic consideration of benthonic fossil assemblages, however, may give certain insights into the nature of the original environment. Growth forms of several types of organisms reflect such factors as water turbulence and other conditions of microenvironment. This is particularly true with red algae (Johnson, 1961) and colonial organisms, such as corals (Vaughan and Wells, 1943) and bryozoans (Stach, 1936; Schopf, 1969). For instance, massive or laminar encrusting forms grow mainly in rough water, whereas delicate branching forms tend to grow only in quiet water. A detailed discussion, however, is beyond the scope of this report.

Certain relationships between benthonic organisms and their substrate have been summarized by Purdy (1964). He noted a direct correlation between deposit feeders and the percentage of mud (silt plus clay) in the sediment. This apparently is due to the higher content in finer sediments of organic material which provides sustenance for deposit feeders. Organic content is higher in fine sediments partly because it largely settles out as fine detritus along with fine terrigenous material in the same quiet environments. Purdy also noted an inverse correlation between suspension feeders and percentage of mud in the sediment. In other words, suspension feeders are more common in the cleaner sands. Several reasons for this are apparent. First of all, suspension feeders are kept to low numbers in muddier environments by high interface turbidity resulting in part from intense deposit-feeding activity that promotes easy resuspension of fine sediment (Rhoads and Waage, 1969). Secondly, a decrease in mud-size material, including organic detritus, would reduce the number of deposit feeders that could be supported, and this would result in a percentage increase in number of suspension feeders in the fauna. More importantly, cleaner sands tend to be better sorted and abraded than muddy sands, and the greater water agitation implied by winnowing would provide more replenishment of suspended nutrients to support a greater overall population of suspension feeders. These general relations of feeding type to grain size of substrate should be obvious in the record, because not only shelled organisms but also sediment grain size is preserved as the substrate becomes rock.

Firmness of substrate.—Certain other aspects of environment, however, may not be as obvious from the rock, but may be suggested by the fossil assemblage. For example, firmness of substrate, which in many cases is independent of grain size, may be interpreted by consideration of types of attachment or mobility represented by members of the benthonic community (fig. 8). Within each major group certain genera and species may require or prefer only a limited portion of the range, but usually only expert paleontologists can distinguish easily the particular forms that may be more sensitive indices to substrate firmness. For our purposes it suffices to know that certain fossil assemblages may suggest a certain firmness of substrate.

For instance, colonial corals, worm tubes, barnacles, encrusting foraminifers, most oysters, bryozoans, red algae, and certain inarticulate brachiopods (craniids) require a hard substrate for attachment; some attach to larger shells. Chitons and limpets (a type of gastropod not shown separately on fig. 8) are mobile, but seem to prefer clinging to hard surfaces. Organisms that are cemented or otherwise strongly attached to hard surfaces can withstand strong water turbulence; thus colonial corals and red algae may suggest the turbulent portions of reef environments, and barnacles, chitons, limpets, and worm tubes are suggestive of wave-battered hard surfaces in the tidal zone along shorelines. Occurrence alone of small organisms that require attachment, including encrusting foraminifers (which are distinguished by the presence of one flattened side), may suggest presence of unpreserved firm features such as plants that are not otherwise apparent in the rock.

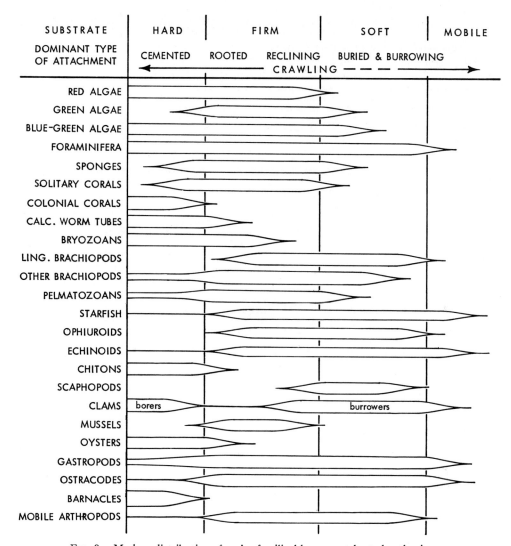

Fig. 8.—Modern distribution of major fossilizable nonvertebrate benthonic groups with respect to firmness of substrate.

A large number of organisms require firm substrate into which they can take root or otherwise fasten without cementing or encrusting. These include most green algae, sponges, solitary corals, pelmatozoans, mussels and many brachiopods, as well as certain members of the groups that prefer hard substrates. Environments indicated by these groups include sediment surfaces that are undergoing little or no deposition and have achieved some degree of induration through compaction and interstitial water loss, and those substrates that contain small hard pieces such as shells which can be used as loci for attachment, particularly of larval stages.

On the other hand, an assemblage of scaphopods, burrowing clams, linguloid brachiopods, certain foraminifers and perhaps some mobile organisms would strongly suggest a soft substrate, particularly in the absence of forms that require hard or firm substrates. Many mobile forms, such as starfish, ophiuroids, echinoids, gastropods, ostracodes and most other arthropods, can crawl about on a wide variety of substrates and thus by themselves suggest little about firmness. An assemblage consisting solely of a few mobile forms, however, may indicate an environment with a mobile substrate, such as an area of rapid deposition or shifting sands. For example, Newell and others (1959) found only a starfish, an echinoid and a mobile clam (one species each) to constitute the entire indigenous shelled fauna on an unstable oolite shoal.

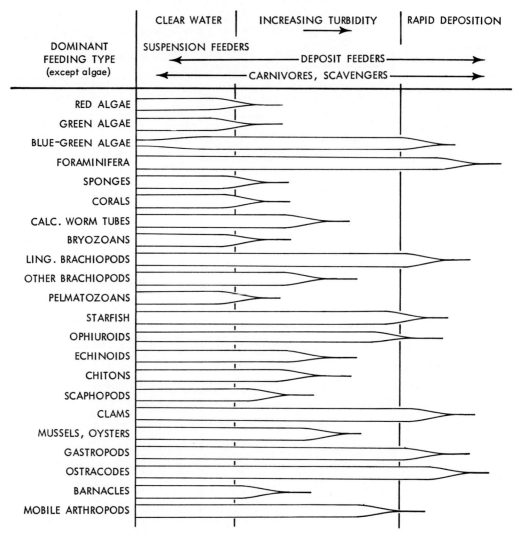

Fig. 9.—Modern distribution of major fossilizable nonvertebrate benthonic groups with respect to turbidity of water and rapidity of deposition.

Turbidity of water.—Consideration of feeding types of benthonic organisms may aid in determination of whether an environment was characterized by clear water or turbid water or was undergoing rapid deposition (fig. 9). Although red and green algae do not feed in the sense of invertebrate animals, they do require nutrient renewal as well as strong sunlight for photosynthesis and they prefer clear water for optimum light penetration. All major organic groups shown on Figure 9 can live in clear water. Moreover, most suspension feeders are restricted to clear water because too much suspended clay and silt (turbidity) will clog and ultimately choke their delicate feeding apparatuses. Thus sponges, corals, bryozoans, and pelmatozoans indicate an environment of clear water containing mainly the microorganisms and organic detritus on which they feed. Members of some suspension-feeding groups, such as tube-secreting worms, brachiopods and certain pelecypods, can withstand moderate amounts of turbidity because they have better methods of disposing of unwanted material or of preventing its ingestion in the first place. All suspension-feeding groups imply a certain amount of current activity for renewal of nutrients. Deposit feeders such as ophiuroids, and certain clams and gastropods can withstand more turbidity because their feeding systems are adapted to straining organic material from sediment. Carnivores and scavengers such as starfish, echinoids, gastropods, ostracodes and other arthropods also can withstand more turbidity, but only to the limit that their respiratory systems are not clogged. Few organisms can exist in a highly turbid environment, and only a few mobile forms of various feeding types such as lingulid brachiopods, burrowing clams, certain snails, ostracodes, foraminifers and starfish can survive in an environment of rapid deposition.

Suggestions concerning nature of the substrate or bottom water are necessarily based on assemblages of benthonic organisms which live in or on the area in question, and of course, are subject to the problem of transportation of fossils into an alien environment, as mentioned previously. Pelagic organisms that occur as fossils include largely nektonic forms such as cephalopods, some modern crinoids, and two groups of clams (limids and pectinids), and planktonic forms such as pteropod snails and several groups of microfossils including radiolarians, silicoflagellates, coccoliths, diatoms and globigerinid foraminifers. After death, the shells of these organisms settle onto all varieties of substrates and consequently yield little information about nature of the substrate or bottom water.

Epiplankton.—Certain members of several groups of benthonic organisms that are able to attach to a firm or hard substrate are known also to attach to planktonic seaweed. Such organisms are thus potentially epiplanktonic and include small encrusting forms of red algae, foraminifers, worms, bryozoans, and perhaps tiny corals, as well as forms attached by an appendage such as certain small brachiopods (see Appendix 1), lepad barnacles, a few small pelmatozoans, and perhaps some mussels. Upon death of only the epiplanktonic organism, those attached by a noncalcified appendage such as the brachiopods, lepad barnacles, and mussels would eventually drop off and settle to the bottom. If the whole mass of weed were to die and decay, then the entire epiplanktonic biota would start settling to the bottom. It is not known how often this occurs, but such a mass of weed is easily washed up and stranded on a beach. In any case, epiplanktonic fossils tell nothing about nature of the sediment substrate, and one must be aware of potential epiplankton when making such interpretations. A few general observations might be helpful. 1) Epiplanktonic fossils are likely to be small because the weed can offer only limited support. 2) They are likely to become scattered on the sediment surface and not particularly noticeable unless they are the only fossils on an otherwise barren substrate. 3) Encrusting forms may come to rest upside down in the sediment and would have no preserved object of attachment. Nevertheless it might be quite difficult to distinguish epiplanktonic elements within an abundant and diverse benthonic biota.

Recognition of a potentially epiplanktonic biota becomes significant in interpretation of bottom conditions if the normal benthonic biota is absent. Oxygenation of the bottom water and sediment surface is indispensible to benthonic life, and is indicated by an assemblage of benthonic fossils and also by trace fossils resulting from organisms living on or in the sediment (Seilacher, 1964). Lack of oxygenation on the sea bottom is suggested by negative evidence, that is, absence of benthonic fossils or organic markings. It would be more strongly suggested, however, by presence of only pelagic (nektonic, planktonic and epiplanktonic) fossils, for these would indicate that surface water was favorable to life while implying that bottom water was not. Presence of only pelagic fossils would also argue against certain other possible causes for apparent lack of benthonic fossils, such as extreme dilution by great amounts of rapidly deposited sediment, for in this case, the pelagic fossils also would be expected to be greatly diluted by sediment and thus not readily noticeable.

Limitations.—Application of biotic considerations to subdivision of the shallow marine environment in the stratigraphic record is not a simple endeavor. Although certain sparsely fossiliferous black shales can be interpreted as products of an unoxygenated substrate because all the fossils that do occur can be attributed to one of the forms of pelagic habitat, evaluation of benthonic fossil communities reveals a much greater complexity of associations and ambiguity of interpretation than is implied in the foregoing discussion of substrate and water-clarity preference of entire organic groups. Furthermore, many other factors contribute their share to control of biotic distribution, and because different independently varying factors affect certain biotic elements more than others, distinctions among fossil assemblages often are blurred, and resulting environmental interpretations must be generalized and often quite conjectural.

Many studies have recognized different shallow marine fossil assemblages within a suite of closely related rock units, and I cannot attempt to cite more than a few, such as the classical work of Elias (1937) on the Lower Permian of the Midcontinent, the works of Johnson (1962) and Moore (1964) on the Midcontinent Pennsylvanian, of Ziegler (1965; Ziegler and others, 1968) on the British Silurian, Stevens (1966) on the Cordilleran Permian, Bretsky (1970) on the

Appalachian Ordovician, Epstein (1969) on the New York Lower Devonian, and Sutton, Bowen and McAlester (1966, 1970) on the New York Upper Devonian. The recent increase in articles shows that recognition and interpretation of different fossil assemblages that represent subdivisions of the shallow marine environment are proceeding at a rapid pace.

SEDIMENTARY MODEL OF A SHALLOW MARINE REGIMEN

Although a certain amount of differentiation of shallow marine regimen can be made by considering lithology and biota alone, and one may examine differentiation based on variation in sedimentary structures and clastic grain size (Allen, 1967; Rich, 1951), the resulting patterns obtained separately may be difficult to comprehend as an integrated whole. Therefore it seems appropriate at this point to synthesize these patterns into a sedimentary model that integrates lithology, biota, certain grain parameters and sedimentary structures into a coherent facies pattern across a shallow sea.

Theoretical Pattern

The general trend of increasing depth away from shoreline in most shallow seas affects waves and currents in a predictable fashion. In turn, differential action of waves and currents on the bottom affects grain size and sorting of the sediment, and causes formation of various sedimentary structures as well as influencing certain previously mentioned factors that control biotic distribution. As a result, a distinct and rather consistent pattern of sedimentary facies, grading laterally into one another perpendicular to shore, is established (fig. 10). This pattern is explainable on theoretical grounds, and is apparent in several modern environments.

The theoretical model (fig. 10A) was described by Irwin (1965) for nondetrital clear-water sedimentation in an epicontinental sea where bottom slope is much gentler than in more commonly studied pericontinental seas. He recognized three major zones characterized by different intensities of water agitation and resulting from the effect of different relative depth on the position at which waves and currents impinge on the bottom and dissipate their energy.

In deeper water below effective wave base, the bottom is sufficiently undisturbed to allow fine sediment to settle out. If the bottom is kept well oxygenated by currents, the biota will be diverse and supply coarser material in the form of shells. Thus, the sediment formed will be mainly mud with sand-size and coarser skeletal material. In the absence of detrital material, the resulting rock type will be shelly calcilutite. Sedimentation in this zone may be very slow for it is dependent solely on detritus derived from the adjacent shoreward zone for the source of mud, because temperature (which generally decreases with depth) would probably be too low for carbonate mud to precipitate chemically (Irwin, 1965). As a corollary, this zone with reduced deposition probably also would be a site of formation of authigenic minerals, such as glauconite, as well as of deposition of fine detrital organic material to form dark-colored sediment and rock.

In the shallower water of the next shoreward zone, effective wave base intersects the bottom surface, and sediments are deposited in water agitated by waves. Little mud can settle out, and thus mainly sand-size and coarser material comes to rest. In the absence of detrital influx this material is largely skeletal debris provided by the diverse biota that is supported by continual replenishment of oxygen and nutrients in this agitated zone. The resulting rock type will be skeletal calcarenite (which here includes what technically could be termed skeletal calcirudite because for our purposes the common origin of the grains is more significant that an arbitrary cutoff point within a gradational size range). The skeletal material may be relatively unfragmented and unabraded, but most is likely to be fragmented and abraded, without necessarily being transported very far, because of the high degree of water agitation that promotes extensive sediment reworking throughout most of the zone. Moreover, the abraded skeletal material is likely to accumulate in cross-bedded, fairly well sorted deposits owing to movement of sediment by traction and extensive winnowing. Reefs develop under certain conditions in this zone as is consistent with the high degree of organic production and their general association with large amounts of skeletal sand. Also, oolite sand forms under certain conditions of supersaturation and water turbulence in shallower waters of the more shoreward extent of this zone.

In the nearshore zone of shallowest water, the environment is again quiet because waves and currents are damped as their kinetic energy is dissipated on the shallowing bottom in the adjacent zone of water agitation. Thus mud again is allowed to accumulate. Its source, at least in part, is shoreward transport of fine products of skeletal disintegration from the adjacent seaward zone. Lime mud also may be derived in part by direct physicochemical precipitation from sea water. This mode of lime-mud formation is presently controversial, but if it occurs at all, it would occur most likely in this zone where water is shallowest and can heat up further to

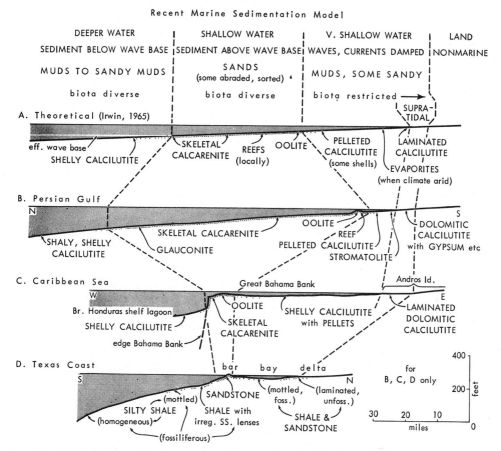

Fig. 10.—Model of Recent shallow marine sedimentation showing theoretical and actual distribution of major sediment types; rock names used for sediments. A) Theoretical epicontinental clear-water sedimentation after Irwin (1965); scale only relative. B) Carbonate sedimentation on south side of epicontinental Persian Gulf. C) Carbonate sedimentation in two parts of pericontinental Caribbean region; diagram combines Great Bahama Bank and British Honduras shelf lagoon. D) Terrigenous detrital sedimentation in northwestern Gulf of Mexico and Texas coast through Aransas bay. Information for B-D summarized from references mentioned in text.

drive off more carbon dioxide, thereby promoting the precipitation reaction more easily than in other parts of the sea. Because very shallow water is subject to reduced circulation and to extremes of salinity as well as temperature, it supports a less diverse and less numerous biota on account of reduced nutrient replenishment and more unstable environmental conditions. Many of the animals that are present produce fecal pellets which commonly become indurated and preserved in this zone. The resulting rock type will be pelleted, sparsely shelly calcilutite. If the climate is arid, and evaporation greatly exceeds compensating fresh-water runoff, evaporites may form in this zone by the process described by Shaw (1964).

At shorelines in intertidal and supratidal environments, temperatures, salinity variation, and periodic desiccation are so extreme that few organisms other than blue-green algae can survive. Here blue-green algal mats strongly influence sedimentation and control or aid in the formation of laminated muds and stromatolitic heads. Birdseye cavities are produced by desiccation and entrapment of gas bubbles. Mud cracks and flat pebbles are produced by periodic alternation of desiccation and inundation. Fine-grained dolomite is produced either directly or syngenetically by evaporitic concentration of Mg-rich water. The greatly reduced fauna leaves few shells and accomplishes only a little vertical burrowing in the sediment. The resulting rock type is mainly barren laminated calcilutite, locally stromatolitic, locally dolomitic, and locally containing mud cracks, flat pebbles, and open, shale-, or spar-filled birdseye structures.

Recent Patterns

Patterns very similar to those predicted on theoretical grounds can be observed in modern carbonate environments of differing topographic expression and geographic setting representing both epicontinental and pericontinental seas. The basic patterns are similar to one another even though the scale of their development is greatly modified by differences in topography. An only slightly modified pattern is also recognizable in modern detrital environments. Three examples are illustrated at the same scale on Figure 10B–D.

Persian Gulf.—The Persian Gulf is a modern epicontinental sea with extensive carbonate sediments extending seaward from the Trucial Coast along the south shore (fig. 10B). Sedimentary information is summarized from Emery (1956); Evans, Kinsman and Shearman (1964); Kinsman (1964b); Illing, Wells and Taylor (1965); and J. L. Wilson (pers. commun. 1965).

In the deeper water along the axis of the gulf, mud is accumulating. It is predominantly carbonate with a subordinate amount of shells and fine terrigenous material that increases northward. Organic content of the sediment is highest in this part of the gulf, and the resulting rock type will be dark, shaly, shelly calcilutite.

These deeper-water muds grade shoreward into a broad belt of skeletal sand. Glauconite occurs in the deeper offshore part of this facies attesting to slowness of sediment accumulation. Rounding of skeletal grains in certain shallower parts of the zone reflects increased water agitation in these areas. Conversely, patches of mud reflecting decreased agitation occur within the sand facies in the protected lee of islands and peninsulas where wave base does not extend as deep. Along the shallowest fringe of this zone are coral reefs and oolite shoals associated with tidal deltas and channel complexes.

Shoreward is a zone of very shallow water, largely in the form of lagoons, in which lime mud is accumulating. The biota consists of small clams, peneroplid foraminifera and small gastropods which provide a large amount of fecal pellets. The resulting rock type will be pelleted calcilutite with a restricted, dominantly molluscan fauna.

Along the shore in the intertidal zone are extensive flats covered with blue-green algal mats exhibiting mud-crack polygons and small stromatolitic heads. Landward is the supratidal flat, or sebkha, which is floored with carbonate mud containing small amounts of fine grained dolomite and zones of gypsum crystals and less commonly anhydrite and celestite. The shoreline sediments will form barren laminated calcilutite ranging from stromatolite to dolomitic and gypsiferous variants.

Caribbean Sea.—Shallower parts of the Caribbean region might be considered an irregular example of a pericontinental sea. Two areas of carbonate sedimentation, the Great Bahama Bank and the British Honduras shelf lagoon, are combined in Figure 10C to better illustrate the entire shallow marine sedimentary pattern because the edge of the Great Bahama Bank desends abruptly to great depths that are below the shallow marine regimen. Information on sedimentary facies is derived from Purdy (1963) and Shinn, Ginsburg and Lloyd (1965) for the Great Bahama Bank, and from Matthews (1966) for the southern British Honduras shelf lagoon.

In deeper water of 60 to 100 feet surrounding local shoals of skeletal sand in the British Honduras shelf lagoon, predominantly lime mud containing shells of the indigenous fauna is accumulating. Terrigenous mud becomes progressively more important in this facies away from the shoals westward toward the main shoreline. Only within 0.1 mile of a shoal does transported skeletal sand derived from the shoal become important in this deeper-water facies. The resulting rock type will be shelly calcilutite grading to calcareous shale toward the detrital source.

The local shoals of southern British Honduras are merely small scale versions of the outer rim of the Great Bahama Bank from the point of view of sedimentary facies, and they provide the link to the greater range of shallower-water carbonate facies developed on the much larger bank. Skeletal sand consisting of diverse endemic invertebrate and algal debris, much of it fragmented and abraded, occurs in a narrow fringe around the rim of the bank. (Reefs occur at a few places in this zone, but are common only on the windward side of the shoal which is not included in Figure 10C). Oolite sand occurs slightly shelfward of the skeletal sand in this zone around much of the bank. The resulting rock types will be abraded skeletal calcarenite and oolitic calcarenite. In the Bahama example, the lime-sand facies is restricted to a much narrower belt than in the Persian Gulf on account of the abrupt bank-edge change in topography which causes nearly all wave and current energy to be dissipated over a very short distance.

The lime sand facies grades eastward into lime mud which accumulates over nearly all this portion of the shallow bank. Not only are waves and currents damped, but this part of the bank also is in the lee of Andros Island and thus is protected from strong, wind-driven waves. (Such waves apparently are strong enough to keep shallow bank water sufficiently agitated to prevent much mud accumulation on the north and south ends of the bank which are unprotected from the strong easterly trade winds; this is merely a subregional modification of the pattern, however, and does not detract from the basic model applying to an enclosed sea.) The shallow-water bank mud contains primarily a molluscan and foraminiferal shelled biota, and fecal pellets are indurated and preserved over a large protion of the environment. Thus the resulting rock type will be shelly, pelleted calcilutite with a somewhat restricted fauna.

Along the east side of the bank lies Andros Island, the western side of which is basically a large tidal flat encompassing both intertidal and supratidal environments. Here pelleted mud is accumulating in the near absence of biota except for algal mats. Much mud that is exposed for long periods of time is being dolomitized to produce barren laminated dolomitic calcilutite, locally with typical associated features such as birdseye and stromatolite.

Texas Gulf Coast.—Although Irwin's model was derived from considerations of nondetrital clear-water sedimentation in shallow seas with a very gentle bottom slope, it is interesting that a similar basic sedimentary pattern with respect to grain size and related parameters is established also in terrigenous detrital environments. Thus, in terms of resulting rock types, the main difference is merely that quartz sandstone replaces calcarenite and shale replaces calcilutite in the model. The similarity exists because detrital sediments react roughly the same as carbonate sediments to relative position of wave base once they are in the basin of deposition, regardless of whether they have been transported as detritus or formed nearly in place, as is most carbonate material. For comparison to the previously described carbonate examples, the detrital facies pattern of the Texas Gulf Coast (fig. 10D) is summarized from Shepard and Moore (1955), Moore and Scruton (1957), and Shepard (1964).

Mud is being deposited both throughout the offshore zone and in protected bays and lagoons. Glauconite occurs only in the offshore mud. Silt is common throughout the offshore mud, and sand is concentrated in nearshore areas. Different structures characterize different offshore areas. As the amount of sand in the sediment decreases progressively offshore, structures range from irregular lenses of sand to sand mottles in the mud. This also may reflect offshore replacement of organic burrowing for mechanical processes as the main mode of reworking the sediment. The farthest offshore region is characterized by mud that appears homogeneous either because of complete organic reworking or absence of sand transport this far.

The steeper slope in this pericontinental setting again causes wave energy to be dissipated over a narrow band, and in this case the cleaner, winnowed sands of the middle zone are thrown up as an actual physical barrier that shelters mud deposition in the bays and lagoons behind the barrier. Bay and lagoonal mud contains sand, particularly in channel bodies, but is poor in silt which apparently is largely transported out into the offshore zone. Peat is forming in marshes around the bays and lagoons. Sediments of the shallow mud zone exhibit much mottling as well as other structures, and the overall sedimentary pattern is more complex on account of the greater number of microenvironments than in the offshore zone. Biota is more diverse in the offshore sediment than in the bays or lagoons, but is nearly lacking in most of the barrier sand because of its littoral or subaerial origin.

River deltas modify the basic sedimentary model in detrital environments. Because deltas may extend either only into the bays or on into the open sea, the poorly fossiliferous, regularly laminated sandstone to siltstone formed in the marine prodelta environment can be inserted into the general model at nearly any point short of the farthest offshore extent. This insertion then will alter the landward facies to the subaerial deltaic regime. Thus the same relative position on Figure 10 of barren laminated prodelta deposits and supratidal carbonate laminites is only coincidental and should not be construed to suggest analogous origin for these two different sediments even though they appear superficially similar with respect to gross sedimentary structure and lack of fossils.

Environmental Summary of Basic Marine Rock Types

Calcilutite and shale.—These two rock types differ mainly in origin and composition of the constituent grains. Both indicate a quiet environment that is sufficiently calm to allow fine material to settle out of suspension. Such an environment may be very shallow and protected from water agitation by a physical barrier, such as a subaerial bar as in the case of a lagoon, or position above mean high tide line as in the case of a supratidal mud flat. Or, the barrier may be the hydrologic effect of a shelf edge or a certain point on a gently sloping bottom where the shallowing floor causes waves to break and dissipate most of their energy before entering shallower water. An environment of mud accumulation may also be relatively deep and protected from water agitation by its position below effective wave base. More detailed characteristics of deeper-water calcilutite, which for our purposes includes rocks termed calcisiltite by some, are discussed at length by Wilson (1969).

The two major environments of mud deposition may be differentiated in some cases by biotic considerations. Biota is generally restricted in very shallow water where salinity and temperature variation are typically high, and circulation which controls nutrient replenishment is reduced. The supratidal portion of this environment is recognized by near absence of fauna and by presence of the typical associated structures mentioned previously. Mud deposited in the deeper portion of the shallow marine regimen is subject to more constant salinity and temperature and thus can support a more diverse biota as long as currents, however slow, replenish nutrients and keep the bottom oxygenated. If bottom water stagnates, however, benthonic organisms will be reduced or eliminated, and remains of pelagic forms will become more noticeable or greatly predominant. In either case, more fine organic matter is incorporated into deeper-water mud because it is less readily oxidized than in typically better aerated shallow water. Thus, at least the calcilutites of deeper origin generally are darker than those of shallow origin.

Abraded calcarenite and clean quartz sandstone.—Calcarenites composed of fragmented, abraded, well-sorted skeletal debris of taxonomically diverse origin and with a primarily spar matrix have nearly the same environmental significance as cleanly washed, well-sorted quartz sandstones. Again, grain provenance and mineralogy are the main differences between the two types. Abraded calcarenite forms in, and clean quartz sandstone ultimately collects in, zones of water agitation above wave base where coarse grains are abraded and fine grains are winnowed. Cross bedding and ripple marking are characteristic of both types. Such clean, abraded sands indicate the zone where waves impinge and dissipate their energy on the sea floor. It lies between the two quiet zones of mud accumulation, one below effective wave base, the other beyond effective wave reach. The agitated zone of sand accumulation may occur at a change in slope which influences breaking of waves or it may occur within a certain range along an evenly shallowing sea floor. Such a zone may be entirely submarine, but under certain conditions, periodic storm waves are powerful enough to pile up the sand to normally subaerial elevations in the form of barrier bars or small islands. In the subaerial position, sand remaining unconsolidated is then often moved about and further winnowed by the wind. Subaerial manifestations of the agitated zone are common in the detrital regime such as along the Texas Gulf Coast with its extensive barrier bars, and are known also in the carbonate regime in the form of lime-sand keys which dot shoals throughout the Bahamas and Caribbean area. Differences in manifestation of the agitated zone are controlled by hydrologic consequences of different sea floor configurations and by different wind and weather patterns for which detailed discussions are beyond the scope of this paper.

Whole-shell calcarenites.—Calcarenites (and calcirudites) composed of skeletal material exhibiting little or no fragmentation, abrasion or sorting, and having only subordinate spar matrix must be contrasted with clean abraded calcarenites for environmental interpretation. Within most quiet mud environments, various types of organisms can proliferate and produce shell material at a faster rate than mud is accumulating. The resulting deposit will be calcarenitic on account of dominance of coarser grains, but is merely a shell accumulation. In contrast to the abraded type, whole-shell calcarenites typically contain much mud and exhibit a variety of sizes of different types of skeletal material giving rise to poor apparent grain sorting. Fragmentation of shells in such a calcarenite can result from activity of scavengers, and some degree of abrasion might result from activity of sediment-ingesting organisms. Rather than being rounded and reduced to somewhat spherical shapes as in an agitated environment of mechanical abrasion, however, organically abraded grains would mainly be comminuted and still inequant and angular, and might best be described as macerated.

Whole-shell calcarenites typically do not exhibit cross bedding or rippling. Moreover, they normally do not have distinct boundaries with surrounding skeletal calcilutite; this reflects their closely related origin as little more than skeletal-rich variants of mud facies. Although such calcarenites often occur as scattered indistinct lenses throughout a calcilutite, some may display substantial lateral and vertical extent. In-situ formation of whole-shell calcarenites, resulting from the autochthonous origin of carbonate grains, exemplifies a major interpretive difference between carbonate and detrital sediments of similar grain size.

More equivocal to interpret would be calcarenites formed by only one type of organism and accumulating with very little mud, because the resulting rock would have a spar matrix and would be well sorted on account of availability of only one size of grain. Such deposits include some stylioline limestones of the Devonian (e.g.,

Genundewa of New York), many crinoidal limestones of the Paleozoic, certain fusulinid limestones of the Upper Paleozoic, and some planktonic foraminiferal and pteropodal limestones of the Mesozoic and Cenozoic. Unless mud-free, well-sorted calcarenites composed of one type of shell display well-developed cross-bedding or consistent grain abrasion (which may be difficult to determine in the case of originally rounded grains such as fusulines), they might indicate nothing more than local proliferation of organisms or lack of dilution of long-term shell accumulation by fine sediment. Lack of dilution, however, might be interpreted by criteria for recognition of diastems, which are elaborated in a later section.

Because mere organic proliferation on any scale may result from factors not otherwise preserved in the record and could happen at any depth with respect to wave base, whole-shell calcarenites as a rock type cannot be fit specifically into the sedimentary model; moreover, they must be considered separately from clean abraded calcarenites for environmental interpretation. Consideration of biotic composition of whole-shell calcarenites may yield helpful environmental information as suggested in previous sections. As brief examples, planktonic foraminiferal calcarenites are most likely deeper-water accumulations, whereas calcarenites composed entirely of small gastropods, ostracodes, or other organisms capable of proliferating under conditions of ecologic stress typical of very shallow water more probably suggest restricted shoreline lagoons. Unabraded, noncross-bedded crinoidal calcarenites so common in the Paleozoic record fully marine conditions on account of the known narrow salinity tolerance of echinoderms and could reflect proliferation of crinoids on such a scale that other organisms were excluded because of lack of extra space or nourishment.

Application to the Stratigraphic Record

The shallow marine sedimentary pattern predicted by the theoretical model and observed in several modern environments can be recognized also in the stratigraphic record. Two examples in the carbonate regime from the Devonian of New York and the Pennsylvanian of Kansas illustrate the vastly different geographic scales at which the model applies (Fig. 11). Many geologists probably are familiar with other examples in the record; possibilities that come immediately to mind are the Lower Cretaceous of Texas in the carbonate regime (Griffith, Pitcher and Rice, 1969), and the Upper Cretaceous of the Rocky Mountains in the detrital regime (Pike, 1947, esp. fig. 3, p. 18; Spieker, 1949).

Lower Devonian, New York.—Carbonate rocks of the Lower Devonian Helderberg Group crop out in central and eastern New York beneath the Catskill sequence. Rickard (1962) and Laporte (1967, 1969) have recognized that several of the named formations representing gross carbonate lithofacies are lateral equivalents of one another and thus represent a lateral as well as vertical succession of carbonate environments. The lateral succession of Helderberg facies and environments (fig. 11A) approximates the geographic scale of the modern examples in Figure 10B–D and was first compared to Irwin's theoretical model by Laporte (1969, fig. 14).

Toward the west end of outcrop the Manlius Formation consists mostly of calcilutite. Some portions are laminated and mudcracked and contain layers of fine-grained dolomite, local birds-eye and very few fossils other than ostracodes. These represent the supratidal environment (Laporte, 1967). Other portions contain algal stromatolites, mud pebble conglomerates and pelleted calcilutite interbedded with skeletal, mostly whole-shell, calcarenite composed chiefly of ostracodes, tentaculites, spirorbid worms and one species each of brachiopod and bryozoan. Scour and fill structures are common, and this facies apparently represents the intertidal environment (Laporte, 1967). Most of the Manlius consists of pelleted calcilutite with local burrows, mostly vertical, and a variety of fragmented and disarticulated skeletal material including stromatoporoids, corals, green algae, snails and other molluscs in addition to more members of the groups present in the intertidal facies. This pelleted calcilutite represents a quiet-water, subtidal, perhaps lagoonal, environment (Laporte, 1967).

Eastward, the Manlius grades into the Coeymans Formation (Rickard, 1962) which is mainly a well-washed, spar-cemented, abraded skeletal calcarenite that exhibits strong crossbedding. It is composed mostly of crinoidal debris, and a great variety of brachiopods, bryozoans, and other groups; it contains many kinds of rugose and tabulate corals that locally form bioherms, often in association with stromatoporoids. The Coeymans represents crinoidal sand bars and local reefy buildups in the zone of greatest water agitation which supported an abundant and diverse suspension-feeding fauna above effective wave base along the margin of the open sea where most wave energy was dissipated (Laporte, 1967).

Farther eastward and southward, the Coeymans grades into the Kalkberg Formation which consists mainly of cherty skeletal calcilutite and passes eventually to shaly skeletal calcilutite of the New Scotland Formation. These formations contain a great variety of fossils of all major

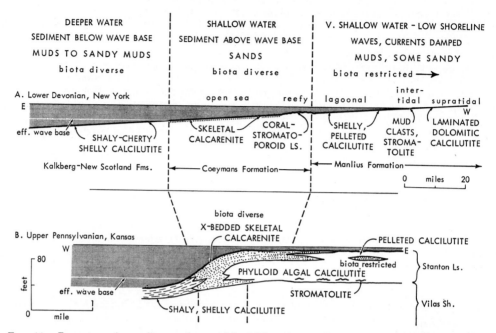

FIG. 11.—Recent marine sedimentation model of Fig. 10 applied to environmental reconstruction of carbonate units in stratigraphic record on two different scales. A) Lower Devonian Helderberg Group in central and eastern New York, after Rickard (1962), and Laporte (1967, 1969). B) Upper Pennsylvanian Stanton Limestone (Lansing Group, Missourian Stage) in central Wilson County, near Benedict, Kansas (author's work); cross section shows present topography on top of Stanton corrected for 20 feet per mile regional westward dip computed along algal calcilutite facies for 7 miles to east.

marine groups except algae. Trilobites and sponges increase in importance, and snails and clams become as common again as in the Manlius. Burrows are more common, shorter, and more nearly horizontal than in the Manlius. The Kalkberg and New Scotland represent a quiet open marine environment of mud deposition that was farthest offshore and below effective wave base, but was subject to enough water circulation to support an abundant and diverse fauna in which bottom-feeding forms were more significant (Laporte, 1969).

Upper Pennsylvanian, Kansas.—The shallow marine sedimentation model applies to the stratigraphic record also on a much smaller geographic scale. Carbonate rocks of the Stanton Formation form an escarpment along outcrop in eastern Kansas. Detailed facies mapping on 7.5-minute quadrangles reveals apparently original topographic features associated with facies changes in the unit. Merely removing the gentle regional westward dip produces the reconstruction illustrated in Figure 11B.

Along the east side of outcrop, the Stanton is primarily calcilutite with large heads of algal stromatolite (first discovered and identified by F. W. Wilson) occurring widespread at the base. Most of this facies is an algal-mound complex consisting of phylloid algal calcilutite with a large amount of spar that fills sheltered voids beneath the large algal blades. Pelleted calcilutite and local zones of skeletal calcarenite occur with algal calcilutite in the upper part of the facies. Biota is restricted and consists mostly of phylloid algae, gastropods, ostracodes, and scattered bryozoans and brachiopods. This facies represents a quiet environment of very shallow water, for the most part choked with algae (Heckel and Cocke, 1969). Intertidal conditions prevailed here during the earliest period of carbonate deposition.

The Stanton grades westward into a cross-bedded skeletal calcarenite facies that occupies a narrow belt along the west side of the low plateau formed by the algal calcilutite facies. Grains in the calcarenite consist of a wide variety of algal and invertebrate debris which generally is abraded and cemented by spar. Many whole or only disarticulated shells occur also, and include echinoderm material, several kinds of bryozoans, brachiopods, corals, fusulinids, and locally clams and sponges. This facies exhibits westward dips up to 10 and 15 degrees along its west side. It apparently represents a fringe of skeletal sand that developed where waves broke and helped support a diverse suspension-feeding fauna around the west edge of an algae-rich shoal.

Westward, the skeletal calcarenite facies dips abruptly beneath an overlying shale. Cores of the Stanton recovered from this western region reveal a thinned sequence of dark shaly calcilutite containing sponges, bryozoans, brachiopods, and echinoderm material. The top of this shaly calcilutite facies lies as much as 50 feet lower than the laterally continuous calcarenite and algal calcilutite facies even after correction for regional dip. Therefore, the shaly calcilutite apparently represents a quiet environment below wave base in a shallow basin west of the shoal.

Because the facies changes in this example coincide with the changes in topography apparent in the limestone today, facies development probably was closely related to variation in bottom topography of the original environment. This illustrates the value of detailed mapping in local well-exposed areas as well as of utilizing sedimentation models as working hypotheses in interpreting such elusive factors as water depth in the stratigraphic record.

MARINE DEPOSITS IN WHICH ORIGIN IS NOT OBVIOUS

Portions of the marine environment are subject to unfavorable extremes of physical factors that control distribution of organisms. When unfavorable conditions become established over a large area, the biota becomes reduced in diversity and often in numbers on a widespread scale and sometimes is eliminated entirely. The resulting deposit will not exhibit the biotic characteristics typical of the normal marine regimen. Such a deposit will be difficult to interpret and may readily give rise to controversy concerning its origin. When confronted by such a case, one must critically evaluate the nature of the few fossils that are found, and the stratigraphic framework as well as lithology, in order to establish a marine origin and determine possible causes for the reduction in biotic elements.

Gray Shales with Sparse Fossils and Low Diversity

An example from the epicontinental regimen for which there appears to be a modern analog is considered.

Thick, gray, "barren" Pennsylvanian shales.—The Upper Pennsylvanian section that crops out in Midcontinent North America from Iowa through eastern Kansas to Oklahoma consists in general of an alternation of limestone and shale formations. The limestone formations with included thin shale members average from 20 to 50 feet thick and typically carry an abundant and diverse marine biota. The intervening shale formations average from 50 to 100 feet thick, and are generally poorly fossiliferous. Most fossils that are found come from the portions immediately adjacent to the intervening limestones, and from a few local thin limy zones within the shale. These shale formations generally display a monotonous sequence of olive-gray to gray, slightly silty shale with thin zones of laminated siltstone that commonly carry macerated plant fragments. Channel sandstones, thin coals with underclays and thin red mudstones occur locally and represent various types of nonmarine environments. Early work recognized presence of both marine and nonmarine environments in many of these thick shales (Moore, 1929). Later, however, the occurrence (even though local) of the nonmarine lithologies, the recognition of the consistent position of the shale formations between definitely marine, predominantly limestone portions of the cyclic vertical sequence, and the general barrenness of these shales caused them to be considered as wholly or predominantly nonmarine deposits (Moore, 1936, p. 29; 1950), a concept that has persisted generally since.

Recent intensive examination of parts of the Kansas outcrop, especially by J. M. Cocke, and of a long core from Iowa by C. D. Conley and myself has revealed the presence of fossils scattered throughout several of these thick shales, particularly the Lane, Bonner Springs and Vilas Formations. Most of the fossils recovered are small, thin-shelled pelecypods and gastropods. Rarer associates include a few echinoderm pieces, articulate brachiopods, and a conularid in the core, and orbiculoid brachiopods, orthocone cephalopods, and possible jellyfish casts on outcrop. In addition, a few thin limy lenses within the shales are basically shell concentrations that yield echinoderm pieces, bryozoans, and brachiopods in addition to pelecypods and gastropods. The small and thin-shelled nature of the molluscs scattered through the shales contributes to their rapid deterioration on outcrop which apparently accounts to a large extent for their previous general neglect.

Preliminary lithologic examination reveals that the zones of laminated siltstone in the Pennsylvanian shales closely resemble the regularly layered structures recognized by Moore and Scruton (1957) in modern prodelta deposits in bays and along the coast of the Gulf of Mexico. Formation and preservation of these structures apparently result from rapid deposition with

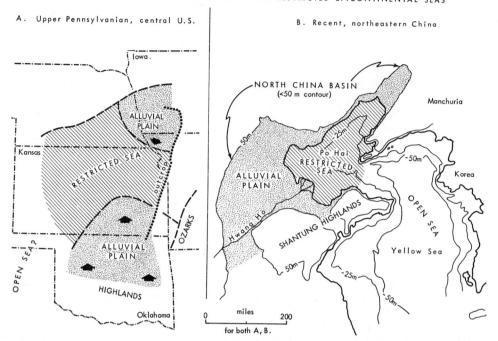

Fig. 12.—Comparison of ancient and modern terrigenous detrital sedimentation in restricted epicontinental seas. A) Hypothetical paleogeographic relations during deposition of thick shale units with restricted faunas in Upper Pennsylvanian of central United States; little detailed work has been done. B) Recent geographic relations of alluvial plain and restricted sea in northeastern China; analogy first suggested by Dunbar and Rodgers (1957, p. 84).

little reworking by organisms. The fauna apparently is reduced by a combination of the high rate of sedimentation and accompanying turbidity and variation in salinity brought about by the fresh-water influx.

The molluscan fauna dominant in the thick Pennsylvanian shales consists primarily of mobile pelecypods and gastropods, both of which can withstand rapid sedimentation and high turbidity better than most other groups. Most of the pelecypods are nuculids and pectinids which today are marine but have members that tolerate lowered salinities at least into the transition range (Ladd, 1951).

Thus lithologic and fossil criteria indicate a restricted marine environment undergoing rapid sedimentation and subject to high turbidity and probably fluctuating but slightly lower salinities for much of these thick, poorly fossiliferous Pennsylvanian shales, an interpretation in general agreement with Weller (1957) and Moore (1929). Hypothetical geographic relations of such a sea is shown in Figure 12A. Probably the scattered cephalopods were nektonic, the orbiculoid brachiopods epiplanktonic, and the jellyfish planktonic; the cephalopods could have entered the restricted sea temporarily, the other two could have drifted in, and none would have been affected by unfavorable bottom conditions. The thin shell lenses containing fossils of less tolerant groups represent times of reduction or cessation of sediment influx when more stable, normal marine conditions returned.

Gulf of Chihli (Po Hai), northeastern China.—Perhaps a very close environmental analogy, first mentioned by Dunbar and Rodgers (1957), to deposition of these thick Midcontinent Pennsylvanian shales exists today in the Gulf of Chihli (Pechili) or Po Hai off the Yellow Sea in northeastern China (fig. 12B). The gulf is about 300 miles long, averages 100 miles wide and encompasses an area of about 30,000 square miles. It occupies a little less than half of an epicontinental sedimentary basin which extends westward as a subaerial alluvial plain drained by the Hwang Ho and other rivers. The similarity of scale is remarkable, as the State of Kansas is just a little larger than the basin, and the part of the Midcontinent Pennsylvanian outcrop trace along which the thick shales occur is just a little longer than the gulf.

Salinity in the Po Hai is reduced to a little over 25 ppt (Grabau, 1931) as a result of fresh-water influx. Thus it is a restricted though still marine environment. The rivers bring in an immense amount of mud and sand that is spread as a subaqueous topset plain over the gulf which nowhere is greater than 160 feet deep and mostly is less than 80 feet deep. By far the dominant elements of the preservable fauna in the Po Hai are pelecypods and gastropods. Other groups are present, but exhibit a greatly reduced number of species, and in most cases fewer individuals, compared to their occurrence in the normal marine Yellow Sea. Crustaceans, barnacles and one species of bryozoan are numerous, starfish are locally common, but only a few echinoids, ophiuroids, brachiopods and corals were found (Grabau, 1931). In addition to reduction in diversity, most individual faunal elements are of smaller size than their compatriots in the Yellow Sea. This apparent dwarfing in the gulf may be due to the lowered salinities or perhaps to the difference in ionic ratios of the salts, as the gulf is enriched in Mg at the expense of Na according to Grabau.

Although most molluscs found in the Pennsylvanian shales are of small size, it is not yet determined if they are smaller than members of the same species in associated, more normal marine deposits. In any case, the faunal composition of the shales is remarkably similar to that in the Po Hai, and the mode of deposition indicated by the lithologic nature of the shales is the same as that apparent today in the gulf. Furthermore, the various nonmarine lithologies present in parts of the shales find their modern counterpart in that portion of the North China basin occupied by the alluvial plain of the Hwang Ho where river channels and swamps are probable locations of channel sand and peat deposits, and well-drained areas potentially could be sites of red soil formation.

To illustrate the delicate balance existing between the marine and nonmarine environments in this regimen, one can assume that Grabau's (1931) figure of about 0.5 cubic km of sediment deposited per year in the gulf by the Hwang Ho alone is the right order of magnitude and make the following computation. This amount is equivalent to a square km of sediment 500 m thick. Assuming 25 m as the average depth of the gulf, 500/25 or 20 sq km of the gulf can be filled each year. As the area of the gulf is only about 70,000 sq km, the present rate of deposition could fill it entirely in 3,500 years if there is no compensating subsidence. On the other hand, subsidence of only 100 feet would flood the entire North China basin and a large area south of the Shantung Highlands (Dunbar and Rogers, 1957). These points are brought up to show the potential transitory nature of either the marine or nonmarine environment in this regimen. Depending on relative rates of subsidence and sediment influx, either environment could become established rather quickly following the other over most of the basin. Thus it becomes less difficult to comprehend the dual restricted marine and nonmarine origin and the probable great complexity of interbedding and facies relations of the respective deposits in the Midcontinent Pennsylvanian shales.

Unfossiliferous Black Shales

Certain black shales are devoid of those fossils and sedimentary structures that are diagnostic for differentiating marine from nonmarine environments. In some cases, however, establishment of a detailed stratigraphic framework allows a reasonable interpretation of marine origin.

General considerations.—Origin of unfossiliferous black shales has long been a subject of controversy. General aspects of the problem are summarized from Strøm (1939) and Dunbar and Rodgers (1957). Black color in these shales is caused by a relatively high content of either unoxidized organic matter or finely divided iron sulphide. Both substances are preserved or formed in large quantities only under reducing conditions which develop in "anoxic" environments where the normal processes of oxygenation have ceased to operate on a scale sufficient for continual oxidation of organic material normally accumulating on the bottom. As soon as oxygen is depleted in the environment, anaerobic bacterial decay of organic matter begins on a large scale, producing hydrogen sulphide which, along with the lack of oxygen, fouls the substrate and bottom waters and makes them inhospitable for benthonic and nektobenthonic life.

One model for development of such an environment is loss of circulation that normally brings oxygenated waters to the bottoms of most modern bodies of water including the oceanic depths. Loss of bottom circulation in modern cases takes place where water bodies, or portions thereof, are barred or silled, and density stratification occurs as water of slightly lower salinity flows in over the denser, normal marine water which is prevented from flowing out of the basin along the bottom by the barrier. This density stratification is stable, and as vertical circulation ceases, bottom stagnation eventually results.

The barrier may be a submarine topographic high, or sill, such as at the mouths of certain Norwegian fiords and the entrance to the Black Sea, or a subaerial bar that cuts off a marginal lagoon or swamp from interchange with the sea. An effective barrier also might be a complex set of little understood nontopographic conditions that prevent outflow or oxygenation of stagnating bottom water. In any case, stagnation could happen at any depth in the marine environment, depending on the level of the barrier, or in freshwater lakes and swamps of the nonmarine regimen. The general end result, loss of bottom life and deposition of black shale, is the same in all cases. Within a silled basin, only water in the deepest part of the basin would stagnate. Thus a full complement of pelagic marine life could occur in the upper reaches of the water, and benthonic forms could live shoreward above the effective level of the barrier.

In order for stagnation that results in fouling to occur in very shallow water, however, sources of oxygenation readily available in near-surface waters through wave- or tide-induced water agitation and planktonic algal photosynthesis must be subdued. Although in broad, shallow, water bodies, wave intensity and tidal exchange are reduced by the damping effect of thinness of the water layer and concommitant increasing effects of bottom friction, shallowness alone probably would not be sufficient to prevent some oxygenation of the bottom water. Indeed, modern shallow-water regions where dark muds are accumulating, such as certain bays along the Baltic Sea, have abundant bottom faunas attesting to occurrence of bottom oxygenation at least on a scale sufficient to support life, if not to completely oxidize all the decaying organic material. This extremely shallow situation is similar to that envisaged by Hallam (1967) for British Jurassic black shales which do contain some bottom-dwelling (though not burrowing) pelecypods and echinoids in addition to many nektonic fossils. The suggestions by Moore (1950) and Zangerl and Richardson (1963), that thick growth of vegetation (either rooted or floating) in nearshore marine marshes or fresh-water swamps would effectively prohibit disturbance of the bottom by winds and waves, also provide an abundant source of organic material. Likewise, perhaps a large unattached floating mat of seaweed similar to that of the modern Sargasso Sea could cover a greater area and prevent agencies of oxygenation from acting on the bottom of a more open shallow sea.

Another model for attaining an oxygen-depleted environment in shallow water, developed by Brongersma-Sanders (e.g., 1966, 1968), involves offshore upwelling of nutrient-rich and oxygen-poor deeper water in response to driving of surface waters seaward by constant strong winds from the adjacent land. Upwelling brings nutrients, especially phosphorus and nitrogen, into the photic zone where they are utilized by phytoplankton which bloom in enormous numbers and produce toxins that cause mass mortality of the neritic fauna, particularly fish (Brongersma-Sanders, 1957). The great amount of decaying organic material both from the mass mortalities and the blooms further depletes the already oxygen-poor water brought up from the oxygen-minimum zone. (See Richards, 1957.) As a result of either periodic or continual upwelling the benthonic fauna is eliminated, and much organic matter accumulates in the sediment, in this case on account of circulation of anoxic water. Although upwelling is known today mainly in pericontinental seas along the west coasts of continents, this model has been extended by Brongersma-Sanders (1966) to epicontinental seas in explaining the origin of the Permian black Kupferschiefer of Germany. High trace element content, association with dry-climate deposits, and well-preserved fish remains of a nature reflecting mass mortality, verify the interpretation of upwelling for formation of this black shale.

Example from the Midcontinent Pennsylvanian. —Recent detailed work by Evans (1967, 1968) on the black Heebner Shale Member of the Oread Limestone provides stratigraphic as well as petrographic evidence that allows the marine origin of a particular type of black shale in an epicontinental regimen to be determined without the aid of a good assemblage of unequivocally marine fossils. The Heebner lies between two limestone members and averages about 6 feet thick along 300 miles of outcrop from Iowa to southern Kansas (fig. 13). About the lower half of the Heebner throughout this area consists of poorly fossiliferous, platy, fissile black shale with numerous thin phosphorite laminae and scattered phosphorite nodules. Fossils in the black shale are mainly conodonts and scattered fish spines, but a few orbiculoid brachiopods and thin-shelled pectinid clams have been found locally. Only the scattered orbiculoids and pectinids would suggest a marine environment by strict modern analogy, and they would not rule out a restricted marine or brackish marsh environment because the pectinids have wide salinity tolerance, and the potentially epiplanktonic orbiculoids could have been attached to floating material that was washed into a marsh. Thus to determine more adequately the depositional environment of this black shale, criteria other than fossils are needed. Evans (1967) has established that the undisturbed nature of the even stratification manifest particularly by the primary phosphorite laminae preclude the former existence of either a root system of attached vegetation or an unpreserved bnrrowiug or bottom feeding fauna. Presence of detrital material

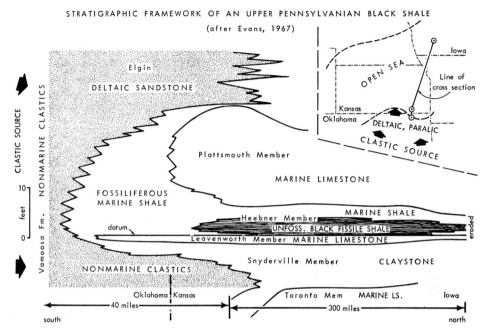

Fig. 13.—Stratigraphic framework of an Upper Pennsylvanian black shale in Midcontinent North America, mostly after Evans (1967). Plattsmouth, Heebner, Leavenworth, Snyderville, and Toronto Members constitute Oread Formation (Shawnee Group, Virgilian Stage).

of only the finest sizes combined with abundance of thin phosphorite laminae which indicate long-term cessation of detrital sedimentation (Goldman, 1922), suggests strongly that the black shale represents a nearly currentless, sediment-starved region where nondeposition prevailed.

Besides making field observations at individual outcrops, Evans established the stratigraphic framework of the black shale lithotope and examined especially the exposed extremities of its extent (fig. 13). Within the Heebner Member, the black shale is overlain by light-colored, fossiliferous marine shale containing coarser quartz silt, and is underlain by either similar shale or a very thin concentration of small brachiopod and pelecypod shells which rests directly upon the underlying marine Leavenworth Limestone Member. Southward, the black shale grades into a thicker section of light-colored shale in northernmost Oklahoma. Because this thicker light shale carries a fauna of brachiopods, gastropods, echinoderms, bryozoans, corals and cephalopods, and grades southward into coarser predominantly nonmarine clastics, it represents a normal marine environment favorable to benthonic organisms between the southern terrigenous detrital source and the region of sediment starvation and oxygen depletion represented by the black shale to the north. Moreover, considering the general paleogeographic distribution of environments indicated by the deposits overlying and underlying the Heebner in the Oread Formation (fig. 13, inset), marine limestones dominate the Kansas sections whereas nonmarine and paralic clastics occupy the Oklahoma sections. Thus, at other times, open marine conditions prevailed in the northern area where black shale is developed in the Heebner.

Because the black shale lithotope indicates relative slowdown of detrital sedimentation as well as lack of carbonate sedimentation, it could not represent encroachment of transitional environments which would imply increased sedimentation necessary to overwhelm the fully marine conditions of the underlying limestone. Furthermore, if the Heebner represented a transitional environment, it should grade directly into the nonmarine sequence of Oklahoma rather than through the fully marine deposits present near the state border. A similar facies change to a fully marine section northward in Iowa is suggested by reappearance of light-colored marine shale below the black facies in exposures just south of the modern erosional termination of outcrop (fig. 13).

Thus the Heebner is inferred to be a marine deposit resulting from extremely slow sedimentation in an oxygen-depleted sea. Lack of evidence of mass mortality such as abundant fish remains and lack of associated dry-climate deposits such as evaporites, each of which would be expected with the upwelling model, suggest that the anoxic conditions responsible for formation of the black Heebner resulted from stagnation of the sea.

Evans (1967) has suggested that stagnation could have been caused by a submarine barrier resulting from sedimentation across the main mouth of the Midcontinent sea to the southwest. If so, the black Heebner represents the deepest part of the silled sea at that time. Depth certainly was not on the order of the present day Black Sea on account of unwarranted tectonic implications, and in fact it need not have been greater than for overlying and underlying marine facies. The sea during black Heebner deposition may be viewed most reasonably as a broad, saucer-shaped basin, perhaps under maximum transgression (as Evans has suggested), in which circulation had ceased over most of the bottom and into which only the finest suspended detritus was spread by surface currents. Nektonic fish and conodont organisms lived only in the aerated surface waters along with epiplanktonic orbiculoids and pectinids (which by modern analogy could have been either similarly attached or free-swimming). Presence of floating seaweed is indicated by the epiplankton, but it is not necessary in great amounts to account for stagnation of the bottom. Only at the extreme edges of this saucer-like basin, as in northern Oklahoma, was detrital sedimentation sufficient to build the bottom up above the level of stagnation into aerated waters where it could support benthonic life. The nature of the barrier across the mouth of the sea to the southwest is obscure. It may have been a shallow sill of clastic sediments as suggested by Evans, or it may have resulted from some sort of nontopographic condition that prevented bottom circulation with the more open ocean to the southwest.

It is appropriate to mention here recent discoveries (1969) of previously unreported fossils in the black Heebner by M. E. Williams of Kansas University. Nautiloid and ammonoid cephalopods and shark remains corroborate a fully marine environment for the black shale, and together with some peculiar arthropods, accentuate the nonbenthonic nature of the faunal assemblage.

Diastems

A type of marine environment that is common in the stratigraphic record but has been little studied is that of nondeposition, which results in a break in the record or diastem. Why diastems are common in the record is well explained by Dunbar and Rodgers (1957, p. 127–134); the paucity of investigation is not surprising when one considers that by definition very little or no sedimentary material is available for study and that a diastem often is not even recognizable in the record. One can, however, study material from thin intervals of sedimentary rock which seem to represent conditions verging on nondeposition, in order to gain insight into how diastems might be recognized more easily in the record.

Certain shale partings in limestone.—Many limestone units exhibit distinct layering in which the limestone layers are separated by a thin shale parting. Such a parting typically weathers in after exposure and thus receives little attention other than as a measuring-unit boundary. One such shale parting in the Toronto Limestone Member of the Oread Formation in the Midcontinent Pennsylvanian was investigated by Troell (1969) in order to justify its use as an internal stratigraphic datum. The thin shale contains corals and encrusting bryozoans which are both suspension feeders requiring clear water and thus indicate relatively slow deposition. Moreover, both require at least firm substrate. From these ecologic considerations, Troell concluded that the shale resulted not from a rapid influx of clastics but rather from cessation of carbonate deposition for a long time with very slow accumulation of suspended detritus, which are conditions more favorable to suspension-feeding organisms.

From purely sedimentologic considerations, I have evaluated similar thin shaly partings in the western extent of the Tully Limestone in the New York Devonian (Heckel, 1966; ms in press). Rock recovered from the partings contains about 3 times the amount of fine noncarbonate detritus and about 3 to 4 times the amount of carbonate skeletal material than the intervening thicker layers of calcilutitic limestone. These partings most likely resulted from near-cessation of lime mud deposition for a long time while the normal slow influx of terrigenous mud and biologic generation of skeletal material continued unchanged and formed a thin concentrated deposit in the absence of dilution by carbonate mud. This explanation is more reasonable than a short-term increase of detrital influx coinciding with a similarly short-term increase in skeletal production while lime-mud deposition remained unchanged. Thus these shaly partings represent near-diastemic conditions at several levels in the Tully Limestone. Moreover, they stand out in contrast to shale partings which are devoid of fossils in the eastern end of the formation, nearer the detrital source. The eastern shale partings more likely represent a rapid influx of fine clastics into an environment where both carbonate mud and skeletal material normally were accumulating slowly.

Concentrations of nondetrital material.—Sedimentary implications of long-term cessation of carbonate deposition in primarily carbonate environments can be similarly applied to long-term cessation of detrital sedimentation in primarily detrital environments. Thus a diastem in a shale sequence might be represented by a thin limestone, particularly a shell concentration like those mentioned previously in thick Midcontinent Pennsylvanian gray shales.

Carried a step further, if deposition of both carbonate and terrigenous detrital material (the dominant sediments in shallow marine environments) ceased for a long time, one should expect concentration on the sedimentary interface of materials that are continually being deposited in small amounts in the marine environment, but normally are so diluted by the dominant sediments that they are hardly noticeable. Nondetrital minerals known to be concentrating today mainly in submarine areas of little or no sedimentation include glauconite, chamosite and phosphorite (Porrenga, 1967; Bromley, 1967). Similarly, manganese nodules also are indicators of slow deposition, but can occur abundantly in fresh-water as well as marine environments (Price, 1967; see also reviews by Twenhofel, 1936; Krumbein, 1942). One of the first to recognize the long time significance of nondetrital mineral concentrations was Goldman (1922) who noted that mixed glauconite and phosphorite concentrations in the record typically contain a wide stratigraphic range of fossils that represent a great thickness of equivalent strata elsewhere. He further concluded that these concentrations probably occurred in a continually marine environment because the surfaces they overlie show no evidence of subaerial weathering or erosion.

Probably the most common material suggesting a diastem is phosphorite which occurs both as fossils (such as conodonts and vertebrate teeth and bones), and as abiotically accreted nodules, laminae, and less commonly, ooids. Most marine vertebrates are nektonic, as presumably also were the conodont organisms, and under conditions short of mass mortality would provide sedimentary material at a fairly constant rate regardless of depositional conditions at the sedimentary interface. Bone beds in the Ohio Devonian apparently are diastemic in origin and may even be lag deposits resulting from submarine erosion (Wells, 1944). Phosphate also is present in particulate form (in dead organisms and feces) in sea water where it is involved in the nutrient cycle of marine life. (See Fox, 1957; Barnes, 1957.) Particulate phosphate that eventually settles out of the cycle to the bottom in areas undergoing negligible sedimentation of other material probably is the source of laminae that occur in black shales like the Heebner. It also may be involved, along with direct precipitation, in the formation of nodules found in certain black shales and other deposits. Direct precipitation probably is responsible for phosphorite ooids that occur in a few deposits.

Other minerals that contain elements present in small amounts in the marine environment, such as barite and sphalerite, also might be expected to be concentrated similarly at marine diastems as a sort of "geochemical residual". This seems to be true in some diastems associated with the Tully Limestone, but it may not be widely applicable because of the tendency of these minerals to occur as secondary fillings and replacements rather than as primary deposits.

Multiple diastems and a submarine unconformity in the Devonian of New York.—In working out detailed stratigraphic correlation within the Tully Limestone of the New York Devonian, I have recognized thin diastemic shale partings between most layers of calcilutite in the western development of the formation. Although these diastems can be recognized in individual outcrops, only establishment of the large-scale stratigraphic framework allowed realization of the full implications of these diastems and their "equivalents" (fig. 14). All diastems within the Tully Limestone join one another westward like tributaries in a stream system to form a major diastem or uncomformity which takes on progressively increasing time value westward as rock units thin to disappearance while approaching and joining it. All rock units involved are demonstrably marine, and there is evidence that the unconformity also is submarine.

The upper Hamilton (Windom Member of the Moscow Formation) consists of gray to dark-gray shale which carries a diverse suite of marine benthonic fossils, some of which define assemblage zones. Cooper (1930) noted that the upper three zones (marked 1, 2, 3 on fig. 14) rise westward and disappear from the top down until the lowest zone (3) occurs at the top of the Hamilton at Lake Erie.

The Tully Limestone consists of well-defined layers of marine calcilutite which are separated by thin diastemic shale partings. In the lower member, these shale partings pass eastward into thicker beds of silty calcilutite and quartz siltstone which represent sediment that accumulated near the detrital source while the diastems were forming westward. In the base of the cal-

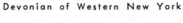
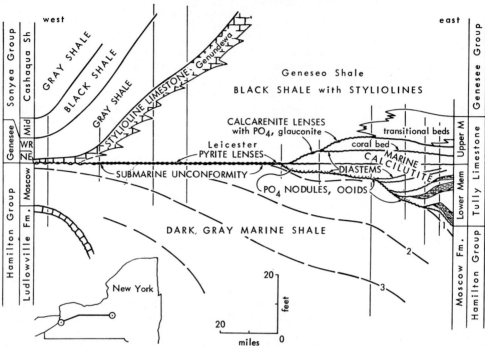

Fig. 14.—Stratigraphic framework of a submarine unconformity in Devonian of western New York. Information from Cooper (1930), Sutton (1951, 1963), Rickard (1964), Heckel (1966; ms. in press) and subsequent field work.

cilutite layers above the Hamilton are local, small, shell concentrations and phosphorite nodules and ooids which strongly suggest submarine nondeposition. Such features occur less commonly in the diastems between the Tully beds, and their greater concentration at the Hamilton contact, particularly westward, attests to the longer duration of nondeposition there. In the upper member of the Tully is the "coral bed", a calcarenitic shaly layer that contains glauconite as well as an extensive development of large corals that further reflects the diastemic nature of this zone on account of the preference of suspension feeders for slow deposition. Scattered between the west ends of both the adjacent calcilutite layers and the overlying Geneseo Shale are locally pyritized calcarenite lenses that contain concentrations of glauconite and phosphorite, particularly conodonts and pieces of bone. In the top of the upper member of the Tully are interbedded layers of poorly fossiliferous dark calcilutite and dark shale that are transitional to the black Geneseo Shale. These transitional beds grade eastward into normal fossiliferous lighter Tully calcilutite and are the only part of the Tully Limestone that actually grades westward into Geneseo Shale. All the lower beds of the formation disappear westward by nondeposition, and exhibit evidence of diastems at their contacts with adjacent shales.

The Geneseo Shale is a fissile platy black shale that carries a greatly reduced marine fauna dominated by styliolines and including similarly pelagic cephalopods and conodonts and potentially epiplanktonic small brachiopods. This fauna is typical of black shale deposited in a stagnant marine basin that supported no benthonic life. Because the Geneseo grades eastward into coarse clastic rocks of the Catskill deltaic sequence, the black shale represents a fondothem (Rich, 1951) in the distal region of fine-sediment deposition away from the detrital source. Above the Geneseo is a sequence of thin dark limestone beds and lenses (Genundewa Limestone) which apparently replace the black shale westward as well. The Genundewa is composed mostly of styliolines and probably represents undiluted longterm accumulation of pelagic skeletons at the edges of, and beyond the farthest extent of, the fine detrital influx that formed the Geneseo. Above the stylioline limestone is an alternation of black shales and dark-gray shales containing small pelecypods in addition to the pelagic fauna. These deposits reflect different degrees of bottom oxygenation in distal environments of subsequent phases of prodeltaic sedimentation (Sutton, 1963).

Between the Geneseo and upper Hamilton west of the Tully Limestone lies the Leicester Pyrite and farther westward the North Evans Limestone. Both are thin lenticular accumulations of invertebrate shells, bone, conodonts, phosphorite ooids, bits of phosphoritized wood and glauconite pellets. In the Leicester Pyrite, most ooids and carbonate shells have been replaced by pyrite and marcasite; authigenic barite and sphalerite occur as matrix and fillings. Many of the pyritized invertebrate shells are dwarfed forms of Hamilton fossils with sessile benthonic elements exhibiting the strongest stunting (Loomis, 1903). Both Loomis (1903) and Fisher (1951), who studied a similar marcasitized assemblage from lower in the Hamilton, feel that the stunting agents were H_2S and/or Fe. L. F. Laporte (personal communication 1969) feels that diminishing oxygenation, which is incipient to formation of H_2S, was the actual stunting agent rather than highly toxic H_2S. In either case, because H_2S and Fe are the substances that eventually combined to replace the fossils, the dwarfed organisms may have been living under incipient stages of the conditions that later caused their mineralogic replacement. If so, later subaerial exposure of this horizon is ruled out because either pyrite or marcasite would have been destroyed under oxidizing conditions. This explanation for stunting of the fauna also suggests that the marine but foul substrate conditions apparent in the overlying black Geneseo Shale began during deposition of the upper Hamilton because the stunted assemblage is probably that of faunal zone 1 (Cooper and Williams, 1935). Unfavorable substrate conditions probably lasted throughout eastward deposition of the Tully, because no uniquely Tully benthonic fossils are known from the pyrite. Thus the diastem occupied by the pyrite spans time encompassing deposition eastward of the upper Hamilton, the entire Tully and the lower Geneseo (fig. 14).

In corroboration of the submarine origin of the major diastem, all adjacent deposits exhibit quiet-water marine facies far away from known aggrading land that is represented by the Catskill redbeds into which all the formations eventually grade farther eastward. One type of deposit associated with the major diastem, the stylioline limestone, represents similarly sediment-starved marine conditions. Furthermore, there are no beach or shoal-water deposits that would be expected to develop around low-lying land if this were a break in the record resulting from subaerial exposure.

Apparently, sedimentation in the Devonian sea over western New York began diminishing during deposition of the upper Hamilton, retreated progressively eastward, and ceased entirely in the region depicted in Figure 14 prior to deposition of the Tully Limestone. Lime-mud deposition of the Tully apparently came in several sporadic waves that spread westward into the area of nondeposition and eventually retreated eastward before being diluted by black terrigenous-mud deposition of the Geneseo which ultimately spread back westward into the sediment-starved sea and covered the diastemic surface. The immediate cause of cessation of sedimentation over the particular area at this time may have been contemporaneous structures, east of the Tully Limestone, that developed during late Hamilton deposition and cut off detrital influx from the east (Heckel, 1965, 1966; ms. in press). These structures not only allowed accumulation of Tully carbonate in a primarily terrigenous detrital deltaic sequence, but also caused a prolonged period of sediment starvation in the region farther west. Sedimentation did not commence again in this area until detrital influx overwhelmed the structures and began its slow westward advance over the limestone and associated diastemic surfaces.

Although previous usage was more general (Twenhofel 1936; Krumbein 1942), the terms unconformity, disconformity, and paraconformity, as discussed by Dunbar and Rodgers (1957), imply that subaerial exposure is responsible for these types of breaks in the record, whereas only the term diastem applies to breaks resulting from submarine nondeposition. Undoubtedly the great majority of diastems in the record are of minor importance in terms of time or change in depositional regimen, but the major diastem described here in the Devonian of western New York is of sufficiently great lateral extent and time duration westward that it approaches the magnitude of many subaerial breaks in the record. In fact, it has angular relations (viewed over a large area) both with the adjacent Geneseo and Tully lithic units and with the Hamilton faunal zones, and it has been referred to previously as an unconformity (Cooper, 1930). Thus, it seems reasonable to refer to it as a submarine unconformity and dispense with the genetic implications of unconformity and related terms that apply to different variations in physical nature of the breaks. Interpretation as to origin of a particular break may be possible only after petrographic and regional stratigraphic study, and only then should a genetic term such as "subaerial" or "submarine" be appended to the descriptive term for the break. In view of this, perhaps many paraconformities and angular unconformities in the record of shallow marine deposition, which generally have been assumed to signify subaerial exposure, should be reinvestigated as to possible submarine origin, as was emphasized earlier by Twenhofel (1936). Commencement of critical reappraisal is indicated by recognition that unconformities of high angular discordance in the Texas Permian and Alberta Devonian may be results of entirely submarine processes (Pray, 1968). Moreover, continuing investigation shows that large areas of extensive modern shallow seas such as the Sahul Shelf and Arafura Sea are actually environments of nondeposition (Tjia, 1966; Van Andel and Veevers, 1967). Although modern shallow-water regions less than 200 feet deep have been subject to marine conditions for only several thousand years, and thus might not be considered significant comparisons to the past, submarine nondeposition could conceivably continue in these areas for enough time to become geologically significant. For an area to receive no noticeable sedimentation, it need only be distant from a detrital source, be subject to currents strong enough only to prevent suspended material from settling, and have a bottom environment unfavorable for organic proliferation on a scale sufficient to produce limestone.

MODERN EPICONTINENTAL SEAS

Most of our knowledge of modern shallow marine sedimentation has been obtained from pericontinental seas between large land masses and the deep ocean. Many of the principles derived from this knowledge are applicable to epicontinental marine sediments that form so much of the stratigraphic record. Nevertheless, the much gentler bottom slopes and extreme shallowness over such broad areas of epicontinental seas necessitate modifications of certain sedimentary principles for them to be applied more appropriately to epicontinental seas. So far, such modification has been mainly theoretical (e.g., Shaw, 1964; Irwin, 1965) because little continental area at present is covered with shallow epicontinental seas, and few of those that do exist have been studied in any detail.

Brief Resume

A quick glance through an atlas reveals six modern seas that can be considered epicontinental in that they are mostly or entirely less than 600 feet deep, yet cover a large area measured from several tens to hundreds of thousands of square miles and are mostly surrounded by continent or large islands. On North America is Hudson Bay, a largely unstudied area subject to a subpolar climate. In the cold temperature portions of Europe is the Baltic-North Sea system whose biota and sediments, mainly terrigenous clastics, are being studied by European geologists (see Donovan, ed., 1968). The Baltic Sea is almost land-locked and largely filled with brackish water (Segerstrale, 1957).

Around southern and eastern Asia lie three epicontinental seas. The Persian Gulf between Arabia and Persia presently is being studied intensively along the south shore where carbonates and evaporites are being deposited (fig. 10B). Extending into the Indonesian islands is the Sunda Shelf comprising the Gulf of Siam, Java Sea, and southwestern part of the South China Sea. Recent large-scale shifting of coastlines in response to both tectonic movement and rapid terrigenous detrital sedimentation in the extremely shallow Indonesian portion of this area has been summarized by Van Bemmelen (1949 p. 298–302). The Yellow Sea between Korea and China has been mentioned previously in conjunction with its gulf, the Po Hai, in the North China Basin (fig. 12). Like the gulf, the Yellow Sea is also mainly a site of terrigenous detrital sedimentation (Niino and Emery, 1961). Little other recent detailed work is available on the Yellow Sea outside of Keulegan and Krumbein (1949), who showed that the gently shelving sea floor configuration along the western shore could effectively damp storm waves such that they do not break at shore, a process that perhaps may not be uncommon in other parts of very shallow seas.

Comparison of Australia-New Guinea Recent to North America Devonian

The epicontinental sea of special interest here is that between the northern coast of Australia and the island of New Guinea (fig. 15B). Comprising the Arafura Sea, Gulf of Carpentaria and Sahul Shelf (part of the Timor Sea), it covers an area between 400,000 and 500,000 square miles that is nowhere greater than 600 feet deep and generally averages about 150 to 250 feet deep. Because it lies between stable cratonic Australia and unstable orogenic New Guinea, it is analogous in position to those shallow seas which lay betwen stable cratons and rising mountain systems and received large amounts of sediment now incorporated throughout the stratigraphic record. Such seas are especially well documented in the Paleozoic. For example, in the Middle to Late Devonian in eastern North America, the rising Acadian Highlands along the present east coast were separated from the North American craton (basically the Canadian Shield and its southwestern appendages) by a broad, probably shallow, epicontinental sea herein termed the Eastern Midcontinent Sea (fig. 15A). The analogy becomes remarkably close when one considers the strikingly similar patterns of sedimen-

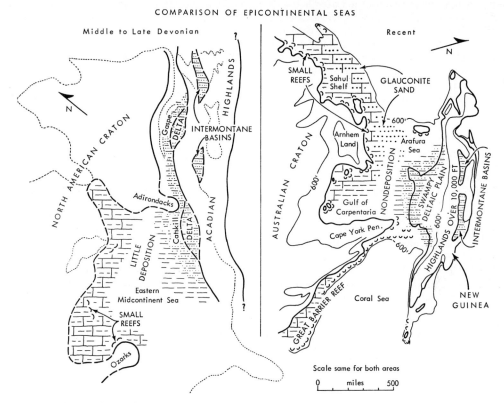

Fig. 15.—Comparison of ancient and modern epicontinental seas. Maps rotated for easier comparison. Standard lithologic symbols apply to both rock and sediment. A) Late Middle to early Late Devonian of eastern North America (anomalous Tully Limestone omitted from Catskill Delta); present features shown by dotted line (counterclockwise from lower right): Florida, Nova Scotia, Newfoundland, northern Quebec, Hudson Bay, and Lake Superior. Information from Clark and Stearn (1960), Sloss, Dapples and Krumbein (1960), and Heckel (1966; ms. in press). B) Recent of Australia and New Guinea; shorelines shown by thick solid line. Information from Van Bemmelen (1949), David (1950), Fairbridge (1950), Phipps (1966), Tjia (1966), and Van Andel and Veevers (1967).

tation that developed at two different times in these two distant regions shown at the same scale in Figure 15.

The backbone of New Guinea is a chain of mountains over 10,000 feet high which is shedding a large amount of terrigenous sediments both into intermontane basins on the north and into the epicontinental sea on the south. The intermontane basins are elongate depressions about 200 to 300 miles long, 30 to 40 miles wide and less than 300 feet above sea level. They are the sites of thick Recent alluvial and paludal deposits of gravel, sand, mud, and decaying vegetation (Van Bemmelen 1949; David, 1950). Neogene and Pleistocene brackish-water and marine deposits, locally including coralline limestone, attest to the recent tectonic activity and concommittant rapid change of depositional regimen in these basins. Sediments shed southward from the New Guinea Highlands have extended the island up to 200 miles into the epicontinental sea as a swampy deltaic plain where

thick Recent alluvial, paludal and lacustrine deposits have accumulated up to perhaps 2000 feet thick (from Van Bemmelen 1949, p. 718). Sediments range from conglomerate to mud and generally become finer-grained southward (David 1950, p. 674). They overlie a much thicker Tertiary sequence which becomes increasingly marine downward into the Miocene (Van Bemmelen 1949, p. 186).

The Devonian Acadian Highlands of eastern North America extended at least 1500 miles from Newfoundland to perhaps North Carolina, a scale remarkably similar to that of the present island of New Guinea (fig. 15). At various times in the Devonian, coarse to fine sediments from these highlands filled nonmarine intermontane basins typically to great thicknesses in Newfoundland and Maine (Clark and Stearn, 1960, p. 114). Clastic sediments from the Acadian Highlands flooded also westward into the epicontinental sea where they formed at least two major deltaic sequences, the Gaspé in the Middle

Devonian and the Catskill in the Late Devonian. Both sequences exhibit general upward trends of transition from gray marine shale and limestone to red-colored, coarser-grained nonmarine sandstone, shale and conglomerate reflecting the aggradation of Acadian land into the Eastern Midcontinent Sea. Volcanics are intercalated in the Gaspé sequence just as they are in parts of the modern deltaic complex of southern New Guinea.

Between the highlands and the craton lie the epicontinental seas which exhibit a remarkably similar lateral succession of sediments. The Recent example has received little study, but from available information, terrigenous sand and mud are accumulating around the deltaic south coast of New Guinea while carbonate sediment is more common around the Australian craton (fig. 15B). Small coral reefs are shown at various places along the Australian coast and around islands by Fairbridge (1950). (But nowhere do these seem to be as thick or extensive as the Great Barrier Reef along the break in slope to oceanic depths off the northeast coast of Australia.) Associated sediment includes coral debris and calcareous mud in the Arafura Sea (Tjia, 1966), Pleistocene cross-bedded calcarenite, Recent green mud and silt, and an area of oolite in the Gulf of Carpentaria (Phipps, 1966; pers. commun., 1969), and mixed skeletal sand and shelly terrigenous mud on the Sahul Shelf (Van Andel and Veevers, 1967). These sediments will produce mainly shaly skeletal limestones containing local reef masses. Over much of this modern epicontinental sea, particularly away from shore, all authors indicate that little or no deposition is taking place. Sediments that are present are being reworked and strongly glauconitized. In addition, there are extensive areas of hard bottom where no sediment has come to rest. This modern surface of nondeposition may be similar to the Devonian surface of probable submarine origin discussed previously.

Sedimentary rock in the Devonian Eastern Midcontinent sea of North America ranges from marine sandstone and shale in the submergent part of the Catskill Delta on the east to shaly limestone along the craton on the west (fig. 15A). Small stromatoporoid-coral reefs occur in Missouri, Iowa, and northern Michigan. The indicated shoreline of the low-lying craton is only general as it probably shifted a great deal in response to slight epeiric movements and changes in sea level. Nevertheless, the Adirondack and Ozark domes seem to have been continually emergent and probably were promontories analogous to Arnhem Land and Cape York Peninsula of modern Australia. Between the carbonate and terrigenous clastic facies of the Devonian sea, few late Middle or early Upper Devonian deposits are known. At least in the western New York portion of this region, this time interval is represented by a submarine unconformity (fig. 14) on which thin lenses of glauconitic skeletal sand have become mostly pyritized. This is analogous to the glauconitic sands scattered over the generally nondepositional surface on much of floor of the modern Arafura Sea. It is not known, however, how far westward the Devonian surface of submarine nondeposition can be traced, or if any part of the eastern Midcontinent region, such as the Cincinnati Arch, were definitely emergent at this time.

In light of the close analogy between sedimentation patterns in the modern, shallow marine shelf between Australia and New Guinea and at least one example of the marine epicontinental deposits that form so much of the stratigraphic record, this modern analog should be a very good laboratory for study of marine epicontinental sedimentation which is only poorly represented on a large scale in modern times. Detailed investigation apparently has just started, but includes published work on the Sahul Shelf by Van Andel and Veevers (1967) and unpublished research in progress on the Gulf of Carpentaria by C. V. G. Phipps of Sydney University (personal communication, 1969).

CONCLUDING COMMENT ON PALEOENVIRONMENTAL INTERPRETATION

In conclusion, much paleoenvironmental inference can be derived from placing known, relatively unambiguous criteria obtained from distinctive portions of a rock unit into their proper position in a sedimentation model. Then, completing stratigraphic environmental reconstructions, such as illustrated in Figures 11–15, is like filling in the last words in a crossword puzzle. Eventually, even the difficult words, or ambiguous facies fall into place if the control is close enough. However, in the case of environmental interpretation of the stratigraphic record, the solution is not necessarily fixed and unique. The problem is open-ended, and the solution merely need be reasonable to our presently accepted frame of reference, in these cases relatively well-tested sedimentation models. But we must take great care not to let the model become the ruling force in our thinking such that we ignore new and perhaps contradictory facts. Quite the contrary, we must continually expand our frame of reference by modifying or augmenting the model with other alternatives, if we see that it is necessary to do so, as we learn more about processes and products of sedimentation in modern environments.

REFERENCES

ALLEN, J. R. L., 1967, Depth indicators of clastic sequences: Marine Geol., v. 5, p. 429–446.
ANDREWS, J. E., SHEPARD, F. P., AND HURLEY, R. J., 1970, Great Bahama Canyon: Geol. Soc. America Bull., v. 81, p. 1061–1078.
BAGNOLD, R. A., 1941, The physics of blown sand dunes and desert dunes: Methuen and Co., London, 265 p.
BALL, M. M., 1967, Carbonate sand bodies of Florida and the Bahamas: Jour. Sed. Petrology, v. 37, p. 556–591.
BANDY, O. L., 1953, Ecology and paleoecology of some California Foraminifera: Jour. Paleont,. v. 27, p. 161–182; 200–203.
———, 1964, General correlation of foraminiferal structure with environment: *in* Approaches to Paleoecology, Imbrie, J., and Newell, N. D., ed., Wiley, p. 75–90.
BARNES, H., 1957, Nutrient elements: Geol. Soc. America Mem. 67, v. 1, p. 297–344.
BATHER, F. A., 1928, The fossil and its environment: Geol. Soc. London, Quart. Jour., v. 84, pt. 2, p. 61–98.
BATHURST, R. G. C., 1966, Boring algae, micrite envelopes and lithification of molluscan biosparites: Geol. Jour., v. 5, p. 15–32.
BAYER, F. M., 1957, Recent Octocorals: Geol. Soc. America Mem. 67, v. 1, p. 1105–1108.
BERNSTEIN, L., 1967, Plants and the supersaline habitat: Univ. Texas Marine Science Inst. Contr. Marine Science, v. 12, p. 242–248.
BRETSKY, P. W., 1970, Upper Ordovician ecology of the central Appalachians: Yale Univ. Peabody Mus. Nat. Hist. Bull. 34, 150 p.
BROMLEY, R. G., 1967, Marine phosphorites as depth indicators: Marine Geol., v. 5, p. 503–509.
BRONGERSMA-SANDERS, M., 1957, Mass mortality in the sea: Geol. Soc. America Mem. 67, v. 1, p. 941–1010.
———, 1966, Metals of Kupferschiefer supplied by normal sea water: Geol. Rundschau (for 1965), v. 55, no. 2, p. 365–375.
———, 1968, On the geographical association of strata-bound ore deposits with evaporites: Mineral. Deposita (Berl.), v. 3, p. 286–291.
BRUUN, A. F., 1957, Deep sea and abyssal depths: Geol. Soc. America Mem. 67, v. 1, p. 641–672.
BULMAN, O. M. B., 1955, Graptolithina: *in* Treatise on Invertebrate Paleontology, R. C. Moore, ed., Part V, 101 p.
———, 1957, Graptolites: Geol. Soc. America Mem. 67, v. 2, p. 987–992.
BUZAS, M. A., AND GIBSON, T. G., 1969, Species diversity: Benthonic Foraminifera in western North Atlantic: Science v. 163, p. 72–75.
CARPELAN, L. H., 1967, Invertebrates in relation to hypersaline habitats: Univ. Texas Marine Science Inst. Contr. Marine Science, v. 12, p. 219–229.
CARTWRIGHT, L. D., JR., 1928, Loss of color of red sandstone upon redeposition: Am. Assoc. Petroleum Geologists Bull., v. 12, p. 72–82.
CASTER, K. E., 1941, The Titusvilliidae; Paleozoic and Recent branching Hexactinellida: Palaeontographica Americana, v. 2, no. 12, p. 1–52 (471–522).
———, 1952, Concerning Enopleura of the Upper Ordovician (Ohio) and its relation to other carpoid Echinodermata: Bull. Am. Paleont., v. 34, no. 141, p. 1–56.
CLARK, T. H., AND STEARN, C. W., 1960, The geological evolution of North America: Ronald Press, New York, 434 p.
CLARKE, J. M., AND RUEDEMANN, R., 1912, The Eurypterida of New York: New York State Mus., Mem. 14, 439 p.
COLE, W. S., 1957, Foraminifera of the Cenozoic: Geol. Soc. America Mem. 67, v. 2, p. 757–762.
COOKE, C. W., 1957, Echinoids: Geol. Soc. America Mem. 67, v. 1, p. 1191–1192.
COOPER, G. A., 1930, Stratigraphy of the Hamilton Group of New York: Am. Jour. Sci., v. 19, p. 116–134, 214–236.
———, 1937, Brachiopod ecology and paleoecology: Rept. Committee on Paleoecology, p. 26–53. National Research Council 1936–1937.
——— AND WILLIAMS, J. S., 1935, Tully formation of New York: Geol. Soc. America Bull., v. 46, p. 781–868.
CRICKMAY, G. W., LADD, H. S., AND HOFFMEISTER, J. E., 1941, Shallow-water *Globigerina* sediments: Geol. Soc. America Bull., v. 52, p. 79–106.
DANGEARD, L. AND GIRESSE, P., 1968, Enseignements geologiques des photographies sous-marines: Bull. Bur. Rech. Geol. Min. Sec. 4, Geol. Gen. & div., no. 2, p. 1–85.
DAVID, T. W. E., 1950, The geology of the Commonwealth of Australia, v. 1; Edward Arnold and Co., London, 747 p.
DEGENS, E. T., 1965, Geochemistry of sediments—a brief survey: Prentice Hall, 342 p.
———, *et al.*, 1957, Environmental studies of Carboniferous sediments Part I: Geochemical criteria for differentiating marine from fresh-water shales: Am. Assoc. Petroleum Geologists Bull., v. 41, p. 2427–2455.
DE LAUBENFELS, M. W., 1955, Porifera: *in* Treatise on Invertebrate Paleontology, Moore, R. C., ed., part E., p. E21–E122.
———, 1957, Marine sponges: Geol. Soc. America Mem. 67, v. 1, p. 1083–1086.
DONOVAN, D. T., ed., 1968, Geology of Shelf Seas: Proc. 14th Inter-Univ. Geol. Congress, Oliver & Boyd, London, 160 p.
DORSEY, G. E., 1926, The origin of the color of redbeds: Jour. Geol., v. 34, p. 131–143.
DUNBAR, C. O., AND RODGERS, J., 1957, Principles of stratigraphy: Wiley, New York, 356 p.
DURHAM, J. W., 1947, Bathymetric distribution of gastropod genera (abs.): Geol. Soc. America, Bull., v. 58, p. 1260.
EAGAR, R. M. C., AND SPEARS, D. A., 1966, Boron content in relation to organic carbon and to paleosalinity in certain British Upper Carboniferous sediments: Nature, v. 209, p. 177–181.
EARDLEY, A. J., 1938, Sediments of Great Salt Lake, Utah: Am. Assoc. Petroleum Geologists Bull., v. 22, p. 1305–1411.

EASTLER, T. E., 1969, Sole marks in non-turbidite sequences (abs.): Geol. Soc. America—Northeastern Section Prog. Ann. Mtg. Albany, New York, p. 15.
EASTON, W. H., 1960, Invertebrate Paleontology, Harper & Brothers, New York, 701 p.
EICHER, D. L., 1969, Paleobathymetry of Cretaceous Greenhorn sea in Eastern Colorado: Am. Assoc. Petroleum Geologists Bull., v. 53, p. 1075–1090.
EKMAN, S., 1953, Zoogeography of the sea: Sedgwick and Jackson, London, 417 p.
ELIAS, M. K., 1937, Depth of deposition of the Big Blue (Late Paleozoic) sediments in Kansas: Geol. Soc. America Bull., v. 48, p. 403–432.
———, Depth of Late Paleozoic sea in Kansas and its megacyclic sedimentation: Kansas Geol. Surv. Bull. 169, v. 1, p. 86–106.
ELLISON, S. P. JR., 1957, Conodonts: Geol. Soc. America Mem. 67, v. 2, p. 993–994.
EMERY, K. O., 1956, Sediments and water of Persian Gulf: Am. Assoc. Petroleum Geologists Bull., v. 40, p. 2354–2383.
EMERY, K. O., STEVENSON, R. E., AND HEDGPETH, J. W., 1957, Estuaries and lagoons: Geol. Soc. America Mem. 67, v. 1, p. 673–750.
EMERY, K. O., TRACEY, J. I., AND LADD, H. S., 1954, Geology of Bikini and nearby atolls: Bikini and nearby atolls: Part 1, Geology: U. S. Geol. Survey Prof. Paper 260-A, 265 p.
EPSTEIN, C. M., 1969, Lithofacies and paleoenvironments of the Kalkberg Formation (Lower Devonian) of central New York, (abs.): Geol. Soc. America—Northeastern Section, Prog. Ann. Mtg., Albany, New York, p. 16–17.
EUGSTER, H. P., 1967, Hydrous sodium silicates from Lake Magadi, Kenya: precursors of bedded chert: Science, v. 157, p. 1177–1180.
———, 1969, Inorganic bedded cherts from the Magadi area, Kenya: Contr. Mineral. & Petrol., v. 22, p. 1–31.
EVANS, G., KINSMAN, D. J. J., AND SHEARMAN, D. J., 1964, A reconnaissance survey of the environment of recent carbonate sedimentation along the Trucial Coast, Persian Gulf: in Developments in Sedimentology, v. 1, Deltaic and Shallow Marine Deposits, L. M. J. U. van Straaten, ed., p. 185–192.
EVANS, J. K., 1967, Depositional environment of a Pennsylvanian black shale (Heebner) in Kansas and adjacent states: Unpub. doctoral dissertation, Rice University, 135 p.
———, 1968, Environment of deposition of a Pennsylvanian "black shale" (Heebner) in Kansas and adjacent states, (abs.): Geol. Soc. America Prog. Ann. Mtg., Mexico City, p. 92–93.
FAGERSTROM, J. A., 1964, Fossil communities in paleoecology: their recognition and significance: Geol. Soc. America Bull., v. 75, p. 1197–1216.
FAIRBRIDGE, R. W., 1950, Recent and Pleistocene coral reefs of Australia: Jour. Geol., v. 58, p. 330–401.
FISCHER, A. G., 1960, Latitudinal variation in organic diversity: Evolution, v. 14, p. 64–81.
FISHER, D. W., 1951, Marcasite fauna in the Ludlowville Formation of western New York: Jour. Paleont., v. 25, p. 365–371.
———, 1962, Small conoidal shells of uncertain affinities: in Treatise on Invertebrate Paleontology, Moore, R. C., ed., Part W, Miscellanea, p. W98–W143.
FLOWER, R. H., 1957, Nautiloids of the Paleozoic: Geol. Soc. America Mem. 67, v. 2, p. 829–852.
FOX, D. L., 1957, Particulate organic detritus: Geol. Soc. America Mem. 67, v. 1, p. 383–390.
FREEMAN, T., 1962, Quiet water oolites from Laguna Madre, Texas: Jour. Sed. Petrology, v. 32, p. 475–483.
FRIEDMAN, G. M., 1969, Recognizing tidal environments in carbonate rocks with particular reference to those of the Lower Paleozoics in the Northern Appalachians (abs.): Geol. Soc. America Northeastern Section, Prog. Ann. Mtg. Albany, New York, p. 20–21.
FUNNELL, B. M., 1967, Foraminifera and radiolaria as depth indicators in the marine environment: Marine Geol., v. 5, p. 333–347.
GARBELL, M. A., 1963, The sea that spills into a desert: Scientific American, v. 209, no. 2, p. 94–100.
GARRISON, R. E., AND FISCHER, A. G., 1969, Deep-water limestones and radiolarities of the Alpine Jurassic: Soc. Econ. Paleontologists and Mineralogists, Spec. Publ. 14, p. 20–54.
GEBELEIN, C. D., 1969, Distribution, morphology, and accretion rate of recent subtidal algal stromatolites, Bermuda: Jour. Sed. Petrology, v. 39, p. 49–69.
GINSBURG, R. N., 1956, Environmental relationships of grain size, and constituent particles of some south Florida carbonate sediments: Am. Assoc. Petroleum Geologists Bull., v. 40, p. 2384–2427.
GOLDMAN, M. I., 1922, Basal glauconite and phosphate beds: Science, v. 56, p. 171–173.
GOLDRING, R., 1964, Trace-fossils and the sedimentary surface in shallow-water marine sediments: in Developments in Sedimentology, v. 1, L. M. J. U. van Straaten, ed., p. 136–143.
GRABAU, A. W., 1931, The Permian of Mongolia: in Natural History of Central Asia, v. 4: Am. Mus. Nat. Hist., New York.
GRIFFITH, L. S., PITCHER, M. G., AND RICE, G. W., 1969, Quantitative environmental analysis of a Lower Cretaceous reef complex: Soc. Econ. Paleontologists and Mineralogists, Spec. Publ. 14, p. 120–137.
GUBER, A. L., 1969, Sedimentary phosphate method for estimating paleosalinities: a paleontological assumption: Science, v. 166, p. 744–746.
GUNTER, G., 1947, Paleoecological import of certain relationships of marine animals to salinity: Jour. Paleont., v. 21, p. 77–79.
———, 1950, Seasonal population changes and distributions as related to salinity of certain invertebrates of the Texas coast, including the commercial shrimp: Texas Univ. Inst. Marine Sci. Publ. v. 1, no. 2, p. 7–51.
HALLAM, A., 1967, An environmental study of the Upper Domerian and Lower Toarcian in Great Britain: Philos. Trans. Roy. Soc. London, Ser. B., Biol. Sci., v. 252, p. 393–445.
HAMBLIN, W. K., 1969, Marine paleocurrent directions in limestones of the Kansas City Group (Upper Pennsylvanian) in eastern Kansas: Kansas Geol. Surv. Bull. 194, pt. 2, 25 p.
HÄNTZSCHEL, W., 1962, Trace Fossils and Problematica: in Treatise on Invertebrate Paleontology, Moore, R. C., ed., Part W, Miscellanea, p. W177–W249.
HARMS, J. C., 1969, Hydraulic significance of some sand ripples: Geol. Soc. America Bull., v. 80, p. 363–396.

HASS, W. H., 1962, Conodonts: *in* Treatise on Invertebrate Paleontology, Moore, R. C., ed., Part W, Miscellanea, p. W3–W98.
HECKEL, P. H., 1965, Apparent structural control of Tully Limestone deposition in the Devonian Catskill delta complex of New York State (abs.): Geol. Soc. America Spec. Paper 87, p. 76.
———, 1966, Stratigraphy, petrography, and depositional environment of the Tully Limestone (Devonian) in New York State and adjacent region: unpub. doctoral dissertation, Rice University, 448 p.
———, Nature, origin, and significance of the Tully Limestone, an anomalous unit in the Catskill Delta, Devonian of New York: Geol. Soc. America Spec. Paper, 138, in press.
HECKEL, P. H., AND COCKE, J. M., 1969, Phylloid algal-mound complexes in outcropping Upper Pennsylvanian rocks of Mid-Continent: Am. Assoc. Petroleum Geologists Bull., v. 53, p. 1058–1074.
HEDGPETH, J. W., 1957, Classification of marine environments: Geol. Soc. America Mem. 67, v. 1, p. 17–28.
HEEZEN, B. C.. AND HOLLISTER, C., 1964, Deep-sea current evidence from abyssal sediments: Marine Geol., v. 1, p. 141–174.
HENBEST, L. G., 1963, Biology, mineralogy and diagenesis of some typical Late Paleozoic sedentary Foraminifera and algal-foraminiferal colonies: Cushman Found, Foram. Research, Spec. Publ. 6, 44 p.
HESSLER, R. R., AND SANDERS, H. L., 1967, Faunal diversity in the deep-sea: Deep Sea Res., v. 14, p. 65–78.
HILL, D., 1948, The distribution and sequence of Carboniferous coral faunas: Geol. Mag., v. 85, p. 122–148.
HOFFMEISTER, J. E., STOCKMAN, K. W., AND MULTER, H. G., 1967, Miami Limestone of Florida and its Recent Bahamian counterpart: Geol. Soc. America Bull., v. 78, p. 175–190.
HOLMES, R. W., 1957, Solar radiation, submarine daylight, and photosynthesis: Geol. Soc. America Mem. 67, v. 1, p. 109–128.
HULSEMANN, J., 1968, Morphology and origins of sedimentary structures on submarine slopes: Science, v. 161, no. 3836, p. 45–47.
ILLING, L. V., WELLS, A. J., AND TAYLOR, J. C. M., 1965, Penecontemporary dolomite in the Persian Gulf: Soc. Econ. Paleontologists and Mineralogists, Spec. Publ. 13, p. 89–111.
IMBRIE, J. AND BUCHANAN, H., 1965, Sedimentary structures in modern carbonate sands of the Bahamas: Soc. Econ. Paleontologists and Mineralogists, Spec. Publ. 12, p. 149–172.
IRWIN, M. L., 1965, General theory of epeiric clear water sedimentation: Am. Assoc. Petroleum Geologists Bull., v. 49, p. 445–459.
JOHNSON, J. H., 1961, Limestone-building algae and algal limestones: Colorado School of Mines Foundation. Johnson Publ. Co., Boulder, 295 p.
JOHNSON, R. G., 1960, Models and methods for analysis of the mode of formation of fossil assemblages: Geol. Soc. America Bull., v. 71, p. 1075–1086.
———, 1962, Interspecific associations in Pennsylvanian fossil assemblages: Jour. Geol., v. 70, p. 32–55.
KEULEGAN, G. H., AND KRUMBEIN, W. C., 1949, Stable configuration of bottom slope in a shallow sea and its bearing on geological processes: Am. Geophys. Union Trans., v. 30, p. 855–861.
KINSMAN, D. J. J., 1964a, Reef coral tolerance of high temperatures and salinities: Nature, v. 202, p. 1280–1282.
———, 1964b, The Recent carbonate sediments near Halat El Bahrani, Trucial Coast, Persian Gulf: *in* Developments in Sedimentology, L. M. J. U. van Straaten, ed., v. 1, p. 185–192.
———, 1966, Gypsum and anhydrite of recent age, Trucial Coast, Persian Gulf: *in* Second Symposium on Salt, J. L. Rau, ed., Northern Ohio Geol. Soc., v. 1, p. 302–326.
KJELLESWIG-WAERING, E. N., 1934, Note on a new eurypterid from the Moscow shales of New York: Am. Jour. Sci., ser. 5, v. 27, p. 386–387.
KOLBE, R. W., 1954, Diatoms from the equatorial Atlantic cores: Repts. Swedish Deep-Sea Expedition., v. 7, Sediment cores from the North Atlantic Ocean, Fasc. 3, p. 151–184.
KRUMBEIN, W. C., 1942, Criteria for subsurface recognition of unconformities: Am. Assoc. Petroleum Geologists Bull., v. 26, p. 36–62.
KRUMBEIN, W. C., AND SLOSS, L. L., 1956, Stratigraphy and sedimentation, W. H. Freeman, San Francisco, 497 p.
KUENEN, P. H., 1958, Turbidity currents as a major factor in flysch deposition: Eclogae Geol. Helvetiae, v. 51, p. 1009–1021.
KUENEN, P. H., AND MENARD, H. W., 1952, Turbidity currents, graded and non-graded deposits: Jour. Sed. Petrology, v. 22, p. 83–96.
LADD, H. S., 1951, Brackish-water and marine assemblages of the Texas coast, with special reference to molluscs: Texas Univ. Inst. Marine Science Publ., v. 2, no. 1, p. 125–163.
———, 1957, Paleoecological evidence: Geol. Soc. America Mem. 67, v. 2, p. 31–66.
LAPORTE, L. F., 1967, Carbonate deposition near mean sea-level and resultant facies mosaic: Manlius Formation (Lower Devonian) of New York State: Am. Assoc. Petroleum Geologists. Bull., v. 51, p. 73–101.
———, 1968, Recent carbonate environments and their paleoecologic implications: *in* Evolution and Environment, Drake, E. T., ed., Yale University Press, New Haven, p. 229–258.
———, 1969, Recognition of a transgressive carbonate sequence within an epeiric sea: Helderberg Group (Lower Devonian) of New York State: Soc. Econ. Paleontologists & Mineralogists, Spec. Publ. 14, p. 98–118.
LECOMPTE, M., 1956, Stromatoporoidea: *in* Treatise on Invertebrate Paleontology, Moore, R. C., ed., Part F, p. F107–144F.
LINEBACK, J. A., 1969, Illinois Basin sediment starved during Mississippian: Am. Assoc. Petroleum Geologists Bull., v. 53, p. 112–126.
LOGAN, B. W., REZAK, R., AND GINSBURG, R. N., 1964, Classification and environmental significance of algal stromatolites: Jour. Geol. v. 72, p. 68–83.
LOHMAN, K. E., 1957, Marine diatoms: Geol. Soc. America Mem. 67, v. 1, p. 1059–1068.
LOOMIS, F. B., 1903, The dwarf fauna of the pyrite layer at the horizon of the Tully Limestone in western New York: New York State Mus. Bull. 69, p. 892–920.
LOOSANOF V. L., 1945, Effects of sea water of reduced salinities upon starfish, *A. forbesi*, of Long Island Sound: Trans. Conn. Acad. Arts & Sci., v. 36, p. 813–835.
LOWENSTAM, H. A., AND MCCONNELL, D., 1968, Biologic precipitation of fluorite: Science, v. 162, p. 1496–1497.

MATTHEWS, R. K., 1966, Genesis of Recent lime mud in southern British Honduras: Jour. Sed. Petrology, v. 36, p. 428–454.
MCALESTER, A. L. AND RHOADS, D. C., 1967, Bivalves as bathymetric indicators: Marine Geol., v. 5, p. 383–388.
MCKEE, E. D., 1964, Inorganic sedimentary structures: *in* Approaches to Paleoecology, Imbrie, J., & Newell, N. D., ed., Wiley, New York, p. 275–295.
MENARD, H. W., 1952, Deep ripple marks in the sea: Jour. Sed. Petrology, v. 22, p. 3–9.
MERRETT, E. A., 1924, Fossil Ostracoda and their use in stratigraphical research: Geol. Mag., v. 61, p. 228–238.
MOORE, D. G., AND SCRUTON, P. C., 1957, Minor internal structures of some recent unconsolidated sediments: Am. Assoc. Petroleum Geologists Bull., v. 41, p. 2723–2751.
MOORE, R. C., 1929, Environment of Pennsylvanian life in North America: Am. Assoc. Petroleum Geologists Bull., v. 13, p. 459–487.
———, 1936, Stratigraphic classification of the Pennsylvanian rocks of Kansas: Kansas Geol. Surv. Bull. 22, 256 p.
———, 1950, Late Paleozoic cyclic sedimentation in central United States: Rept. 18th Internat. Geol. Cong., London, pt. 4, p. 5–16.
———, 1964, Paleoecological aspects of Kansas Pennsylvanian and Permian cyclothems: Kansas Geol. Surv. Bull. 169, v. 1, p. 287–380.
———, LALICKER, C. G., AND FISCHER, A. G., 1952, Invertebrate fossils, McGraw-Hill, New York. 766 p.
MORRIS, R. C., AND DICKEY, P. A., 1957, Modern evaporite deposition in Peru: Am. Assoc. Petroleum Geologists Bull., v. 41, p. 2467–2474.
MULLER, G., 1969, Sedimentary phosphate method for estimating paleosalinities: limited applicability: Science, v. 163, p. 812–813.
MYERS, E. H., AND COLE, W. S., 1957, Foraminifera: Geol. Soc. America. Mem. 67, v. 1, p. 1075–1082.
NELSON, B. W., 1967, Sedimentary phosphate method for estimating paleosalinities: Science, v. 158, p. 917–920.
NEWELL, N. D., IMBRIE, J., PURDY, E. G., AND THURBER, D. T., 1959, Organism communities and bottom facies, Great Bahama Bank: Am. Mus. Nat. Hist. Bull., v. 117, p. 117–228.
———, PURDY, E. G., AND IMBRIE, J., 1960, Bahamian oolitic sand: Jour. Geol., v. 68, p. 481–497.
NIINO, H. AND EMERY, K. O., 1961, Sediments of shallow portions of East China Sea and South China Sea: Geol. Soc. America Bull., v. 72, p. 731–762.
ODUM, H. T., CLANTON, J. E., AND KORNICKER, L. S., 1960, An organizational hierarchy postulate for the interpretation of species-individual distributions, species entropy, ecosytem evolution, and the meaning of a species-variety index: Ecology, v. 41, p. 395–399.
OKULITCH, V. J., 1943, North American Pleospongia: Geol. Soc. America Spec. Paper 48, 112 p.
OLSEN, S. 1944, Danish Charophyta, chronological, ecological and biological investigations: Det K. Danske Vidensk., Biol. Skr., bind 3, nr. 1.
OSBURN, R. C., 1957, Marine Bryozoa: Geol. Soc. America Mem. 67, v. 1. p. 1109–1112.
PARKER, F. L., 1948, Foraminifera of the continental shelf from the Gulf of Maine to Maryland: Bull. Mus. Comp. Zool., v. 100, p. 213–241.
PEARSE, A. S., 1928, On the ability of certain marine invertebrates to live in diluted sea water: Biol. Bull., v. 54, p. 405–409.
———, 1950, The emigrations of animals from the sea: Sherwood Press, Dryden, N. Y., 210 p.
———, AND GUNTER, G., 1957, Salinity: Geol. Soc. America, Mem. 67, v. 1, p. 129–158.
PETERSON, M. N. A., AND VON DER BORCH, C. C., 1965, Chert: modern inorganic deposition in a carbonate-precipitating locality: Science, v. 149, p. 1501–1503.
PEVEAR, D. R., 1966, The estuarine formation of United States Atlantic coastal plain phosphorite: Econ. Geol. v. 61, p. 251–256.
PHIPPS, C. V. G., 1966, Gulf of Carpentaria (Northern Australia): p. 316–324 *in* The Encyclopedia of Oceanography, R. W. Fairbridge, ed., Reinhold, New York, 1021 p.
PHLEGER, F. B., 1951, Ecology of Foraminifera, northwest Gulf of Mexico. Pt. 1. Foraminiferal distribution: Geol. Soc. America Mem. 46, 88 p.
———, AND EWING, G. C., 1962, Sedimentology and oceanography of coastal lagoons in Baja California: Geol. Soc. America Bull., v. 73, p. 145–182.
PICARD, M. D., AND HIGH, L. R., 1968, Sedimentary cycles in the Green River Formation (Eocene), Uinta Basin, Utah: Jour. Sed. Petrology, v. 38, p. 378–383.
PIKE, W. S., 1947, Intertonguing marine and nonmarine Upper Cretaceous deposits of New Mexico, Arizona, and southwestern Colorado: Geol. Soc. America Mem. 24, 103 p.
PLAYFORD, P. E. AND COCKBAIN, A. E., 1969, Algal stromatolites: deepwater forms in the Devonian of Western Australia: Science, v. 165, p. 1008–1010.
PLIČKA, M., 1968, *Zoophycos*, and a proposed classification of sabellid worms: Jour. Paleont., v. 42, p. 836–849.
PORRENGA, D. H., 1967, Glauconite and chamosite as depth indicators in the marine environment: Marine Geol. v. 5, p. 495–501.
POTTER, P. E., AND PETTIJOHN, F. J., 1963, Paleocurrents and basin analysis: Academic Press, New York, 326 p.
PRAY, L. C., 1968, Basin-sloping submarine (?) unconformities at margins of Paleozoic banks, west Texas and Alberta (abs.): Geol. Soc. America Prog. Ann. Mtgs. Mexico City, p. 243.
PRICE, N. B., 1967, Some geochemical observations on manganese-iron oxide nodules from different depth environments: Marine Geol. v. 5, p. 511–538.
PURDY, E. G., 1961, Bahamian oolite shoals: *in* Peterson, J. A. and Osmond, J. C. eds., Geometry of sandstone bodies: Am. Assoc. Petroleum Geologists, Tulsa, p. 53–62.
———, 1963, Recent calcium carbonate facies of the Great Bahama Bank: Jour. Geol., v. 71, p. 334–355, 472–497.
———, 1964, Sediments as substrates: *in* Approaches to Paleoecology, J. Imbrie & N. D. Newell, ed., Wiley, New York, p. 238–271.
RAYMOND, P. E., 1927, The significance of red color in sediments: Am. Jour. Sci., v. 240, p. 658–669.
REINECK, H.-E., 1969, Tidal flats (abs.): Am. Assoc. Petroleum Geologists Bull., v. 53, p. 737.

Rhoads, D. C., 1966, Missing fossils and paleoecology: Discovery, magazine of the Peabody Mus. Nat. His., v. 2, p. 19–22.
———, 1967, Biogenic reworking of intertidal and subtidal sediments in Barnstable Harbor and Buzzards Bay, Massachusetts: Jour. Geol., v. 75, p. 461–475.
———, AND WAAGE, K. M., 1969, Sediment control of faunal distribution patterns in Late Cretaceous marginal marine deposits of South Dakota (abs.): Am. Assoc. Petroleum Geologists Bull., v. 53, p. 737–738.
RICH, J. L., 1951, Three critical environments of deposition and criteria for recognition of rocks deposited in each of them: Geol. Soc. America Bull., v. 62, p. 1–20.
RICHARDS, F. A., 1957, Oxygen in the ocean: Geol. Soc. America Mem. 67, v. 1, p. 185–238.
RICHTER-BERNBURG, G., 1955, Über salinare sedimentation: Zeitschrift der deutschen Geologischen Gesellschaft, v. 105, p. 593–645.
RICKARD, L. V., 1962, Late Cayugan (Upper Silurian) and Helderbergian (Lower Devonian) stratigraphy in New York: N. Y. State Mus. & Sci. Service Bull. 386, 157 p.
———, 1964, Correlation of the Devonian rocks in New York State: New York State Mus. & Sci. Service, Geol. Surv., Map and Chart Series: no. 4.
RIEDEL, W. R., AND HOLM, E. A., 1957, Radiolaria: Geol. Soc. America, Mem. 67, v. 1, p. 1069–1072.
RODGERS, J., 1957, The distribution of marine carbonate sediments: a review: Soc. Econ. Paleontologists & Mineralogists, Spec. Publ. 5, p. 2–14.
ROEHL, P. O., 1967, Carbonate facies, Williston Basin and Bahamas: Am. Assoc. Petroleum Geologists Bull., v. 51, p. 1979–2032.
ROSSMAN, R., AND CALLENDER, E., 1968, Manganese nodules in Lake Michigan: Science, v. 162, p. 1123–1124.
RUDWICK, M. J. S., 1965, Ecology and paleoecology: *in* Moore, R. C. ed., Treatise on Invertebrate Paleontology. Part H. Brachiopoda, p. H199–H214.
RUEDEMANN, R., 1925, Siluric faunal facies in juxtaposition: Pan-Am. Geol., v. 44, p. 309–312.
RUSNAK, G. A., 1960, Some observations of Recent oolites: Jour. Sed. Petrology, v. 30, p. 471–480.
SCHENCK, H. G., 1928, The biostratigraphic aspect of micropaleontology: Jour. Paleont., v. 2, p. 158–165.
SCHMALZ, R. F., 1968, Formation of red beds in modern and ancient deserts: discussion: Geol. Soc. America Bull., v. 79, p. 277–280.
———, 1969, Deep-water evaporite deposition: a genetic model: Am. Assoc. Petroleum Geologists Bull., v. 53, p. 798–823.
SCHOPF, T. J. M., 1969, Paleoecology of ectoprocts (Bryozoans): Jour. Paleont., v. 43, p. 234–244.
SCOTT, G., 1940, Paleoecological factors controlling the distribution and mode of life of Cretaceous ammonoids in the Texas area: Jour. Paleont., v. 14, p. 299–323.
SCRUTON, P. C., 1953, Deposition of evaporites: Am. Assoc. Petroleum Geologists Bull., v. 37, p. 2498–2512.
SEGERSTRALE, S. G., 1957, Baltic Sea: Geol. Soc. America Mem. 67, v. 1, p. 751–802.
SEILACHER, A., 1964, Biogenic sedimentary structures: *in* Approaches to Paleoecology, Imbrie, J., and Newell, N., ed., Wiley, New York, p. 296–316.
———, 1967, Bathymetry of trace fossils: Marine Geol., v. 5, p. 413–428.
SHAW, A. B., 1964, Time in stratigraphy: McGraw-Hill, New York, 365 p.
SHEPARD, F. P., 1964, Criteria in modern sediments useful in recognizing ancient sedimentary environments: *in* Developments in sedimentology, v. 1, Deltaic and shallow marine deposits, L. M. J. U. van Straaten, ed., p. 1–25.
———, AND MOORE, D. G., 1955, Central Texas Coast sedimentation: characteristics of sedimentary environment, recent history, and diagenesis: Am. Assoc. Petroleum Geologists Bull., v. 39, p. 1463–1593.
SHINN, E. A., 1968a, Practical significance of birdseye structures in carbonate rocks: Jour. Sed. Petrology, v. 38, p. 215–233.
———, 1968b, Burrowing in Recent lime sediments of Florida and the Bahamas: Jour. Paleont., v. 42, p. 879–894.
———, GINSBURG, R. N., AND LLOYD, R. M., 1965, Recent supratidal dolomite from Andros Island, Bahamas: Soc. Econ. Paleontologists & Mineralogists, Spec. Publ. 13, p. 112–123.
SHROCK, R. R., AND TWENHOFEL, W. H., 1953, Principles of invertebrate paleontology: McGraw-Hill, New York, 816 p.
SLOSS, L. L., 1953, The significance of evaporites: Jour. Sed. Petrology, v. 23, p. 143–161.
———, DAPPLES, E. C., AND KRUMBEIN, W. C., 1960, Lithofacies maps: An atlas of the United States and southern Canada: Wiley, New York, 108 p.
SOHN, I. G., 1957, Ostracodes of the post-Paleozoic: Geol. Soc. America, Mem. 67, v. 2, p. 937–941.
SPIEKER, E. M., 1949, Sedimentary facies and associated diastrophism in the Upper Cretaceous of central and eastern Utah: Geol. Soc. America Mem. 39, p. 55–81.
STACH, L. W., 1936, Correlation of zoarial form with habitat: Jour. Geol. v. 44, p. 60–65.
STANLEY, D. J., 1968, Graded bedding-sole marking-graywacke assemblage and related sedimentary structures in some Carboniferous flood deposits, eastern Massachusetts: Geol. Soc. America Spec. Paper 106, p. 211–239.
STEVENS, C. H., 1966, Paleoecologic implications of early Permian fossil communities in eastern Nevada and western Utah: Geol. Soc. America Bull., v. 77, p. 1121–1130.
———, 1968, Variability of Pennsylvanian marine fossils correlated with depth and distance from shore (abs.): Geol. Soc. America Prog. Ann. Mtg., Mexico City, p. 291.
STRØM, K. M., 1939, Land-locked waters and the deposition of black muds: *in* Recent marine sediments, P. D. Trask, ed., Am. Assoc. Petroleum Geologists, Tulsa, p. 356–372.
SUTTON, R. G., 1951, Stratigraphy and structure of the Batavia Quadrangle: Proc. Rochester Acad. Science, v. 9, p. 348–408.
———, 1963, Correlation of Upper Devonian strata in south-central New York: Pennsylvania Geol. Surv., 4th ser., Bull. G 39., p. 87–102.
———, 1969, Sedimentary structures and their environmental significance in the marine Catskill Delta of New York (abs.): Geol. Soc. America-Northeastern Section, Prog. Ann. Mtg., Albany, New York, p. 58.

———, Bowen, Z. P., and McAlester, A. L., 1966, Multiple-approach environmental study of the Upper Devonian Sonyea Group of New York (abs.): Geol. Soc. America Prog. Ann. Mtg. San Francisco, p. 214.
———, Bowen, Z. P., and McAlester, A. L., 1970, Marine shelf environments of the Upper Devonian Sonyea Group of New York: Geol. Soc. America Bull., v. 81, p. 2975–2992.
Swinchatt, J. P., 1969, Algal boring: a possible depth indicator in carbonate rocks and sediments: Geol. Soc. America Bull., v. 80, p. 1391–1396.
Tasch, P., 1963, Fossil content of salt and associated evaporites: in Symposium on Salt, A. C. Bersticker, ed., Northern Ohio Geol. Soc. p. 96–102.
Taylor, W. R., 1954, Sketch of the character of the marine algal vegetation of the shores of the Gulf of Mexico: U. S. Fish and Wildlife Serv., Fishery Bull. 89, p. 177–189.
Teichert, C., 1943, The Devonian of Western Australia: Am. Jour. Sci., v. 241, p. 69–94, 167–184.
———, 1958, Cold- and deep-water coral banks: Am. Assoc. Petroleum Geologists Bull., v. 42, p. 1064–1082.
Tjia, H. D., 1966, Arafura Sea: p. 45–47, in The Encyclopedia of Oceanography, R. W. Fairbridge, ed., Reinhold, New York, 1021 p.
Tressler, W. L., 1957, Marine Ostracoda: Geol. Soc. America Mem. 67, v. 1, p. 1161–1164.
Troell, A. R., 1969, Depositional facies of Toronto Limestone Member (Oread Limestone, Pennsylvanian), subsurface marker unit in Kansas: Kansas Geol. Surv. Bull. 197, 29 p.
Trueman, E. R., 1964, Adaptive morphology in paleoecological interpretation: in Approaches to Paleoecology, Imbrie, J. & Newell, N. D., ed., John Wiley, New York, p. 45–74.
Twenhofel, W. H., 1936, Marine unconformities, marine conglomerates and thicknesses of strata: Am. Assoc. Petroleum Geologists Bull., v. 20, p. 677–703.
Van Andel, T. H., and Veevers, J. J., 1967, Morphology and sediments of the Timor Sea: Australia Bureau of Mineral Resources: Geol. & Geophys. Bull. 83.
Van Bemmelen, R. W., 1949, The Geology of Indonesia, v. 1A, General Geology: Gov. Printing Office: Martinus Nijhoff, agents, The Hague, 732 p.
Van Houten, F. B., 1968, Iron oxides in red beds: Geol. Soc. America Bull., v. 79, p. 399–416.
Vaughan, T. W., and Wells, J. W., 1943, Revision of the suborders, families, and genera of the Scleractinia: Geol. Soc. America Spec. Paper 44, 363 p.
Walker, C. T., and Price, N. B., 1963, Departure curves for computing paleosalinity from boron in illites and shales: Amer. Assoc. Petroleum Geologists Bull., v. 47, p. 833–841.
Walker, R. G., and Sutton, R. G., 1967, Quantitative analysis of turbidites in the Upper Devonian Sonyea Group, New York: Jour. Sed. Petrology, v. 37, p. 1012–1022.
Walker, T. R., 1959, Fossiliferous marine redbeds in Minturn Formation (Des Moines) near McCoy, Colorado: Am. Assoc. Petroleum Geologists Bull., v. 43, p. 1069–1071.
———, 1967, Formation of red beds in modern and ancient deserts: Geol. Soc. America Bull., v. 78, p. 353–368.
Walton, W. R., 1964, Recent foraminiferal ecology and paleoecology: in Approaches to Paleoecology, Imbrie, J., and Newell N. D., ed., Wiley, New York, p. 151–237.
Weller, J. M., 1957, Paleoecology of the Pennsylvanian Period in Illinois and adjacent states: Geol. Soc. America Mem. 67, v. 2, p. 325–364.
———, 1960, Stratigraphic principles and practice, Harper & Brothers, New York, 725 p.
Wells, J. W., 1944, Middle Devonian bone beds of Ohio: Geol. Soc. America Bull., v. 55, p. 273–302.
———, 1957a, Corals: Geol. Soc. America Mem. 67, v. 1, p. 1087–1104.
———, 1957b, Corals: Geol. Soc. America Mem. 67, v. 2, p. 773–782.
———, 1967, Corals as bathometers: Marine Geol., v. 5, p. 349–365.
Wilson, J. L., 1969, Microfacies and sedimentary structures in "deeper water" lime mudstones: Soc. Econ. Paleontologists & Mineralogists Spec. Publ. 14, p. 4–19.
Woodring, W. P., 1957, Marine Pleistocene of California: Geol. Soc. America Mem. 67, v. 2, p. 589–598.
Yochelson, E. L., 1961, The operculum and mode of life of Hyolithes: Jour. Paleont. v. 35, p. 152–161.
Zangerl, R., and Richardson, E. S., Jr., 1963, The paleoecological history of two Pennsylvanian black shales: Chicago Nat. Hist. Mus. Fieldiana: Geol. Mem., v. 4, 352 p.
Ziegler, A. M., 1965, Silurian marine communities and their environmental significance: Nature, v. 207. p. 270–272.
———, et al., 1968, The composition and structure of Lower Silurian marine communities: Lethaia, v. 1, p. 1–27.

APPENDICES

1. *Environmental Tolerances of Living Members of Preservable Organic Groups.*

Information in the following resume is summarized from the Treatise on Marine Ecology and Paleoecology (H. S. Ladd, ed. 1957, Geological Society of America Memoir 67), the Treatise on Invertebrate Paleontology (R. C. Moore, ed.) and textbooks on invertebrate paleontology (e.g. Moore, Lalicker and Fischer, 1952; Shrock and Twenhofel, 1953; Easton, 1960). Reference citation to these sources is kept minimal.

Calcareous algae.—Calcareous algae comprise a diverse group of carbonate-secreting and -binding aquatic plants which contribute much material to modern sediments and were responsible for a substantial amount of rock and fossils in the stratigraphic record. (See Johnson, 1961.) Certain tiny forms bore into carbonate skeletal debris and leave the grains with characteristic rotten-appearing dark edges that may be responsible for "micrite envelopes" noticable on some carbonate grains in the record (Bathurst 1966; Swinchatt, 1969.) The single most important factor controlling distribution of algae is sunlight, which is necessary for their metabolic process of photosynthesis. Therefore algae live only in the upper, well-lit photic zone of water bodies. Depth of light penetration is reduced by turbidity, and luxurance of algal growth is inversely proportional to turbidity of the water. Under the best conditions in the clearest ocean water, only about 10 percent of the total incident light energy from a zenith sun penetrates to 110 feet, and less than 1 percent penetrates to 300 feet (Holmes, 1957). As a result most algae are restricted to depths of less than 200 feet. Abundant algal boring seems restricted to depths of less than 120 feet (Swinchatt, 1969). Although calcareous algae generally are marine, some forms are adapted to salinities ranging from fresh water to salt lakes. Algae live in waters of all temperatures, but are most diverse in tropical waters. Most algae need a stable substrate for attachment.

Red algae (Rhodophyta) are represented mainly by the family Corallinaceae and include a variety of encrusting massive to digitate types and bushy, articulated forms. Encrusting forms require a hard surface such as rock, coral and shells. Some bladelike forms grow free on sand or mud substrates. Growth form can be correlated to some extent with microenvironment. Red algae grow most profusely in the marine sublittoral zone just below low tide level where light is strong and water agitation at least moderate. Red algae have been found living at depths as great as 450 feet in the tropical Pacific, but decrease in size and abundance below about 70 feet. Although they are most diverse in tropical seas, some genera grow abundantly and form extensive banks in temperate and even polar waters. Red algae show less tolerance to salinity changes than do other types of calcareous algae, and the overwhelming majority are exclusively marine, although a few fresh water types are known. Marine forms seem to tolerate some freshening of water near estuaries or after heavy rainfall in shallow places, and some species of the subfamily Melobesieae survive salinities appreciably above that of normal sea water for a short time. Salinity tolerance range for red algae is 18 to 54 ppt, but the range that can be tolerated for longer times, or within which growth could be maintained, would be substantially narrower (Bernstein, 1967).

Green algae (Chlorophyta) include two carbonate-secreting families, Codiaceae and Dasycladaceae, which are mainly segmented and unsegmented, erect branching plants. Members of both groups live attached to stable sandy or muddy substrate by a blade or central stem. Some codiaceans are crustose or nodular. All modern dasycladaceans and most codiaceans are restricted to shallow warm tropical waters where certain genera grow luxuriantly and form "meadows" in at least slightly agitated water from below low-tide level to depths of 180 feet. Within this range different taxa exhibit different optimum depths. Although many noncalcified green algae live in fresh water, codiaceans and dasycladaceans are exclusively marine. They seem to tolerate lower salinities and salinity variation to a greater extent than red algae, however, as Ginsburg (1956) found green algae extending farther away from the normal marine reef tract into Florida Bay where salinity variation is greater. Likewise, several types of green algae, but no reds, occur in brackish bays and lagoons along the Texas coast (Taylor, 1954).

Charophytes, classified by some in the Chlorophyta, are bushy branching plants in which the distinctive female oogonia become strongly calcified and are preserved as fossils. Most modern charophytes are restricted to fresh water; a few species have a wider salinity tolerance and inhabit brackish waters up to 18 ppt and possibly to 26 ppt (Olsen, 1944). Except for scattered specimens that may be carried down estuaries into the sea, charophyte oogonia are indicative of nonmarine environments where locally they may be sufficiently abundant to form freshwater limestones.

Blue-green algae (variously termed Myxophyceae, Schizophyta, Cyanophyta) occur in a great variety of environments from fresh water to marine to hypersaline. They are especially well developed in fresh water and inhabit the range from cold arctic lakes to hot springs where they influence formation of calcareous tufa and travertine. Blue-green algae inhabiting marine environments are primarily unicellular filamentous forms which, along with chlorophytes of similar nature, form mucilaginous mats in low-lying, broad intertidal and supratidal flats. These mats can withstand extreme salinity variation and periodic long-term desiccation. Fine sediment, usually carbonate, that is carried by tides periodically through these environments, adheres to the sticky mats and eventually forms distinctive laminated deposits known as algal stromatolites which are organism-induced sedimentary structures with no involvement of skeletal secretion. Various forms of Recent stromatolites have environmental significance (Logan, Rezak and Ginsburg, 1964). In general, flat to crinkly laminae locally exhibiting desiccation cracks or small connected domes are characteristic of the supratidal or high intertidal zone; discrete laminated domes, or "cabbage heads" termed *Cryptozoon* and *Collenia*, indicate an intertidal environment with a substantial tidal range and continual drainage between the heads. Stromatolites developed around discrete loose nuclei such as mud chips or shell fragments form spheroidal or ellipsoidal, laminated algal balls termed oncolites, which occur in the low intertidal to subtidal zone where water agitation is more constant. In the Upper Paleozoic, a variety of similar laminated structures form uniform coatings around small sand-size grains (*Osagia*) to uneven, irregular partial coatings around typically larger grains and sometimes extending onto the adjacent sediment surface (*Ottonosia*) often involve not only blue-green algae, but also encrusting foraminifers (Henbest, 1963), and occasionally red algae and other types of encrusting organisms. Minute tubules (*Girvanella*) representing secretions around blue-green algal filaments sometimes are preserved in these coatings. Because these coated grains depend on algal activity often in conjunction with water agitation, they indicate the shallow subtidal environment. Association of other less tolerant organisms indicates more nearly normal marine salinity. Although blue-green algae today seem rare in the marine regimen outside of marginal to shallow subtidal environments, some Paleozoic stromatolites apparently lived to depths of about 150 ft (Playford and Cockbain, 1969).

Protistid microfossils.—Several groups of unicellular microscopic organisms, classified as protistids because of difficulty of assignment as either plant or animal, secrete tests that often are preserved as fossils. Most forms are planktonic and therefore achieve wide distribution. Some groups provide much of the sediment in deep oceans today.

Radiolarians are exclusively marine planktonic organisms that secrete siliceous tests. They live from surface waters to great depth, and abandoned tests sink to the bottom and form extensive deposits in modern oceans where undiluted by other sediment (Riedel and Holm, 1957). Radiolarian ooze is most common at depths greater than 12,000 feet because planktonic carbonate material generally dissolves at these depths and no longer masks the siliceous sediment. (See Garrison and Fischer, 1969.) Radiolarians, however, also occur in shallow marine deposits including beach sands, but in these situations they are largely masked by other material.

Silicoflagellates also secrete siliceous tests, and as they too are exlusively marine and planktonic, they commonly are found associated with radiolarians.

Coccoliths are minute calcite tests secreted by organisms classified by some as golden-brown algae. They are planktonic and predominantly marine. Although living only in the photic zone, coccoliths settle into oceanic depths of generally less than 12,000 feet where they constitute a fair proportion of calcareous oozes.

Diatoms are phytoplankton that secrete minute siliceous tests and live in the photic zone throughout marine, brackish, and fresh waters. Different genera and species of diatoms exhibit definite environmental restriction related to salinity, and because many living species occur in Tertiary sediments, diatom assemblages are useful for differentiating marine, brackish and fresh-water environments in the Cenozoic (Lohman, 1957).

Foraminifera.—Most foraminifers inhabit marine environments, but some live in brackish or hypersaline water, and one family inhabits fresh water (Shrock and Twenhofel, 1953). Most also are benthonic and browse the substrate for food. Some encrust shells or vegetation and can be recognized by presence of one flattened side of the test.

Globigerinids and related genera are planktonic and populate the open oceans in great numbers. Perhaps 35 percent of the present ocean bottoms are covered with oozes that consist mainly of globigerinid foraminifers and contain appreciable numbers of tests of other planktonic organisms. Although concentrations of planktonic organic remains are characteristic of deep oceanic sediments, globigerinid foraminifera can be carried into and deposited in shallow water (Crickmay, Ladd and Hoffmeister, 1941). Because they do not survive for long in shallow seas (Myers and Cole, 1957), the number of globigerinids in the sediment decreases markedly as shoreline is approached (see Walton, 1964), and their occurrence in nearshore sediments normally is greatly overshadowed by shallow-water forms.

Benthonic foraminifers exhibit definite patterns of distribution that can be correlated with salinity, temperature and depth, and this information has been applied to environmental interpretation of Tertiary rocks. (See Cole, 1957; Walton, 1964; Funnell, 1967.) It is not known how much of the apparent correlation with depth reflects pressure, which is directly related to depth, or some factor not directly dependent on water depth. Several studies suggest that "depth biofacies" may be controlled mainly by temperature which, although generally decreasing with depth, is still an independent variable (Parker, 1948; Phleger, 1951; Bandy, 1953). Nevertheless, depth zonation is very distinct in the Java Sea where there is little temperature difference with depth (Myers and Cole, 1957). General effects of environment on foraminiferal morphology have been summarized by Bandy (1964).

Walton's (1964) study of modern foraminiferal distribution in the northeastern Gulf of Mexico shows that in nearshore marginal environments, the foraminiferal fauna is dominated by simple agglutinated forms. On the least marine fringes of these environments, thecamoebinids (small agglutinated organisms that may not be true foraminifers) take over dominance and populate fresh-water environments. The few calcareous foraminifers that occur in marginal environments tend to be smaller and thinner-shelled than their counterparts in the offshore open marine environment, perhaps because of increased difficulty of calcium carbonate precipitation in hyposaline waters. They also possess preservable chitinous inner linings which are lacking in their offshore counterparts. The region transitional to normal marine water is dominated by a limited number of calcareous genera. Shallow but open marine environments extending to about 600 feet in depth exhibit several faunal associations dominated by a greater number of calcareous forms. Arenaceous forms that are present are subordinate and possess a more complex internal structure than those occurring in marginal environments. Open marine environments deeper than 600 feet are dominated by different calcareous genera and increasing numbers of planktonic forms.

Sponges.—Sponges live attached to a solid substrate in areas that experience slight to moderate water agitation. They cannot tolerate covering by silt and therefore flourish mainly in areas of slow deposition. Because of these ecologic restrictions, sponges typically exhibit only local distribution and abundance in otherwise favorable regimens. Of the three classes of sponges, two are restricted to rather definite environmental regimens (deLaubenfels, 1957). Calcisponges, which secrete only calcareous spicules and spicular skeletons, live only in normal marine shallow water less than 350 feet deep and are most abundant at depths less than 30 feet. Hyalosponges secrete only siliceous spicules and skeletons and live only in deeper marine water, most abundantly below 600 feet (Caster, 1941), but extending upward locally to less than 300 feet where sheltered underneath permanent ice. Demosponges secrete siliceous spicules and/or spongin which is not preserved in the record. Most live in marine waters of all depths, but a few live in intertidal, brackish and even fresh water. Those that live in fresh water secrete spicules similar to some occuring in the hyalosponges (deLaubenfels, 1955).

Corals.—All hydrozoan, anthozoan and alcyonarian coelenterates that have calcareous skeletons are considered corals because ecologically they are similar (Wells, 1957a). All corals are strictly marine and, although tolerating salinities from 27 to 48 ppt, find their optimum from 34 to 36 ppt. Corals do not tolerate influxes of fresh water for even short periods to time. The few hydrozoans that live in fresh water do not secrete calcareous skeletons. Because they are sessile but feed on small nektonic or suspended organisms, corals need consistent water movement and good circulation. Few corals can withstand much turbidity of water. Coral growth form is variable and is strongly influenced by intensity of water agitation in the microenvironment. (See Vaughan and Wells 1943, p. 61.) Ecologically, corals may be grouped into hermatypic and ahermatypic forms.

Ahermatypic corals include many anthozoans and all hydrozoans. Most are relatively small solitary forms. Those that are colonial produce small encrusting to massive or ramose colonies in which corallites are relatively widely spaced and number of corallites per area of colony is low. Colonies require hard substrate for attachment, but many solitary forms can live on soft substrates. Ahermatypic corals tolerate a wide range of temperatures and live at all depths attaining their best development between 600 and 1800 feet along margins of the continental shelves where a few species form deep water coral banks (Teichert, 1958). Different genera have different depth and temperature ranges, and fossil assemblages can be used to suggest depositional depth of Cenozoic units (Wells, 1967). Most alcyonarians are considered ecologically as ahermatypic, but some reef-dwelling forms are known (Bayer, 1957).

Hermatypic or *reef corals* include all anthozoan hexacorals that possess in their tissue symbiotic unicellular algae (zooxanthellae), which provide an efficient mechanism for absorbing metabolic waste products as well as aid in calcification, thereby promoting rapid and profuse coral growth. Because of this, hermatypic corals can form very large colonies with great numbers of closely-packed corallites per unit area. As a result of this dependence on the symbiotic algae, however, hermatypic corals are restricted to the photic zone and attain their best development in depths less than 60 feet. These corals also require high temperatures and are not found outside tropical waters. Because of the great size attainable by hermatypic corals, they form the solid framework of modern reefs in association with red algae that function primarily as frame and sediment binders. This sort of shallow water tropical reef is distinguished by great faunal diversity as well as by the two types of algae (only one preservable) which are responsible for their environmental restriction.

Most *Paleozoic corals* probably are more comparable to modern ahermatypic types because of the similar preponderance of solitary forms and generally small size of many colonial forms. This does not necessarily mean that their maximum development occurred at greater depths, however, because modern ahermatypic forms are relegated there to a large extent because of superior competition from faster-growing hermatypic forms in shallow water. It is not known exactly when the symbiotic relationship between hermatypic hexacorals and the zooxanthellae was established (Wells, 1957b), but the early development of hexacoral reefs suggests origin at least in the early Mesozoic. Although they probably belong to a different lineage, it is possible that some of the Paleozoic colonial corals with closely packed corallites had an equivalent relationship with similar algae and thereby were similarly restricted to shallow marine environments. Some Silurian and Devonian reefs show evidence of living in warm shallow water (Wells, 1967). Even so, Paleozoic colonial corals did not commonly inhabit highly agitated reef environments like their modern hexacoral counterparts (Wells, 1957b). Instead, the now-extinct stromatoporoids apparently filled most of the ecologic niches involved in building potentially wave resistant structures. Nevertheless, distinct Carboniferous coral assemblages in correlation with lithology seem to define various environments ranging from reeflike conditions and clear shallow seas to murky, perhaps deeper waters (Hill, 1948), which is not unlike the environmental range of modern corals.

Worms.—Only polychaete annelids produce hard parts that are preserved commonly in the stratigraphic record. Serpulid and spirorbid polychaetes that form calcareous tubes live primarily in marine environments, but some forms inhabit brackish and even fresh water (Shrock and Twenhofel, 1953). Tube-secreting worms need a firm substrate for attachment and often encrust other shells. They live through a large range of depths but seem more common in shallow water where serpulids locally grow attached to one another so profusely that they form small reefs. Burrows and trails are formed in sediment by several types of worms. (See text on sedimentary structures.)

Bryozoans.—Members of this group that secrete calcareous skeletons are restricted to marine environments. They have greater tolerance of brackish water than do corals, and a few genera withstand salinities less than 20 ppt (Osburn, 1957). Bryozoans occur through a wide range of temperatures and depths but the great majority of genera live in warmer waters less than 1200 feet deep. As sessile colonial organisms, bryozoans require a firm substrate for attachment and normally do not live on unstable sand or mud substrates. They often encrust shells of other organisms and also commonly are attached to floating seaweed. Because they are suspension feeders, bryozoans prefer clear, at least slightly agitated water, and cannot withstand much turbidity. Both substrate requirement and low tolerance for turbidity inhibit their development in areas of rapid deposition. Bryozoan colonies possess forms ranging from encrusting to massive, ramose, flabellate, and fenestrate, and type of zoarial form has been correlated with depth and degree of agitation in the environment (Stach, 1936; Schopf, 1969).

Brachiopods.—All living forms of both inarticulate and articulate brachiopods are marine, sessile, benthonic suspension feeders, and these general characteristics apparently obtained throughout the history of the phylum (Rudwick, 1965). All living brachiopods are intolerant of lower salinities; even *Lingula*, which lives in the intertidal zone, closes tightly and suspends life activities during influxes of brackish or fresh water, and there is no evidence that any fossil brachiopod has become adapted to a truly nonmarine environment. Modern brachiopods tolerate wide ranges of temperature and depth both as a group and in many cases as individual genera. One of the few living exceptions is *Lingula* which seems restricted to depths less than 120 feet.

Brachiopods live on either hard or soft substrate. A few cement their shells to a hard object, but most are attached by their pedicle to something firm. This can be a rock substrate, or a shell or firm noncalcified organism lying on a soft sand or mud bottom. Some brachiopods can anchor the pedicle directly into a soft substrate. Strophomenoids generally lost the pedicle and lay free on the bottom with modifications of either broad shape or spines (productids) to keep the organism from sinking into the mud. Possessing mobility on account of their extendable pedicle, some lingulids live in burrows. Because brachiopods can attach to firm but noncalcifed objects that normally would not be preserved, certain lingulids, acrotretids and rhynchonellids without other benthonic biota in black shales that represent foul substrates probably were epiplanktonic on floating vegetation (Rudwick, 1965).

Although as suspension feeders, some brachiopods prefer clear water, many forms can tolerate a large amount of turbidity because of high development of their feeding-rejection mechanism or merely the capability of closing their shells for a time (Rudwick, 1965). As sessile organisms, however, they cannot tolerate rapid accumulation of sediment, except for *Lingula* which has a pedicle capable of moving it up through the mud. Outside of those forms (craniids, oldhaminids) cemented to the substrate, or the ponderous pentamerids, few brachiopods can withstand intense water turbulence. Although all brachiopods require some movement of water for feeding, their ability to set up strong, individual feeding currents would allow them to thrive in very quiet environments unfavorable to other suspension feeders. Much of the variation in shell morphology can be related to response to problems of attachment, feeding, and stabilization in the microenvironment (Cooper, 1937).

Echinoderms.—All echinoderms are exclusively marine organisms, but some of the more mobile forms occasionally invade lagoons and estuaries where salinities are brackish (Cooke, 1957). Although certain starfish can withstand water of only 3 ppt salinity for a very short time, they die in sustained salinities of less than 20 ppt (Loosanof, 1945). All living groups of echinoderms exist through a wide range of temperatures and inhabit depths from shoreline to 34,000 feet.

Pelmatozoans are represented today only by crinoids which are suspension feeders living attached to the substrate by a stalk, at least in early life stages. Many modern crinoids leave their place of attachment to become free-swimming adults. Crinoids that remain stalked require a firm substrate, although a few in deeper water possess long slender cirri that stabilize attachment in a muddy substrate. The extinct blastoids and cystoids probably lived in a manner similar to stalked crinoids, thus would have required clear water with good circulation as well as a firm substrate. Although most stalked crinoids now inhabit deeper water, it is likely that they and their extinct relatives formerly were well distributed through shallow water as well. Of two minor pelmatozoan groups, carpoids may have been vagrant bottom-feeders (Caster, 1952), and edrioasteroids lived mainly attached to other objects. Some small pelmatozoans may have attached to floating seaweed (Bather, 1928).

Eleutherozoans are vagrant benthonic forms well represented in modern seas. Starfish are voracious predators that are viable on any substrate. A starfish and an echinoid are the only skeletal forms, other than a mobile pelecypod upon which the starfish feeds, which inhabit the extremely unstable substrate of modern oolite shoals (Newell *et al.*, 1959). Ophiuroids and holothurians are mainly deposit feeders and inhabit substrates rich in organic matter. Echinoids are scavengers as well as predators and seem to prefer clear over muddy water. Whereas regular (round) echinoids occur at all depths, greatly flattened forms live mainly in shallow water (Cooke, 1957).

Molluscs.—This group comprises a large variety of organisms that have radiated into many different environments. Three small primitive groups are exclusively marine. *Monoplacophorans* are found today only in deep water, but probably inhabited shallower environments in the past. *Chitons* are bottom feeders that live mainly on rocky surfaces. *Scaphopods* are suspension feeders that live partially submerged in soft sediment. Chitons and scaphopods live at all depths from shoreline and shoals to several thousand feet.

Cephalopods living today also are exclusively marine, although squids will enter bays and estuaries where salinities range from 20 to 30 ppt (Gunter, 1950). There is no evidence in faunal associations that any type of cephalopod now extinct was other than marine. Modern cephalopods are nektonic or nektoplanktonic carnivores and scavengers in shallow to relatively deep waters and therefore are little influenced by substrate. Among the great variety of ancient cephalopods, however, nautiloids ranged from planktonic to nektobenthonic and strictly benthonic (Flower, 1957). Ammonoids seem to have been more generally nektonic or nektobenthonic, but display bathymetric zonation of genera in the Cretaceous of Texas (Scott, 1940). Living cephalopods with only an internal shell (coleoids) are widespread in modern seas, but *Nautilus*, the only living cephalopod with an external shell, is confined to the tropical Pacific. It is unknown if the multitude of extinct forms with external shells (nautiloids and ammonoids), were similarly restricted to warm waters, but if so, flotation of the gas-filled shell after death until waterlogged might distribute them widely to alien environments. Some workers, however, feel that flotation of shells in many cases was negligible, and Teichert (1943) shows that goniatites apparently inhabited only an offshore environment avoiding both the coast and barrier reef in the Devonian of Western Australia.

Pelecypods inhabit marine, brackish and fresh water with equally great numbers of individuals, although many more genera are strictly marine than strictly nonmarine. Within this range of adaptation, specific genera inhabit specific environments of different salinities. Unfortunately, there appears to be no major functional morphologic differences inherent in the shell that adapts them to different salinities. Therefore ancient environments cannot be deduced from shell morphology alone outside of trusting occurrences of certain genera of known environment back into the record. As a group, pelecypods are mainly benthonic and inhabit all depths from intertidal to at least 34,000 feet. They also exist through a wide range of temperatures. Pelecypods range from filter feeders to sediment ingesters and scavengers and have radiated into perhaps a broader set of substrate environments than any other benthonic group. Most pelecypods are mobile clams that crawl across or burrow through soft substrate. Burrowing clams are adapted for their way of life by possessing a retractable siphon. Fossil burrowers can be distinguished from nonburrowers by presence of a deep pallial sinus (indentation in the mantle attachment line) in the interior of the shell, and by presence of a gape between an end of the valves when closed; both features are adaptations to accommodate the retractable siphon (Trueman, 1964). Burrowers and many bottom feeders can withstand turbidity, and some burrowers can live in areas of rapid deposition. Some clams bore into rock for protection, and others that have developed diminutive valves bore into and ingest wood. All these types of mobile clams have two well-developed adductor muscle scars in each valve. Other pelecypods are adapted for a sessile existence; these are suspension feeders that require cleaner waters than burrowers and cannot tolerate rapid deposition. Mussels (mytilids) are attached to a firm substrate by the threadlike organic byssus. Many forms are attached only when young, and a few such as the scallop (*Pecten*) become free-swimming in adulthood. A few groups cement one valve to a hard substrate; oysters cement to one another and locally form extensive reefs in brackish bays today, and the extinct rudistids accomplished similar feats in marine environments in the Cretaceous. Forms cemented to a hard substrate have developed valves of unequal size with the large attached valve on the bottom and the small movable valve above serving as a sort of operculum. In many types of attached sessile pelecypods, such as oysters and mussels, one adductor muscle is lacking or greatly reduced, and these attached forms are distinguished from mobile forms by having only one major muscle scar in each valve (Trueman, 1964).

Gastropods have adapted to terrestrial life as well as the entire range of aquatic habits. In the sea, they range from high tide level to nearly 18,000 feet in depth. Most genera inhabit shallow marine waters less than 600 feet deep, but there is a secondary concentration at about 3500 to 4500 feet (Durham, 1947). As with pelecypods, certain genera of gastropods have been used as indicators of environments with specific salinity and depth ranges in the Cenozoic. Most gastropods are benthonic snails that move slowly over both hard and soft substrate; some forms burrow in mud or sand. Certain genera are known to inhabit certain types of substrates, but the only such forms that are recognized easily are the cap-shaped limpets which live mainly clinging to hard rock. Most snails are bottom-feeders that consume plants or decomposing organic matter, and a few are carnivores that prey on clams and oysters. Thus, mobile gastropods are not as strongly affected by turbid water or rapid deposition as sessile suspension feeders would be.

Pteropods are a group of pelagic gastropods that locally are abundant on the surface of the open sea. Some pteropods secrete tiny cone-shaped shells that sink and accumulate in substantial amounts to form pteropod ooze at places on the deep ocean floor.

Arthropods.—These also comprise a large variety of organisms that have become adapted to many different environments. Aside from the extinct trilobites and eurypterids (discussed later), only crustaceans have left much of a fossil record.

Ostracodes are minute, calcareous, bivalved crustaceans that inhabit all aquatic environments. Most genera are marine but many are euryhaline and inhabit brackish or hypersaline lagoons, and some inhabit fresh water. Because many genera are confined to specific ranges of salinity tolerance, they are useful for paleoenvironmental interpretation back through the Mesozoic when many of the extant genera arose. (See Sohn, 1957.) As a group, ostracodes live through a wide range of temperatures and exist from shallow pools to marine depths of 14,000 feet. Deeper water forms tend to be thin-shelled and unornamented. Ostracodes are primarily swimmers and crawlers, but a few are planktonic.

Some are suspension feeders, others are bottom feeders, and many are voracious scavengers. Certain genera prefer different kinds of substrate; some live clinging to seaweed and therefore are independent of substrate (Tressler, 1957).

Barnacles are exclusively marine crustaceans that live attached to a hard surface such as rocks or shells. They are suspension feeders and seem to prefer turbulent water. Although most abundant in the intertidal zone, some live to depths of 12,000 feet. Some attach to floating logs and thus could be deposited on soft substrates. Because the shell consists of several pieces that might become disarticulated upon settling in such a case, however, complete fossil barnacles would most likely imply a hard substrate.

Branchiopods are small mobile, calcareochitinous, mostly bivalved crustaceans many of which resemble tiny clams. A few modern forms (some cladocerans) are marine, but most forms (conchostracans and other cladocerans) inhabit fresh or brackish water environments, and a few forms (anostracans or brine shrimp) live in highly saline inland waters. Thus their potential significance for suggesting such restricted ancient environments far outweighs their scattered fossil record.

Malacostracan crustaceans include the common crabs, shrimps, lobsters, and crayfish and live throughout the salinity range from marine to fresh water. Fossils usually are rare, and only forms related to modern genera of restricted tolerance are useful in paleoenvironmental interpretation.

Merostomes are aquatic chelicerates (arachnoids) of which the only living representative is the xiphosuran *Limulus*, the horse-shoe crab. Today, *Limulus* is a bottom feeder that lives on soft sandy substrates in shallow marine water. Experimentally, however, the organism can survive at salinities as low as 9 ppt for several weeks (Pearse, 1928), and thus exhibits the wide salinity tolerance characteristic of many arthropods. Many ancient merostomes are determined from their biotic associates to have inhabited brackish or fresh-water environments.

Other major arthropod groups such as spiders, scorpions, insects, centipedes and millipedes are today all nonmarine air-breathing organisms. Their fossil record is scattered and mostly the result of preservation under special conditions, such as in tar pits, amber, tree-trunk fillings, and volcanic ash falls in fresh-water lakes. Few such organisms are preserved after transport into a marine environment.

2. *Inferred Environments of Habitat of Extinct Fossil Groups.*

Most information in the following resume was obtained from the same general sources as Appendix 1.

Archaeocyathids.—Exclusively Cambrian, vase-shaped, carbonate-secreting archaeocyathids resemble both calcisponges and corals and lived in association with brachiopods, gastropods, trilobites and algae (Okulitch, 1943). As sessile benthonic organisms, archaeocyathids probably were suspension feeders and therefore required relatively clear water with slow deposition as well as attachment to a firm substrate.

Stromatoporoids.—These hydrozoan- and sponge-like organisms inhabited reefy environments where they exhibit many different growth forms from encrusting to ramose, domal and massive. Although occurring with most major groups, stromatoporoids lived particularly in close association with corals and algae in the Paleozoic and with hexacorals and rudistid clams in the Mesozoic (LeCompte, 1956). As sessile suspension feeders they probably required attachment to a firm substrate, low turbidity and slow deposition.

Hyolithids.—The Paleozoic hyolithids were probably virtually sedentary, though unattached, benthonic organisms (Yochelson, 1961) that lived associated with most major organic groups but apparently avoided reefy environments.

Trilobites.—Members of this diverse and abundant Paleozoic arthropod group occur with all other major groups in a variety of marine environments. Most were mobile benthonic forms with both crawling and swimming ability, and a few may have been pelagic. Different types probably ranged from suspension feeders to burrowing bottom feeders; most probably preferred soft substrates but several inhabited reefy environments.

Conularids.—Occurring with most major organic groups, the Paleozoic conularids seem most common in dark shales with similarly phosphatic inarticulate brachiopods, conodonts and fish remains (Moore, Lalicker, and Fischer, 1952). Conularids may have been weakly attached to the substrate or perhaps nektonic.

Tentaculites.—These small but thick-walled conical shells of the Middle Paleozoic occur with a diverse group of associates. When in large numbers, however, tentaculites are associated primarily with ostracodes, conodonts, brachiopods, bryozoans and pelecypods (Fisher, 1962). They are considered to have been nektobenthonic in habit.

Styliolines.—These tiny, thin-walled, conical shells of the Middle Paleozoic often occur with a diverse group of associates. Styliolines also are found in great numbers with few associates, particularly in dark shales where they may form limestone lenses. Styliolines originally were thought to be pteropod snail shells which they closely resemble. Although now taxonomically distinct, styliolines are considered to be similarly pelagic on the basis of shell form and structure (Fisher, 1962).

Conodonts.—Associated with most marine organic groups in the Paleozoic, conodonts seem most common with inarticulate brachiopods, ostracodes, gastropods, cephalopods, pelecypods, and fish (Ellison, 1957; Hass, 1962), and may occur alone in black shales. Very little is known of the problematical organism that produced these toothlike phosphatic fossils, but it most likely was pelagic in view of wide distribution of the taxa.

Graptolites.—This group of Paleozoic chitinous colonial organisms includes both sessile, attached forms (dendroids) and planktonic forms (graptoloids) which achieved worldwide distribution (Bulman, 1955, 1957). Graptolites sometimes are associated with other groups, but often occur alone in black shales. One notable occurrence of dendroids with brachiopods and worms in a small shale lens in a predominantly coralline and crinoidal limestone (Ruedemann, 1925) suggests that limited lithologic distribution may be primarily a matter of differential preservation.

Eurypterids.—In contrast to the preceding groups, this Paleozoic group of merostome arthropods is not exclusively associated with marine forms, but seems to have emigrated from the marine environment during the course of its fossil history. (See Clarke and Ruedemann, 1912.) Ordovician eurypterids occur mainly with marine faunas of cephalopods, brachiopods, trilobites and graptolites in dark shales. Some Silurian eurypterids are associated with similar marine faunas, but others are found in restricted environments. They may occur alone or with limulids, scorpions and myriapods (the latter two now air-breathing arthropods) in thin shales thought to represent brackish-water lagoons in shoreline paralic sequences. Or, they may occur with a restricted assemblage locally exhibiting great numbers of individuals of one or two taxa of ostracodes, linguloid brachiopods, gastropods, clams, and some peculiar arthropods in shales and fine-grained laminated carbonates known as waterlimes; both rock types occupy generally an intermediate stratigraphic position between marine limestones and evaporites and apparently record transitional environments with highly fluctuating salinites trending toward the hypersaline side. Only a few Devonian eurypterids have marine associates (Kjelleswig-Waering, 1934), and most occur in what are considered estuarine deposits in redbed sequences. Carboniferous and Permian eurypterids are known only from coal-bearing sequences or redbeds where they are associated with branchiopods, ostracodes or other crustaceans, and often only with leaves of land plants.

3. *Significance of Individual Stratification Features*

Most information in the following resume of features appearing on Figure 6 but not discussed in the text is summarized from Potter and Pettijohn (1963), McKee (1964), Moore and Scruton (1957), and Hulsemann (1968). Reference citation to these sources has been kept minimal.

Lamination.—Sedimentary laminae reflect either intermittency of deposition or slight changes in sediment provenance or in velocity of the transporting medium. Preservation of laminae is inversely related to activity of burrowing infauna. Thus lamination characterizes those environments unfavorable to benthonic organisms such as stagnant marine basins, lagoons, or lakes, areas above water level like the supratidal, alluvial, or subaerial regimes, and areas of very rapid deposition as along delta fronts and on some submarine slopes. Conversely lack of lamination in sedimentary rocks may reflect intense burrowing activity in either marine, lagoonal, estuarine, or lacustrine environments, where the bottom is oxygenated, and deposition is slow.

Homogeneous rock.—Structureless deposits may result from complete bioturbation, from availability of only one sediment type, or from such rapid deposition in delta fronts that no sorting is accomplished.

Mottling.—Mottled sediments and disturbed stratification seem to result mostly from incomplete homogenization by burrowers which leave remnants of original structure or introduce new sediment of visibly different characteristics.

Horizontal bedding.—Flat parallel bedding reflects intermittent settling of particles from suspension and can occur throughout a wide range of environments from marine to intertidal, estuarine, alluvial, lacustrine, and even subaerial (McKee, 1964). Because of its origin, this type of bedding is most common in muds; but it is apparent also in some subaerial interdune sand and turbidity-current sand, both of which also are deposits that settle from suspension.

Graded bedding.—Vertical decrease of grain size within a single layer reflects settling of a wide range of grain sizes from suspension as current velocity diminishes. Graded bedding is typically associated with turbidity currents. In order to carry sand-size material in suspension such currents require initiating slopes greater than 1 in 250 (Kuenen, 1958). Because averages slopes in pericontinental seas range from 1 in 500 to 1 in 2500, and those in epicontinental seas are even less, turbidity currents are active mainly on the continental slope and carry material into the deep sea. Turbidites typically occur in repetitive sequences of relatively coarse-grained, graded layers with sharp lower contacts alternating with layers of fine mud that settled from suspension between times of rapid current flow (Shepard, 1964). Turbidites also tend to be more poorly sorted than other graded sand deposits (Kuenen and Menard, 1952). Because turbidity-current deposits are basically dependent only on slope, they may denote a sense of increasing depth. But they do not necessarily reflect absolute depth because they can occur on steep slopes in relatively shallow water such as off the coast of southern California. Turbidites would suggest local basins with relatively steep slopes if found in an epicontinental regime. (See Walker and Sutton, 1967).

Vertically graded bedding *per se* is not confined to turbidites and could occur in any environment where material of a wide range of grain size is suspended, then deposited rapidly when velocity of the transporting medium is abruptly impeded. Thus alluvial deposits resulting from periodic flooding are graded (Stanley, 1968). Graded bedding has been reported in some marsh deposits (McKee, 1964) and could occur also in shallow seas as a result of sediment suspension by storm waves (Kuenen and Menard, 1952).

Cross-bedding.—Inclined stratification within an individual sedimentation unit results from transportation of silt, sand and gravel by traction. Cross-bedding is common in shallow marine, littoral, fluvial and aeolian deposits, and it has been reported also in presumably deep-water graywacke turbidites (Potter and Pettijohn, 1963). Several types of cross-bedding recognized by McKee (1964) may be more characteristic of certain environments than of others. *Low-angle* ($<12°$) simple or planar cross bedding is most common in beach and offshore bar deposits. (This occurs also in alluvial fans, but is extremely variable in such poorly sorted deposits which exhibit little similarity to typically better sorted marine sand and gravel.) Tabular planar cross bedding of *moderate angle* (gen. 18–28°), sometimes called torrential cross bedding, is most typical of delta foresets building into most any type of water body, but occurs also in some alluvial sand-plain deposits and offshore bars. Moderate angles of cross bedding are found also in beaches consisting largely of irregular grains such as shell material. *High-angle* (24 to 34°) wedge-planar cross bedding is most common in aeolian dune deposits but can occur also in delta foresets. *Trough* cross-bedding and *festoon* patterns reflect scour and fill and are most common in fluvial deposits, but occur also in channels in tidal flats, backshore beaches, alluvial fans, and locally in dunes.

Ripple marks.—Rippling reflects movement of fluid and is known from aeolian, fluvial, littoral and marine environments. Possible distinction between aqueous and aeolian ripple marks is based on differences in mode of transport of particles (Bagnold, 1941). Wind moves grains by saltation, and the coarser grains which resist bombardment by other grains remain on the crest of the ripple. Water moves grains by traction, and the coarser grains typically come to rest in or near the troughs after rolling down the downcurrent slope. Symmetrical ripple marks reflect water oscillation, and asymmetrical marks reflect unidirectional currents which orient the steeper face toward the downcurrent direction. Further distinctions between current versus wave control of ripple formation and methods of estimation of water velocities and wave periods which can yield information on size of the water body and other aspects of depositional environment have been presented recently by Harms (1969). Although once considered indicative of shallow water, an increasing variety of ripple marks have been found in deep water (Menard, 1952; Heezen and Hollister, 1964; Dangeard and Giresse, 1968).

Megaripples with wave lengths on the order of several tens of feet and amplitudes up to 2 or 3 feet occur in marine calcilutitic limestones in the Pennsylvanian of Kansas (Hamblin, 1969; personal observation). Large ripplelike features of similar or greater scale, presumably resulting from current flow, can be seen in Bahamian lime sands from the air. (See Imbrie and Buchanan, 1965, p. 171.) Possible causes for symmetrical megaripples in mud, however, have been little discussed to my knowledge. Perhaps effects of large-scale standing waves in the shallow marine environment would be a fruitful area for investigation. (See Ball, 1967, p. 564.)

Sole markings.—Occurring on the underside of sandstone beds that overlie finer grained deposits, sole markings include spoon-shaped *flute casts*, resulting from current scour, and continuous linear *groove casts* and related discontinuous bounce and prod casts resulting from even and uneven dragging of objects along the bottom by currents. Flute casts record current direction and thus might give a sense of increasing depth. Groove casts and related features record sense though not usually direction of current flow. All types of sole markings indicate moderate to strong currents and are characteristic of slope and turbidity-current deposits. Flute casts, however, have been recorded from shelf deposits, stream-cut channels, and flood-plain deposits where they are associated with mud cracks and raindrop prints (Stanley, 1968; Eastler, 1969). Groove casts are known in shallow water and on tidal flats (Reineck, 1969) and probably also can be formed in nonmarine channels. Because sole markings are most abundant in turbidite sequences where typical alternations of sandstone and shale layers offer the best conditions for their preservation, Potter and Pettijohn (1963) state that in great numbers they are suggestive of the slope to deep-water environment. It would seem, however, that these features could be formed in any shallow, nonslope environment subject to strong currents, but their preservation in the record is dependent upon alternation of the proper rock types. Whereas loose sand might easily shift and collapse immediately into a scour mark, mud is more likely to retain the shape. Above the scour, only sediment that eventually indurates into coherent slabs will record the mark as a cast for geologists to observe. The alternation of mud and muddy sand characteristic of turbidite sequences is greatly favorable to such preservation. Perhaps many scour markings are formed also in shallow water muds, but are poorly preserved in crumbly sands or silty shales not sufficiently different from the underlying mud to part easily along the bedding surface.

Deformational structures.—Postdepositional, prelithificational features such as *load casts* and ball-and-pillow structures result from vertical movement arising from unequal loading and subsequent sinking of a denser sand layer into plastic water-logged mud. Thus they are controlled by the juxtaposition of sedimentary layers of particular consistencies, and because sand overlies mud in a great variety of environments, loading features indicate little about salinity or depth of water.

Slump and flow structures result from lateral movement of unindurated sediment by gravity on a slope or by sudden rapid current flow on a level surface such as during a flash flood. They are controlled by sediment instability and can occur in a wide range of environments, although they seem more common in areas with steep slopes such as delta fronts and the continental slope. Direction of increasing depth is indicated when direction of flow can be determined from the structures.

SUBMARINE CHANNEL DEPOSITS, FLUXOTURBIDITES AND OTHER INDICATORS OF SLOPE AND BASE-OF-SLOPE ENVIRONMENTS IN MODERN AND ANCIENT MARINE BASINS

DANIEL J. STANLEY
Division of Sedimentology, Smithsonian Institution, Washington, D.C. 20560

AND

RAFAEL UNRUG
Department of Geology, Jagellonian University, Cracow, Poland

ABSTRACT

Submarine slopes are distinctive depositional environments because of their gradients and their setting between sediment source locales at their upper level and areas more favorable for preservation on basin floors beyond. The model for slope sedimentation must incorporate such factors as structural framework, gradient, dissection, type and rate of sediment input, and transport processes. Slopes are most often envisioned as inclined planes reflecting depositional instability, i.e., temporary resting places for sediments during their passage to depositional sites in deeper, more distal environments. Sediments are, however, preserved in a multiple set of slope subenvironments, and there are criteria for recognizing most of these.

The dominant slope assemblage comprises fine-grained pelagic deposits, hemipelagic materials influenced by bottom current activity (and often entirely reworked by benthic organisms), turbidites with their mixed faunas and mineralogy, contorted slumped units, and large allochthonous slices. Distinguishing paleoslope from paleobasin deposits is generally difficult because of the merging of sedimentary facies in the two environments. Marine geological investigations have shed light on the three-dimensional geometry, vertical-lateral facies relationships, sedimentary properties and processes on modern slopes. These studies, combined with investigations of paleoslope deposits in certain Cretaceous and Tertiary flysch basins in the Alps and Carpathians, allow more precise paleogeographic interpretations.

Channel deposits, coarse fills of submarine canyons and valleys answering the description of *fluxoturbidites*, are important in this respect. The shoe-string bodies migrate downslope and incise pelagic and bottom-current transported units and turbidite bundles. These fills of former sedimentary funnels are well developed on lower slopes, subsea fans and rises, and can be traced well into basins; they need not necessarily be geographically proximal to source. Associated primary structures indicate that traction and grain-flow processes are significant in the deposition of submarine channel units.

In association with these channels, wedges of pebbly mudstone often occur at the base of the slopes and, where concentrated, signal the position of a predominent decrease in gradient. Large, often rounded, blocks and boulders enrobed in contorted mud suggest conditions of rapid sedimentation, as off river mouths or along rapidly eroded coastlines where materials are periodically moved across narrow shelves and then *en masse*, on relatively steep slopes, in and between canyons. The assemblage of channel, pebbly mudstone, and slump deposits define paleoslope and basin trends, pin-point source input along ancient basin margins, and serve to distinguish lateral from longitudinal dispersal patterns.

ACKNOWLEDGMENTS

Many of the ideas outlined in this article developed from discussions and, in some cases, joint studies with colleagues. We are indebted to many individuals. Among others should be mentioned S. Dzulynski, C. Gehin, M. Gennesseaux, Y. Gubler, J. B. Hersey, C. Hubert, G. Kelling, M. Ksiazkiewicz, J. Lajoie, G. V. Middleton, E. J. Mutti, G. Pautot, W. F. Ryan, N. Silverberg, R. Walker, and F. C. Wezel. Various aspects of the work (D.J.S.) on modern slopes (off east coast of the U.S., Mediterranean) and paleoslopes (Maritime Alps, Carpathians) have been supported since 1966 by the Smithsonian Research Foundation, the U.S. National Academy of Sciences, the U.S. Coast Guard, the National Geographic Society and the SACLANT Research Centre, La Spezia. The Polish Academy of Sciences is thanked for funds and generous long-term support of field work (R.U.) in the Carpathians.

INTRODUCTION

Problems related to provenance, to the location of point-source and sediment entry into a basin, to sediment transport and dispersal paths, and to the physiographic and tectonic framework are inherent to most paleobasin analyses. Equally important is the recognition of paleoslope position and attitude. Delineating slopes is an essential aspect of regional paleogeographic reconstruction because of the very specific setting of basin slopes between sediment source locales near or beyond the shelf break and zones more favorable for the accumulation of sediment on basin plains beyond. *Gradient* is the most distinctive characteristic of a slope. This inclined surface is visualized as a temporary resting surface for sediments during their transport from the shelf-break to final depositional sites in deeper, less steep environments beyond. Depositional instability, sediment failure and downslope movement are generally associated with this outer margin environment.

The importance of accurately defining slopes in paleobasin studies has long been realized (summary in Potter and Pettijohn, 1963). There are two major difficulties: (a) even at present marine slopes remain one of the enigmatic and least studied sedimentological provinces, and (b) marginal facies of most former deep-sea basins tend to be only partially preserved in the geological record. Erosion and tectonic displacement have frequently resulted in the truncation, removal, or cover of preexisting terrace deposits. Thus, only fragmentary evidence of former slopes, most often in the form of a few isolated and tectonically disturbed outcrops and corehole horizons, is available to geologists.

More comprehensive marine geological investigations of modern slope environments, made possible through the development of oceanographic techniques not available twenty or even ten years ago, now afford an opportunity to better assess critical large and small-scale sedimentary features most likely to be of use in recognizing ancient slope deposits. Much has been made of the point that the Quaternary eustatic oscillations have left a marked imprint on the physiography and sediments of the world's continental margins, and that modern shelves and slopes are not representative of those of past periods. It can be readily demonstrated, however, that certain dominant aspects of Quaternary slope topography and of selected modern slope deposits can be recognized in preserved ancient basin margin units.

Several distinct slope facies and the lithologic assemblages related to each of them are described in following sections. Of potential value to field geologists and stratigraphers concerned with ancient marine basins are coarse submarine channel deposits occasionally noted to interfinger with relatively well stratified finer-grained units. Deep-sea channel fills, a still poorly defined type of slope and base-of-slope (a catch-all term for continental rise, basin apron and subsea fan) facies, merits serious attention. Emphasis is thus placed on the description and interpretation of different types of shoe-string and wedge-shaped bodies, most of which serve as valuable paleogeographic indicators and some of which are undoubtedly of economic importance.

INADEQUACY OF SOME PREVIOUSLY HELD CONCEPTS

Twenty years ago, Rich (1950, 1951) summarized in a series of popular papers criteria for recognizing slope deposits. The lithologic criteria of rock units formed in the slope environment, which Rich referred to as the *clinoform*, were selected on the basis of field observations and information of slope sedimentation available at that time. Many of the lithologic features that he and other workers chose as valid slope (or *clino*) criteria were based on observations of selected ancient formations believed to represent slope facies and on inferred sedimentary transport processes (fig. 1).

Most of the key features of the slope assemblage that Rich and others long before him have cited have proven to be valid ones. Few would doubt, for instance, that the slope environment is a zone of reworking and instability as attested by slumping and larger scale gravity sliding. Furthermore, the predominant sediment type on the slopes of most basins is mud (silt-clay mixtures), sole marking orientations tend to be consistent (current sense is most often normal or subnormal to the strike of the slope), and the skeletal remains of organisms tend to be uncommon (although evidence of bioturbation is frequently observed).

On the other hand, Rich (1950, 1951) called attention to an assemblage comprising relative thin, even/or uniform bedded strata permitting long distance correlation, and units with a marked absence of ripple marks and cross-bedding, and a sparsity of fossils suggesting conditions of sterility. These parameters do not generally conform with observations of modern slope environments. Furthermore, the presence of sole markings, intrastratal flowage and crumpling, and rhythmic alternation of bedding which were inferred to be key features of slopes are now known to be present in a host of other environments, including some without gradients and others that are nonmarine (Dott and Howard, 1962; Stanley, 1968; and others).

Geologists have most often visualized the *clinoform* as a smooth surface, convex-up near the shelf edge or outer margin break, and concave-upward at the base of the slope (fig. 1). The great topographic complexity of this environment, generally corrugated and only rarely resembling a simple inclined plane, could hardly have been imagined by early workers. Although the upper boundary of the slope is generally sharply defined at the shelf break, the lower slope generally merges imperceptibly with the basin floor. It is not surprising, therefore, that base-of-slope facies are often confused with basin floor deposits [Rich's *fondothem* (his fig. 1)], and vice versa.

Furthermore, information acquired during the past decade on sediment transport mechanics also require a reassessment of the slope lithological assemblage. The wave-base concept, as originally envisioned for instance, has seriously

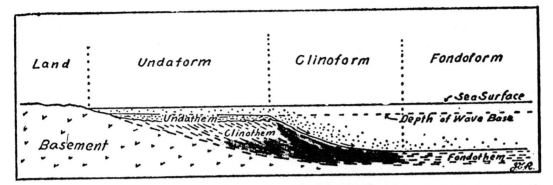

FIG. 1. Sketch illustrating definitions proposed by Rich (1951). The slope environment (clinoform) extends between the platform at or above wave-base (undaform) and the basin plain (fondoform). Clinothem facies are the slope deposits. Arrows indicate density (turbidity) currents; stippling indicates muddy water.

been questioned (Dietz, 1963b); as such it does not provide a satisfactory explanation for the origin and development of continental slopes as shown on figure 1. Investigations during the 1960's have also resulted in a reassessment of the role of turbidity currents. Although turbidity currents are undoubtedly an important factor in basin sedimentation, they are only one of several possible transport mechanisms that can be called upon to explain the assemblage of sedimentary structures of slope and deep-sea deposits. Visual observations and direct measurements of the sea floor beyond the shelf-break and the water mass above it indicate a striking diversity of transport processes. Bottom currents and associated sediment motion by traction, even at great depths, are once again receiving the attention they deserve.

Detailed morphological studies invariably show that most submarine slopes comprise several distinct environments. It is reasonable to assume that each of these lesser-order morphological provinces comprises a characteristic lithological assemblage which could be recognized if encountered in the fossil record. This complexity of modern marine slopes, whether they be large continental slopes at the junction between continental margins and major deep open ocean basins or much smaller-scale basin slopes extending into small and relatively shallow, land-encircled troughs, requires closer examination.

SUBMARINE SLOPE ENVIRONMENTS
Modern Slopes in General

The slope s.l. is defined as the relatively steeply inclined (generally from 3 to 6°) portion of the sea floor lying seaward of the usually sharp shelf-break which serves as its upper boundary (Dietz and Menard, 1951). Continental slopes bounding ocean basins are physiographic provinces of the first order on the face of the earth. They can be visualized as steep escarpments about 3.5 km high. Modern continental slopes bordering deep sea floors or trenches have a total length exceeding 110,000 km. The lower portion of a slope, generally less steep, merges gradually with the near-horizontal basin or abyssal plain. Slopes and the subsea fans, coalescing fans, rises and aprons at their base comprise a zone of transition between ocean basins and continents; these submerged inclined regions occupy 10 to 15 percent of the total earth's surface.

No single form or origin characterizes all slopes. It is as hazardous, therefore, to generalize about submarine slopes as it is to generalize about slopes exposed on the earth's surface. A model for slope sedimentation must incorporate, for instance, tectonic framework, gradient, degree of dissection, type and rate of sediment input at its upper surface, and transport mechanisms active between the shelf-break and the deeper sea floor beyond. The characteristics of the water mass overlying a slope is also an essential parameter. Variance of even one or two of these factors would alter the geometry and lithological make-up of a slope deposit.

The diverse origin of slopes has long been a subject of discussion (Gardiner, 1915; Cotton, 1918; Emery 1950, Kuenen, 1950; Dietz, 1952; Guilcher, 1958; Shepard, 1963, Emery, 1965a, 1968; Worzel, 1968; Emery et al., 1970). Most geologists and geophysicists agree that slopes, for the most part, have a diastrophic origin, i.e., continental slopes are ultimately structural with sedimentary modifications. Shepard (1963, p. 310) has cited the general straightness of slopes, angular changes in trend, steepness, the association with earthquake belts and deep trenches, and exposure of bedrock as evidence for tectonic origin. Hypotheses proposed to explain the formation of slopes include, among others, faulting (Emery, 1960; Heezen, 1962; Scholl et al., 1966; Hoskins, 1967), damming of outer margins (Emery, 1968), progradation (Uchupi and Emery, 1967), slumping and gravity sliding tectonics (Rona and Clay, 1967; Andrews, 1967; Uchupi, 1968a, b; Stride et al., 1969) downwarping of continental margins (Veach and Smith, 1939; Bourcart, 1949; Heezen and Drake, 1963), rifting (Carey, 1958; Dietz, 1963a), catastrophic events such as asteroid impact (Harrison, 1960), and development as the flank of an accretionary folded belt or *orogen*, usually a collapsed continental rise prism (Dietz, 1963a; Dietz and Holden, 1966). The concept of seafloor spreading has enjoyed growing popularity during the past few years. *Perhaps slopes bounding ocean basins are a logical consequence of the separation of continental plates.* The validity and relative importance of this and the other theories in light of crustal data and deep-sea drill cores now available, exciting subjects, are not discussed here.

Most slopes, subsequently to their structural formation, have been modified to some degree by processes that cause either erosion or progradation, a depositional build-up of the outer margin. The wave-built and wave-cut terrace hypotheses are discussed by Shepard (1963), Dietz (1963b) and others. The secondary sedimentological modifications of slopes are of special interest for we are concerned with physical and biological criteria likely to be recognized in the geological record. Slope modification and sedimentation is most notable in areas such as those off large rivers and deltas, or where ice and fluvio-glacial transport mechanisms have been important. In such areas, progradation of the slope and build-up of subsea fans, aprons and rises (these tend to on-lap on lower continental slopes) is a common phenomenon. Slopes resulting from carbonate build-ups on submerging reef platforms, steep volcanic oceanic islands and other variants of depositional slope development are not elaborated upon here.

Outer Margin Topography

There is a close and direct relationship between submarine slope topography and the associated depositional geometry and lithology. Dominant physiographic characteristics of the outer margin province are highlighted in Figure 2. For convenience of this discussion, the outer margin province is subdivided into outermost shelf, shelf-break, upper and midslope, and lower slope and base-of-slope environments.

Continental slopes are narrow, range from 5 to 35 km in width, and have a straight to curving trend. Slopes in smaller marine basins are frequently less wide. The greatest gradients generally occur just below the shelf-break on upper and midslope reaches. The typical continental slope is not precipitious, averaging 4° 17' for the first 1800 m of descent (Shepard, 1963, p. 298). Modern slopes off stable coasts lacking major rivers average about 3°; those off young mountain range coasts average 4.6°; slopes off fault coasts with narrow shelves average 5.6°. Steep slopes in excess of 10° occur on the walls of submarine canyons and occasionally on the margins of smaller basins, particularly those off narrow and tectonically active mountainous coasts such as off Southern California (Emery and Terry, 1956; Emery, 1960). A minimum gradient of 1:40 has been used to define the continental slope province (Heezen et al., 1959). On the Atlantic margin off North America, the continental slope lies between the shelf-break, at about 100 to 180 m, and the more gentle upper rise margin at depths of 1400 to 3200 m. Gradients of the continental rise below the slope proper range from about 1:40 to 1:1000; depths of the rise off eastern North America range from 1400 to 5000 m. The presence of steep scarps on otherwise gentle fan or rise surfaces result from faulting or slumping.

Present-day marine slopes are generally spectacularly dissected surfaces almost ubiquitously cut by submarine valleys and gullies as illustrated in a comprehensive survey compiled by Shepard and Dill (1966). These linear and sinuous depressions, generally incised perpendicularly to the strike of slopes, have been charted for over a century (Dana, 1863; Buchanan, 1887; and others). Leadline surveys and, since the 1930's, echo sounding and precise navigation techniques (Veatch and Smith, 1939; Buffington, 1964; and others) have brought out the density and complexity of these and other sea floor irregularities. Regional physiographic surveys (Heezen and Tharp, 1962, 1964, 1968; Pratt, 1967; Uchupi, 1968) and most detailed charts indicate that smooth slopes are a rarity indeed, even in areas of high sedimentation (Shepard and Dill, 1966). Relief between canyon axis and adjacent canyon walls is greatest in the uppermost reaches of the slope and tends to decrease downslope. Frequently, the most intensely dissected sectors are the mid and lower stretches of the slope.

Earlier workers attempted to find analogies between submarine and mountain stream valley systems (Veatch and Smith, 1939). Although the relief of both types may be comparable in some respects, marine drainage patterns are generally less dense than those in subaerial environments (Shepard and Dill, 1966, their figure 76, and others). Submarine valley patterns on the outermost shelf and upper and midslope reaches com-

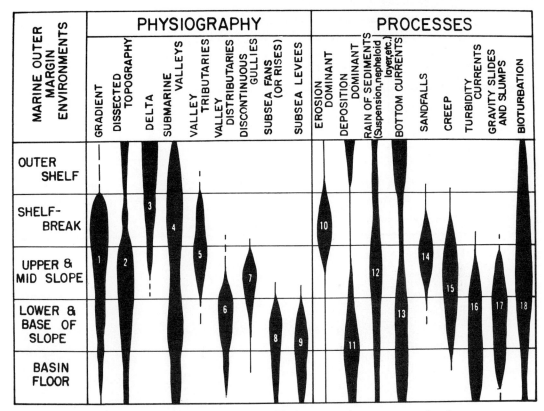

Fig. 2.—Some dominant physiographic features and processes associated with marine outer margin environments.

monly display distinct tributary systems. Major distributary systems occur at and beyond the base of the slope of subsea fans, aprons and rises. Levees are frequently noted bordering submarine valleys beyond the base of the slope well within the basin (Buffington, 1952; Shepard and Buffington, 1968). The associated distributary system, submarine levees, and subsea fans (Menard, 1960), aprons (Emery, 1960) and rises (Menard, 1955), in fact define lower and base-of-slope environments (fig. 2).

Outer Margin Processes

The slope is a physiographic province in which erosion and depositional accretion are in delicate balance. Degree of crustal mobility, eustatic sea-level changes and related transgressions and regressions, rates of sedimentation, gradient, morphological relation to the shelf-break and coast, sediment transport processes and bioturbation are among the critical controlling factors. The interplay of several of the above parameters (fig. 3) determines whether a slope serves as a temporary depositional surface (a discontinuity in the geological record) or as a horizon of permanent continuous deposition and progradation. Seismic profiling and direct observation of the bottom with submersibles and television surveys show that in probably a majority of cases erosion is a dominant factor in the shelf-break and uppermost slope environments (fig. 4); progradation is prevalent on mid-slope, but is best developed at the base of the slope and on the basin plain beyond (fig. 2).

Slope erosion predominates in areas where seismic activity is prevalent, where the outer shelf is narrow and shallow or subaerially exposed (such as during periods of low sea level stands), where gradients are high, where high sedimentation rates result in increased pore water pressure, and where sediment failure frequently occurs by slumping and other mass gravity processes (fig. 3). Examples of slopes that border tectonically active basins include those off Southern California and other circum-Pacific sectors, the Mediterranean and the Caribbean. Continuous seismic profiles made along slopes in these areas reveal locally thin and discontinuous unconsolidated sediment build-up on older deposits (Scholl et al., 1966; Moore, 1966; Curray, 1966; Ryan et al., 1970; and others). It must be noted, however, that

SEDIMENTATION PATTERNS ON THE SLOPE PROPER	FACTORS INFLUENCING SLOPE DEPOSITION						
	TECTONIC ACTIVITY	EUSTATIC SEA LEVEL CHANGES	GRADIENT	GEOGRAPHIC SETTING OF SLOPE IN RELATION TO SHELF AND COAST	RATE AND AMOUNT OF SEDIMENTATION	SEDIMENT TRANSPORT PROCESSES	BIOTURBATION
PROGRADATION AND PERMANENT ACCUMULATION OF DEPOSITS ON SLOPE	Stable tectonic setting; low level of seismicity	High sea level stands; outer shelf submerged	Relatively low	Shelf-break in deep water; little or no coastal influence	Low sedimentation rates; or large supply accumulation with low pore water pressure	Pelagic and hemipelagic sedimentation dominant	Low rate of sediment turnover
SLOPE SERVES AS TEMPORARY SEDIMENTATION SURFACE: EROSIONAL PROCESSES DOMINANT	Unstable tectonic setting; frequent earthquakes, faulting, etc.	Lower sea level stands; outer shelf subaerially exposed, or at shallow depth	Relatively high	Shelf-break shallow and close to coast; influence of seasonal floods and wave and tidal currents	High sedimentation rates; or sediment accumulating with high pore water pressure	Sediment failure by mass-gravity processes dominant	Rapid and thorough reworking of bottom sediment

FIG. 3.—Diagram showing interplay of factors influencing slope sedimentation: deposition versus erosion.

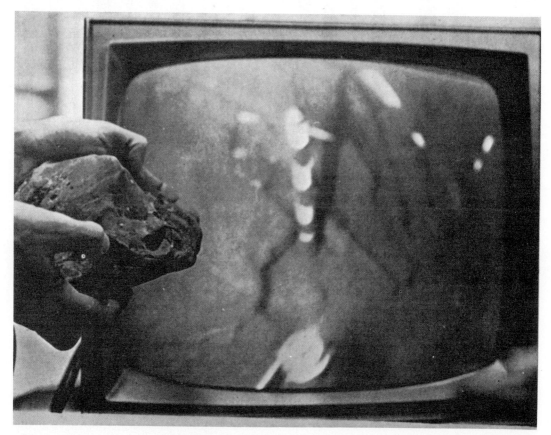

FIG. 4.—Sandstone sample of Pleistocene age in foreground was dredged on the west wall in the upper reaches of Wilmington Canyon. Television monitor in background shows the massive rock outcrop at a depth of approximately 110 m where this sample was obtained. This cliff was subaerially exposed during the Wisconsin low stand of the sea. Penetrometer and compass (7 cm diameter) provides scale. Erosion has exposed these older rocks (see cross-section profiles in Fig. 21) which contribute rock-fall deposits occasionally dredged near the canyon axis.

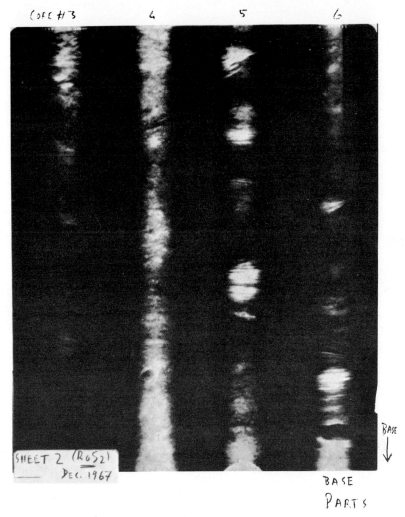

Fig. 5.—X-radiographs of unsplit cores (diameter approximately 6 cm) collected on the upper continental rise (2875 m) southeast of Delaware Bay. Note distinct alternations of current laminated silt and sandy silt (light) and bioturbate muddy (dark) strata (after Stanley, 1970b).

similar observations are also made in more tectonically stable areas such as off Western Europe (Curray et al., 1966; Stride et al., 1969) and on the Atlantic margin off North America (Moore and Curray, 1963; Rona and Clay, 1967; Uchupi and Emery, 1967; Knott and Hoskins, 1968; Kelling and Stanley, 1970).

No mechanism of sediment transport active on slopes is necessarily restricted to submarine slopes or even to the marine environment. Although knowledge of the diverse transport mechanisms is still sketchy, we can outline a characteristic grouping or assemblage of dominant processes that defines the slope province (fig. 2). Sediment movement involving both traction and suspension as a result of bottom currents associated with storm and normal wave and tidal activity has long been recognized as a dominant factor in deep marine environments. In recent years, visual techniques allowing direct observation of bottom currents (underwater camera, television, SCUBA and submerisble dives) have supplemented current meter measurements and empirical predictions of water mass movement. There is sufficient evidence that the ocean floor, even at great depths, is at least temporarily affected by moving water masses; current velocities ranging from less than 2 cm to perhaps as high as 1 m per second (Heezen and Hollister, 1964; Emery and Ross, 1968) have been measured.

Many workers have emphasized the predominance of *relict* sediments on continental margins and particularly present day shelves

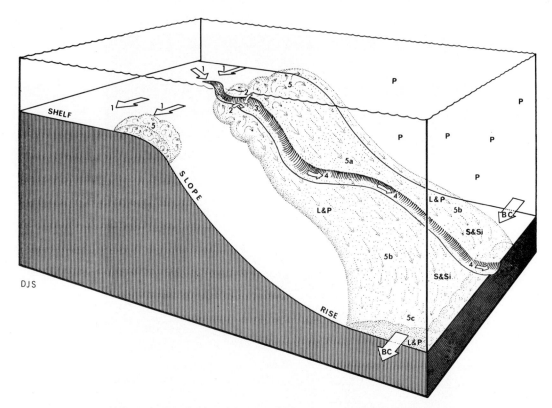

Fig. 6.—Schematic diagram, based on evidence discussed in text, showing low density turbidity current flow on the slope and rise in the vicinity of Wilmington Canyon. Stirring of the outer shelf (5) by breaking storm and internal waves brings fines into suspension; these are moved seawardly by water masses onto the rise (5a) and beyond. Contour-flowing bottom currents (arrow BC) deflect turbid, suspension-rich (fines include lutite and pelagic fraction, L & P) downslope-moving water mass (5b, 5c). Sand and silt (S & Si) of this and earlier flows are also deflected, but are moved laterally along the bottom shorter distances than the suspended fine fraction in the water mass. Concurrently, sand on the shelf-edge moved by bottom currents (open arrow 1) is spilled over and trapped in the upper reaches of the canyon heading on the shelf (2). Slumping down the canyon walls (3) toward the axis is also prevalent. The canyon serves as a funnel (4) in the transfer of sediments to the continental rise and abyssal plain. (After Stanley, 1970b).

(Emery, 1965b). However, bottom photographs and cores indicate that there are very few areas not in some fashion affected by the movement of water masses flowing above the bottom. The strength of currents varies with geographic siting: they are frequently stronger on the outermost shelf and on the rise beyond the base of the slope proper. In most areas where they have been measured, currents have sufficient velocity to transport materials of at least silt and clay size. The deep contour-following currents (moving along the general strike of the slope) of the type noted on the Atlantic continental margin (Swallow and Worthington, 1961; Heezen et al., 1966; Schneider et al., 1967, Heezen, 1968; Stanley and Kelling, 1969) may well occur along the slope and rise in other areas. Such water mass movements are produced by differences in density (salinity and temperature controlled) or by differences in surface level or both. The Western Boundary Under Current in the northwest Atlantic, flowing at velocities ranging to 18 cm per second (fig. 6), is competent to transport most textural types encounted on the continental rise (Fox and Heezen, 1968). Geostrophic currents of this type are recognized in other ocean basins as well [i.e., the Mediterranean, for instance according to Gostan (1967) and Ryan et al., (1970)].

Seaward dispersion of fine terrigenous material in the water column accounts for much of the finer-grained sediments that veneer the slope and comprise a large fraction of the prograding apron, fan, and rise deposits. However, a simple pelagic "rain of sediment" process does not explain the predominance of alternating clay and laminated silt units forming the surficial sediment section of the slope and rise. Noteworthy, then, is the recent discovery of thick suspended sediment-rich layers (ranging from 200 to 950 m in thickness) in the water mass above the continental slope, rise and abyssal plain off the east coast of the U.S. (Eittreim *et al.*, 1969). This type of suspension-rich horizon has been called a *nepheloid* layer by Ewing and Thorndike (1965). Concentrations may be such as to induce downslope flow and result in the lateral transfer of significant amounts of fine-grained inorganic and organic matter into the deep ocean basins. One centrifuged sample of 200 liters collected at a depth greater than 4000 m contained 0.50 g of suspended lutite and organisms (Ewing and Thorndike, 1965). Presumably much of the lutite originates from erosion on the outer continental shelf and uppermost slope (Lyall *et al.*, 1971), but more data are needed to substantiate this. Where measured in the north and south Atlantic, the nepheloid layer tends to follow bottom topography and appears to be permanent and widespread. Downslope flow of nepheloid layers may be deflected by topography, Coriolis flow, and the deep bottom currents as suggested by Heezen (1968).

The origin of such nepheloid layers, their composition and regional extent, and the amount of pelagic organisms entrained and reworked downslope are just several of the questions that need to be resolved. It is apparent that the effect of deep currents on concentrated suspensions of the type described above may account for the deposition of significant amounts of hemipelagic (terrestrial plus pelagic components) sediments (*hemipelagites*) accumulating in the base-of-slope and basin plain environments. Workers have, for instance, suggested that this type of interplay may have resulted in the local accumulation of huge sediment drifts and the development of features such as the Blake-Bahamas Outer Ridge (Heezen, 1968).

The mass of literature during the last two decades has emphasized the role of turbidity currents. Displaced shallow water fauna and lithologies occurring in deep environments were noted over a century ago (Bailey, 1854). There are still few well documented cases in the modern oceans (cf. Heezen and Ewing, 1955). Even the now-classic Grand Banks example has recently been reinterpreted as a massive slump (Heezen and Drake, 1964). The relative role of turbidity currents in the downslope transfer of sediments in modern ocean basins has recently been reevaluated by Hubert (1964), Heezen *et al.* (1966), Schneider *et al.* (1967), Kuenen (1967) and others. One would conclude, on the basis of these and other recent studies, that turbidity current deposition represents only one of several important processes active on slopes, and that it is, in some cases, only of relatively minor significance. Moore (1966) in his evaluation of basin-filling process on the California continental borderland suggests that purely pelagic and hemipelagic deposition have been of little importance volumetrically in slope progradation and basin fill. Resuspension at the shelf-break *and* low-density, low-velocity turbidity currents are selected as the prime moving forces carrying large volumes of silt and clay-sized particles downslope to these basin plains. Evidence for this sequence of events is available for the area off the Middle Atlantic States (Lyall *et al.*, 1971). A bottom photography—underwater TV— bottom-sampling program in the vicinity of the Wilmington Canyon (Stanley and Kelling, 1969, Stanley, 1970b) indicates that (1) bottom currents on the outer shelf are responsible for the movement of silt and sand into the head of the canyon (fig. 2); (2) bottom currents with velocities ranging from 10 to 25 cm per sec. were noted to depths of 500 m within the canyon and long-term transport trends are down the canyon walls toward the axis and down-axis; (3) a relatively dense layer of suspended matter in the water column above the slope and rise hampers observation of the bottom (Lyall *et al.*, 1971, their fig. 2); (4) bottom currents on the upper rise tend to move sediment along a southwesterly direction, i.e., along the contours (Stanley and Kelling, 1969); (5) cores collected on the lower slope and rise tend to consist of distinctly alternating current-laminated silt and bioturbate mud (fig. 5).

The schema, shown in figure 6, shows suspended sediment scoured by storm wave and possibly breaking wave activity or other as yet unidentified phenomena on the outer shelf. These resuspended sediments have actually been sampled in the area of the Wilmington Canyon (fig. 7). Low density turbidity currents carry these resuspended fines (silt and clay) downslope in suspension, tend to move the somewhat coarser sediment (including sand) on the bottom by traction, and locally scour the bottom as indicated by bottom photographs. At this time, however, it would be difficult to distinguish a nepheloid deposit [*nephelite?*] from a low density, slow-moving turbidity current deposit (turbidite). Perhaps these two mechanisms may actually be related in some way. As suggested by the figure 6, deflection of such downslope-current activity by deep contour-flowing currents may well occur on the rise. The alternation of current laminated (and occasionally graded) silt and sand and mud in cores (fig. 5) can thus be explained.

Coarser sediment tends to be trapped in canyons, submarine valleys and other depressions. These are *channelized* and then funneled downslope by sand flows and creep, at least in

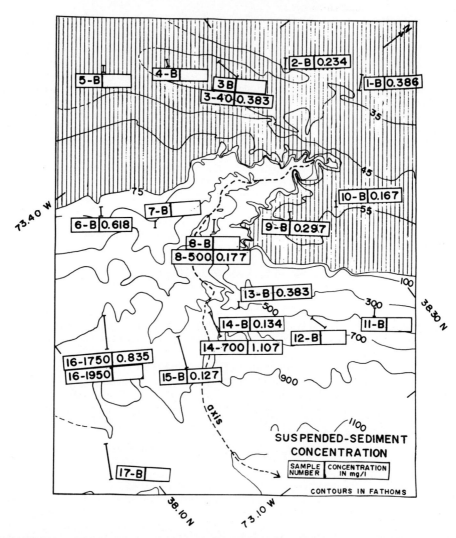

Fig. 7.—Chart showing location of 17 water sample stations (depicted by bar symbol) and total concentration of suspended particulate matter in mg/l (12 samples) in the vicinity of Wilmington Canyon. The first number represents the station number; the letter B indicates near-bottom samples. The second number at stations 3, 8, 14 and 16 designates the depth of sample within the water column, in meters. Data detailed in Lyall et al. (1971).

the upper reaches of the canyon (Dill, 1964). Evidence for channelization includes natural levees, and submarine valleys migrating on giant submarine fan surfaces.

Bottom and seismic profiling indicate that in most regions gravity slides and slumps are volumetrically the most important sedimentation processes affecting slopes. Mechanics of submarine slumping are reviewed by Moore (1961), Dott (1963), and others. The presence of mass gravity sediment failure is recognized on the basis of slide-type topography (Pratt, 1967; Stanley and Silverberg, 1969), continuous seismic profiling showing disrupted reflectors (Emery et al. 1970; Kelling and Stanley 1970, their fig. 12), reworking of shallow water and outer shelf facies downslope (fig. 8), and contorted stratification in cores (Stanley and Silverberg, 1969). Failure of this type is particularly prone in areas of high sedimentation, such as seaward of the Mississippi (Shepard, 1965) and Rhône Deltas (Duboul-Razavet, 1959; Menard et al., 1965), on slopes having received high amounts of fluvio-glacial sediments (Stanley and Silverberg, 1969) in higher latitudes, off reefs in warm oceans, and particularly off steep and tectonically active slopes (Fairbridge, 1947). Slumping was particularly important during glacial low stands when rivers debouched large volumes of sediments directly onto the upper continental slopes. Discontinuous wedge-shaped and often contorted slumps and slices tend to be the base-of-slope facies *par excellence*. Large 'allochthonous' slices are ubiquitous and numerous examples in the modern oceans are cited (Arkhanguelsky, 1927; Moore, 1961; Buffington and Moore, 1963, Uchupi, 1967, 1968; Gorsline and Emery, 1959; Hoskins, 1967; Rona and Clay, 1967; Stride *et al.*, 1969; Kelling and Stanley, 1970).

Bioturbation, the reworking of sediments and disturbance of primary physical structures by organisms, is important, but perhaps nowhere as critical in affecting sedimentation as in the slope environment. In this environment richly endowed with diverse bottom-dwelling animals, the feeding, nesting, burrowing and other activities change the physical properties of sediments. By altering the original grain-to-grain texture and affecting changes in the pore water content of sediments resting on inclined surfaces, organisms are quite probably responsible—at least to some degree—in bringing about the failure of slope deposits and triggering their downslope movement (Stanley, 1970a).

RECOGNIZING SLOPE DEPOSITS

Slope Deposits in General

Submarine slopes have been examined from the point of view of morphology, tectonic setting and origin, but surprisingly few detailed analyses of slope deposits proper are available. An example of ancient slope deposits described in the

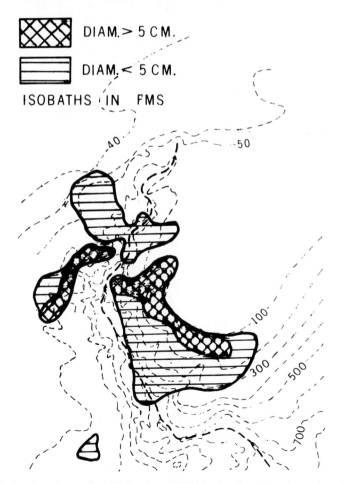

Fig. 8.—Distribution of reworked Pleistocene stiff blue clay fragments in the head of Wilmington Canyon. Distribution patterns pin-point the presence of rock exposures and erosion by rock fall and slumping along the upper walls of the canyon. See also Fig. 21.

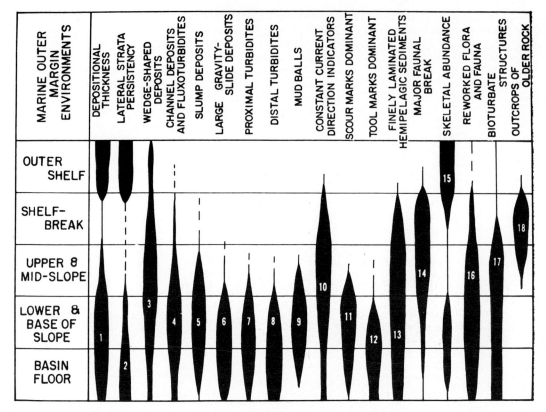

Fig. 9.—Diagnostic lithological assemblage in submarine outer margin environments.

literature (Rich, 1950) is the Silurian flysch sequence at Aberystwyth, Wales. The remarkable lateral persistency of thin sedimentation units in this area (Wood et al., 1967) would preclude an upper to midslope origin. The thin and very regular alternations of clay and silt strata which Rich illustrated would now be recognized as more distal basin apron and plain deposits.

With the possible exception of *in situ* benthic forms with known restricted depth ranges, and and associated organic tracks and markings ("lebensspuren"), there are few parameters truly unique to submarine slopes. Even the fauna examined beyond the shelf-break must be interpreted carefully for such assemblages frequently consist of mixed neritic-pelagic forms as a result of downslope displacement. Recent continuous seismic surveys and more intensive coring programs, however, are beginning to bring to light the three dimensional nature and petrological characteristic of slope deposits. In this and following sections, a reevaluation of the more obvious criteria used to define submarine slope deposits is attempted (fig. 9).

The subbottom configuration of the Atlantic margin off North America that has received intensive examination during recent years (Moore and Curray, 1963; Hoskins, 1967; Rona and Clay, 1967; Uchupi and Emery, 1967; Knott and Hoskins, 1968; Rona, 1969; Emery et al., 1970, and others) provides some insight into the style of sedimentation active on a relatively stable outer continental margin. The internal structure of this sedimentary terrace (shelf and slope) is one of outbuilding (progradation) of slope deposits and upbuilding by shelf and paralic facies on a subsiding base (Moore and Curray, 1963). Locally along the outer Atlantic margin, as in other areas, there has been considerable modification of slope deposits by canyon cutting and large gravity slides (Archanguelsky, 1927; Curray et al., 1966; Hoskins, 1967; Scholl et al., 1968; Stanley and Silverberg, 1969; Stride et al., 1969). Erosion is also prevalent in areas of particularly high sedimentation, such as off the Mississippi Delta where there is a marked increase in the sediment pore-water pressure within the pro-delta sections (Shepard, 1955; Moore, 1961).

Not all upper-midslope environments, however, need show a predominance of erosion, as is attested by the thick prograding sequences off Norfolk, Virginia, the Northwest Gulf of

Mexico, and off west-central Mexico (Moore and Curray, 1963). Profiles across certain of these areas show reflectors that extend from the slope into the continental rise (Uchupi and Emery, 1967). In many areas also the uppermost slope appears smooth and undissected as a result of the prograding sequences; an example on the margin off Nova Scotia is depicted in Stanley and Silverberg (1969).

In order to establish a sedimentation model applicable to the middle Atlantic slope, Rona (1969, p. 1960) outlines factors that need be considered: (a) the thickness of sediment deposited on the outer continental shelf and upper continental rise are approximately equal; (b) the seaward thickening of sedimentary strata underlying the coastal plain and continental terrace suggests that the basement is subsiding along a hinge line landward of the continental shelf; (c) the sedimentary regime of the continental slope is distinct from that of the continental shelf, but related to that of the continental rise; and (d) deposition and erosion are probably periodic, and the relative importance of each can vary along a single continental margin sector. Two sedimentation models are proposed by Rona (1969, his figs. 10 and 11): in one model (fig. 10A), there is a sequence of simultaneous deposition on the continental shelf, slope and rise; in the other (fig. 10B), deposition on the shelf is shown to alternate with deposition on the slope and rise.

As described earlier, the balance between erosion and deposition would vary somewhat from region to region depending upon differences of such factors as proximity to source and sedimentation rates, the amount of slope inclination, bioturbation, and structural mobility of the outer margin (fig. 3).

Upper and Midslope Deposits

It is probable that outer-shelf to upper-slope deposits are occasionally buried under transgressive-regressive sequences or by a tectonic cover that protects such deposits from further erosion. When preserved in the rock record, such upper-slope sections are likely to be distinguished by a lack of laterally persistent strata, stratal truncation, angular unconformities, and alternating topset and foreset strata. (See, for instance, seismic profiles in Knott and Hoskins, 1968; and Kelling and Stanley, 1970.) These features and shelf-edge benches, terraces and buried channels commonly noted in profiles result from alternating fluvial and marine forces during eustatic sea level oscillations. Seaward dipping strata of the continental terrace, consisting of sandstones and shales, extend to, and are frequently truncated at, the slope where they are veneered by younger acoustically transparent

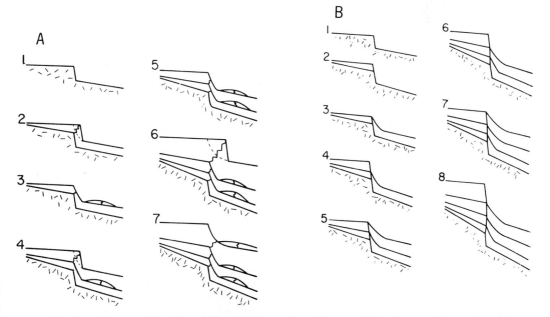

Fig. 10.—Sequencial models of the middle Atlantic continental terrace (after Rona, 1969). A, deposition on the continental shelf, slope, and upper rise is simultaneous; distinct sedimentary facies are separated by wavy line at shelf break. B, deposition on continental shelf alternates with deposition on slope plus upper continental rise.

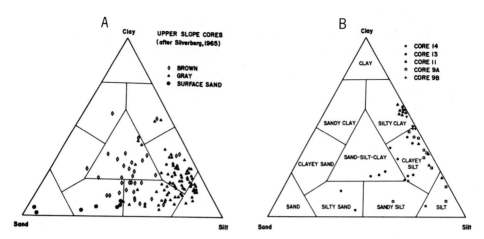

Fig. 11.—Textural analysis of Nova Scotian continental slope and rise sediments plotted on triangular diagrams. Some silty and sandy sediments with mud matrix (note, for instance, those plotted in the sand-silty-clay sector) may be likened to modern graywackes. Most samples occupying this area of the triangle are of Pleistocene age (open lozenges in A, solid dots of Core 13 in B), after Stanley (1970b).

fine-grained deposits. Not unfrequently such strata actually crop out in areas of high relief such as in submarine canyons (fig. 4) (Roberson, 1964) and ridges. Basement rocks consisting of older sedimentary units and volcanic and crystalline series also are exposed locally on the slope (Curray, 1966; Scholl et al., 1966; Stride et al., 1969). The thickness of the accumulating prograding sequences of strata dipping seaward (parallel to the slope surface) is generally less pronounced at the uppermost slope than at the base of the slope.

The shelf-break and uppermost slope is frequently associated by a rather abrupt diminution of grain size (older authors referred to "mud-line") such as that noted along much of the outer east U.S. margin. The associated marked faunal discontinuity at the shelf-break is due, in large part, to temperature changes (Sanders and Hessler, 1969). Biomass studies reveal that slopes are distinguished by a rapid decline in animal density and a benthic fauna of high diversity. A distinct faunal distribution is observed on the slope southeast of Delaware Bay (Stanley and Kelling, 1969). Beyond the shelf-break there is a definite increase in percent of reworked and displaced flora and fauna. The predominance of bioturbate structures preserved on the slope is in part related to the considerably higher amount of finely laminated lutite and silt available to burrowing organisms in this environment than on the outermost shelf.

Current markings such as scour marks on the upper to midslope proper tend to show a fairly constant current orientation. In some cases these are oriented downslope (Owen and Emery, 1967), while in other areas they trend subparallel to contours (Emery and Ross, 1968). Submarine valley, canyon head and valley tributary facies associated with slope deposits have ocassionally been recognized in the fossil record (Whitaker, 1962).

Lower Slope and Base-of-Slope Deposits

Lower slope and base-of-slope environments are the sites of deposition *par excellence* as confirmed by most seismic surveys across the outer margins. Sediments deposited in these deep-water sites are known to be preserved in the rock record. Increased thickness of onlapping depositional sequences, lateral persistency of subparallel strata and the concentration of slump units and large-gravity slices resting disconformably upon mid and lower-slope strata characterize these deeper facies (fig. 9).

Seismic reflection profiles show that the continental rise southeast of New England, for instance, is a sedimentary apron 1.6 km thick consisting of conformable layers that lap up on the slope between 1200 to 2000 m below sea level (Hoskins, 1967). Sediments accumulating on this marginal base have by-passed the slope (the slope is veneered by only 0.4 km of unconformable fine-grained sediment) to form the onlapping continental rise wedge. Individual reflectors can be traced for up to 15 km and, in many cases, sandy horizons mineralogically appear to be derived from the outer edge of the shelf and slope (Hubert and Neal, 1967). In recent years, attention has been called to the importance of bottom currents in deep water, and some workers have suggested that such currents are the most probable means by which late glacial and postglacial sands are moved on outer margins (Hubert, 1964, Hubert and Neal, 1967; Heezen, 1968). Some workers also affirm that turbidites are concentrated on the abyssal plain and not on the rise wedge which is molded by moving water masses (geostrophic currents) (Heezen et al., 1966 and Heezen, 1968).

Cores collected on basin aprons and plains, however, do reveal the presence of graded micaceous silty and sandy units (Ericson et al., 1952; Stanley et al., 1970); these are occasionally plant-rich. Most workers would still agree that the turbidity current hypothesis best explains the assemblage of structures and texture of laterally persistant silt and sand units. In the base-of-slope environments (Gorsline and Emery, 1959; Moore, 1966; Kuenen, 1967), a higher proportion of coarser and thicker *proximal* turbidites (after Walker, 1966a, b; 1967) occurs near the base of the slope, while thinner, finer-grained *distal* turbidites occur in the basin plains. Textural analyses indicate that some of these Quaternary graded units comprise sand-silt-mud admixtures (fig. 11) that call to mind the immature "dirty sand" or graywacke texture of ancient flysch sandstones.

Split cores occasionally reveal that the base of these units have scour and tool marks (Stanley, 1967) of the type found on the soles of graded beds in ancient flysch and other facies (fig. 12). These graded beds are generally interbedded within much thicker mud units. Most of the muds, when examined with X-ray techniques (Hamblin, 1962; Bouma, 1964; Stanley and Blanchard, 1967) appear to consist of alternating silt and clay laminae (fig. 5). These current laminated silts (low-density, turbidity current deposits or normal bottom current deposits?) alternate with finer-grained hemipelagic deposits that may have been deposited from suspension (nepheloid deposits?) and then reworked by organisms after deposition.

A detailed seismic-coring survey of the Alboran Sea in the Western Mediterranean (fig. 13) reveals thin turbidites 2 to 5 cm thick which can be traced continuously for a distance of more than 65 km from the lower part of a basin slope across the basin apron and an entire basin plain (Stanley et al., 1970). A similar phenomenon has been observed elsewhere in the Mediterranean: in the Tyrrhenian Sea (Ryan et al., 1965)

FIG. 12.—Submarine slope facies in the French Maritime Alps (Annot Sandstone, upper Eocene-lowermost Oligocene age). A, arrow points to incipient soft sediment failure within one sand unit (approximately 50 cm thick); B, small slump overfold in thick sandstone unit about 4 m thick; C, groove cast at base of sandstone unit (hammer gives scale); D, frondescent, groove, slide and other sole markings on an overturned sandstone block.

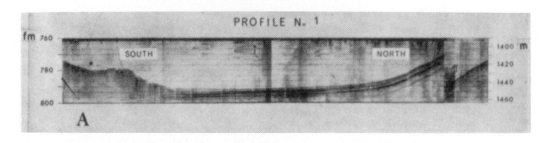

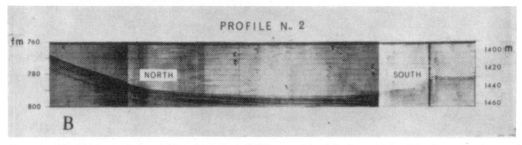

Fig. 13.—Two high-resolution subbottom traces (PGR records) with about 10 fm (18 m) penetration made across the Western Alboran Basin (between Morocco and Spain, southeast of the Strait of Gibraltar, see Fig. 40). This basin is about 1500 m deep, and is 35 km long and 22 km wide. Note that reflectors [these prove to be thin sand layers in cores (see Fig. 39A)] are continuous from the base of the slope, across the basin apron and plain. Alternating nature of mud and sand strata calls to mind typical flysch sequences preserved in the rock record. Traces courtesy of SACLANTCEN, La Spezia, Italy.

and in the Adriatic Sea (van Straaten, 1968). Some of the graded units are composed in part, or entirely, of shell and foraminifera swept up and concentrated by the flows during the movement downslope (Kuenen, 1964a, b). Shell, coarse sand, and granule-size material originating in neritic or upper slope environments are occasionally concentrated at depth as discontinuous lenses and lag deposits by turbidity and bottom currents.

To the assemblage of turbidites, *hemipelagites* and bottom current transported deposits (*tractionites*) that fill sedimentary *ponds* (Hersey, 1965a, b) should be added the volumetrically important mass-gravity deposits: localized slump blocks and large gravity slices (fig. 14). There is ample evidence of rotational slumping and gravitational gliding on a large scale revealed by the bottom morphology and the presence of discontinuous stratigraphic series noted on subbottom profiles across the lower slope, fan and/or rise provinces (Rona and Clay, 1967; Emery *et al.*, 1970). The smaller slump units occasionally consist of contorted sediment as revealed in cores (Stanley and Silverberg, 1969) and call to mind *olisthanites* and related slump units in ancient marine flysch basins (de Raaf, 1968; Stanley, 1970b). The presence of large mounds at the base of gullied slopes and slump-scar valleys (Gorsline and Emery, 1959; Buffington and Moore, 1963; Uchupi, 1967; Pratt, 1967) attests to the predominance of this type of deposit on submarine aprons, fans and rises.

Large out-of-place slices, the downslope movement of which was triggered by depositional oversteepening or earthquakes, call to mind gravity tectonics as recognized in ancient marine basins. Particularly noteworthy in this respect are immense sheets, some of them more than 400 m thick, over 100 km long and even greater width that have been detected in the base-of-slope off the western Atlantic (Heezen and Drake, 1964). Profiles obtained southeast of Cape Hatteras, North Carolina, reveal a composite of allochthonous units consisting of discontinuous strata with bedding planes nearly parallel with the water-sediment interface. These deposits are tens of kilometers wide and hundreds of meters thick and have apparently been moved into place by gliding (Rona and Clay, 1967). Rona and Clay suggest that such strata may have glided for distances up to about 100 km across a near-horizontal abyssal plain surface. Gravity sliding, more than any other phenomenon, is thus invoked to explain the erosion of the slope proper, and, at the same time, the build up of basin aprons and fans, and the development of large morphological features such as the Blake Outer Ridge (Andrews, 1967).

PEBBLY MUDSTONE: A DIAGNOSTIC SLOPE INDICATOR

The pebbly mudstone facies is one of the most useful parameters for distinguishing slope deposits (fig. 9). The origin of till-like subaqueous pebbly mudstones, admixtures of sand and pebbles dispersed in a mud matrix (fig. 15) has been discussed by Dorreen (1951), Crowell (1957), Ksiazkiewicz (1958), Stanley (1961), Dott (1963), Unrug (1963), Scott (1966) and others. Among the best exposures described are those in flysch and flysch-like formations in the Alps, Apennines, Carpathians, Andes, and coastal ranges of Oregon and California. Pebbly mudstones in most cases reflect gravity-triggered slope movement and represent "arrested flows which just surpassed their liquid limits, but regained cohesion thixotropically before a true turbidity current could form" (Dott, 1963, p. 126). There appears to be a gradation between viscous fluid flow, represented by turbidity current suspensions, and plastic mass flow. However a pebbly mudstone origin would preclude considerable lateral movement as a turbulent suspension generally associated with turbidity

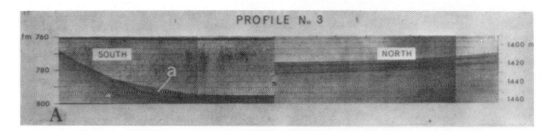

FIG. 14.—A, a high-resolution subbottom trace (PGR record) with about 10 fm (18 m) penetration made across the Western Alboran Basin between Morocco and Spain. Note the presence of a large slice (see a) that has slid onto the basin apron at the base of the slope. This wedge covers an area approximately 100 km². This allochthonous packet consists of slightly deformed mud and silt strata is revealed in cores; structures are not unlike those of the soft sediment deformation illustrated in Fig. 12B. Trace courtesy of SACLANTCEN, La Spezia, Italy. B, Lower Miocene ("Vaqueros stage") section of California state highway 1 about 1500 m NW of Point Mugu (see Crowell, 1957, his fig. 11).

Fig. 15.—Pebbly mudstone strata in the French Maritime Alps (uppermost Eocene Annot Sandstone formation) and Polish Carpathians. A, typical pebbly mudstone horizon about 5 m thick interbedded with proximal turbidites (see top of photo). Rounded clasts from sand to boulder size (a, metamorphic boulder 60 cm long) are dispersed in contorted mud matrix. This wedge-shaped unit at La Boucharde locality was deposited at the base of a basin slope, probably a subsea fan surface. B, large boulder of gneiss on a lenticular pebbly mudstone unit in the Polish Carpathians (Silesian Beskid, Janoska creek locality). C, D, contorted mudstone enclosed in a thick amalgamated submarine channel sand unit (fluxoturbidite), Contes (A.-M.) France. E, slump unit (a) consisting of contorted mud 8 m thick below amalgamated submarine channel sands at the Contes locality (Stanley, 1967). F, thin vertical rip-up clast of alternating silt-shale flysch (arrow) enclosed in a coarse channel sand, Menton (A.-M.) France. G, contorted mudstone covered by a coarse sandy channel deposit at the Contes locality.

current flows. Unrug (1959), Dott (1963) and others infer that in many cases turbidity currents have resulted from mass flows. The fact that in most sections pebbly mudstones are less common than turbidities could be used as supporting evidence. Triggering of mass movements leading to pebbly mudstone formation includes earthquakes, overloading due to extremely rapid sediment build-up (such as off deltas) and oversteepening and undercutting of slopes.

Crowell (1957) has suggested that lenticular, poorly sorted matrix-rich strata, often interbedded between undisturbed graded beds (fig. 15A), are formed by slumping in a "turbidity current environment." The conditions of pebbly mudstone origin as outlined by Crowell (1957, p. 1004) can be paraphrased as follows: (a) coarse sediment is deposited on a mud surface; (b) gravel lobes sinking into the underlying mud, set off downslope movement of gravel and mud together; (c) during downslope movement, pebbles become dispersed throughout the mud; (d) if the flow is relatively fluid, the resulting deposit develops as a uniform layer of pebbly mudstone; (e) if the flow is viscous and plows into the substratum, slabs of underlying units are ripped up and incorporated in the mass as slump overfolds and the resulting deposit is lenticular (fig. 15).

Most workers agree that the presence of a slope and an underlayer of soft mud is essential in the formation of pebbly mudstones. Definition of the mechanism whereby gravels are transported onto a mud surface is more of a problem. Crowell (1957) emphasizes turbidity currents as a means of transporting gravels into a muddy site; earthquakes are, in turn, called upon to initiate the movement of mud and gravel downslope. Recent observations made in the Mediterranean, off the Provençal coast of France, shed new light on this matter.

The presence of fluvial source inputs along land-enclosed, mobile basins such as this provide a key to the problem. Dives (D.J.S.) made in 1967 and 1968 off the Var and Paillon Rivers at Nice revealed large (10 to 50 cm diameter) subangular to rounded carbonate cobbles and boulders on a sloping mud bottom (fig. 16). The initial conditions for pebbly mudstone formation are thus met. Mud is derived from the normal suspended load discharged by rivers at the coast during most of the year. Extensive plumes of suspended matter at the delta mouth are recognized from the air (fig. 17); large volumes of sediment in the water column seriously reduce visibility under water and hamper the diver's observation of the bottom in this area. Gravels, the bed-load of rivers and torrents draining the adjacent Maritime Alps, are transported to the coast where they form small, coarse-textured deltas such as the one at the mouth of the Var (fig. 17A). Abrasion during lateral near-shore transport of pebbles and cobbles along the coast (the coarse-textured strandlines belt or 'cordon litoral') accentuates their rounding.

The extremely narrow coastal platform onto which the gravels are transported is bounded by 3° to 6° slopes. The lateral transport of gravels directly onto the slope is a seasonal event, i.e., occuring during flood stages and during periods of strong near-shore current (stormwave) activity. Once transported beyond the river mouth and onto the mud-covered slope, gravels apparently begin a journey which extends to the basin plain at depths exceeding 2400 m (Bourcart, 1964).

A most interesting observation was made during a dive to about 25 m at the mouth of the Paillon River at the Baie des Anges (fig. 16B). Large cobbles and some boulders (fig. 16C, D) rest on steep (5° to 10°) slopes near the head of the Paillon Canyon (fig. 16A). When touched lightly, rounded blocks were made to slide on the slick mud surface for distances of 10 to 30 cm. It is quite probable that the sudden introduction of additional coarse sediment onto this already unstable mud-gravel surface during flood stages would result in oversteepening, almost immediate failure, and a mass transfer of the unstable deposits downslope. Cores collected off the Provençal coast frequently penetrate mixtures of mud and gravels (fig. 18). Evidence that pebbles move rapidly downslope to the basin apron and plain is provided by oil- and tar-coated pebbles in dredge hauls (tar accumulates on pebbles at the strand line) and by direct visual observation of the bottom (Gennesseaux, 1962a, 1962b, 1966). An excellent sequence of photographs made in the canyons and adjacent gullies by Gennesseaux (1966) shows that bottom currents, as indicated by ripple and scour marks, are almost inevitably present in pebble- and cobble-rich zones along the slope. Bourcart (1964) suggests that pebbles are transported into place by slumping and possibly by movement of individual pebbles by rotation (creep?).

Occasionally much larger rock masses several meters or more in diameter are noted resting on the mud slope (fig. 16E). These angular blocks of broken cliff material ("rock-fall deposits") are not unlike those found in certain wildflysch facies of the Alps, Apennines, Newfoundland, and other mobile belts. The spectacularly large allochthonous blocks resting on the contorted (soft-sediment deformation)

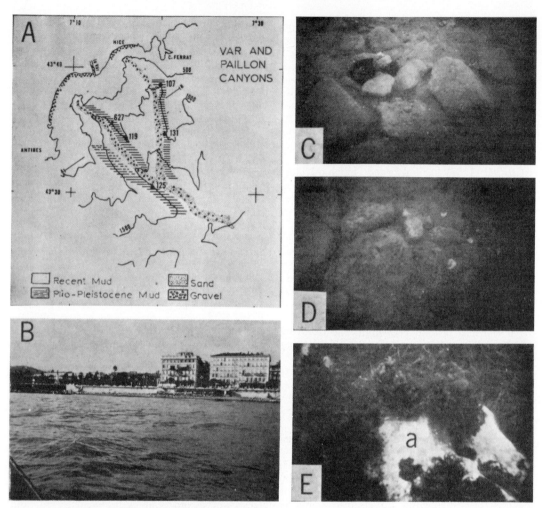

Fig. 16.—Modern pebbly mudstone formation along the Provencal coast of France near Nice (Alpes-Maritimes). A, distribution of recent fluvially transported sand and gravel in the "cordon littoral" and coarse sediment in the Var and Paillon Canyons. Older Plio-Pleistocene mud units are exposed in the canyons (after Gennesseaux, 1962b). B, Paillon river (arrow) outlet and pebble beach, Baie des Anges at Nice. C, D, boulder (maximum diameter 50 cm) and mud admixtures at diving locality shown in B. Depth of about −25 m. E, large encrusted blocks of Jurassic limestone (an example of submarine rock fall) and *Poseidonia* growth near the Cap Ferrat. Depth about −15 to −20 m.

pebbly mudstone facies of the Haymond formation (Marathon Basin) at the base of Housetop Mountain, Texas locality (McBride, 1966) may have had, in fact, a similar origin. Massive slides, perhaps triggered by earthquakes, could have facilitated the transfer of these large masses and their substrate together basinward. A large gneiss boulder resting on a pebbly mudstone horizon in the Janoska creek section of the Polish western Carpathians (fig. 15B) probably has a similar origin. In this region of the Carpathians, pebbly mudstones form lenses several meters thick and from 700 to 1000 m in width. The pebbly mudstone units frequently contain angular fragments of siltstone rip-up clasts; the lower surface of these lenses display irregular erosional scour-like surfaces (fig. 19). Inspection of these deposits indicates that, in some cases, pebbly mudstone lenses are composite in origin: interbedded siltstone layers have been partially eroded by successive mass-gravity flows transporting a mixture of mud, sand, and gravel along the bottom. Erosional features preserved at the base of surfaces serves to distinguish mass-gravity transported tilloids from poorly sorted (mud to boulder), ice-rafted, glacio-marine slope deposits of the type found on slopes in higher latitudes (Brundage et al., 1967; Stanley and Cok, 1968).

In present-day ocean basins, pebbly mudstone-like deposits are associated with submarine canyons (Stanley, 1967) and, frequently, with subsea fans. Aprons along the coastal side of basin floors on the California borderland consist of discontinuous and contorted sediments, some of them including sand and gravel pockets in mud (Gorsline and Emery, 1959; Shepard et al., 1969). Pebbly mudstones in ancient marine basins are also distributed at the base of relatively steep slopes, in apron sequences (Unrug, 1963, Stanley and Bouma, 1964) and in subsea fans (Walker, 1966). Sand grains and crystalline pebbles in pebbly mudstone units are frequently rounder and have a different composition than the sand-gravel fraction of submarine channel deposits (Unrug, 1963) of the type described in the following sections. One explanation is that the coarse clastic fraction eventually dispersed in mud matrix remains longer in coastal environ-

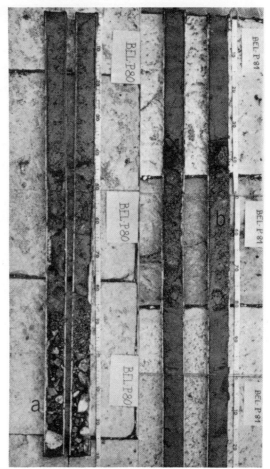

FIG. 18.—Cores collected in the Golfe de la Napoule off Cannes, France. Core BEL-P80, 43°30'8 N Lat., 7°01'7 E Long., depth: −95 m. The upper 75 cm consists of pebbly, muddy sand (mode: 0.80 mm). Sand-sized fragments of lamellibranchs and bryozoans are common. The lower portion of the core (a) contains subangular limestone and dolomitic pebbles (Jurassic) derived from the Iles de Lerins area. This coarse horizon may represent an ancient (Wurm?) strandline.

Core BEL-P81, 43°30'5 N Lat., 6°57'9 E Long., Depth: −90 m. The upper 35 cm of beige Holocene muds cover a 45 cm thick section of muddy, pebbly sand (b). This coarse, nongraded horizon (probably a slump mass) resembles the pebbly mudstone facies (see Fig. 15A) often found in the rock record. It covers stiff gray mud of probable Pliocene age in the lower part of the core. Photos courtesy of G. Pautot, Station de Géodynamique Sous-Marine, Villefranche-sur-Mer, France.

FIG. 17.—A, aerial photograph oriented east toward the Var delta and the city of Nice on the French Mediterranean coast. The light semicircular area seaward of the delta (a) delineates the concentrated suspended sediment zone in time of normal summer effluence (June, 1966). B, closer view of the sharp boundary separating suspended sediment-rich water (a) from the bluer coastal water mass. Sailing vessel in lower part of photo gives scale. Rivers such as this are important point-sources of sediment that eventually accumulate in the deep Mediterranean basins.

ments before its transfer basinward than sand and gravel trapped directly by canyon heads and funnelled rapidly downslope.

Mapping the pebbly mudstone distribution in a paleobasin serves to draw inferences on: (a) the position of source input along the basin mar-

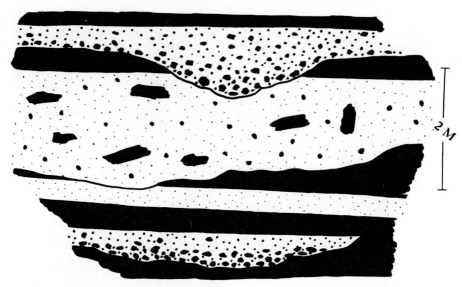

Fig. 19.—Three pebby mudstone units showing moderate erosion of interbedded dark siltstone (shown in black). This is a field sketch of a large lenticular amalgamated pebbly mudstone section in the Silesian Beskid area, Bystra Creek, Kamesznica, Poland.

gin (including river mouths and small deltas); (b) the mineralogy and morphology of the source terrain being drained by the fluvial network that carrier terrigenous sediments to the basin; (c) the nature of the platform and relation of the slope to the coastline; and (d) the breaks in slope (i.e., within the slope-apron-basin plain complex). In all cases known to the authors, pebbly mudstone deposits serve as indicators of lateral introduction and transport of sediments into a basin. This facies, unlike some slumps that are initiated within the basin trough in the manner suggested by Kuenen (1967), is a most valuable criterion for distinguishing transverse basin slope sedimentation.

SHOE-STRING BODIES IN DEEP MARINE ENVIRONMENTS

A survey of the world's modern slope province shows that a nondissected outer margin is indeed a rarity (an excellent region-by-region review is presented in Shepard and Dill, 1966). Thus, to the turbidite-hemipelagite-pebbly mudstone-slump assemblage (fig. 9) that comprises the slope facies can be added channel deposits that fill different types of submarine valleys. Although this type of deposit is limited volumetrically and restricted geographically, it is one of the most diagnostic criteria of submarine slopes. The submarine channel-fan complex is a most important type of petroleum reservoir as demonstrated by recent exploratory work in California, Carpathians, southwestern France, and other areas.

The importance of submarine valleys in outer margin sedimentation is generally recognized. Most workers now believe that a large proportion of shallow marine sediments found in the deep sea have been transported downslope via submarine canyons (Shepard, 1965a; Heezen et al., 1964, 1966; Martin and Emery, 1967). This funneling effect (see fig. 6) has resulted in the development of enormous sediment wedges at the base of slopes in the form of submarine fans and coalescing lenses such as the continental rise of eastern North America, and in the filling (ponding) of deep basins (Gorsline and Emery, 1959; Hersey, 1965; Ryan et al., 1965, 1970, Stanley et al., 1970).

The term *submarine canyon* pertains to a steep-walled valley that cuts into a continental slope and frequently heads on a continental shelf [*submarine valley* has a much broader connotation and refers to almost any variety of seafloor depression (Shepard and Dill, 1966)]. To date canyon sedimentation has been examined most closely in the upper reaches of modern canyons and, to a lesser extent, in base-of-slope areas beyond canyon mouths, i.e., primarily on submarine fans. There is still much uncertainty as to the details of sediment facies and processes active within deep submarine valleys proper, although recent techniques including deep submersibles (Shepard, 1965b, Shepard et al., 1969), precision navigation (Buffington, 1964), underwater television (Stanley, Fenner et al., 1969), and high-resolution subbottom profiling (Roberson, 1964, Moore, 1966; Shepard et al., 1969; Kelling and Stanley, 1970) facilitate work in this environment.

It can be demonstrated that submarine valley deposits are distinct and clearly recognizable lithofacies in the rock record. Because of the relatively fixed position of a slope depression, the resulting canyon deposit tends to be massive and have a high thickness: width ratio. Like the valleys themselves, deposits trend parallel or subparallel to the slope dip. On the slope proper, primary dips of 5° to 10° or more are not uncommon along the axis; greater dips (to 30°) occur along the uppermost slope and along steep flanks, while slopes of less than 1° occur on basin aprons. In short, the most obvious characteristic of submarine valley fills, regardless of texture or composition, is their downslope trending, straight to sinuous, shoe-string "channel" geometry.

MODERN SUBMARINE CHANNEL DEPOSITS

Like fluvial and some shelf channel deposits, submarine valley fills tend to be coarser than the sediments (in this case, slope and fan deposits) with which they interfinger. But textural composition, stratification, assemblage of sedimentary structures, and, in particular, *geometry* and *geographic position*, serve to distinguish canyon deposits from other types of channel deposits.

There are numerous investigations detailing morphology of modern submarine valleys and their sediment fill (among others, Kuenen, 1950; Shepard, 1951, 1965a; Bourcart, 1959; Dill, 1964; Gennesseaux, 1966; Shepard and Dill, 1966; Shepard et al., 1969) The Wilmington Canyon (Stanley and Kelling, 1969) and canyons off Nova Scotia (Stanley, 1967; Stanley and Silverberg, 1969) cited earlier also provide information on canyon sediment entrapment and downslope funneling.

Processes active in moving sediments downslope via canyons differ somewhat from those of fluvial and shelf channels. Down-flank slumping is probably the dominant process active in the canyons studied. The evidence for this includes sudden changes of topography (Shepard, 1951), and the occurrence of contorted mud and sand strata (sometimes overturned and oriented down-flank toward the axis or downchannel) mud clasts, pockets of sand, and small to large sand spheroids, sometimes with mud centers. Sediments which move down-flank along the steep walls (fig. 20) merge with those transported down the canyon axis resulting in a diagnostic assemblage of petrologic features.

Grain size sometimes decreases somewhat

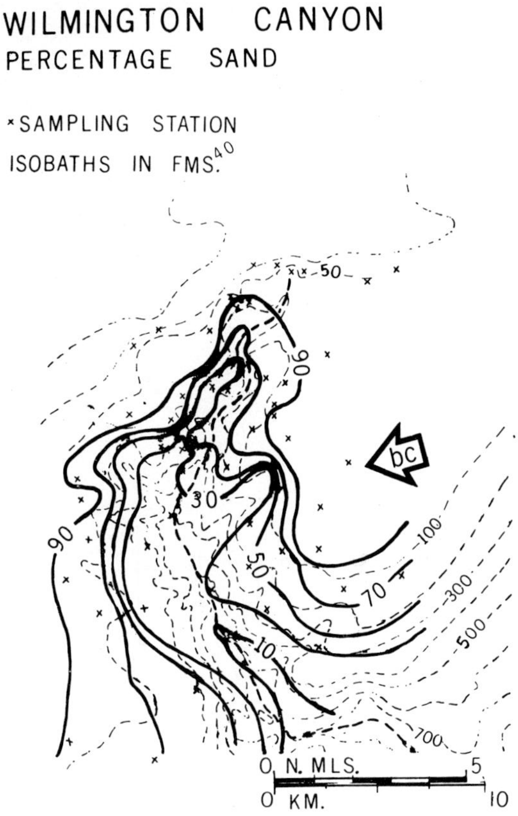

FIG. 20.—Chart showing percent of sand in the surficial sediment cover of Wilmington Canyon and adjacent outer shelf margin. Note that sand draping from the shelf to the canyon axis ("spillover") is more extensive on the east wall. This reflects predominant postglacial sediment transport toward the west (see also Fig. 21) on the outer shelf near the canyon head. Heavy dotted line represents canyon axis.

down-axis; more often than not, however, this phenomenon is generally not clearly evident. Sorting is consistently poor. Mineralogical composition, like texture, is extremely variable downslope; there is generally a close mineralogical correlation with sediments on the shallow adjacent shelf. Skeletal remains, some of which are of shallow origin, include microfossils and shells indicating downslope entrapment. Some fauna also appears to be reworked from older formations that crop out along canyon walls (fig. 21). Submarine weathering and erosion are undoubtedly important processes. Muds are locally green to black, rich in organic matter, and occasionally reek of H_2S. Plant debris is not uncommon. Disturbed stratification and contorted laminations, probably produced by burrowing organisms, are also noteworthy.

Submarine valleys are generally wide (1 to

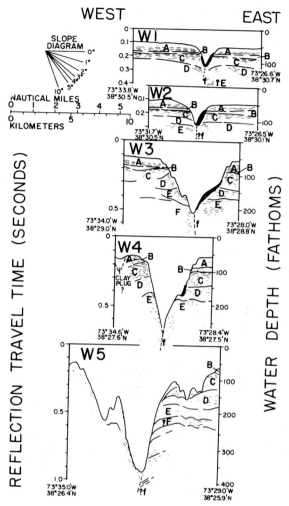

Fig. 21.—Seismic profiles across Wilmington Canyon showing the presence of axial faulting (f), offset of strata on opposite sides of the canyon, exposure of bedrock in the upper canyon reaches, and veneer of recent sediment (in black) on the east wall of the canyon. This coincides with the textural chart (Fig. 20) showing sand spilling over the shelf-break and draping onto the east wall (after Kelling and Stanley, 1970 their Fig. 5).

10 km or more), straight to sinuous depressions on the steeper part of the slope; their winding or meandering is most pronounced on submarine fans (here they are called fan-valleys), basin aprons, continental rises (fig. 22), and plains (Hand and Emery, 1964). Fan-valley walls are relatively steep (>3°) although showing considerably less relief than canyon walls. Fan-valley walls are also asymmetrical and do not include consolidated rock (Shepard, 1965a; Shepard and Buffington, 1968). Valleys generally lack tributaries but often have distributaries and are occasionally braided (Normark and Piper, 1969). Transverse profiles show a development of natural levees (fig. 23) often better developed on one side of the fan valley (Shepard, 1965). In these respects they resemble river channels crossing deltas and valleys crossing alluvial fans (Shepard and Buffington, 1968). The nature of a particular modern channel system is probably most closely dependent on postglacial changes of sea level (fig. 24) and the amount of sedimentation (Normark and Piper, 1969). When rapid sedimentation occurs, shifting and braided channels develop near the apex of the fan; on fans where there is slower or finer sedimentation (such as the well-studied La Jolla fan valley), a single relatively straight fan valley tends to develop (fig. 24).

The idealized section across a submarine fan or apron generally depicts a wedge-shaped depositional build-up (Gorsline and Emery, 1959; Menard, 1960). This is not always the case, however, as exemplified by the La Jolla fan. This feature examined with seismic equipment shows that the inner fan is formed of a thin cover of unconsolidated sediment over faulted older rocks; a section over 1000 m of Quaternary sediments with buried channels (fig. 25) forms the outer fan and adjacent San Diego trough (Shepard et al., 1969).

The flanks of canyons and valleys are not strictly parallel and tend to widen near the canyon mouth at the base of the slope. Complex sets of tributary valleys entering at the canyon head and along the walls provide deposits which merge with the major canyon fill. The longitudinal profiles of submarine channels, particularly fan valleys, tend to develop a profile in which a dynamic equilibrium exists between erosion and deposition not unlike those of river channels (Normark and Piper, 1969). Both headward erosion of canyons and increased deposition on the lower fan or apron frequently accompany a rise in sea level (Gennesseaux, 1962a). The erosion of the upper part of a fan valley accompanying a rise in sea level is shown in Figure 24. Another hypothesis, that of combined axial downcutting and rim up-building, has been used to explain the enlargement of canyons in their vertical dimensions (von der Borch, 1969).

Processes believed to affect sedimentation in submarine valleys are varied. The turbidity current hypothesis (Kuenen, 1950; Shepard, 1963) and slumping (Shepard, 1951; Stanley, 1967) have been recognized as the two dominant processes for a good many years. However, the laterally discontinuous nature of sand and mud strata, the amalgamation of sand or sand and silt beds that comprise thick coarse units, the

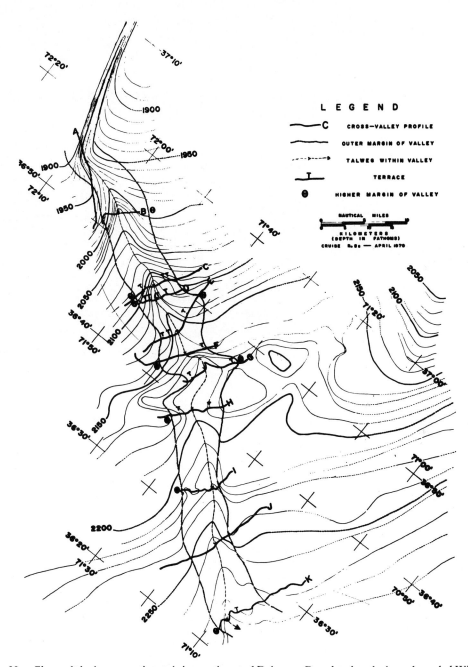

Fig. 22.—Chart of the lower continental rise southeast of Delaware Bay showing the levee-bounded Wilmington Valley trending downslope. Note talweg migrating within the valley. Wilmington Valley serves to channelize sediment between the upper continental rise and the Hatteras Abyssal Plain. Channel bank overflow results in the build-up of the rise wedge in this region.

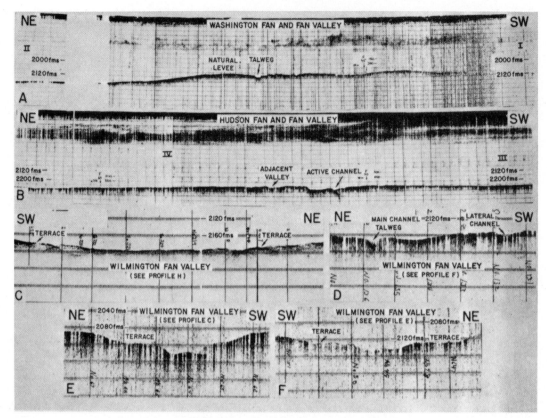

Fig. 23.—Bathymetric profiles (PESR) made parallel to isobaths on the lower continental rise east off the Middle Atlantic States. A, levee-bounded Washington Valley on crest of mega-fan formed as a result of downslope sediment transport. B, Hudson Valley and adjacent valley meandering on mega-fan surface. C, D, E, F, profiles across Wilmington Valley (see Fig. 22 for position) showing main channel talweg, subsidiary (distributary?) channels, and terraces, all indicative of predominant downslope sediment channelization.

abundance of laminated sand, the presence of imbricated pebbles (Gennesseaux, 1966), and erosion shadows and scour depressions all suggest the importance of traction transport, most likely as a result of bottom currents. Breaking internal waves or tides or both may influence such bottom currents (Lyall et al., 1971). Gravity creep and sand flows (Dill, 1964) are of considerable importance in the canyon head and walls where high gradients are prevalent. Bottom current measurements in canyons show currents capable of transporting sand [to 20 cm/sec. in the Wilmington Canyon (Fenner et al., 1971); to 34 cm/sec. in the La Jolla Canyon (Shepard and Marshall, 1969). Bottom flow in some instances appears related to the tidal cycle; although both up-canyon and down-canyon currents are measured, in most cases examined to date there is a net downslope movement. Additional factors affecting water movement are storm and breaking internal waves. Sediment dynamics on slopes and in canyons continues to be a promising area of research.

The presence of graded beds (26 percent in the La Jolla fan; 59 percent of the sand units show parallel lamination and 41 percent have current-ripple cross-lamination according to Shepard et al., 1969) and bank spill-over and natural levee deposits along channels beyond the base of submarine slopes (Komar, 1969) indicate that the turbidity current hypothesis cannot be readily discounted although there is a tendency to do so at present. The presence of low density muddy clouds on slope and base-of-slope environments (fig. 6) may account for the covering mud layer observed in many canyons (Shepard et al., 1969; Stanley and Kelling, 1969).

Evidence from coring programs indicate that individual strata (=sedimentation units) in canyons generally cannot be traced for any distance. Strata are lenticular; slump masses occur as isolated pods oriented down-flank or down-channel. The base of many beds, particularly those composed of cleaner sands and gravels, is concave-up but irregular due to cut-and-fill structures (Stanley, 1967). The zone of maxi-

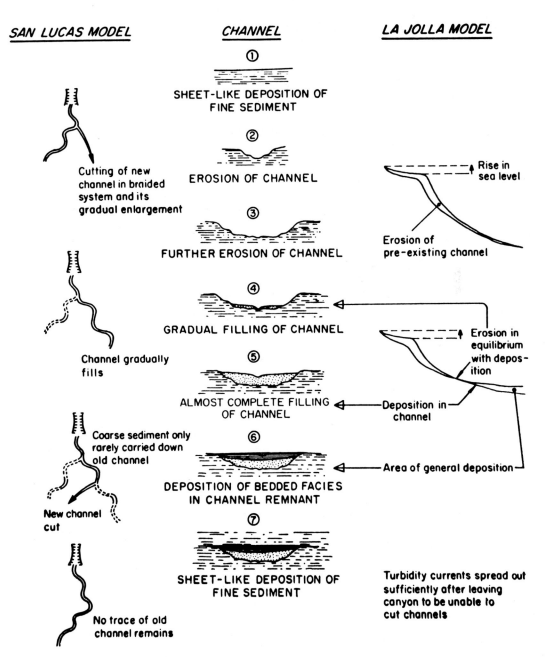

Fig. 24.—Generalized model for erosion and filling of deep-sea fan valleys, after Normark and Piper (1969) The development of a braiding system of channels is illustrated by the San Lucas valley system; the series of depressions is due to the incomplete filling of older channels. The middle column shows corresponding cross-sectional profiles. That rising sea level leads to incision of the upper part of the fan valley is depicted by the La Jolla fan.

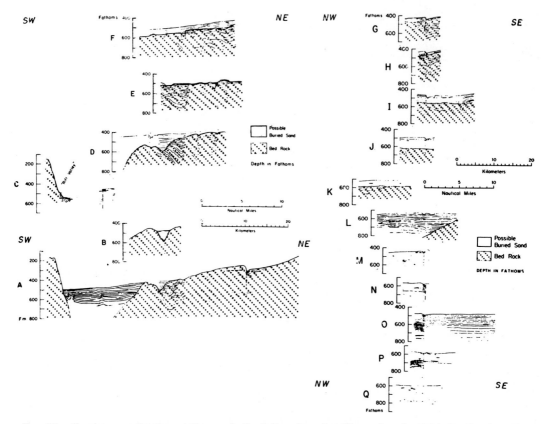

Fig. 25.—Continuous reflection profiles on the La Jolla subsea fan (Shepard et al., 1969). Profiles A to E are roughly parallel with the fan-valley axis; profiles F to Q are transverse to the axis. Note the "possible buried sand" localized in fan deposits (? former abandoned or fan-valley deposits). Analogous sand deposits in the ancient rock record may serve as excellent petroleum reservoirs. (See text.)

mum bed thickness appears to coincide closely with the position of the canyon axis; here, successive beds, especially of sand, are "fused" or amalgamated so that bed thickness is unusually pronounced. Graded and cross-laminated silt and sand strata occur as thin, discontinous sheets as do interbedded muds. Coarse sediments, including gravels, occur as isolated channel tongues in the axis of the main and tributary canyons (fig. 16) and occasionally drape the canyon flanks as well (figs. 8, 20). Gravel units tend to be more localized than the sands with which they interfinger.

Gravel and boulder concentrations are found near the head and the base of flanks as well as in the axis. Downslope imbrication is occasionally observed. That gravels often lie directly above a thick mud or sand substratum suggests that the coarse fraction is being transferred downslope and is only temporarily in the position where it was sampled.

Well-defined graded beds of silt and sand are not a predominant lithology in submarine channel deposits such as those of the Gully Canyon off Nova Scotia (Stanley, 1967). Coring indicates that where they occur, graded units are limited areally; they cannot be correlated from core to core. This is in contrast with an increased proportion of graded turbidites in submarine fans and aprons beyond the canyon mouth and basin plains.

Thin cross-laminated units, some probably deposited by normal bottom currents, are much more common in canyons. Bottom photographs near the head of the Wilmington and the Gully Canyons show ripple marks oriented normal to the axes. Scour marks behind irregularities also emphasize the role of bottom currents in these and other canyons in moving material downslope (Gennesseaux, 1966; Shepard et al., 1969).

Observations point to rapid but intermittent sediment transport, and to the general instability of canyon fills. There are, however, localized areas of relatively slow sediment accumulation as suggested by mottling and sections totally reworked by organisms (Stanley and

Kelling, 1968). Subbottom profiles in submarine valleys also indicate locally important fine-grained sediment-filled pockets or "ponds" consisting of acoustically transparent hemipelagic accumulations. These fines become proportionately more important beyond the base of the slope on fans.

ANCIENT SUBMARINE CHANNEL DEPOSITS AND FLUXOTURBIDITES

Canyon fills, although temporary features localized in time and space, are undoubtedly preserved in the rock record. A significant change of conditions, such as crustal downwarping or a major eustatic fluctuation of sea level causing burial under prograding sediments, would aid in the preservation of such deposits. It has been previously suggested that many of the as yet enigmatic linear belts and localized wedges of ancient coarse sediment similar to those described in the previous section (particularly in flysch basins) are actually submarine valley deposits (Stanley, 1961, 1967). Some suggested fossil canyon and submarine valley deposits include, among others, those recorded in the Alps (Stanley, 1961, 1967; Bosellini, 1967), Apennines (Mutti and deRosa, 1968), the Carpathians in Poland (Unrug, 1963) and Czechoslovakia (Marschalko, 1964), Israel (Neev, 1960), Rhodes (Mutti, 1969), Sicily and Tunisia (Wezel, 1968), Spain (vanHoorn, 1969), Great Britain (Kelling, 1962; Kelling and Woolards, 1969; Whitaker, 1962; Walker, 1966), California (Sullwold, 1960 a, b; Martin, 1963; Bartow, 1966; Dickas and Payne, 1967; Normark and Piper, 1969), Delaware Basin (Jacka et al., 1968), and Gulf Coast of the United States (Bornhauser, 1948).

Noteworthy are the particularly well exposed coarse, shoestring-shaped bodies, generally interfingering with finer grained deposits, that have been interpreted as submarine canyon and valley fill deposits in the Maritime Alps (Stanley, 1961, 1967; Stanley and Bouma, 1964) and in the Carpathians (Unrug, 1963). These ancient channels have a number of characteristics in common with the axis fill of some modern canyons (Wilmington and The Gully Canyons, for instance, on the Atlantic margin). Such units of Cretaceous to Tertiary age in the Carpathians and Maritime Alps have been called *fluxoturbidites*. This term was introduced by Kuenen (1958, p. 332) for deposits that are "characterized by coarse grain, thick bedding, poor development of grading, of sole markings, and of shales." Kuenen inferred that such units originated from slides down steep slopes which did not develop fully into turbidity currents. The source of sediment for this type of deposit was envisioned to be close by. A more detailed description of such deposits in the Polish Carpathians was given by Dzulynski et al. (1959, p. 1114):

> In this type the grain size is large, and the beds tend to be less muddy. The bedding is thick and rather irregular, and the shales between are silty to sandy and thin or even absent. Current sole markings are scarce, load casting is more common, and coarse current bedding of somewhat variable direction is encountered. Indications of slumping are found, and grading is absent, repetitive, irregular, or even inverted, and irregular lenses of coarse grain occur in a few places inside the beds. These sandstones may occur as large lenses between normal flysch or shales. In other cases the material or direction of supply contrast with those of the normal surrounding flysch of the same age.

Dzulynski et al. (1959) and Kuenen (1964a) suggested that such beds—or a section composed largely of such units—appear transitional between turbidites and slides, and that the mechanism involved may be related to both turbidity current flow and sliding. Unfortunately too little is known about the transport mechanisms involved and, understandably, few geologists have been satisfied with a genetic term such as fluxoturbidite. That the term be abandoned has been advocated because it has been applied to a bed, to a facies and to a process (Walker, 1967, p. 38):

> First, it is unsuitable as a bed name. No type of bed can be clearly defined as a fluxoturbidite, and no clear difference can be demonstrated between some fluxoturbidite beds and proximal turbidites, and between other fluxoturbidite beds and slide conglomerates. Second, it is unsuitable as a facies name because it is insufficiently descriptive. Better descriptive names would include turbidite-slump facies, or turbidite-conglomerate facies. Third, it is unsuitable as a genetic term because it attempts to classify a very poorly understood mechanism of transport, or series of mechanisms.

A recent reexamination of deposits in the Carpathians and Maritime Alps described in the published literature as fluxoturbidites reveals that there is one diagnostic parameter common to all: *their geometry*. These fluxoturbidites (*"fluxies"* as they are sometimes called in the field) are lenticular or tonguelike and display, at least to some degree, a characteristic "shoestring" aspect. They are cut in thinner bedded and finer grained alternating base-of-slope, basin apron and basin plain facies. The geometry and assemblage of structures of the Annot, Istebna and Ciezkowice units that we (the authors were joined by St. Dzulynski) examined in the field all suggest a submarine channel origin. We thus propose that thick coarse tongues of this type on outer submarine margins be referred to as submarine valley or submarine channel deposits. These terms minimize to some degree the objectionable mixing of descriptive characteristics and genetic (process) implications [the latter discussed by Stauffer (1967), Walker (1967), Aalto and Dott (1970) and others] that admittedly are still poorly understood. The term submarine channel emphasizes the descriptive (mappable) aspect of these deposits rather than the transport mechanisms.

We do not imply that all units described as fluxoturbidites in the literature are so clearly channelized. For instance somewhat less lenticular blanket sand bodies characterize much of the Numidian Flysch of Tertiary age that crops out in Southern Italy and North Africa. These deposits are interpreted as sheet sandflows that have moved downslope and spread across paleofans and thus have only been partially channelized (F. Wezel, personal communication).

On the other hand, most of the fluxoturbidites described in the French and Italian Maritime Alps and Carpathians bear out the contention that sands and finer-textured sediments are trapped and channelized in neritic or bathyal environments and moved downslope, sometimes for great distances, to basin and abyssal plains by traction, mass-gravity and also in some cases by suspension displacement. In many flysch formations channel deposits and units described as fluxoturbidites generally comprise only a relatively small volume of the total base-of-slope and rise wedge. However, their greater resistance to erosion (perhaps due to the lower percentage of intercalated pelitic beds) than the finer slope, apron and basin sediments with which they are interbedded results in a somewhat anomalously high proportion of preserved channel deposits in the rock record.

SUBMARINE CHANNEL DEPOSITS (FLUXO-TURBIDITES) IN THE FRENCH ALPS

Canyons and channels were of equal or perhaps even greater importance in the downslope transfer of sediments at certain times in the past than they are today. Because of their potential significance in interpreting the geologic record, it is worthwhile to formulate a model for canyon sedimentation and to establish criteria for recognizing types of submarine valley deposits in the ancient record.

Outcrop localities of the Tertiary Annot Sandstone in the French Maritime Alps at Annot (Stanley, 1961; Stanley and Bouma, 1964), at Contes (Stanley, 1967) and at Menton (Stanley, 1970b and this study) offer an unusual opportunity because submarine valley deposits are oriented almost parallel to depositional dip and are well exposed (fig. 26A). The well preserved and exposed channel facies sequence from upper to lower slope can be examined in far greater detail than is presently possible in modern canyons. Bundles of strata may be followed continuously for up to 8 km so that a fairly complete three-dimensional interpretation may be constructed.

Most of these fossil submarine canyons and related submarine channel fills superficially resemble meandering fluvial channels or fingerlike distributary channels of deltaic environments. However, alluvial and submarine shelf channels can be distinguished by crevasse-splay deposits (Allen, 1965), "spill-over fingers" (Sedimentation Seminar, 1969), associated floodplain, backswamp and other associated facies. Erosion exposing a section cutting transversely across a submarine canyon fill shows a bundle of strata, generally coarse grained, that is wedge-shaped (figs. 27, 29). The base of the canyon wedge at Annot is generally concave-up truncating older underlying and lateral units (fig. 29 B,E).

Particularly characteristic of submarine slope channel deposits (when they can be followed in longitudinal profiles) is the tendency for bundles of beds to thicken downslope (fig. 29A). This thickening occurs near the base of the slope where the gradient decreases and deposition exceeds erosion as submarine valleys begin to migrate across submarine fans. This is documented at localities north of Contes (Alpes-Maritimes) and Annot (Basses-Alpes) where paleocanyons extend downslope toward deep (bathyal depth) marine flysch basins (Stanley and Bouma, 1964).

The fingerlike accumulation of friable Upper Eocene sandstones near Contes (fig. 28) is oriented NNW-SSE and occupies about 17 km^2; the stratigraphic thickness at Contes, exceeds 350 m at some points. This area was selected for detailed study, as were those at Annot (fig. 29) and Menton (figs. 26, 27), because the thick accumulation of massive, coarse, wedge-shaped beds are usually well exposed and relatively (by Alpine standards) undeformed structurally. Interbedded shales are thinner in these localities than in the basin apron (fig. 29A) and basin plain facies of the Annot Sandstone flysch formation cropping out to the north (Stanley and Bouma, 1964). Paleogeographic reconstruction indicates that shelf and emergent areas existed to the south, and that linear deposits accumulated in canyon or channel depressions on a paleoslope dipping north (fig. 26A) toward the basin plain (Stanley, 1961; Stanley and Mutti, 1968).

Shape and thickness of strata, sedimentary structures, and textural and mineralogical data were collected at about 70 localities within the

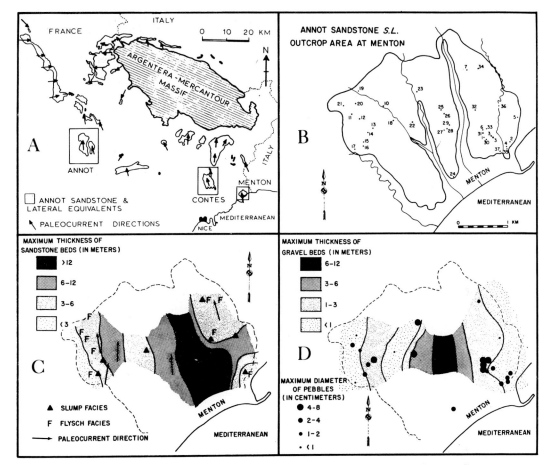

FIG. 26.—A, map showing outcrop localities of the Annot Sandstone *s.l.* in the French and Italian Maritime Alps. Arrows indicate paleocurrent directions in the Tertiary marine basin. Enclosed areas at Annot, Contes, and Menton are sites of probable submarine valley deposits. B, large-scale map of outcrop locality at Menton, France. Numbers refer to selected measured sections. C, map showing north-south trend of sandstone strata, maximum thickness and dominant northward paleocurrent directions. D, map showing north-south trend of maximum gravel bed thickness and maximum pebble diameter. Coarse materials were channelized northward into the Annot Flysch Basin.

Contes outcrop area. It now appears that the presently isolated linear belt of sandstone coincides closely with the major axis of a depression cut into shales of the underlying Eocene Marnes Bleues Formation. A similar situation is found at Annot (Stanley, 1961) and at Menton (Stanley, 1970a).

In cross-section, the bundles of strata in these three localities are coarse-grained, wedge-shaped and truncate older underlying units (fig. 29E). In longitudinal section, the total thickness of beds thins or wedges out toward the uppermost slope and head; the strata thicken toward the base of the paleoslope (subsea fan or apron environments in fig. 29A). The beds appear to consist of irregularly stratified, coarse-grained sand, particularly at the base of canyon walls, near tributary canyons, and in the canyon axis. Sand units may be thin (<1 m) or thick (>16 m); most thick sand strata are actually composite in origin and lack interbedded muds (fig. 27). The thick superimposed strata of sand, such as those at Contes, Annot, and Menton, actually represent a fusion or amalgamation of rapidly deposited sedimentation units. Pockets of gravel, consisting of crystalline and mud pebbles, and slump masses are incorporated within sand strata. The base of thick sand units is generally irregular, sometimes revealing scour-and-fill structures and other sole-markings. Scour-and-fill and channel structures are commonly observed *within* the thick fluxoturbidite sand and gravel beds.

Scour structures (fig. 30B), shale clasts, low angle cross-stratification and large sand spheroids (fig. 30A) within channelized beds indicate

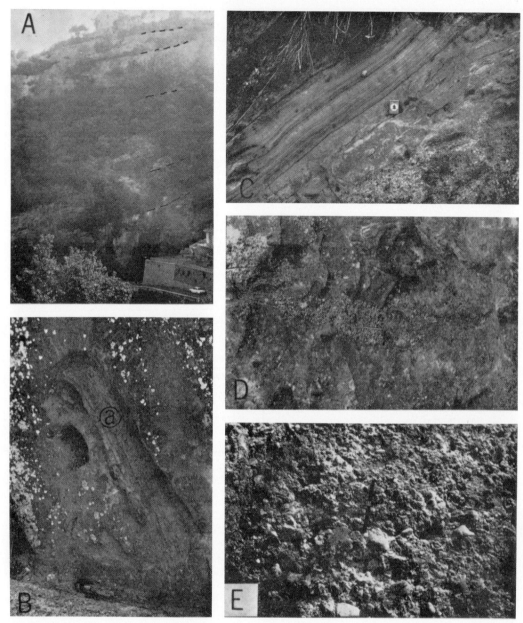

Fig. 27.—Thick, poorly consolidated submarine valley deposits of Annot Sandstone at Menton, France. A, coarse "fluxoturbidite" sequence at section 3 (see B, Fig. 26). Thickness of composite sandstone strata range from several to more than 20 m in thickness. B, contorted "rip-up" sandstone clast (a) (>1 m length in sandstone stratum. C, horizontally laminated sandstone (matchbox gives scale) section above gravel-rich sandstone bed. D, pebble tongue (a) in coarse sand layer (pencil gives scale). E, pebbly mudstone horizon between thick sandstone beds.

that bottom current activated flows were capable of submarine erosion (Stanley, 1964). These coarse units may have been deposited, in part, by watery slides and sand flows of the type recently observed in canyons on the west coast of North America (Dill, 1964). In summary, the coarse-grained, poorly sorted sands generally show no, or poorly developed, graded bedding, and display the stratification features and primary sedimentary structures of deposits called "fluxoturbidites" described in the previous section. They have many of the "proximal turbidite" characteristics summarized by Walker (1967); however, their position within distal turbidite bundles well within the basin cannot be overlooked. These "shoe-string" bodies, in the Annot Sandstone formation display all of the features of "channelized" sediments and, because of their excellent exposure, should serve as one type example for fossil submarine valley deposits.

SUBMARINE CHANNEL DEPOSITS (FLUXOTURBITES) IN THE POLISH CARPATHIANS

Channel deposits in flysch formations, earlier thought to be an unusual or abnormal facies (Dzulynski et al., 1959), have been recognized in the Polish Carpathians (Unrug, 1963) where they are, in fact, not an uncommon lithofacies. As in the Annot Sandstones, their most obvious characteristic is a lenticular sand body geometry, strongly contrasting with the more typical flysch sandstone sheets in which individual beds show a greater constancy of bed thickness, grain size, and sedimentary structures over larger areas.

Typical examples of channelized deposits are provided by Lower Eocene and Paleocene formations of the Silesian Series of the Carpathian Flysch, exposed on the south slopes of the Silesian Beskid range (Polish Western Carpathians). The stratigraphic succession pertinent to this discussion includes:

Hieroglyphic Beds (Middle Eocene)—green shales and thin-bedded sandstones;
Black Siltstones (Paleocene—Lower Eocene) with two channelized sandstone units;
Ciezkowice Sandstone (Lower Eocene);
Upper Istebna Sandstone (Paleocene); and pebbly mudstone lenses below the Upper Istebna sandstone;
Lower Istebna Sandstone (Upper Senonian).

The Ciezkowice sandstone consists of mappable lenticular sand bodies ranging in width from c. 750 m to c. 3000 m (fig. 31). The thickness of the individual sandstone lenses ranges up to c. 100 m. Because of the uniform dip and stratigraphic cover in the described area the length of the lenticular sand bodies cannot be determined. Paleocurrent analyses, however, show that the clastic material was transported from south to north.

In the central part of the lenticular bodies, the sandstone is pebbly and coarse grained. Bedding is indistinct, and the thickness of beds often exceeds 10 m. Such beds contain irregularly distributed lenses of conglomerate and show repeated graded bedding. Large rip-up clasts of

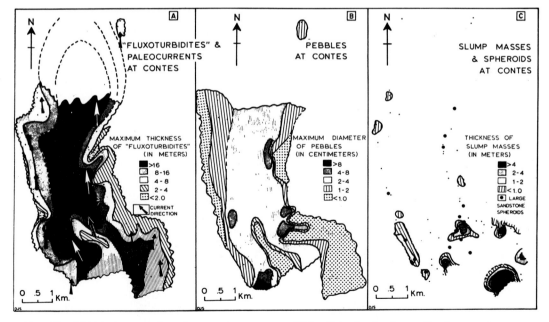

FIG. 28.—Lithofacies maps based on measurements at 70 localities in the Annot Sandstone outcrop near Contes, France (after Stanley, 1967). Note that contours of maximum bed thickness (A) and maximum pebble diameter (B) bring out a dominant north-south depositional trend of the submarine channel. Paleocurrents indicate dominant sediment transport toward the north. Isolated pockets of gravel (B) and of slump masses (C) indicate the importance of lateral dispersal trends and down-flank sedimentation along the former canyon walls. Laterally introduced sediments merged with those of the major axial trend oriented toward the north. An example of spheroids (in C) shown in Fig. 30A.

Fig. 29.—Selected sections of the grès d'Annot Formation, a marine basin slope and basin sandy flysch of late Eocene-early Oligocene age exposed in the French Maritime Alps. A, thick (over 600 m) section of a base-of-slope (probably submarine apron) facies (bundle of turbidites) at the Lauzanier Lake locality on the French-Italian border. B, the Annot Sandstone in its type locality near Annot (Basses-Alpes) France (locality is shown in map A, Fig. 26). This photo oriented toward the north along the Coulomp River shows the thick, coarse, poorly stratified and poorly graded beds of sandstone fluxoturbidites lying upon the older Priabonian "Marnes Bleues" formation. The thick sandstone once filled a submarine valley that funneled shallow water clastics downslope northward (toward right of photo) into a basin deeper than 1000 m. C, D, typical fluxoturbidites (channelized deposits), lenticular stratum >20 m thick consisting of amalgamated sandstone, near Contes (A-M). E, sandstone fill (a) of the submarine canyon at Annot (see also photo B) that was cut in the older (Eocene) Marnes Bleues (blue marls). The canyon deposits shown in B and E are to be visualized as cast fillings while the blue marls (b), much less resistant to erosion, are the mold.

Fig. 30.—A, large mudstone-nucleus sandstone spheroids within amalgamated submarine channel (fluxoturbidite) deposit at Contes (B-A) France (after Stanley, 1964). Arrow points to gravel-filled erosional channel within the thick (>15 m) sandstone stratum. Hammer gives scale. B, erosional base of graded sandy gravel stratum (arrow) near Menton (A-M). Hammer gives scale.

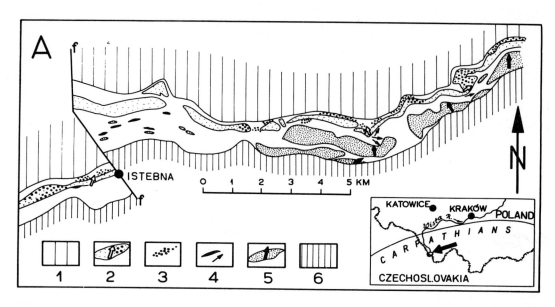
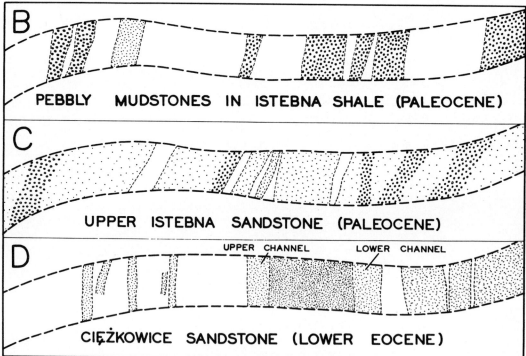

Fig. 31.—Geological map of the southern slopes of the Silesian Beskid range, Western Polish Carpathians, showing coarse lenticular north-south trending sand bodies (B, C, and D) embedded in black siltstones. 1, Lower Istebna Sandstone (Upper Senonian); 2, lenses of Upper Istebna Sandstone (Paleocene) in black siltstones; 3, conglomerate lenses within the Upper Istebna Sandstone and pebbly mudstone lenses within the black siltstones; 4, red shales; 5, lenses of Ciezkowice Sandstone within black siltstones; and 6, Hieroglyphic beds (Middle Eocene).

black siltstone and black shale are common and disseminated throughout the beds. These thick complex beds are locally separated by thin intercalations of black sandy siltstones to silty shales (fig. 32A).

Towards the margins of the lenses there is a marked decrease of bed thickness and grain size. The sandstones become medium bedded and medium grained, and the beds are separated by black silt layers. In spite of the small thickness of beds and absence of coarse-grained material there is strong localized erosion at the base of the beds, causing marked changes of bed thickness.

The large sandstone lenses form two discontinuous stratigraphic horizons. In the areas where the large sandstone lenses wedge out there are smaller lenses composed of fine-grained sandstones up to 100 m wide and a few tens of meters thick. The detrital material of these small lenses is similar to the material forming the marginal parts of the large lenses. The channelized sandstone lenses of the Ciezkowice Sandstone occur within black siltstones and silty shales containing abundant plant detritus and alternating with very thin-bedded fine-grained sandstones. Red shales form small local intercalations. The sandstone lenses of the lower horizon are embedded in black siltstones, while those of the upper horizon are covered directly by green shales and thin-bedded sandstones of the Hieroglyphic Beds (Middle Eocene). The nature of the exposures unfortunately does not permit observation of the contact of sandstones and siltstones at the base of the large sandstone lenses; exposure of the marginal sections indicates that the lower boundary of the sandstone units is erosive.

Similar lenticular sand bodies form the Upper Istebna Sandstone (fig. 31). The thickness of sandstone lenses ranges to 150 m. In those zones where the Upper Istebna Sandstone has its maximum thickness, the underlying black siltstones are entirely eroded by sandstone units resting directly upon the Lower Istebna Sandstone. Here, the lower boundary of the sandstone lenses is clearly erosive. The sandstone bodies contain discrete mappable lenses of conglomerate up to a few tens of meters thick and up to 1000 m wide, which form locally as many as three distinct superposed horizons. In the sandstones grading is often absent, and if present, of repetitive character. Large-scale cross-bedding (fig. 33) is relatively frequent. The sand bodies are embedded in black siltstones and silty shales with rare thin beds of ankeritic siderites.

A different type of channelized deposit consists of thick-bedded and coarse-grained sandstone forming large sets of lenticular bodies. The exposures available in the Polish Carpathians indicate that the lenticular sand bodies were several tens of meters thick and their width ranged to a few hundred metres. The lower surface of the sandstone lenses is erosive, cutting across underlying strata and forming the marginal parts of other lenticular bodies. Within the lenses, especially in their central part, erosion causes the amalgamation of beds which are complex and show in some cases, multiple grading. Fine-grained intercalations, if present, consist of sandy silts with abundant carbonized plant detritus (fig. 34 B, C)

A typical example of such sets of channelized sand bodies is illustrated by the Lower Istebna Beds (Upper Senonian) which were deposited near the *center of the sedimentary basin* (fig. 34A). The lenticular sand bodies, interpreted here again as the fill of erosive channels, are both superposed and interfingering. Some of the components of the complex lenticular beds filling the channels contain abundant clasts of siltstones and shales. There are no sole markings at the base of the sandstone lenses other than large load casts (fig. 32A)

Locally, such channelized sandstone series contain numerous beds of pebbly mudstones up to 10 m thick, alternating with the amalgamated sandstone beds. Pebbly mudstones may form up to 40 percent of all coarse-grained beds. Further away from the source of clastics, at the base of the slope, channelized sandstone deposits sometimes pass into nonchannelized sheet or blanket-shaped beds. Sole markings appear on the basal surfaces of the sandstone beds and armored mud balls form regular horizons near the top of the beds. Here the bedding becomes more regular and the siltstone intercalations are more continuous although some complex amalgamated beds still occur.

CHANNEL DEPOSITS AND BASE-OF-SLOPE ZONATION

Neritic and outer shelf deposits are most often totally absent in ancient marine basins (such as flysch basins) that have undergone considerable syn- and post-depositional tectonic modification. It is possible nevertheless, to arrive at some coherent interpretation of paleoslope configuration, proximity of source area, location of delta point sources along basin margins and rate of sedimentation by carefully mapping ancient channel deposits.

The presence of ancient submarine channel deposits of the type described in preceding sectors generally pin-points a lateral (or adjacent) source of coarse clastics entering a marine basin. The texture of particularly coarse units also suggests a high-standing land area and a high rate of sedimentation on a relatively nar-

Fig. 32.—A, a minor channel (marginal part of a lenticular sand body), filled with medium-grained sandstone, cut in black siltstones with laminae of fine-grained sandstone. Ciezkowice Sandstone (lower Eocene), Silesian Beskid, Janoska creek. B, Lower Istebna Sandstone (Upper Senonian) in the outer zone. A slide plane (dashed line) is visible in the central part of the thick sandstone stratum. Arrows show basal surfaces of beds with large sole markings within fused stratum. Exposure on the castle hill at Dobczyce.

Fig. 33.—Small outcrop of Lower Istebna Sandstone in a creek bed in Polish Carpathians showing high angle cross-stratification of gravelly sand. This type of material was channelized within a submarine valley.

row shelf platform. Oversteepening and periodic failure of the rapidly accumulating sediments, perhaps as a result of seasonal flooding, result from these conditions. This high rate of sedimentation seems to be localized, i.e., most probably *seaward of deltas* built by rapidly flowing rivers carrying coarse sediments along their steep gradients from mountains to the coast. This interpretation is suggested by the typical association of coarse-grained sediment shoe-string bodies and abundant plant detritus (fig. 34) as noted earlier in the Maritime Alps and Carpathian examples.

The lateral variation of Istebna Beds and the Ciezkowice Sandstone of the Silesian Series of the Polish Carpathian flysch permit three geomorphic zones to be distinguished on the outermost margin bordering a marine basin filled with coarse detrital material:

1. An inner zone, in which elongate sand bodies (lenticular in cross-section and consisting of thick amalgamated beds) fill channels cut in silt and finer grained sandstone units. These thick, coarse-grained sand bodies represent the fill of submarine valleys dissecting the lower part of a continental or insular slope.

2. A middle zone, situated at a greater distance from the source area of clastic material, consists of superposed and interfingering lenticular amalgamated sandstone deposits. This zone coincides with the upper submarine fan where canyons open into submarine valleys of less relief at a geographic point near the decrease in slope.

3. An outer zone, in which the lenticular character of the sandstone bodies is less conspicuous and erosion less pronounced. The assemblage of sedimentary features suggests deposition of a reasonably large percent of sandstone beds by turbidity currents.

This zonation established on the basis of sand body geometry of the lithologies with which they interfinger and of associated sedimentary structures reflects an environmental continuum: submarine valley→upper part of submarine fan with channels→lower (distal) part of submarine fan with less marked channel development. This pattern calls to mind the disposition of modern channels in the San Diego Trough off Southern California (Hand and Emery, 1964) and the importance of similar ancient channel complexes in subsea fan deposits in the Delaware Basin (Jacka *et al.*, 1968). These and other examples, such as those of the French Alps, Carpathians, California (Sullwold, 1960a, b) and Great Britain (Kelling and Woollards, 1969; Walker, 1966) shed light on the varieties of channel configuration and the relation of such channels to base-of-slope sedimentation (figs. 35, 36). Examples of submarine levee deposits in the rock record are cited by Jacka *et al.* (1968).

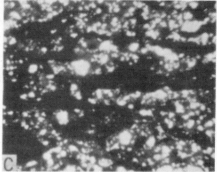

Fig. 34.—A, amalgamated and interfingering sandstone lenses (fluxoturbidite) of a submarine channel [Lower Istebna sandstone (Upper Senonian) at quarry on Roznow Lake, Polish Carpathians]. B, plant-rich laminated sandstone sample collected near axis of channel shown in A. Note oriented plant fragments on top surface. C, thin section (natural light) showing laminations of quarts (white) and plant (black) matter. This channel probably funneled deltaic sediments introduced at the basin margin downslope toward the basin plain.

Both ancient and modern examples indicate that channels provide a means for coarse-grained material to move away from continental margins and penetrate far into a basin, i.e., across basin aprons and well into basin plains. The thickness and the extent of sand bodies is notably greater in the upper part of submarine fans (zone of deceleration) than on slopes. This indicates that coarse detrital material is "funneled" or channelized along submarine valleys, for the most part by-passing the proximal marginal sector of the basin. As one might well expect, the thickness to width ratio of shoe-string and lenticular sandstone bodies tends to diminish in the lower distal sector of submarine fans and in basin troughs.

COMPARISON OF MEDITERRANEAN AND ANNOT BASIN MARGINS

A comparison of sedimentation patterns on selected modern and fossil outer margins calls attention to the role of sediment entry into basins. For such a comparison to be most useful, topographic setting and sedimentation patterns of the selected modern margin should approximate in as many ways as possible the inferred paleogeographic configuration of the selected ancient sedimentary basin. It is appropriate, therefore, to discuss sediment entry into the modern Mediterranean Sea and onto margins of an ancient marine flysch basin, the Annot Sandstone Basin. The latter occupied much of what is now the French and Italian Maritime Alps in late Eocene time.

The selection of acceptable basin analogs is obviously of prime importance, and whether this can be done at all has been seriously questioned. It has been stated that "present-day ocean basins in which nearly all deep-sea sands have been found are obviously very different from the geosynclinal basins of alpine orogens" (Kuenen, 1964a, p. 19). One would have to agree that the configuration of troughs in the modern Mediterranean is not, by any means, identical with that of the Annot Basin or even perhaps any other similar ancient furrows. There are, nevertheless, a significantly large number of topographic, structural and sedimentological characteristics of this ancient flysch sea that recall geographic and geologic setting of certain modern Mediterranean basins.

The particular geographic setting of the almost completely landlocked and silled deep basins of both eastern and western Mediterranean insures their serving as excellent sediment ponds, i.e., ultimate catchments for sediments, particularly coarser "gravitational" terrigenous and tephra deposits (Hersey, 1965a). Over half of the Mediterranean floor lies at depths greater than 1000 m (fig. 37), and several small basins in the Ionian Sea are deeper than 4000 m (Hersey, 1965b).

The Annot Sandstone flysch formation of Eocene age crops out in a region covering about 6000 km² south, west and north of the Argentera-Mercantour Massif and north of the Maures-Esterel Massif (fig. 26). Sandstone and shale outcrops comprise basin slope, apron and the basin plain facies (the latter covering little more than 3,000 km²). The Tyrrhenian abyssal plain east of Sardinia, north of Sicily and west of the Italian boot (covering an area of 2900 km²) is comparable in size, inferred depth and shape to the Annot basin plain.

Schematic diagrams in Figure 38 serve to highlight the reconstructed paleogeography of the Annot Basin toward the end of Eocene time (after Stanley, 1961; Stanley and Bouma, 1964). In diagram 1, paleoslopes (represented by hachures) are shown dipping north, east and west into a basin with a depth estimated at 1000 to 2000 m. The basin lay west of the Argentera-Mercantour Massif (a) and the Dôme de Barrot (b), and north of the Maures-Esterel Massif (c). The position of the slopes delineates the irregular shape of the furrow in which were deposited thick (500–600 m) sequences of well-stratified sands and muds. The Annot Basin does not display a simple elongate trough configuration of the type frequently depicted for flysch seas. Although the early Tertiary Alpine sea (Tethys), like the Mediterranean, was long and arcuate (Stanley, 1965), all of the basins within it were not necessarily long narrow furrows, nor oriented parallel to the sea as a whole. The basins were partially isolated topographically as are, say, the Balearic basin from the Tyrrhenian or Ionian basins.

Large gravity slide blocks and conglomerates concentrated at the base of slopes and the presence of north-trending canyons suggest paleoslope gradients of at least 3°, and, locally, in excess of 5°. Modern Mediterranean basins are

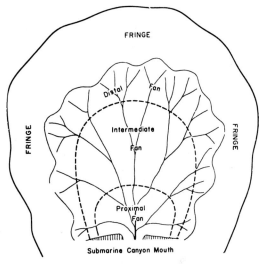

Jacka et al., 1968

Fig. 35.—Idealized plan of a deep sea fan showing distributary pattern of fan valley complex beyond the canyon mouth at the base of the slope (after Jacka et al., 1968).

similarly bordered by slopes that generally range from about 4° to 10° (Ryan et al., 1965, 1970). Much gentler gradients occur in the Adriatic Sea and in regions off deltas (a declivity of 1°26' of foreset beds of the Nile Cone has been calculated by Emery et al., 1966).

Diagram 2 (fig. 38) subdivides the Annot formation s.l. into a flysch facies (I) deposited at the base of the slope and on the basin floor (fig. 29), and a nonflysch facies (II) deposited on the narrow and shallow margins. This "narrow shelf-deep large basin" configuration is also typical of much of the modern Mediterranean (fig. 37).

Arrows in diagram 3 (fig. 38) indicate predominant paleocurrent directions measured in the southern sector of the Annot Basin. Sediments were transported northward away from the Maures-Esterel Massif and what is now deep ocean in the present Mediterranean area (Stanley and Mutti, 1968). The east-west hachured bands, becoming progressively lighter toward the north, symbolize the thinning and fining of turbidites and other sand strata basinward, i.e., toward the north. The slope is less than 10 km wide.

Four major source areas (A, B, C, D) provided material to the basin as shown in diagram 4, figure 38. Only a small (northwest) sector of the now-exposed Argentera massif (A) served as a point source at the end of Eocene time. Sediments derived from sources A and B were brought together and mixed in that part of the

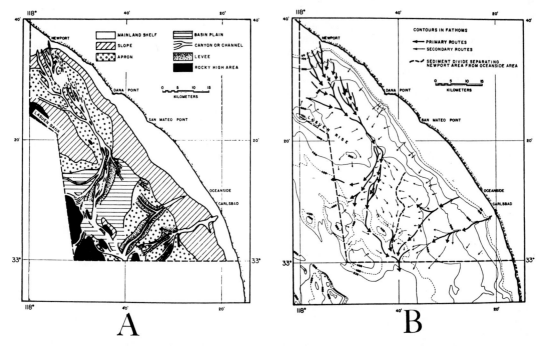

Fig. 36.—A, geomorphic provinces at the north end of San Diego trough; note channels cutting across basin slopes and aprons and extending to basin plain; B, inferred sediment dispersal in large part controlled by the submarine valley system (modified after Hand and Emery, 1964).

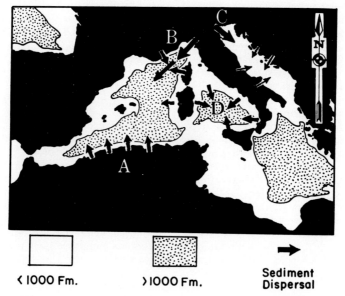

Fig. 37.—Schematic diagram showing different types of sediment dispersal patterns on selected margins of the western Mediterranean (after Stanley, 1970b). A, dominant LATERAL source off North Africa; B, LATERAL AND DISTAL source in the Ligurian Sea; C, dominant LONGITUDINAL source in the northern Adriatic Sea; D, MULTIPLE SOURCE in the Tyrrhenian Sea. Explanation in text.

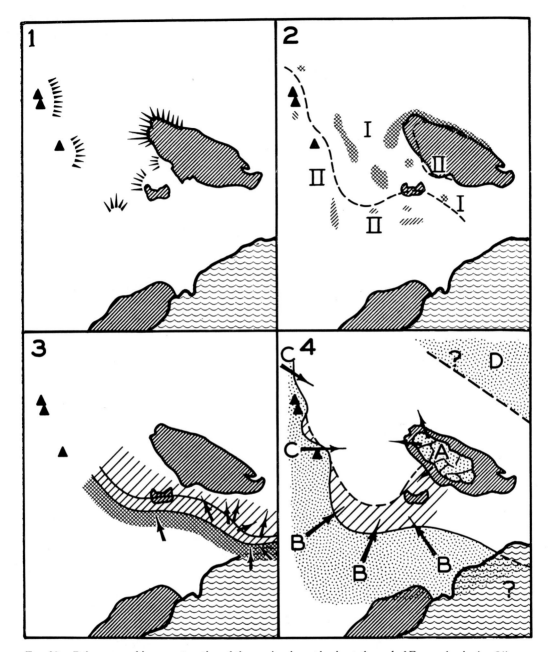

Fig. 38.—Paleogeographic reconstruction of the marine Annot basin at the end of Eocene-beginning Oligocene time (after Stanley and Bouma, 1964). A, Argentera-Mercantour Massif; B, Dôme de Barrot; C, Maures-Esterel Massif. Map showing outcrop localities is shown in fig. 26. Explanation in text.

basin delineated by the hachures. Source area B, lying to the south was, most likely, part of the once-subaerially exposed Esterel-Tyrrhenide land mass that extended in the area now beneath the Mediterranean south of Nice (Kuenen, 1959; Stanley and Mutti, 1968).

The modern Mediterranean slopes, like those of the Annot Basin, are incised by canyons and gullies, the density of which approximates the fluvial drainage on land. The close spacing of major canyons cut into the slope off the French margin is noted on detailed charts (Bourcart, 1959). Such canyons on the margins of the modern Mediterranean can be visualized as sedimentary "funnels" that channelize and transfer coarse sediments from the narrow margins to the basin floor. Probable fossil canyon localities have been mapped at Annot (Stanley, 1961), Contes (Stanley, 1967) and Menton (enclosed areas in A, fig. 26). The N-S oriented channel at Menton (figs. 26, 27) funneled coarse sand and gravel toward the north, as did the canyons at Annot and Contes (figs. 28, 29). Canyon deposits merge on the basin floor at the base of the slope and form thick wedges that are probable analogs of modern subsea fans of the type found in the Mediterranean. These thick wedges of alternating sandstone units (many of them graded) and finer-grained hemipelagites, comprising the bulk of the diagnostic flysch facies, thin basinward. It is possible to trace laterally bundles of sand-shale sequences along a continuous outcrop exposure for a distance of approximately 8 km. In the modern Mediterranean, PGR soundings also reveal reflectors (cores show these to be thin sand layers) that can be traced from the base of slopes across basin floors (fig. 13). Diagnostic ash and sand layers in cores collected in certain basins like the Tyrrhenian Sea (Ryan et al., 1965) and the Alboran Sea (Stanley et al., 1970) confirm the continuous and extensively wide coverage of strata across the nearly flat (less than 1:1000 gradient) floors.

INTERPRETING LATERAL VERSUS
LONGITUDINAL SOURCE INPUT

The Annot Sandstone formation is defined as a coarse, sand-rich flysch, and this facies generally connotes rising cordilleras supplying materials seasonally to coasts and seaward to adjacent margins. That these geographic conditions are not always met is illustrated by several ancient flysch belts (Kuenen, 1958; Dzulynski et al., 1959) and by certain sectors of the modern Mediterranean. Aspects of the problem as to whether deep-sea flysch type sediments are derived from adjacent-lateral or from distal-longitudinal sources, as discussed by Kuenen (1958), can be examined in the Mediterranean. A possible example of a large, distant source area can be found in the eastern Mediterranean where large volumes of sediment are transported by the Nile draining an area of 1,150,000 sq. mi. Another example of a dominantly longitudinal transport condition is also encountered in the Adriatic Sea that receives much of its fill from the Po delta at its northwest sector (fig. 37c).

Sediment transport from adjacent land areas is prevalent in the present Mediterranean as it was during early Tertiary flysch deposition in the Alpine Sea (Stanley, 1965). In the area off west North Africa, for instance, material from the narrow Atlas belt is transported northward across the narrow shelf and into the Balearic Basin. Sediment dispersal in this region (fig. 37A) has been discussed by Heezen and Ewing (1955) and Rosfelder (1955). Frequently, both lateral and longitudinal sources are responsible for providing materials to a basin. A possible example is the Ligurian-northeast Balearic Basin area where sediments from the Provençal coast of France and from Corsica are added to those from the Genoa-Liguria area of Northwest Italy (fig. 37B). The multiple source area of the Tyrrhenian Sea (fig. 37D) shows schematically the numerous point sources including Sardinia, Sicily, Calabria, and main Italian peninsula. The Alboran Sea east of the Strait of Gibraltar is also an excellent example displaying a multiple source pattern (Figs. 39, 40). This multiple sediment dispersal pattern, with both adjacent and more distant source areas, most closely approximates that of the Annot Basin.

The probability of internal ("intrabasinal") sediment sources is encountered in some ancient marine basins such as the Polish flysch Carpathians (the Silesian Cordillera during Cenomanian and Turonian time (Dzulynski et al., 1959; Unrug, 1968) and Rumanian Carpathians in Barremian-Aptian time (Dumitriu and Dumitriu, 1968). It has been suggested that slumps originate on basin slopes or even within basin plains (Kuenen, 1967). The partially emerged Corsican-Sardinian and Balearic blocs are important internal contributors in the western Mediterranean. Less obvious but equally important intrabasinal source areas in the eastern Mediterranean are the numerous, smaller, deep-submerged ridges lying along the Crete-Cyprus arc [this complex is known as the Mediterranean Ridge (Ryan et al., 1970)]. Small islands such as those in the Aegean region are local contributors. Similar examples occur in ancient basins. In the Annot Basin, for example, the northwest sector of the now-exposed Argentera-Mercantour Massif was a small island that provided a limited sediment volume to the adjacent deep (diagram 4, fig. 38).

The major point sources along the modern Mediterranean are the deltaic areas where large rivers debouche large volumes of materials directly on the narrow basin margins. The two most important rivers in the Mediterranean are the Nile with its annual discharge of 60,000,000 tons and the Rhône with an annual discharge of 45,000,000 tons. Surprisingly, the close relation of deltas and thick base-of-slope clastic wedges is too often overlooked in the study of flysch. There is ample evidence that much of the supply for flysch sequences in times past was provided by fluvial sources. Noteworthy, for instance, is the fact that the thickest marine clastic se-

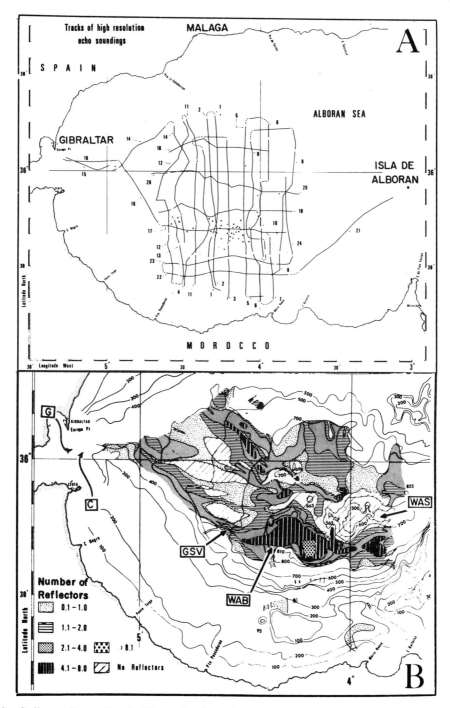

Fig. 39.—Sediment dispersal in the Alboran Sea (after Stanley et al., 1970). A, chart showing track lines where high-resolution subbottom profiles (PGR) were obtained and cores (see dots) were collected. B, chart showing distribution of sand layers in the upper 18 m of sediment (see figures 13 and 14). Note that the number of sand units is irregular: the greatest number are concentrated in the Western Alboran Basin (WAB). Sand is also channelized downslope in the Gibraltar (G) and Ceuta (C) canyons and in the Gibraltar Submarine Valley (GSV). Tongues of sand also drape basinward from the Spanish and Moroccan margins. The Western Alboran Seamount (WAS) chain appears free of sand layers (no reflectors noted).

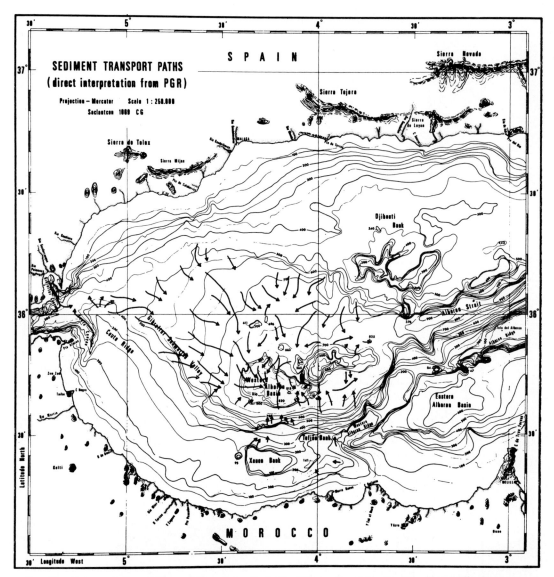

Fig. 40.—Interpretation of the sand dispersal paths in the Alboran Sea (after Stanley et al., 1970). This type of gravity-induced multiple source sedimentation in a modern deep basin is analogous in certain respects to sediment dispersal in some ancient flysch basins (see figure 38D).

quences most often have not accumulated in the basin center and distal abyssal plains (Kuenen, 1959), but near the base of the slope. This is evident in the modern Mediterranean in such areas as the Rhône fan (Menard et al., 1965) and the Nile Cone (Emery et al., 1966). The type of material deposited in deep water often has a "shallow-water," and even a nonmarine appearance as in the case of plant-rich sands (fig. 34). This type of shallow-water lithology is commonly encountered in modern deep-sea cores collected in the Mediterranean.

In most ancient basins (particularly flysch basins) studied to date, it has been demonstrated that faulting is contemporaneous with subsidence and deposition (Shelton, 1968) and is, in fact, an essential criterion affecting outer margin sedimentation. Faults that were active during the filling of the Annot Basin at the end of Eocene time have been mapped (Stanley, 1961). Contemporaneous tectonic activity in the form of earthquakes, faulting and volcanism is concurrent with sedimentation in much of the Mediterranean (Ryan et al., 1970). Submerged ridges, that need not necessarily be large or continuous, are topographic obstacles resulting in the "barrière-en-creux" effect) that obstruct the lateral dispersal of sediments from one furrow to another (Stanley and Kelling, 1968). Sediment failure along such ridges can result in the formation of flank breccias. How much intrabasinal material these ridges supply or how significant a role they actually play in sedimentation needs to be clearly evaluated in the case of the Mediterranean.

OUTER MARGIN DISPERSAL PATTERN IN THE MEDITERRANEAN: EXAMPLES APPLICABLE TO THE GEOLOGICAL RECORD

Charts showing the general regional surficial sediment distribution in the Mediterranean (U.S. Naval Oceanographic Office, 1965, figs. V-9, V-10) indicate that slopes and basins seaward of the narrow shelves are essentially covered with a gray-olive mud except in areas off large deltas or island platforms. Examination of surface and intermediate current patterns indicate a general counter-clockwise pattern in the Mediterranean as a whole (Nielson, 1912). This counter-clockwise movement is particularly pronounced in the Balearic, Tyrrhenian, Adriatic and Aegean Seas as well as at the eastern end of the Mediterranean. Vertical mixing and passage of water masses from basin to basin, sometimes at considerable velocities (locally in excess of 4 knots) insures that basins are not entirely closed sedimentological systems.

That fines in suspension tend to undergo some lateral transport is indicated by studies of the regional clay mineral distribution. Although rivers drain diverse and mineralogically distinct source terrains in southern Europe and North Africa, there appears to be a considerable degree of homogenization of the fine terrigenous fraction, at least in the western Mediterranean. The wide distribution of volcanic glass shards related to two major eruptions in the eastern Mediterranean (Ninkovich and Heezen, 1967) also reflects the influence of water mass movement in laterally redistributing materials settling from the atmosphere.

A diverse assemblage of gravel and sand plus mud, and occasionally bedrock, often indicative of local or adjacent source areas, floors shelves and coastal areas, island platforms, and submarine canyons and gullies. Slope and deep basin areas off deltas also tend to receive tongues of coarser materials. Cores collected in most Mediterranean basins (Bourcart, 1964; Ryan et al., 1965, 1970; Emery et al., 1966; Stanley et al. 1970; and others) indicate that mud generally veneers a highly variable topography and lithology consisting of alternating sequences of pebbles, sands, silts, clays and, occasionally, tephra materials.

A small amount of material is derived from reworked Pliocene and older formations (Gennesseaux, 1962b) eroded along the basin slopes (see fig. 16). By far the most important source, however, is fluvially transported terrigenous material introduced at delta point sources along the margins (fig. 17). The amount of material transported seaward in this fashion increases considerably during flood stage influx. As might be expected, fines carried to coastal areas are moved greater distances in basins than the coarse fraction which tends to be funneled downslope and deposited more proximally to basin margins. Thus while the distribution of fine silts and clays is most closely related to water-mass movements, the distribution of the cobble to silt fraction is essentially topographically controlled (i.e., channelized downslope in canyons and gullies, for instance). Similar coarser deposits preserved in marine sequences would thus be of obvious importance in recognizing lateral dispersal systems.

Biomass and ecological studies (Vaissière, 1964; Vaissière and Carpine, 1964; Fredj, 1964; Pérès, 1967; and others) have pinpointed local marginal contributions as reflected by the distribution of biogenic material. Such studies reveal the normal *in situ* living distribution and displaced assemblages on the basin margins and slopes. Shallow water organisms are commonly encountered in deep-sea cores indicating the importance of reworking and downslope transport processes. In this respect, organic fragments of *Poseidonia* are particularly useful basin margin source indicators. *Poseidonia* growth areas off the Provençal coast of France, examined by Nesteroff (1965) and others, provide organic materials that become an integral part of the sediment upon decomposition (Bianchi, 1968).

The effects of wave- and tide-generated currents are most noticeable in shallow coastal areas (Nesteroff, 1965) where fluvially transported cobbles are concentrated in the "cordon littoral." At greater depths, sand and gravel also tend to be concentrated (Allan, 1966) where the water mass movement is channelized and where bottom current velocities are intensified in such areas as the Straits of Gibraltar and Messina. Bedrock outcrops are locally exposed in such environments. Bottom currents at depth can develop sediment build-ups in areas such as the Messina Cone (Ryan and Heezen, 1965). Evidence of a deep bottom current-sediment "pile-up" phenomenon has not been recognized in the Tertiary Annot Sandstone Basin but conceivably may have been of some importance.

Mechanisms most frequently related to basin slope and rise-fan sedimentation include the mass-gravity processes (Dott, 1963), i.e., large gravity slides, slumps, and turbidity currents. Evidence that these catastrophic processes are important factors in modern Mediterranean sedimentation is best provided by continuous seismic profiles and cores (figs. 13, 14). In various Mediterranean basins, such as those bordering the mid-Mediterranean Ridge, large "sedimentary klippe" and gravity slide masses are recognized. These have moved downslope on basin margins and off the flanks of submerged and emerged ridges well into basins (Ryan et al., 1970). These allochthonous slices, some of them hundreds of meters thick and covering large areas, resemble deposits in fossil basins (such as those in the Apennines) that have been called "olistoliths" and "olistostromes" (Beneo, 1956). There is also ample evidence of subaerial and submarine failure such as rock falls. On narrow cliffed coastal margins off the French Provençal coast, for instance, large limestone blocks, occasionally 10 m³ or larger lie at the base of emerged and submerged cliffs (fig. 16). Observations from submersibles show that such rock masses can eventually work themselves downslope to considerable depths by creep and processes related to failure of metastable sediments on relatively steep slopes. Earthquake epicenters concentrated near basin margins undoubtedly serve as major triggering mechanisms. The recent detection of fresh-water nappes emerging along basin margins (Bellaiche et al., 1966) and the resulting formation of "soft layers" in cores might well also explain periodic downslope failure on slopes. Visual observation of a major underwater avalanche in the Mediterranean has been recorded (Dugan, 1967).

Contorted sediment strata commonly observed in cores collected on basin slopes and aprons indicate that small-scale slumping of material downslope is a most frequent mechanism. Similar sequences of contorted mud and cobble admixtures (i.e., the Swiss "wildflysch" and pebbly mudstone facies described by Crowell, 1957) are a common facies in the rock record (fig. 15). A logical site of origin for this type of deposit is the upper slope, particularly near fluvial sites, where fine to coarse admixtures are transported directly onto narrow basin margins (Ryan et al., 1965). An example off the Var and Paillon Rivers, where admixtures of large subangular to rounded carbonate blocks and mud (fig. 16C, D) are moved rapidly downslope to considerable depths has been discussed in an earlier section. The Var and Paillon canyons heading nearshore also trap, funnel and channelize coarse materials downslope by bottom traction-related processes; they recall the ancient valleys of Annot, Menton and Contes.

It is remarkable how the geometry and structures of sand-pebble deposits of the Var and Paillon Canyons resemble those of the Tertiary Annot Basin. As an example, stiff Pliocene clays exposed on the walls of the modern Var and Paillon canyons are eroded and rip-up clasts are incorporated in the canyon fills. An ancient analog of this may be the large clasts of contorted strata (fig. 27B) in the Menton sandstone channel fill: this is best explained by erosion of partially consolidated materials and the addition of such clasts to the sand fill of the main channel.

The near-horizontality of coarse sediment layers traced across large sectors of basin floors beneath the thin veneer of mud (fig. 13) requires transport mechanisms that tend to distribute sediments evenly. Graded sand layers (Bourcart, 1964; Emery et al., 1966 and others) encountered in Mediterranean cores collected in areas of cable breaks (Heezen and Ewing, 1955) strongly suggest that transport by turbidity currents has been an important mechanism during the Quaternary as it was during times of basin margin deposition in the past. Modern turbidites have been found on slopes as well as on basin floors, and their texture and structures resemble those observed in flysch (Hersey, 1965b; Ryan et al., 1965; Stanley et al., 1969). The presence of graded beds suggests that possible matrix-rich sands, so commonly found associated with geosynclinal belts are being deposited in the modern Mediterranean. Modern Mediterranean graded beds, however, tend to be less rich in mud content than ancient turbidites and texturally could, at best, be considered "incipient graywackes."

It is probable that silts and sands, once transferred into a Mediterranean basin, can be redistributed laterally somewhat by other types of bottom-hugging ("contour") currents. The role of nepheloid layers and other possible mechanisms not sufficiently investigated in this area cannot be ruled out. Geostrophic currents operative in regions such as the Balearic Basin (Gostan, 1967) might produce some winnowing and redeposition of fines in deeper and quieter environments, and concentration of clean laminated sands. Such currents could well be responsible for rippled sand observed at the base of the slope in some areas.

A relatively large volume of coarse to fine sediment admixtures seasonally transported directly onto a narrow outer shelf and steep upper slope results in the build-up of unstable depositional wedges that are prone to fail periodically. The frequency of earthquakes in much of the Mediterranean practically insures the downslope movement of even the coarsest sediment to base-of-slope environments. High resolution subbottom profile and core programs in areas such as the Alboran Sea show that tongues of sand drape downslope from the surrounding margins toward basin plains (fig. 39). These sands, introduced from multiple point sources around such semi-enclosed basins (fig. 40) call to mind the dispersal pattern inferred for the Tertiary Annot Sandstone Basin (diagram 4, fig. 38). It is probable that mud, like sand, is ponded in basin plains by turbidity currents (Stanley et al., 1970). These hemipelagic *mud turbidites* need not only be introduced by mass failure along the basin margin: seasonal introduction during flood influx would best explain the relatively high average rates of muddy sediment accumulation (20 to 40 cm per 1000 years) in most Mediterranean basins. However, in the Mediterranean, as in ancient flysch and other marine basins, turbidity currents are one, but only one, of several mass-gravity mechanisms. Continuing concurrent examination of modern and ancient environments is clearly a most rewarding means for interpreting sedimentation dispersal patterns and depositional characteristics of outer continental margins.

REFERENCES

AALTO, K. R. AND DOTT, R. H., JR., 1970, Late Mesozoic conglomeratic flysch in southwestern Oregon, and the problem of transport of coarse gravel in deep water: *in* J. Lajoie, (ed.), Flysch Sedimentology in North America: Geol. Assoc. Canada Sp. Pap. 7, pp. 53–65.
ALLAN, T. D., 1966, Underwater photographs in the Strait of Gibraltar: Mem. NATO-SACLANT ASW Res. Centre, 19 p.
ALLEN, J. R. L., 1965, A review of the origin and characteristics of recent alluvial sediments: Sedimentology, v. 5, p. 89–191.
ANDREWS, J. E., 1967, Blake Outer Ridge: development by gravity tectonics: Science, v. 156, p. 642–645.
ARCHANGUELSKY, A. D., 1927, On the Black Sea sediments and their importance for the study of sedimentary rocks: Soc. Naturalistes Moscou Bull., v. 35, p. 276–277.
BAILEY, J. W., 1854, Deep soundings from the Atlantic Ocean: Am. Jour. Sci., v. 17, p. 176–178.
BARTOW, J. A., 1966, Deep submarine channel in upper Miocene, Orange County, California: Jour. Sed. Petrology, v. 36, p. 700–705.
BELLAICHE, G., LEENHARDT, O., AND PAUTOT, G., 1966, Sur l'origine des niveaux de vase liquide dans les carottes sous-marines prélevées au carottier à piston: C. R. Acad. Sci. Paris, v. 263, p. 808–811.
BIANCHI, A. J. M., 1968, Aperçu sur la distribution de certains groupements fonctionnels bactoriens au niveau de trois stades d'évolution de feuilles de Posidonies: Rec. Trav. St. Mar. End. Bull., v. 43, p. 351–357.
BORNHAUSER, M., 1948, Possible ancient submarine canyon in southwestern Louisiana: Am. Assoc. Petroleum Geologists Bull., v. 32, No. 12, p. 2287–2294.
BOSSELLINI, A., 1967, Frane sottomarine nel giurassico del bellunese e del friuli: Accad. Nz. dei Lincei, v. XLII, p. 465–467.
BOUMA, A. H., 1964, Sampling and treatment of unconsolidated sediments for study of internal structures: Jour. Sed. Petrology, v. 34, p. 349–354.
BOURCART, J., 1949, Géographie du Fond des Mers, Etudes du Relief des Océans: Payot, Paris, 307 p.
———, 1959, Morphologie du précontinent des Pyrénées à la Sardaigne: *in* La Topographie et la Géologie des Profondeurs Océaniques, J. Bourcart, (ed.): Centre Nat. Rech. Scientifique, Paris, p. 33–50.
———, 1964, Les sables profonds de la Méditerranée occidentale: *in* Turbidites, Bouma, A. H., and Brouwer, A., (eds.): Developments in Sedimentology, v. 3, Elsevier, p. 148–155.
BRUNDAGE, W. L., JR., C. L. BUCHANAN, AND R. B. PATTERSON, 1967, Search and serendipity: *in* Deep-Sea Photography, Hersey, J. B., (ed.): Baltimore, Johns Hopkins Press, p. 75–87.
BUCHANAN, J. Y., 1887, On the land slopes separating continents and ocean-basins, especially those on the west coast of Africa: Scottish Geographical Mag., v. 3, p. 217–238.
BUFFINGTON, E. C., 1952, Submarine "natural levees": Jour. Geol., v. 60, p. 473–479.
———, 1964, Structural control and precision bathymetry of La Jolla submarine canyon, California: Marine Geol., v. 1, p. 44–58.
———, AND MOORE, D. G., 1963, Geophysical evidence on the origin of gullied submarine slopes, San Clemente, California: Jour. Geol., v. 71, p. 356–370.
CAREY, S. W., 1958, A tectonic approach to continental drift, *in* Continental Drift, A Symposium, Carey, S. W., (ed.): Hobart, Univ. Tasmania, p. 177–355.
COTTON, C. A., 1918, Conditions of deposition on the continental shelf and slope: Jour. Geol., v. 26, p. 135–160.
CROWELL, J. C., 1957, Origin of pebbly mudstones: Geol. Soc. America Bull., v. 68, p. 993–1010.
CURRAY, J. R., 1966, Geologic structure on the continental margin from subbottom profiles, northern and central California: Geol. of Northern California Bull. 190, p. 337–342.
———, MOORE, D. G., BELDERSON, R. H., AND STRIDE, A. H., 1966, Continental margin of western Europe: slope progradation and erosion; Science, v. 154, p. 265–266.
DANA, J. D., 1863, A Manual of Geology: Philadelphia, 798 p.
DICKAS, A. B., AND PAYNE, J. L., 1967, Upper Paleocene buried channel in Sacramento Valley, California: Am. Assoc. Petroleum Geologists Bull., v. 51, p. 873–882.
DIETZ, R. S., 1952, Geomorphic evolution of continental terrace (continental shelf and slope): Am. Assoc. Petroleum Geologists Bull., v. 36, p. 1802–1819.
———, 1963a, Collapsing continental rises: an actualistic concept of geosynclines and mountain building: Jour. Geol., v. 71, p. 314–333.
———, 1963b, Wave-base, marine profile of equilibrium, and wave built terraces: a critical appraisal: Geol. Soc. America Bull., v. 74, p. 971–990.

———, AND HOLDEN, J. C., 1966, Deep-sea deposits in but not on the continents: Am. Assoc. Petrol. Geologists Bull., v. 50, p. 351–362.
———, AND MENARD, H. W., 1951, Origin of abrupt change in slope at continental shelf margin: Am. Assoc. Petroleum Geologists Bull., v. 35, p. 1994–2016.
DILL, R. F., 1964, Sedimentation and erosion in Scripps submarine canyon head: *in* Papers in Marine Geology, Shepard Commemorative Volume, Miller, R. L., (ed.): The MacMillan Co., New York, p. 23–41.
DORREEN, J. M., 1951, Rubble bedding and graded bedding in Talara formation of Northwestern Peru: Am. Assoc. Petroleum Geologists Bull., v. 35, p. 1829–1849.
DOTT, R. H. JR., 1963, Dynamics of subaqueous gravity depositional processes: Am. Assoc. Petroleum Geologists Bull., v. 47, p. 104–128.
———, AND J. K. HOWARD, 1962, Convolute lamination in non-graded sequences: Jour. Geol., v. 70, p. 114–121.
DUBOUL-RAZAVET, C., 1959, Contribution à l'étude géologique et sédimentologique du delta du Rhône: Mem. 71-1, Soc. Géol. France, 234 p.
DUGAN, J., 1967, Taxis to the deep: *in* World Beneath the Sea: National Geographic Soc., Washington, p. 143–163.
DUMITRIU, M. AND DUMITRIU, C., 1968, Quelques aspects paléogeographiques des Carpates orientales Roumaines d'après l'étude des paléocourants: Rev. Géograph. Phys. Géol. Dynamique, v. 10, p. 13–20.
DZULYNSKI, S., KSIAZKIEWICZ, M., AND KUENEN, PH. H., 1959, Turbidities in flysch of the Polish Carpathian Mountains: Bull. Geol. Soc. Am., v. 70, p. 1089–1118.
EITTREIM, S., EWING, M., AND THORNDIKE, E. M., 1969, Suspended matter along the continental margin of the North American Basin: Deep-Sea Res., v. 16, p. 613–624.
EMERY, K. O., 1950, A suggested origin of continental slopes and of submarine canyons: Geol. Mag., v. 87, p. 102–104.
———, 1960, Basin plains and aprons off southern California: Jour. Geol., v. 68, p. 464–479.
———, 1965a, Characteristics of continental shelves and slopes: Am. Assoc. Petroleum Geologists Bull., v. 49, p. 1379–1384.
———, 1965b, Geology of the continental margin off eastern United States: *in* Submarine Geology and Geophysics, Whittard, W. F., and Bradshaw, R., (eds.): Colston Papers, v. 17, Butterworths, London, p. 1–20.
———, 1968, Shallow structure of continental shelves and slopes: Southeastern Geol., v. 9, p. 173–194.
———, HEEZEN, B. C., AND ALLAN, T. D., 1966, Bathymetry of the eastern Mediterranean Sea: Deep-Sea Res. v. 13, p. 173–192.
———, AND ROSS, D. A., 1968, Topography and sediments of a small area of the continental slope south of Martha's Vineyard: Deep-Sea Research, v. 15, p. 415–422.
———, AND TERRY, R. D., 1956, A submarine slope off southern California: J. Geol. v. 64, p. 271–280.
———, UCHUPI, E., PHILLIPS, J. D., BOWIN, C. O., BUNCE, E. T., AND KNOTT, S. T., 1970, Continental rise off eastern North America: Am. Assoc. Petroleum Geologists Bull, v. 54, p. 44–108.
ERICSON, D. B., EWING, M., AND HEEZEN, B. C., 1952, Turbidity currents and sediments in North Atlantic: Am. Assoc. Petroleum Geologists Bull., v. 36, p. 489–511.
EWING, M., AND THORNDIKE, E. M., 1965, Suspended matter in deep-ocean water: Science, v. 147, p. 1291–1294.
FAIRBRIDGE, R. W., 1947, Coarse sediments on the edge of the continental shelf: Am. Jour. Science, v. 245, p. 146–153.
FENNER, P., KELLING, G., AND STANLEY, D. J. 1971, Bottom currents in Wilmington Canyon, Eastern U.S.A.: Nature, v. 229, p. 52–54.
FOX, P. J., AND HEEZEN, B. C., 1968, Abyssal anti-dunes: Nature, v. 220, p. 470–472.
FREDJ, G., 1964, Contribution à l'étude bionomique de la Méditerranée occidentale. La région de Saint-Tropez: du cap Taillat au cap de Saint-Tropez (région A1): Bull. Inst. Océanogr. Monaco, v. 63, 55 p.
GARDINER, J. S., 1915, Submarine slopes: Geographical Jour., v. 45, p. 202–216.
GENNESSEAUX, M., 1962a, Les canyons de la baie des Anges, leur remplissage sédimentaire et leur role dans la sédimentation profonde: C. R. Acad. Sci. Paris, v. 254, p. 2409–2411.
———, 1962b, Travaux du Laboratoire de Géologie Sous-Marine concernant les grands carottages effectués sur le précontinent de la région Niçoise: *in* Océanographie Géologique et Géophysique de la Méditerraneé Occidentale: Colloques Internationaux C.N.R.S. Villefranche, Avril 1961, p. 177–181.
———, 1966, Prospection photographique des canyons sous-marins du Var et du Paillon (Alpes-Maritimes) au moyen de la Troïka: Rev. Géograph. Phys. Géol. Dynamique, v. 8, p. 3–38.
GORSLINE, D. S., AND EMERY, K. O., 1959, Turbidity-current deposits in San Pedro and Santa Monica Basins off southern California: Geol. Soc. America Bull., v. 70, p. 279–290.
GOSTAN, Z., 1967, Etude du courant géostrophique entre Villefranche-sur-Mer et Calvi: Cahiers Océanogr., v. 19, p. 329–345.
GUILCHER, A., 1958, Coastal and submarine morphology: Methuen & Co., Ltd., London, 274 p.
HAMBLIN, W. K., 1962, X-ray radiography in the study of structures in homogenous sediments: Jour. Sed. Petrology, v. 32, p. 201–210.
HAND, B. M., AND EMERY, K. O., 1964, Turbidites and topography of north end of San Diego Trough, California: Jour. Geol., v. 72, p. 526–552.
HARRISON, E., 1960, Origin of the Pacific Basin: a meteorite impact hypothesis: Nature, v. 188, p. 1065–1067.
HEEZEN, B. C., 1962, The deep sea floor: *in* Continental Drift, Runcorn, S. K., (ed.): Academic Press, p. 235–288.
———, 1968, The Atlantic continental margin: Univ. Missouri Rolla Jour., p. 5–25.
———, AND DRAKE, C. L., 1963, Gravity tectonics, turbidity currents, and geosynclinal accumulations in the continental margin off eastern North America: Univ. Tasmania Symp. p. D1–D10.
——— AND ———, 1965, Grand Banks Slump: Am. Assoc. Petroleum Geologists Bull., v. 48, p. 221–233.
———, AND EWING, M., 1955, Orleansville earthquake and turbidity currents: Am. Assoc. Petroleum Geologists Bull., v. 39, p. 2505–2514.

———, AND HOLLISTER, C. D., 1964, Deep-sea current evidence from abyssal sediments. Marine Geol., v. 1, p. 141–174.

———, AND RUDDIMAN, W. F., 1966, Shaping of the continental rise by deep geostrophic contour currents: Science, v. 152, p. 502–508.

———, MENZIES, R. J., SCHNEIDER, E. D., EWING, W. M., AND GRANELLI, N. C. L., 1964, Congo submarine canyon: Am. Assoc. Petroleum Geologists Bull., v. 48, p. 1126–1149.

———, AND THARP, M., 1962, Physiographic diagram of the South Atlantic Ocean, the Caribbean Sea, the Scotia Sea, and the eastern margin of the South Pacific Ocean (chart 1: 11,000,000): Geol. Soc. America Chart.

——— AND ———, 1964, Physiographic diagram of the Indian Ocean, the Red Sea, the South China Sea, the Zulu Sea, and the Celebes Sea (chart 1: 11,000,000): Geol. Soc. America Chart.

——— AND ———, 1968, Physiographic diagram of the North Atlantic Ocean (revised), (chart 1: 5,000,000): Geol. Soc. America Chart.

——— AND ———, AND EWING, M., 1959, The floors of the oceans. The North Atlantic: Geol. Soc. America Spec. Paper 65, 122 p.

HERSEY, J. B., 1965a, Sedimentary basins of the Mediterranean Sea: *in* Submarine Geology and Geophysics, Whittard, W. F., and Bradshaw, R., (eds.): Proc. 17th Symposium Colston Res. Soc. (April 5–9, 1965), Butterworths, London, p. 75–91.

———, 1965b, Sediment ponding in the deep-sea: Geol. Soc. America Bull., v. 76, p. 1251–1260.

HOSKINS, H., 1967, Seismic observations on the Atlantic-continental shelf, slope, and rise southeast of New England. Jour. Geol., v. 75, p. 598–611.

HUBERT, J. F., 1964, Textural evidence for deposition of many western north Atlantic deep-sea sands by ocean bottom currents rather than turbidity currents. Jour. Geol., v. 72, p. 757–785.

———, AND NEAL W. J., 1967, Mineral composition and dispersal-patterns of deep-sea sands in the western North Atlantic petrologic province: Geol. Soc. America Bull., v. 78, p. 749–772.

JACKA, A. D., BECK, R. H., GERMAIN, L. ST., AND HARRISON, S. C., 1968, Permian deep-sea fans of the Delaware Mountain group (Guadalupian), Delaware Basin: *in* Guadalupian facies, Apache Mountain area, west Texas. Permian Basin Section, Soc. Econ. Paleont. Mineralogists Publ. 68-11, p. 49–90.

KELLING, G., 1962, The petrology and sedimentation of Upper Ordovician rocks in the Rhinns of Galloway, southwest Scotland: Trans. Royal Soc. Edin., v. 65, p. 107–137.

———, AND STANLEY, D. J. 1970. Morphology and structure of Wilmington and Baltimore Canyons, Eastern United States: Jour. Geol., v. 78, p. 637–660.

———, AND WOOLLARDS, M. A., 1969, The stratigraphy and sedimentation of the Llandoverian rocks of the Rhayader district, *in* The Pre-Cambrian and Lower Palaeozoic Rocks of Wales, Wood, A. (ed.): Univ. Wales Press, p. 255–282.

KNOTT, S. T., AND HOSKINS, H., 1968, Evidence of Pleistocene events in the structure of the continental shelf off the northeastern U. S.: Marine Geol., v. 6, p. 5–43.

KOMAR, P. D., 1969, The channelized flow of turbidity currents with application to Monterey Deep-Sea Fan Channel: Jour. Geophys. Res., v. 74, p. 4544–4558.

KSIAZKIEWICZ, M., 1958, Submarine slumping in the Carpathian Flysch. Ann. Soc. Géol. Pologne, v. 28, p. 123–151.

KUENEN, PH. H., 1950, Marine Geology: John Wiley and Sons, New York, 568 p.

———, 1958, Problems concerning source and transportation of flysch sediments: Geologie en Mijnbouw, v. 20, p. 329–339.

———, 1959, Age d'un bassin Méditerranéen: *in* La Topographie et la géologie des Profondeurs Océaniques: Colloques Internationaux C.N.R.S., Nice-Villefranche, Mai 1958, p. 157–162.

———, 1964a, Deep-sea sands and ancient turbidites: *in* Turbidites, A. H. Bouma and A. Brouwer (ed.), Developments in Sedimentology 3: Amsterdam, Elsevier Publ. Co., p. 3–33.

———, 1964b, The shell pavement below oceanic turbidites: Marine Geol., v. 2, p. 236–246.

———, 1967, Emplacement of flysch-type sand beds: Sedimentology, v. 9, p. 203–243.

LYALL, A. K., STANLEY, D. J., GILES, H. N., AND FISHER, A., JR., 1971, Suspended sediment and transport at the shelf-break and on the slope, Wilmington Canyon area, Eastern U.S.A.: Mar. Tech. Soc. Jour., v. 5, p. 15–27.

MARSCHALKO, R., 1964, Sedimentary structures and paleocurrents in the marginal lithofacies of the Central-Carpathian flysch: *in* Turbidites—Developments in Sedimentology, 3, Bouma A. H., and Brouwer, A., (eds.): Elsevier, Amsterdam, p. 106–126.

MARTIN, B. D., 1963, Rosedale channel-evidence for Miocene submarine erosion in great valley of California: Am. Assoc. Petroleum Geologists Bull., v. 47, p. 441–456.

MARTIN, B. D., AND EMERY, K. O., 1967, Geology of Monterey Canyon, California: Am. Assoc. Petroleum Geologists Bull., v. 51, p. 2281–2304.

MCBRIDE, E. F., 1966, Sedimentary petrology and history of the Haymond Formation (Pennsylvania), Marathon Basin, Texas: Bur. Econ. Geol. Texas, Rept. Invest. 57, 101 p.

MENARD, H. W., JR., 1955, Deep Sea channels, topography, and sedimentation: Am. Assoc. Petroleum Geologists Bull., v. 39, p. 236–255.

———, 1960, Possible pre-Pleistocene deep-sea fans off central California: Geol. Soc. America Bull., v. 71, p. 1271–1278.

———, SMITH, S. M. AND PRATT, R. M., 1965, The Rhône deep-sea fan: *in* Submarine Geology and Geophysics (W. F. Whittard and R. Bradshaw, Eds.): Proc. 17th Symposium Colston Res. Soc. (April 5–9, 1965), Butterworths, London, p. 271–285.

MOORE, D. G., 1961, Submarine slumps: Jour. Sed. Petrology, v. 31, pp. 343–357.

———, 1966, Structure, litho-orogenic units, and postorogenic basin fill by reflection profiling: California continental borderland (Ph.D. thesis): U. S. Navy Elect. Lab., San Diego, 151 p.

———, AND CURRAY, J. R., 1963, Sedimentary framework of continental terrace off Norfolk, Virginia and Newport, Rhode Island: Am. Assoc. Petroleum Geologists Bull., v. 47, p. 2051–2054.

Mutti, E., 1969, Sedimentologie delle Arenarie di Messanagros (Oligocene-Aquitaniano) nell'isola di Rodi: Mem. Soc. Geol. Italiana, v. 8, p. 1027–1070.

———, and de Rosa, E., 1968, Caratteri sedimentologici delle Arenarie di Ranzano e della Formazione di Val Luretta nel basso Appennino di Placenza: Riv. Ital. Paleont., v. 74, p. 71–120.

Neev, D., 1960, A pre-Neogene erosion channel in the southern coastal plain of Israel: Israel Ministry Devel. Geol. Survey Bull., 25, Oil Div. Paper 7.

Nesteroff, W., 1965, Recherches sur les sédiments marins actuels de la région d'Antibes: Ann. Inst. Océanogr. Monaco, v. 43, 136 p.

Nielson, J. N., 1912, Hydrography of the Mediterranean and adjacent waters: Danish Oceanogr. Exped. 1908–1910, Rept. 1, p. 77–191.

Ninkovich, D. and Heezen, B. C., 1965, Santorini Tephra: in Submarine Geology and Geophysics, Whittard W. F., and Bradshaw, R., (eds.): Butterworth; London, p. 413–452.

Normark, W. A. and Piper, D. J. W., 1969, Deep-sea fan-valleys, past and present: Geol. Soc. America Bull., v. 80, p. 1859–1866.

Owen, D. M., and Emery, K. O., 1967, Current markings on the continental slope: in Deep-Sea Photography, Hersey, J. B., (ed.), p. 167–172.

Pérès, J. M., 1967, Les bioceonose benthiques dans le système phytal: Rec. Trav. Mar. End. Bull., v. 42, p. 3–113.

Potter, P. E., and Pettijohn, F. J., 1963, Paleocurrent and Basin Analysis: Academic Press, Inc., Publishers, Berlin, 296 p.

Pratt, R. M., 1967, The seaward extension of submarine canyons off the northeast coast of the United States: Deep-Sea Research, v. 14, p. 409–420.

de Raaf, J. F. M., 1968, Turbidites et associations sédimentaires apparentées: I and II. Koninkl: Nederl. Akad. Van Wetenschappen Proc., v. 71, p. 1–23.

Rich, J. L., 1950, Flow markings, groovings, and intra-stratal crumplings as criteria for recognition of slope deposits, with illustrations from Silurian rocks of Wales: Am. Assoc. Petroleum Geologists Bull., v. 34, p. 717–741.

———, 1951, Three critical environments of deposition, and criteria for recognition of rocks deposited in each of them: Geol. Soc. America Bull., v. 62, p. 1–20.

Roberson, M. I., 1964, Continuous seismic profiler survey of Oceanographer, Gilbert, and Lydonia submarine canyons, Georges Bank: J. Geophysical Research, v. 69, p. 4779–4789.

Rona, P. A., 1969, Middle Atlantic continental slope of the United States: deposition and erosion: Am. Assoc. Petroleum Geologists Bull., v. 53, p. 1453–1465.

———, and Clay, C. S., 1967, Stratigraphy and structure along a continuous seismic reflection profile from Cape Hatteras, North Carolina, to the Bermuda rise: J. Geophysical Research, v. 72, p. 2107–2130.

Rosfelder, A., 1955, Carte provisoire au 1/500.000e de la marge continentale algérienne; Note de présentation: Publ. Serv. Cart Géol. Algerie (N.S.) v. 5, p. 57–106, plus map.

Ryan, W. B. F. and Heezen, B. C., 1965, Ionian Sea submarine canyons and the 1908 Messina turbidity current: Geol. Soc. America Bull., v. 76, pp. 915–932.

———, Stanley, D. J., Hersey, J. B., Fahlquist, D. A., and Allan, T. D., 1970, The tectonics and geology of the Mediterranean Sea: in The Sea, Maxwell, (ed.): Interscience Publishers, John Wiley & Sons, New York, v. 4, Part II, p. 387–492.

———, Workum, F., Jr., and Hersey, J. B., 1965, Sediments on the Tyrrhenian abyssal plain: Geol. Soc. America Bull., v. 76, p. 1261–1282.

Sanders, H. L., and Hessler, R. R., 1969, Ecology of the deep-sea benthos: Science, v. 163, p. 1419–1424.

Schneider, E. D., Fox, P. J., Hollister, C. D., Needham, H. D., and Heezen, B. C., 1967, Further evidence of contour currents in the Western North Atlantic: Earth and Planetary Sci. Letters, v. 2, p. 351–359.

Scholl, D. W., Buffington, E. C. and Hopkins, D., 1966, Exposure of basement rock on the continental slope of the Bering Sea: Science, v. 153, p. 992–994.

———, 1968, Geologic history of the continental margin of North America in the Bering Sea: Marine Geol., v. 6, p. 297–330.

Scott, K. M., 1966, Sedimentology and dispersal pattern of a Cretaceous flysch sequence, Patagonian Andes, Southern Chile: Am. Assoc. Petroleum Geologists Bull., v. 50, p. 72–107.

Sedimentation Seminar, 1969, Bethel Sandstone (Mississippian) of Western Kentucky and south-central Indiana, a submarine-channel fill: Kentucky Geol. Surv. Rept. Invest. 11, 24 p.

Shelton, J. W., 1968, Role of contemporaneous faulting during basinal subsidence: Am. Assoc. Petroleum Geologists Bull., v. 52, p. 399–413.

Shepard, F. P., 1951, Mass movements in submarine canyon heads: Am. Geophys. Union Trans., v. 32, p. 405–418.

———, 1955, Delta-front valleys bordering the Mississippi distributaries: Geol. Soc. America Bull., v. 66, p. 1489–1498.

———, 1963, Submarine Geology, (2nd Ed.): Harper & Row, Publishers, 557 p.

———, 1965a, Importance of submarine valleys in funneling sediments to the deep sea: Progress in Oceanography, v. 3, p. 321–332.

———, 1965b, Submarine canyons explored by Cousteau's Diving Saucer: in Geology and Geophysics, Whittard, W. F., and Bradshaw, R., (eds.): Colston Papers, v. 17, Butterworths, London, p. 303–309.

———, and Buffington, E. C., 1968, La Jolla submarine fan-valley: Marine Geol., v. 6, p. 107–143.

———, and Dill, R. F., 1966, Submarine Canyons and other sea valleys: Rand McNally & Co., Chicago, 381 p.

———, and von Rad, U., 1969, Physiography and sedimentary processes of La Jolla submarine fan and fan-valley, California: Am. Assoc. Petroleum Geologists Bull., v. 53, p. 390–420.

———, and Marshall, N. F., 1969, Currents in La Jolla and Scripps Submarine Canyons. Science, v. 165, p. 177–178.

SILVERBERG N., 1965, Reconnaissance of the Upper Continental Slope off Sable Island Bank, Nova Scotia: unpublished M.Sc. thesis, Dalhousie Univ., 79 p.
STANLEY, D. J., 1961, Etudes sédimentologiques des grès d'Annot et de leurs équivalents latéraux: Inst. Franç. Pétrole, Ref. *6821*, Société des Editions Technip, Paris, 158 p.
——, 1963, Nonturbidites in flysch-type sequences: their significance in basin studies (Abstract): Geol. Soc. America, Prog. Ann. Mtg., p. 155A–156A.
——, 1964, Large mudstone-nucleus sandstone spheroids in submarine channel deposits: Jour. Sed. Petrology, v. 34, p. 672–676.
——, 1965, Heavy minerals and provenance of sands in flysch of central and southern French Alps: Am. Assoc. Petroleum Geologists Bull., v. 49, p. 22–40.
——, 1967, Comparing patterns of sedimentation in some modern and ancient submarine canyons: Earth and Planetary Sc. Letters, v. 3, p. 371–380.
——, 1968, Graded bedding—sole marking—graywacke assemblage and related sedimentary structures in some Carboniferous flood deposits, Eastern Massachusetts: *in* Late Paleozoic and Mesozoic Continental Sedimentation, Northeastern North America, Klein, G. deV., (ed.): Geol. Soc. America Spec. Paper 106, p. 211–239.
——, 1970a, Bioturbation and sediment failure in some submarine canyons: *in* IIIrd European Symposium on Marine Biology, Vie et Milieu, Supplement 22, v. 2, p. 541–555.
——, 1970b, Flyschoid sedimentation on the outer Atlantic margin off Northeast North America: *in* Flysch Sedimentology in North America, Lajoie, J., (ed.): Geol. Assoc. Canada Spec. Paper 7, p. 179–210.
——, AND BLANCHARD, L. R., 1967, Scanning of long unsplit cores by X-radiography: Deep-Sea Res., v. 14, p. 379–380.
——, AND BOUMA, A. H., 1964, Methodology and paleogeographic interpretation of flysch formations: a summary of studies in the Maritime Alps: *in* Turbidites, Bouma A. H., and Brouwer, A., (eds.): Developments in Sedimentology, 3, Elsevier, p. 34–64.
——, AND COK, A. E., 1968, Sediment transport by ice on the Nova Scotian Shelf: *in* Symposium on Ocean Sciences & Engineering of the Atlantic Shelf: Trans. Mar. Tech. Soc., Delaware Valley Section, Philadelphia, p. 109–125.
——, FENNER, P., KELLING, G., AND SWIFT, D., 1969. Underwater television as a tool for mapping the outer continental margin: Abstracts with Programs for 1969, Geol. Soc. America, p. 56–57.
——, GEHIN, C. E., AND BARTOLINI, C., 1970, Flysch-type sedimentation in the Alboran Sea, Western Mediterranean: Nature, v. 228, p. 979–983.
——, AND KELLING, G., 1968, Sedimentation patterns in the Wilmington Submarine Canyon area, *in* Symposium on Ocean Sciences and Engineering of the Atlantic Shelf: Trans. Mar. Tech. Society, Delaware Valley Section, Philadelphia, p. 127–142.
——, AND KELLING, G., 1969, Photographic investigation of sediment texture, bottom current activity, and benthonic organisms in the Wilmington submarine canyon: U. S. Coast Guard Oceanogr. Report 22, 95 p.
—— AND ——, 1970, Interpretation of a levee-like ridge and associated features, Wilmington submarine canyon, eastern U.S.A.: Geol. Soc. America Bull., v. 81, p. 3747–3752.
——, AND MUTTI, E., 1968, Sedimentological evidence for an emerged land mass in the Ligurian Sea during the Paleogene: Nature, v. 218, p. 32–36.
——, AND SILVERBERG, N., 1969, Recent slumping on the continental slope off Sable Island Bank, southeast Canada: Earth and Planetary Sci. Letters, v. 6, p. 123–133.
STAUFFER, P. H., 1967, Grain flow deposits and their implications, Santa Ynez Mountains, California: Jour. Sed. Petrology, v. 37, p. 487–508.
STRIDE, A. H., CURRAY, J. R., MOORE, D. G., AND BELDERSON, R. H., 1969, Marine geology of the Atlantic continental margin of Europe: Philosophical Trans. Royal Soc. London, v. 264, p. 31–75.
SULLWOLD, H. H., JR., 1960a, Tarzana fan, deep submarine canyon of late Miocene age, Los Angeles County, California: Am. Assoc. Petroleum Geologists Bull., v. 44, p. 433–457.
——, 1960b, Turbidites in oil exploration: *in* Geometry of Sandstone Bodies, Peterson, J. A., and Osmond, J. C., (eds.): Tulsa, Am. Assoc. Petroleum Geologists Bull., p. 63–81.
SWALLOW, J. C., AND WORTHINGTON, L. V., 1961, An observation of a deep countercurrent in the Western North Atlantic: Deep-Sea Res., v. 8, p. 1–19.
UCHUPI, E., 1967, Slumping on the continental margin southeast of Long Island, New York: Deep-Sea Research, v. 14, p. 635–639.
——, 1968a, Atlantic continental shelf and slope of the United States—Topography: U. S. Geol. Survey Prof. Paper 529-C, 30 p.
——, 1968b, Tortugas terrace, a slip surface?: Geol. Survey Prof. Paper 600-D, p. 231–234.
——, AND EMERY, K. O., 1967, Structure of continental margin off Atlantic coast of United States: Am. Assoc. Petroleum Geologists Bull., v. 51, p. 223–234.
UNRUG, R., 1959, On the sedimentation of the Lgota beds, Bielsko area, Western Carpathians: Ann. Soc. Géol. Pologne, v. 29, p. 197–225.
——, 1963, Istebna beds—a fluxoturbidity formation in the Carpathian Flysch: Ann. Soc. Géol. Pologne, v. 33, p. 49–92.
——, 1964, Turbidites and fluxoturbidites in the Moravia-Silesia Kulm zone: Acad. Pol. Sci. Bull. Ser. Sc. Geol. Geograph., v. 12, p. 187–194.
——, 1968, The Silesian cordillera as the source of clastic material of the flysch sandstones of the Beskid Slaski and Beskid Wysoki ranges, Polish Western Carpathians: Ann. Soc. Géol. Pologne, v. 38, p. 81–164.
U. S. NAVAL OCEANOGRAPHIC OFFICE, 1965, Oceanographic Atlas of the North Atlantic Ocean: NAVOCEANO Pub. 700, 71 p.
VAISSIÈRE, R., 1964, Contributions à l'étude bionomique de la Méditerranée occidentale. Généralités I: Bull. Inst. Océanogr. Monaco, v. 63, 12 p.
——, AND I. CARPINE, 1964, Contributions à l'étude bionomique de la Méditerranée occidentale. Compte rendu de plongées en soucoupe plongeante SP 300 (région A1): Bull. Inst. Océanogr. Monaco, v. 63, 14 p.

VAN HOORN, B., 1969, Submarine canyon and fan deposits in the upper Cretaceous of the south-central Pyrenees, Spain: Geologie en Mijnbouw, v. 48, p. 67–72.

VAN STRAATEN, L. M. J. U., 1968, Turbidites, ash layers, and shell beds in the bathyal zone of the southern Adriatic Sea: Revue Géographie Physique Géologie Dynamique, v. 9, p. 219–240.

VEATCH, A. C. AND SMITH, P. A., 1939, Atlantic submarine valleys of the United States and the Congo Submarine Valley: Geol. Soc. America Spec. Paper 7, 101 p.

VON DER BORCH, C. C., 1969, Submarine canyons of southeastern New Guinea: Seismic and bathymetric evidence for their mode of origin: Deep-Sea Res., v. 16, p. 323–328.

WALKER, R. G., 1966a, Deep channels in turbidite-bearing formations: Am. Assoc. Petroleum Geologists Bull., v. 50, p. 1899–1917.

———, 1966b, Shale grit and Grindslow shales: transition from turbidites to shallow water sediments in the Upper Carboniferous of northern England: Jour. Sed. Petrology, v. 36, p. 90–114.

———, 1967, Turbidite sedimentary structures and their relationship to proximal and distal depositional environments: Jour. Sed. Petrology, v. 37, p. 25–43.

WEZEL, F. C., 1968, Osservazioni sur sedimenti dell' Oligocene-Miocene inferiore della Tunisia settentrionale: Mem. Soc. Geol. Italy, v. 7 p. 417–439.

WHITAKER, J. H., 1962, The geology of the area around Leintwardine, Herefordshire: Quart. Jour. Geol. Soc. London, v. 68, p. 319–351.

WOOD, A., SMITH, A. J., AND CUMMINS, W., 1967, The Silurian graywackes of Wales: Excursion A1 notes, International Sedimentological Congress, Great Britain, 6 p.

WORZEL, J., 1968, Survey of continental margins: *in* Geology of shelf Seas, Donovan, D. T., (ed.): Edinburgh, Oliver and Boyd, p. 117–152.